Partial differential equations: time-periodic solutions

Otto Vejvoda

in collaboration with

Leopold Herrmann / Vladimír Lovicar / Miroslav Sova
Ivan Straškraba / Milan Štědrý

1982

Martinus Nijhoff Publishers
The Hague / Boston / London

Distributors:

for the United States and Canada

Kluwer Boston, Inc.
190 Old Derby Street
Hingham, MA 02043
USA

for all other countries

Kluwer Academic Publishers Group
Distribution Center
P.O. Box 322
3300 AH Dordrecht
The Netherlands

Library of Congress Cataloging in Publication Data

Vejvoda, Otto.
Partial differential equations, time-periodic solutions.

Includes index.
1. Differential equations, Partial--Numerical solutions. I. Title.
QA374.V38 1982 515.3'53 82-18957
ISBN 90-247-2772-3

ISBN 90-247-2772-3

Joint edition published by
MARTINUS NIJHOFF PUBLISHERS
P.O.B. 566
2501 CN The Hague
The Netherlands

and

SNTL, PUBLISHERS OF TECHNICAL LITERATURE
Prague 1, Spálená 51
Prague, Czechoslovakia

Printed in Czechoslovakia

Preface

As far as the number of new results and quoted papers is concerned the present book may be considered a monograph. However, it also has some features of a textbook. Firstly, it proceeds from concrete problems to abstract ones, and secondly, all considerations and procedures are presented in much detail when met for the first time (such very elementary expositions can be found especially at the beginning of Chapters III and V). Finally, the authors focus their attention on elementary problems which can be dealt with by relatively simple methods. The authors hope that all this will make it possible also for an applied or technical research worker with some mathematical training to read this book. Naturally, the reader is supposed to be familiar with some basic notions from mathematical analysis, functional analysis and theory of partial differential equations. Also, the arguments and procedures which are repeated in the book are presented more briefly when met again, the reader being expected to become gradually more thoroughly acquainted with them.

The authors have tried to provide a complete bibliography of all relevant publications (their number reaches about 500) from the theory of time-periodic solutions to non-linear partial and abstract differential equations whose origin may be put in the early thirties of this century.

About 360 papers are closely connected with the problems studied in Chapters III – VII and each of them (provided it was available to the authors) is, at least briefly, characterized in the text. The others are listed in the Bibliography of papers on related topics. The authors would be very grateful for any information concerning papers which might have been neglected in the bibliography.

The book was written by a group of mathematicians from the Department of the theory of partial differential equations of the Institute of Mathematics, Czechoslovak Academy of Sciences: L. Herrmann (Chap. I, Chap. III except § 6, Chap. IV except § 3, Chap. VI), V. Lovicar (Chap. I, § 5 of Chap. III, §§ 4, 5 of Chap. VII), M. Sova (Chap. I, §§ 1, 2, 3 of Chap. VII), I. Straškraba

(Chap. II, §§ 1, 2 of Chap. III, § 3 of Chap. IV, §§ 4, 5 of Chap. VII), M. Štědrý (Chap. V), O. Vejvoda (Chaps. II–VII). The numbers of paragraphs and chapters given in parentheses indicate the parts of the book on which the efforts of the respective co-author were concentrated in the first place. However, it should be pointed out that each of the authors also collaborated on parts other than those the detailed preparation of which was his primary task. In preparing § 6 of Chap. III the authors were helped by J. Barták and J. Neustupa. When applying some results and methods from number theory the help of B. Diviš and B. Novák was of great importance. The book could hardly have appeared without the more than ten years' activity of the seminar on the theory of partial differential equations of evolution; the members have been (in addition to all those mentioned above as co-authors) P. Filip, J. Havlová, O. Horáček, J. Kopáček, M. Kopáčková, N. Krylová, J. Pešl, H. Petzeltová, V. Šťastnová and V. Vítek. The authors wish to express their gratitude to W. S. Hall from the University of Pittsburgh, USA, and to their colleagues S. Fučík, J. Jarník, A. Kufner and K. Rektorys for reading carefully the manuscript and offering valuable advice and many interesting suggestions. Last but not least, we are indebted to R. Pachtová for completing carefully and promptly her tedious task of typewriting the manuscript.

Contents

Introduction

In this book we deal with partial differential evolution equations, which means that one of the independent variables (as a rule denoted by t) represents time. We are interested in periodic solutions with respect to the time variable. The period of solutions is denoted by ω if not specified otherwise.

In Chapters III–VI we investigate successively periodic solutions of several fundamental types of equations of mathematical physics, namely: *the heat equation*

$$u_t(t, x) - u_{xx}(t, x) + c\, u(t, x) = g(t, x)\,, \quad (t, x) \in \mathbf{R} \times I\,, \tag{1}$$

the telegraph equation

$$u_{tt}(t, x) + a\, u_t(t, x) - u_{xx}(t, x) + c\, u(t, x) = g(t, x)\,, \quad (t, x) \in \mathbf{R} \times I\,, \tag{2}$$

the wave equation

$$u_{tt}(t, x) - u_{xx}(t, x) = g(t, x)\,, \quad (t, x) \in \mathbf{R} \times I\,, \tag{3}$$

and *the beam equation*

$$u_{tt}(t, x) + u_{xxxx}(t, x) + a\, u_t(t, x) + c\, u(t, x) = g(t, x)\,, \quad (t, x) \in \mathbf{R} \times I\,, \tag{4}$$

where a, $c \in \mathbf{R}$, $|a| + |c| > 0$ in (2), and $I = \mathbf{R}$ or $I = (0, l)$. In the case of a bounded interval the solutions of the equations of the first three types are subject to some of the following boundary conditions:

a) *the Dirichlet boundary conditions*:

$$u(t, 0) = h_0(t)\,, \quad u(t, l) = h_1(t)\,, \quad t \in \mathbf{R}\,, \tag{5}$$

b) *the Newton boundary conditions*:

$$\begin{aligned} &u_x(t, 0) + \sigma_0\, u(t, 0) = h_0(t)\,, \\ &u_x(t, l) + \sigma_1\, u(t, l) = h_1(t)\,, \quad t \in \mathbf{R}\,, \end{aligned} \tag{6}$$

(if $\sigma_0 = \sigma_1 = 0$ then they are called *the Neumann conditions*),

c) *the combined boundary conditions*:

$$u_x(t, 0) + \sigma_0 u(t, 0) = h_0(t), \quad u(t, l) = h_1(t), \quad t \in \mathbf{R}. \tag{7}$$

As a rule, in addition to the above-mentioned problems, we study corresponding n-dimensional and weakly non-linear problems. As usual, a problem is called *weakly non-linear* if it involves a so-called small parameter ε in such a way that the problem reduces to a linear one if ε vanishes. For instance, instead of the equation (1) we can consider the equation

$$u_t(t, x) - u_{xx}(t, x) + c\, u(t, x) = g(t, x) + \varepsilon F(u, \varepsilon)(t, x), \quad (t, x) \in \mathbf{R} \times I, \tag{8}$$

where F is, in general, a non-linear operator. It is easy to see that this equation can be transformed to one with $g = 0$, provided the linear case has been already dealt with.

We study *non-autonomous* problems. The *autonomous* ones (that is, those whose data, coefficients, right-hand sides etc., do not depend explicitly on the time variable so that the period of the required solution represents, in general, an additional unknown) are mentioned only in Comments to the corresponding chapters.

The complexity of solving a given periodic problem depends considerably on the properties of the corresponding homogeneous problem. A problem is said to be *non-critical* (or *non-resonant*) if the corresponding homogeneous problem has only the trivial solution (in other words, the null-space of the corresponding differential operator is trivial). Otherwise it is said to be *critical* (or *resonant*). For example, the problem

$$\begin{aligned} &u_t(t, x) - u_{xx}(t, x) + c\, u(t, x) = g(t, x), \quad (t, x) \in \mathbf{R} \times (0, \pi), \\ &u(t, 0) = h_0(t), \quad u(t, \pi) = h_1(t), \quad t \in \mathbf{R}, \\ &u(t + \omega, x) - u(t, x) = 0, \quad (t, x) \in \mathbf{R} \times (0, \pi), \end{aligned} \tag{9}$$

where g, h_0, h_1 are ω-periodic in t, is critical or non-critical for $c = -1$ or $c = 0$, respectively, for the corresponding homogeneous problem

$$\begin{aligned} &u_t(t, x) - u_{xx}(t, x) + c\, u(t, x) = 0, \quad (t, x) \in \mathbf{R} \times (0, \pi), \\ &u(t, 0) = u(t, \pi) = 0, \quad t \in \mathbf{R}, \\ &u(t + \omega, x) - u(t, x) = 0, \quad (t, x) \in \mathbf{R} \times (0, \pi) \end{aligned}$$

has in the latter case only the trivial solution, whereas in the former case it has a one parameter family of solutions $d \sin x$, $d \in \mathbf{R}$ (see III: Sec. 1.1). Of course, even if the null-space is trivial the solution of the original problem still need not be easy (see V: Sec. 6.3).

In the critical case the inhomogeneous linear problem has a solution only if the right-hand sides satisfy some orthogonality conditions just as in the theory of Fredholm integral equations of the second kind. For instance, in the problem (9) with $c = -1$ the functions g, h_0, and h_1 have to satisfy the condition

$$\int_0^\omega \int_0^\pi g(\tau, \xi) \sin \xi \, d\xi \, d\tau + \int_0^\omega (h_0(\tau) + h_1(\tau)) \, d\tau = 0 .$$

Investigating a weakly non-linear problem

$$u_t(t, x) - u_{xx}(t, x) + u(t, x) = \varepsilon F(u, \varepsilon)(t, x), \quad (t, x) \in \mathbf{R} \times (0, \pi),$$
$$u(t, 0) = \varepsilon X_0(u)(t), \quad u(t, \pi) = \varepsilon X_1(u)(t), \quad t \in \mathbf{R},$$
$$u(t + \omega, x) - u(t, x) = 0, \quad (t, x) \in \mathbf{R} \times (0, \pi),$$

we find that this condition turns into a certain condition on the operators F, X_0, and X_1, which is called the *bifurcation equation.* (The same terminology is used in similar cases.) The bifurcation equations represent conditions on certain free data of the solution, for example, on its initial data.

There are a great many methods, both special and general, for investigating the existence of periodic solutions to partial differential equations. They can be divided roughly into two large classes, those of direct and of indirect methods. By a *direct method* we mean one that makes it possible from the very beginning to look for a solution in the space of periodic functions. On the other hand, when we use an *indirect method,* we find first a solution to some related problem (for example, the corresponding initial value or initial boundary-value problem), and only then do we try to determine some "free" data (for example, the initial data) so as to obtain a periodic solution of the original problem.

The search for a solution in the form

$$u(t, x) = \sum_{j=-\infty}^{\infty} u_j(x) \, e^{ijvt} \quad (v = 2\pi/\omega)$$

is an example of a method belonging to the first class. In what follows this method will be called the *time-Fourier method,* briefly the *t-Fourier method.*

Another representative of this group is the method called here the *time-space-Fourier method,* briefly the *(t, s)-Fourier method.* In this case the solution is looked for in the form

$$u(t, x) = \sum_{j=-\infty}^{\infty} \sum_{k=1}^{\infty} u_{jk} \, e^{ijvt} \, v_k(x) \quad (v = 2\pi/\omega),$$

where $\{v_k\}_{k=1}^{\infty}$ is a base in an appropriate space. This base is formed, as a rule, by the eigenfunctions of the self-adjoint elliptic operator that appears in the problem. For example, in the problem (9) with $h_0 = h_1 = 0$ we choose $v_k(x) = \sin kx$, which represent the eigenfunctions of the eigenvalue problem

$$-v_{xx} = \lambda v\,, \quad v(0) = v(\pi) = 0\,.$$

One of the best-known indirect methods is the *Poincaré method*. Suppose that there exists a *solution operator* S that associates uniquely with the vector of the initial data φ the value of the solution and its derivatives at a time t, that is,

$$S(\varphi)(t, x) = \left(u(t, x), \frac{\partial}{\partial t} u(t, x), \ldots, \frac{\partial^{n-1}}{\partial t^{n-1}} u(t, x)\right), \tag{10}$$

where n is the order in t of the differential equation. The solution of the corresponding periodic problem reduces to the investigation of the equation

$$\varphi(x) = T(\varphi)(x) \tag{11}$$

(that is to finding the fixed points of the operator T) where $T(\varphi)(x) = S(\varphi)(\omega, x)$ and ω is the period of the required solution. T is often called the *translation operator*. For instance, in the case of the equation (3) with the initial conditions $u(0, x) = \varphi_0(x)$, $u_t(0, x) = \varphi_1(x)$, the solution operator S is given by

$$2S(\varphi_0, \varphi_1)(t, x) = \Big(\varphi_0(x + t) + \varphi_0(x - t) + \int_{x-t}^{x+t} \varphi_1(\xi)\, d\xi + \int_0^t \int_{x-t+\tau}^{x+t-\tau} g(\tau, \xi)\, d\xi\, d\tau\,,$$

$$\varphi_0'(x + t) - \varphi_0'(x - t) + \varphi_1(x + t) + \varphi_1(x - t) + \int_0^t [g(\tau, x + t - \tau) + g(\tau, x - t + \tau)]\, d\tau\Big)$$

and the equation (11) becomes

$$(\varphi_0(x), \varphi_1(x)) = S(\varphi_0, \varphi_1)(\omega, x)\,.$$

Note that in this book the term "Poincaré method" has a more specialized sense than that mentioned above. Namely, we use it only when the solution of the corresponding initial value or initial boundary-value problem is used in a "closed" form as, for example, in the case just described.

Another indirect method is the *space-, Fourier method*, briefly *s-Fourier method,* which is well-known for solving initial boundary-value problems. We look for a solution in the form

$$u(t, x) = \sum_{k=1}^{\infty} u_k(t)\, v_k(x)$$

where $\{v_k\}_{k=1}^{\infty}$ is a base, and we try to determine the initial vector φ so that u_k are periodic.

The Ficken-Fleishman method (see IV: Sec. 4.2) provides another example of an indirect method. Of course, there are procedures that have some features of both direct and indirect methods, for example *the Günzler method* (see V: § 2) or the *two-time method* (see VII: Sec. 5.5).

In this book a number of problems are solved by several different methods. This is justified by the fact that the method influences substantially the choice of spaces to which the solutions and the right-hand sides belong. The most suitable method depends on the character of the problem. Of course, in some cases it may be purely mathematical reasons that lead us to the choice of a certain method. For instance, the (t, s)-Fourier method has the obvious advantage of reducing the task of finding periodic solutions in a linear case to the examination of a system of linear algebraic equations, whereas the t- and s-Fourier methods transfer the problem to boundary-value problems for ordinary (or partial) differential equations. Further, when we expand a solution with respect to a variable ξ, we are led to consider the solution in this variable in a Hilbert space, because these spaces offer more convenient convergence criteria. On the other hand, the Poincaré method in our sense or the Günzler method enables us to work in Banach spaces of continuous functions.

In § 3 of Chapter IV *the "singularly" perturbed problem*

$$\begin{aligned} &\mu\, u_{tt}(t, x) - u_{xx}(t, x) + u_t(t, x) + c\, u(t, x) = g(t, x)\,, \quad (t, x) \in \mathbf{R} \times (0, \pi)\,, \\ &u(t, 0) = u(t, \pi) = 0\,, \quad t \in \mathbf{R}\,, \qquad (12) \\ &u(t + \omega, x) - u(t, x) = 0\,, \quad (t, x) \in \mathbf{R} \times (0, \pi) \end{aligned}$$

is investigated by means of the (t, s)-Fourier method, and the relation between u^μ and u^0 is studied in detail, where u^μ and u^0 are the solutions to (12) and (9) (with $h_0 = h_1 = 0$), respectively.

The results for linear problems are stated in such a form that they allow us to formulate easily the results for the corresponding weakly non-linear problems with help of the Banach contraction principle in non-critical cases and with help of the implicit function theorem in critical ones.

In Chapter VI the Faedo-Galerkin method, one of the methods most frequently employed for solving strongly non linear problems, is used to examine the non-linear vibrations of thin plates. As this is the single application of the method in the present book, we omit the description of its theoretical background. In II: [13] the reader will find a number of applications of this method to initial boundary-value and periodic problems.

In Chapter VII abstract differential equations are treated. Unfortunately, lack of space compelled us to restrict ourselves to the description of the t-Fourier method (in a version that was prepared by SOVA for a lecture in Novosibirsk in 1970). This method is applied to the equation

$$M\left(\frac{\mathrm{d}}{\mathrm{d}t}\,u(t)\right) + A\,u(t) = g(t) + \varepsilon\,F(t, u(t))\,, \quad t \in \mathbf{R}\,,$$

where $M(\xi)$ is a polynomial of degree n in ξ and A is a closed (or normal or self-adjoint) operator. The (t, s)-Fourier method has been recently examined by HERRMANN (VII: [40]). The paper by STRAŠKRABA and VEJVODA (VII: [62]), where the spectral resolution of the operator A is employed, is perhaps closest from the formal point of view to the Poincaré method. If A has a point spectrum only, then, as far as the actual calculation is concerned, their method coincides with the s-Fourier method.

In Chapter I some fundamental results, concepts, and notation from functional analysis are stated. The same applies to Chapter II with regard to the theory of partial differential equations. It is hardly necessary to point out that the material presented in these two chapters represents a very special choice of topics adapted for the purposes of this book. These chapters are intended to enable the reader to check easily the formulation of those theorems that are quoted in the main text. At the first reading the reader is supposed to get acquainted with their contents only cursorily.

The last paragraphs of each chapter are usually devoted to comments. Papers whose content is close to the material treated in the main text are commented on in the form of remarks. The length of comments does not necessarily correspond to the importance of the paper. Some very important papers make use of analytical tools too complicated to be explained adequately. It should be mentioned that even in comments the reader will meet some notions that are not treated in the first two auxiliary chapters. To facilitate a comparison of our comment with the original paper we keep occasionally the author's notation even when it differs from that used in our text. We usually state not all but only the principal assumptions of the article in question. Exceptionally we modify the author's assertion to avoid the necessity of introducing new concepts. We have tried our best to comment on every paper dealing with time-periodic solutions to non-linear partial differential equations, at least briefly. The page where a paper is mentioned is given in parentheses after the corresponding bibliographical entry. As the theory of time-periodic solutions to partial differential equations has continued to develop very intensively since the book was handed over to the publishers we try to face this situation by supplementing the Bibliography by Addenda of 50 titles of publications which have appeared during the last three years.

Every chapter has its own list of references. References to Chapters III – VII include only papers on periodic solutions. References to publications of an auxiliary character are concentrated in the bibliographies to Chapters I and II. Besides, a bibliography of papers on related topics is added. It contains papers that are not so close to the main text. Let us emphasize that papers dealing with almost periodic solutions are not quoted, since the methods of studying them are quite different from those described in this book. To help the reader in finding papers of his own interest, the references to Navier-Stokes equations and to the autonomous wave equations in the bibliographies to Chapters III and V, respectively, are quoted separately. By including several papers on stability of solutions to partial and abstract differential equations in the bibliography of papers on related topics we wish to compensate for the fact that the investigation of stability of solutions is left aside because of lack of space, although a certain degree of stability is always required to guarantee the existence of a periodic motion in a physical system. The titles of journals are abbreviated as in the Mathematical Reviews. To transliterate the Cyrillic alphabet we have also used the system of the Mathematical Reviews.

Let us conclude our Introduction by explaining our method of referring to previous results (which has already been used above): if the number of a theorem, a section, a formula, a reference etc. is given without further specification, then it refers to the same chapter in which the quotation appears. If the quotation concerns another chapter, then its number is given first, for example, III: (1.1.1) denotes the formula (1.1.1) of Chapter III. If a system of equations is denoted by the same number, say (1.1.1), then $(1.1.1_n)$ denotes the n-th equation in this system.

Chapter I

Preliminaries from functional analysis

§ 1. Linear spaces and operators

1.1. *Basic notation.*

We use standard set-theory terminology and notation. Here we recall some useful symbols and concepts.

$\mathbf{N}$ – the set of all positive integers,
$\mathbf{Z}$ – the set of all integers,
$\mathbf{Z}^+$ – the set of all non-negative integers,
$\mathbf{R}$ – the set of all real numbers,
$\mathbf{R}^+$ – the set of all non-negative numbers,
$\mathbf{C}$ – the set of all complex numbers,
$\mathbf{S}$ – $\mathbf{R}$ or $\mathbf{C}$,
$\mathbf{R}^n$ – $\mathbf{R} \times \ldots \times \mathbf{R}$ (n factors),

$\{x \in X;\ P(x)\}$ – the subset of X containing all elements that have the property $P(x)$,

$[a, b]$ – the closed interval $\{x \in \mathbf{R};\ a \leqq x \leqq b\}$,

(a, b) – the open interval $\{x \in \mathbf{R};\ a < x < b\}$,

$[a, b)$, $(a, b]$ – semiclosed intervals.

If M is a subset of a metric space X, then we denote by $\operatorname{cl} M$ or $\overline{M}$ the closure of M in X and by ∂M its boundary.

A non-void subset $\Omega \subset \mathbf{R}^n$ is called a *domain* if $\operatorname{cl}(\Omega \setminus \partial\Omega) \supset \Omega$.

By a *ball* in a metric space X $(= (X, d))$ with centre at $u \in X$ and radius $\varrho > 0$ we mean the set

$$B(u, \varrho; X) = \{v \in X;\ d(u, v) \leqq \varrho\}\,.$$

1.2. *Banach spaces*

Let B be a Banach space (over $\mathbf{S}$). Two norms $\|\cdot\|_1$ and $\|\cdot\|_2$ on B are said to be *equivalent* ($\|\cdot\|_1 \sim \|\cdot\|_2$), if there are $c_1 > 0$ and $c_2 > 0$ such that

$$c_1\|u\|_2 \leqq \|u\|_1 \leqq c_2\|u\|_2\,, \quad u \in B\,.$$

If M is a subset of B, then we denote by $\operatorname{lin} M$ the set of all elements $u \in B$ of the form $u = \sum_{j=1}^{n} \alpha_j v_j$, where $n \in \mathbf{N}$, $\alpha_j \in \mathbf{S}$, and $v_j \in M$.

By $\mathscr{D}(A)$ ($\mathscr{R}(A)$) we denote the *domain* (the *range*) of a mapping (operator or function) A from a Banach space B_1 to another Banach space B_2. We write $A : \mathscr{D}(A) \to B_2$ or $A : \mathscr{D}(A) \subset B_1 \to B_2$. If $\operatorname{cl} \mathscr{D}(A) = B_1$, we say that A is *densely defined*. The restriction of A to a set $M \subset B_1$ is denoted by $A \mid M$. If A is linear, then by the *null-space* of A we mean the set $\mathscr{N}(A) = \{u \in \mathscr{D}(A); Au = 0\}$.

For any linear operator $A : \mathscr{D}(A) \subset B_1 \to B_2$ we define on $\mathscr{D}(A)$ the so-called *graph norm*

$$\|u\|_{\mathscr{D}(A)} = (\|u\|_{B_1}^2 + \|Au\|_{B_2}^2)^{1/2}\,.$$

$\mathscr{D}(A)$ is a Banach space with respect to this graph norm if and only if A is *closed*, that is, whenever $u_n \in \mathscr{D}(A)$, $u_n \to u$ and $Au_n \to z$, then $u \in \mathscr{D}(A)$ and $Au = z$.

By $\mathscr{L}(B_1, B_2)$ we denote the Banach space of all continuous linear operators (or, what is the same, of all bounded linear operators) $A : B_1 \to B_2$.

THEOREM 1.2.1 (*Banach closed graph theorem, see* [43], *p.* 77). *Let B_1 and B_2 be Banach spaces. If $A : B_1 \to B_2$ is a closed linear operator, then $A \in \mathscr{L}(B_1, B_2)$.*

THEOREM 1.2.2 (*Banach theorem on continuity of the inverse operator, see* [43], *p.* 77). *Let B_1 and B_2 be Banach spaces. If $A \in \mathscr{L}(B_1, B_2)$ and A is a one-to-one operator onto B_2 ($\mathscr{R}(A) = B_2$), then $A^{-1} \in \mathscr{L}(B_2, B_1)$.*

1.3. *Hilbert spaces*

A normed linear space K is called an *inner* (or *scalar*) *product space* if there exists a function $f : K \times K \to \mathbf{S}$ such that

(1) $f(\alpha u, v) = \alpha f(u, v)$,
(2) $f(u + v, w) = f(u, w) + f(v, w)$,

(3) $f(u, v) = \overline{f(v, u)}$ (the bar here denotes the complex conjugate),
(4) $f(u, u) = \|u\|^2$

for all $u, v, w \in K$, $\alpha \in \mathbf{S}$.

The function f is uniquely determined and is called the inner (scalar) product on the space K. We write $f(u, v) = \langle u, v \rangle$ or $\langle u, v \rangle_K$.

The above properties imply that

$$|\langle u, v \rangle| \leqq \|u\| \, \|v\| \,, \quad u, v \in K$$

(*the Schwarz inequality*).

A complete inner product space is called a *Hilbert space*. Evidently, it is an inner product Banach space.

Two elements u and v of an inner product space K are called *orthogonal* if $\langle u, v \rangle = 0$. If M is an arbitrary subset of K, then by the *orthogonal complement* of M in K we mean the set

$$M^{\perp} = \{u \in K; \langle u, v \rangle = 0 \quad \text{for every} \quad v \in M\} \,.$$

THEOREM 1.3.1 (*see* [33], *p.* 79). *Let H be a Hilbert space and M a closed subspace of H. Then there exists a unique pair of mappings P and Q such that $P : H \to M$, $Q : H \to M^{\perp}$, and*

$$u = Pu + Qu \,, \quad u \in H \,.$$

Moreover, P and Q are linear; if $u \in M$, then $Pu = u$, $Qu = 0$, and if $u \in M^{\perp}$, then $Pu = 0$, $Qu = u$.

The operators P and Q $(= I - P)$ are called the *orthogonal projections* of H onto M and $M^{\perp}$, respectively.

1.4. *Fourier series*

Let K be an inner product space. A system $\{e_j\}_{j \in \mathbf{J}}$ ($\mathbf{J}$ is an index set, possibly uncountable) from K is called *orthogonal* in K if $\langle e_j, e_k \rangle = 0$, $j, k \in \mathbf{J}$, $j \neq k$, and *orthonormal* in K if, moreover, $\|e_j\| = 1$, $j \in \mathbf{J}$.

If K is separable (which is the case we shall deal with in this book), then any orthonormal system in K is at most countable. In most of our applications it is $\mathbf{J} = \mathbf{Z}$ or $\mathbf{J} = \mathbf{N}$.

THEOREM 1.4.1 (*see* [7], *p.* 151). *Let $\{e_j\}_{j \in \mathbf{J}}$ be an orthonormal system in a separable Hilbert space and $\{\alpha_j\}_{j \in \mathbf{J}}$ a system of complex numbers. Then the series $\sum_{j \in \mathbf{J}} \alpha_j e_j$ is convergent if and only if $\sum_{j \in \mathbf{J}} |\alpha_j|^2 < \infty$. Moreover,*

$$\Big\langle \sum_{j \in \mathbf{J}} \alpha_j e_j, e_k \Big\rangle = \alpha_k \,, \quad k \in \mathbf{J} \,.$$

THEOREM 1.4.2 (*see* [7], *p.* 152). *If* $\{e_j\}_{j\in \mathbf{J}}$ *is an orthonormal system in a separable inner product space* K, *then the following statements are equivalent*:

(*a*) *the set of all finite linear combinations of members of* $\{e_j\}_{j\in \mathbf{J}}$ *is dense in* K, *that is* $\mathrm{cl}\,\mathrm{lin}\,\{e_j\}_{j\in \mathbf{J}} = K$;

(*b*) *for any* $u \in K$,

$$u = \sum_{j\in \mathbf{J}} \langle u, e_j\rangle\, e_j\ ;$$

(*c*) *for any* $u \in K$,

$$\|u\|^2 = \sum_{j\in \mathbf{J}} |\langle u, e_j\rangle|^2$$

(*the Parseval identity*).

If one of the conditions (a)–(c) is satisfied, then $\{e_j\}_{j\in \mathbf{J}}$ is said to be an *orthonormal base* (or an *orthonormal complete system*) *in* K. An orthonormal system $\{e_j\}_{j\in \mathbf{J}}$ in a Hilbert space is an orthonormal base if and only if $\langle u, e_j\rangle = 0$ for any $j \in \mathbf{J}$ implies that $u = 0$. Any Hilbert space has an orthonormal base.

The series in (b) is called the *Fourier series of* u *relative to* $\{e_j\}_{j\in \mathbf{J}}$, and the numbers $\langle u, e_j\rangle$ are called the *Fourier coefficients of* u *relative to* $\{e_j\}_{j\in \mathbf{J}}$.

1.5. *Self-adjoint operators*

Let K be an inner product space and $A : \mathscr{D}(A) \subset K \to K$ a linear operator. If A is densely defined, then there exists a unique operator A^* from K to K, called the *adjoint* of A, that is characterized by the following properties:

(1) $\langle Au, v\rangle = \langle u, A^*v\rangle$, $u \in \mathscr{D}(A)$, $v \in \mathscr{D}(A^*)$,

(2) $v \in \mathscr{D}(A^*)$ if and only if there is a $w \in K$ such that $\langle Au, v\rangle = \langle u, w\rangle$ for any $u \in \mathscr{D}(A)$.

THEOREM 1.5.1 (*see* [31], *p.* 252). *Let* H *be a Hilbert space and* $A : \mathscr{D}(A) \subset H \to H$ *a linear operator. If* A *is densely defined and closed, then the adjoint* A^* *is densely defined and* $(A^*)^* = A$. *Moreover*, $\mathrm{cl}\,\mathscr{R}(A) = \mathscr{N}(A^*)^\perp$ *and* $\mathrm{cl}\,\mathscr{R}(A) + \mathscr{N}(A^*) = H$.

A linear operator $A : \mathscr{D}(A) \subset K \to K$ is said to be *formally self-adjoint* (or *symmetric*) if $\langle Au, v\rangle = \langle u, Av\rangle$, $u, v \in \mathscr{D}(A)$. Let K be a Hilbert space. Then A is said to be

self-adjoint if it is densely defined and $A = A^*$,

normal if it is densely defined, closed and $A^*A = AA^*$.

If A is densely defined, then A is formally self-adjoint if and only if it is the restriction of A^*. Every self-adjoint operator is normal. If A is self-adjoint, then also A^n is self-adjoint for any $n \in \mathbf{N}$. If A is self-adjoint and one-to-one, then A^{-1} is densely defined and self-adjoint. An operator A is normal if and only if it is densely defined, that is if $\mathscr{D}(A) = \mathscr{D}(A^*)$, and $\|Au\| = \|A^*u\|$ for all $u \in \mathscr{D}(A)$.

A formally self-adjoint operator is said to be *bounded below* (*non-negative, positive, positive definite*) if

$$\langle Au, u\rangle \geqq c\|u\|^2, \quad u \in \mathscr{D}(A)$$

with a suitable $c \in \mathbf{R}$ ($c = 0$, $c = 0$ and $\langle Au, u\rangle = 0$ only if $u = 0$, $c > 0$).

Every formally self-adjoint operator in a Hilbert space that is bounded below has a self-adjoint extension.

THEOREM 1.5.2 *Let H be a Hilbert space and $A : \mathscr{D}(A) \subset H \to H$ a self-adjoint positive definite operator. Then $\mathscr{D}(A)$ equipped with the norm $\|Au\|$ (which is equivalent to the graph norm) is a Hilbert space. The inner product is given by $\langle Au, Av\rangle$.*

1.6. *Spectral properties of operators; compact operators*

Let B be a Banach space over $\mathbf{C}$ and let $A : \mathscr{D}(A) \subset B \to B$ be a linear closed operator.

The set of all $\lambda \in \mathbf{C}$ for which the operator $\lambda - A$ is one-to-one and $(\lambda - A)^{-1} \in \mathscr{L}(B, B)$ is called the *resolvent set* of the operator A and is denoted by $\varrho(A)$. The set $\sigma(A) = \mathbf{C} \setminus \varrho(A)$ is called the *spectrum of A*. The spectrum $\sigma(A)$ consists of three disjoint sets:

the point spectrum $\sigma_p(A) = \{\lambda \in \mathbf{C};\ \mathscr{N}(\lambda - A) \neq \{0\}\}$,
the continuous spectrum $\sigma_c(A) = \{\lambda \in \mathbf{C};\ \mathscr{N}(\lambda - A) = \{0\}$, $B = \operatorname{cl} \mathscr{R}(\lambda - A) \neq \mathscr{R}(\lambda - A)\}$,
the residual spectrum $\sigma_r(A) = \{\lambda \in \mathbf{C};\ \mathscr{N}(\lambda - A) = \{0\}$, $\operatorname{cl} \mathscr{R}(\lambda - A) \neq B\}$.

The elements of $\sigma_p(A)$ are called *eigenvalues* of A. If $\lambda \in \sigma_p(A)$, then the non-trivial elements of $\mathscr{N}(\lambda - A)$ are called *eigenvectors* (or *eigenfunctions*) if B is a function space) of A corresponding to the eigenvalue λ. The dimension of $\mathscr{N}(\lambda - A)$ is called the *multiplicity* of λ.

The spectrum $\sigma(A)$ is a closed set; the void set or the whole $\mathbf{C}$ are not excluded. If $A \in \mathscr{L}(B, B)$, then $\sigma(A)$ is a non-void compact set.

THEOREM 1.6.1 (*see* [32], *p.* 434). *Let H be a Hilbert space and $A : \mathscr{D}(A) \subset \subset H \to H$ a linear operator. If A is self-adjoint, then $\sigma(A) \subset \mathbf{R}$ and, in particular, A has only real eigenvalues. If A is a normal operator such that $\sigma(A) \subset \mathbf{R}$, then A is self-adjoint.*

Let us recall that an operator (possibly non-linear) F from a Banach space B_1 to another Banach space B_2 is called *compact* if for every bounded subset $M \subset \mathscr{D}(A)$ the set $F(M)$ is relatively compact (that is, any sequence in $F(M)$ contains a convergent subsequence in B_2).

THEOREM 1.6.2 (*see* [32], *p.* 231). *Let H be a Hilbert space and let $A \in \mathscr{L}(H, H)$ be normal and compact. Then*

(1) *there exists an orthonormal base $\{e_j\}_{j \in \mathbf{J}}$ in H of eigenvectors of A with the corresponding system of eigenvalues $\{\lambda_j\}_{j \in \mathbf{J}}$, that is, $Ae_j = \lambda_j e_j$, $j \in \mathbf{J}$,*

(2) *the set $\{j \in \mathbf{J};\ |\lambda_j| \geqq \gamma\}$ is finite for every $\gamma > 0$,*

(3) *the set $\{k \in \mathbf{J};\ \lambda_k = \lambda_j\}$ is finite for every $j \in \mathbf{J}$ for which $\lambda_j \neq 0$.*

If, in particular, the operator A is self-adjoint, then the numbers $\{\lambda_j\}_{j \in \mathbf{J}}$ are real.

Let us modify these results to fit a situation that occurs very often in the theory of partial differential operators.

THEOREM 1.6.3 (*see* [32], *p.* 231). *Let H be an infinite-dimensional Hilbert space and $A : \mathscr{D}(A) \subset H \to H$ a one-to-one normal operator such that $A^{-1} \in \mathscr{L}(H, H)$ is compact. Then*

(1) *the space H is separable,*

(2) *there exists a countable orthonormal base $\{e_j\}_{j \in \mathbf{J}}$ in H of eigenvectors of A with the corresponding eigenvalues $\{\lambda_j\}_{j \in \mathbf{J}}$,*

(3) *the set $\{j \in \mathbf{J};\ |\lambda_j| \leqq c\}$ is finite for every $c > 0$.*

If, in particular, A is a self-adjoint operator, then the numbers $\{\lambda_j\}_{j \in \mathbf{J}}$ are real.

The next theorem is useful when solving equations of the type $Au - \lambda u = g$.

THEOREM 1.6.4 (*see* [39], *pp.* 279–281). *Let H be a separable Hilbert space, $A \in \mathscr{L}(H, H)$. Suppose that there exists an orthonormal base $\{e_j\}_{j \in \mathbf{J}}$ in H of eigenvectors of A with the corresponding eigenvalues $\{\lambda_j\}_{j \in \mathbf{J}}$. Then*

$$Au = \sum_{j \in \mathbf{J}} \lambda_j \langle u, e_j \rangle \, e_j \quad \textit{for all} \quad u \in H\,,$$

$$Au = \sum_{\substack{j \in \mathbf{J} \\ \lambda_j \neq \lambda}} \lambda_j \frac{\langle \lambda u - Au, e_j \rangle}{\lambda - \lambda_j} \, e_j + \lambda \sum_{\substack{j \in \mathbf{J} \\ \lambda_j = \lambda}} \langle u, e_j \rangle \, e_j$$

for all $u \in H$ and $\lambda \in \mathbf{S}$,

$$u = \frac{1}{\lambda}(\lambda u - Au) + \frac{1}{\lambda} \sum_{\substack{j \in \mathbf{J} \\ \lambda_j \neq \lambda}} \lambda_j \frac{\langle \lambda u - Au, e_j \rangle}{\lambda - \lambda_j} e_j + \sum_{\substack{j \in \mathbf{J} \\ \lambda_j = \lambda}} \langle u, e_j \rangle e_j$$

for all $u \in H$, $\lambda \in \mathbf{S}$, $\lambda \neq 0$.

Let H be a separable Hilbert space and $A : \mathscr{D}(A) \subset H \to H$ a self-adjoint operator. Suppose that there exists an orthonormal base $\{e_j\}_{j \in \mathbf{J}}$ in H of eigenvectors of A with the corresponding eigenvalues $\{\lambda_j\}_{j \in \mathbf{J}}$. If A is non-negative (positive, positive definite) then $\lambda_j \geqq 0$ ($\lambda_j > 0$, $\lambda_j \geqq c > 0$) for all $j \in \mathbf{J}$. Assume that A is non-negative. The *powers* A^α for $\alpha > 0$ are defined as follows:

(1) $u \in \mathscr{D}(A^\alpha)$ if and only if $\sum_{j \in \mathbf{J}} \lambda_j^{2\alpha} |\langle u, e_j \rangle|^2 < \infty$,

(2) $A^\alpha u = \sum_{j \in \mathbf{J}} \lambda_j^\alpha \langle u, e_j \rangle e_j$, $u \in \mathscr{D}(A^\alpha)$.

A^α is the unique self-adjoint operator for which $A^\alpha e_j = \lambda_j^\alpha e_j$, $j \in \mathbf{J}$. A^α is non-negative, positive, or positive definite in accordance with the corresponding properties of A. If $\alpha > 0$, $\beta > 0$, then $\langle A^{\alpha+\beta} u, v \rangle = \langle A^\alpha u, A^\beta v \rangle$, $u \in \mathscr{D}(A^{\alpha+\beta})$, $v \in \mathscr{D}(A^\beta)$.

1.7. *Embeddings; negative norms*

Let B_1 and B_2 be Banach spaces. If $B_1 \subset B_2$ and the unit ball $B(0, 1; B_1)$ of B_1 is a bounded (relatively compact) subset of B_2, then we say that B_1 *is embedded* (*compactly embedded*) *in* B_2. If, moreover, B_1 is a dense subset of B_2, then B_1 is said to be *densely embedded* or *densely compactly embedded* in B_2.

Let H_1 be a Hilbert space that is densely embedded in a Hilbert space H_0. We define on H_0 the so-called *negative norm* $\|\cdot\|_{-1}$ by the identity

$$\|u\|_{-1} = \sup \{\langle u, v \rangle; \|v\|_{H_1} \leqq 1\}.$$

The linear space H_0 becomes an inner product space with respect to this negative norm, and the corresponding completion of H_0 is denoted by $H_{-1}(H_0, H_1)$.

THEOREM 1.7.1 (*cf.* [43], *p.* 98 *or* [30], *pp.* 166–167). *Let H_0 and H_1 be two Hilbert spaces. If H_1 is densely embedded in H_0, then (we write H_{-1} instead of $H_{-1}(H_0, H_1)$)*

(*a*) *there exists a unique continuous function $g : H_1 \times H_{-1} \to \mathbf{S}$ such that $g(u, v) = \langle u, v \rangle_{H_0}$ for any $u \in H_1$ and $v \in H_0$,*

(*b*) *there exists an isometric isomorphism T_0 of H_{-1} onto H_1 such that $\langle u, T_0 v \rangle_{H_1} = g(u, v)$ for any $u \in H_1$ and $v \in H_{-1}$.*

1.8. *Differentials*

Let X and Y be Banach spaces and $\Omega \subset X$ an open set. A mapping $F : \Omega \to Y$ is called *Gâteaux differentiable* (*G-differentiable*) *at a point* $x \in \Omega$ if there exists an operator $T \in \mathscr{L}(X, Y)$ such that

$$\lim_{\tau \to 0} \frac{1}{\tau} (F(x + \tau h) - F(x)) = Th$$

for any $h \in X$.

The operator T is called the *Gâteaux derivative* (*G-derivative*) *of* F *at* x and is denoted by $DF(x)$.

Further, a mapping $F : \Omega \to Y$ is called *Fréchet differentiable* (*F-differentiable*) *at a point* $x \in \Omega$ if there exists an operator $T \in \mathscr{L}(X, Y)$ such that

$$\lim_{h \to 0} \frac{\|F(x + h) - F(x) - Th\|}{\|h\|} = 0 .$$

The operator T is called the *Fréchet derivative* (*F-derivative*) *of* F *at* x and is denoted by $F'(x)$.

THEOREM 1.8.1 *If* $F : \Omega \to Y$ *is F-differentiable at a point* $x \in \Omega$, *then it is also G-differentiable at this point and* $F'(x) = DF(x)$.

THEOREM 1.8.2 (*see* [29], *p.* 6). *Let* $F : \Omega \to Y$ *be G-differentiable at every point* $x \in \Omega$. *If* $tx_2 + (1 - t) x_1 \in \Omega$ *for any* $t \in [0, 1]$, *then* $\|F(x_1) - F(x_2)\| \leqq \sup_{0 \leqq t \leqq 1} \|DF(tx_2 + (1 - t) x_1)\| \|x_1 - x_2\|$.

THEOREM 1.8.3 (*see* [29], *p.* 7). *If* $F : \Omega \to Y$ *is G-differentiable in* Ω *and* $DF : \Omega \to \mathscr{L}(X, Y)$ *is continuous at a point* $x \in \Omega$, *then* F *is F-differentiable at* x *and* $DF(x) = F'(x)$.

This theorem enables us to use the symbol F' for the continuous G-derivative of an operator F.

THEOREM 1.8.4 *If* $F : \Omega \to Y$ *is F-differentiable at a point* $x \in \Omega$, *then* F *is continuous at this point.*

THEOREM 1.8.5 (*see* [29], *p.* 3). *Let* X, Y, *and* Z *be three Banach spaces,* Ω *an open set in* X, Λ *an open set in* Y, $F : \Omega \to \Lambda$, *and* $G : \Lambda \to Z$. *If the mapping* F *is F-differentiable at a point* x *and* G *is F-differentiable at the point* $F(x)$, *then the composite mapping* $G \circ F$ *is F-differentiable at* x *and* $(G \circ F)'(x) = G'(F(x)) \circ F'(x)$.

THEOREM 1.8.6 *If $A \in \mathscr{L}(X, Y)$, then the mapping A is F-differentiable at all points $x \in X$ and $A'(x) = A$ for all $x \in X$.*

In some cases we shall consider *partial Gâteaux* or *Fréchet derivatives* in the following sense: Let $X_1, X_2, \ldots, X_n$, and Y be Banach spaces, Ω an open set in $X_1 \times X_2 \times \ldots \times X_n$ and $F : \Omega \to Y$. Then F is called G- or F-differentiable at a point $(x_1, x_2, \ldots, x_n) \in \Omega$ with respect to x_j if the function $F_0(x) = F(x_1, x_2, \ldots, x, \ldots, x_n)$ is respectively G- or F-differentiable at x_j. This derivative is then denoted by $D_{x_j}F\ (x_1, x_2, \ldots, x_n)$ or $F'_{x_j}(x_1, x_2, \ldots, x_n)$, respectively.

§ 2. Function spaces

2.1. *Spaces of smooth functions*

Let $\Omega \subset \mathbf{R}^n$ be a domain, B a Banach space.

As usual, D^α $(\alpha \in (\mathbf{Z}^+)^n)$ denotes the partial differential operator

$$D^\alpha (= D^\alpha_x = D^{\alpha_1}_{x_1} \ldots D^{\alpha_n}_{x_n}) = \frac{\partial^{|\alpha|}}{\partial x_1^{\alpha_1} \ldots \partial x_n^{\alpha_n}}$$

$(|\alpha| = \alpha_1 + \ldots + \alpha_n)$.

Let $k \in \mathbf{Z}^+$, $\gamma \in (\mathbf{Z}^+)^n$. By $C^k_{\mathrm{loc}}(\Omega; B)$ $(C^{\gamma_1,\ldots,\gamma_n}_{\mathrm{loc}}(\Omega; B)$, $C^{(\gamma_1,\ldots,\gamma_n)}_{\mathrm{loc}}(\Omega; B))$ we denote the linear space of all (vector-valued) functions $u : \Omega \to B$ that have continuous partial derivatives $D^\alpha u$ for all $\alpha \in \Gamma_1$ (Γ_2, Γ_3), where

$$\Gamma_1 = \{\alpha \in (\mathbf{Z}^+)^n;\ |\alpha| \leqq k\}\,, \tag{2.1.1}$$

$$\Gamma_2 = \{\alpha \in (\mathbf{Z}^+)^n;\ \alpha = (0, \ldots, \alpha_j, \ldots, 0)\,,\ \alpha_j \leqq \gamma_j\,,\ j = 1, \ldots, n\}\,, \tag{2.1.2}$$

$$\Gamma_3 = \{\alpha \in (\mathbf{Z}^+)^n;\ \alpha_j \leqq \gamma_j, j = 1, \ldots, n\}\,. \tag{2.1.3}$$

By $C^k(\Omega; B)$ $(C^{\gamma_1,\ldots,\gamma_n}(\Omega; B)$, $C^{(\gamma_1,\ldots,\gamma_n)}(\Omega; B))$ we denote the Banach space of all functions $u \in C^k_{\mathrm{loc}}(\Omega; B)$ $(C^{\gamma_1,\ldots,\gamma_n}_{\mathrm{loc}}(\Omega; B)$, $C^{(\gamma_1,\ldots,\gamma_n)}_{\mathrm{loc}}(\Omega; B))$ for which $D^\alpha u(\Omega)$ is a bounded subset of B for all $\alpha \in \Gamma_1$ (Γ_2, Γ_3). The norm is given by

$$\sum_{\alpha \in \Gamma_j} \sup_{x \in \Omega} \|D^\alpha u(x)\|_B\,.$$

Let us note that if Ω is not an open set, then the statement "u has a continuous derivative $D^\alpha u$" means that there exists a uniquely determined function v_α that is continuous on Ω and such that its restriction to $\Omega \setminus \partial\Omega$ coincides with $D^\alpha u$. (This case occurs if $D^\alpha u$ is uniformly continuous and bounded on $\Omega \setminus \partial\Omega$.) Then we regard $D^\alpha u$ on Ω as v_α.

Further, we write $C^\infty(\Omega; B) = \bigcap_{k=0}^{\infty} C^k(\Omega; B)$, while $C_0^\infty(\Omega; B)$ means the subspace of $C^\infty(\Omega; B)$ of all functions u whose *supports* $\mathrm{cl}\,\{x \in \Omega;\ u(x) \neq 0\}$ are compact subsets of $\Omega \setminus \partial\Omega$.

We write $C_{\mathrm{loc}}(\Omega; B)$ (and $C(\Omega; B)$) instead of $C^0_{\mathrm{loc}}(\Omega; B)$ (and $C^0(\Omega; B)$).

If $B = \mathbf{R}$, we write $C^k_{\mathrm{loc}}(\Omega)$ instead of $C^k_{\mathrm{loc}}(\Omega; \mathbf{R})$ and similarly in the other cases.

2.2. *Spaces of integrable functions*

Let $\Omega \subset \mathbf{R}^n$ be a measurable set and B a Banach space.

Let $p \in [1, \infty)$. By $L_p(\Omega; B)$ we denote the Banach space of all functions $u : \Omega \to B$ that are (strongly) measurable and for which $\|u(\cdot)\|_B^p$ is a Lebesgue integrable function on Ω (actually, u represents an equivalence class of functions that coincide almost everywhere). The norm is given by

$$\left(\int_\Omega \|u(x)\|^p \,\mathrm{d}x\right)^{1/p}.$$

Let us note that if B is separable, then the strong measurability of u means that $\langle u^*, u(\cdot)\rangle$ is a Lebesgue measurable function for all u^* from the dual space of B.

By $L_\infty(\Omega; B)$ we denote the Banach space of all (classes of) functions $u : \Omega \to B$ that are (strongly) measurable and essentially bounded in Ω. The norm is given by

$$\sup_{x\in\Omega} \operatorname{ess} \|u(x)\|_B .$$

If $B = \mathbf{R}$, we write $L_p(\Omega)$ instead of $L_p(\Omega; \mathbf{R})$.

THEOREM 2.2.1 (*see* [31], *p.* 51). *Let $\Omega \subset \mathbf{R}^n$ and $\Lambda \subset \mathbf{R}^m$ be measurable sets. If $\{\varphi_j\}_{j\in\mathbf{J}}$ and $\{\chi_k\}_{k\in\mathbf{K}}$ are orthonormal bases in $L_2(\Omega; \mathrm{S})$ and $L_2(\Lambda; \mathrm{S})$ respectively, then the system of functions $\{\varphi_j(x)\,\chi_k(y)\}_{j\in\mathbf{J},k\in\mathbf{K}}$ is an orthonormal base in $L_2(\Omega \times \Lambda; \mathrm{S})$.*

THEOREM 2.2.2 *Let $\Omega \subset \mathbf{R}^n$ be a measurable set and H a Hilbert space. If $\{\varphi_j\}_{j\in\mathbf{J}}$ is an orthonormal base in $L_2(\Omega; \mathrm{S})$, then for all $u \in L_2(\Omega; H)$*

$$(a)\ \sum_{j\in\mathbf{J}} \int_\Omega \overline{\varphi_j(\tau)}\, u(\tau)\,\mathrm{d}\tau\, \varphi_j = u\,,$$

$$(b)\ \sum_{j\in\mathbf{J}} \left\| \int_\Omega \overline{\varphi_j(\tau)}\, u(\tau)\,\mathrm{d}\tau \right\|^2 = \|u\|^2_{L_2(\Omega;H)}\,.$$

THEOREM 2.2.3 *Let $\Omega \subset \mathbf{R}^n$ be a measurable set and H a Hilbert space. If $\{\varphi_j\}_{j\in\mathbf{J}}$ is an orthonormal base in $L_2(\Omega; \mathrm{S})$ and $\{e_k\}_{k\in\mathbf{K}}$ an orthonormal base in H, then the system $\{\varphi_j e_k\}_{j\in\mathbf{J},k\in\mathbf{K}}$ is an orthonormal base in $L_2(\Omega; H)$.*

2.3. *Sobolev spaces. (Spaces of differentiable functions)*

Let $\Omega \subset \mathbf{R}^n$ be a measurable domain and B a Banach space.

We say that a locally integrable function $u : \Omega \to B$ has a *generalized derivative* $D^\alpha u$ $(\alpha \in (\mathbf{Z}^+)^n)$ if there exists a locally integrable function v_α such that

$$\int_\Omega D^\alpha \varphi\, u \,\mathrm{d}x = (-1)^{|\alpha|} \int_\Omega \varphi v_\alpha \,\mathrm{d}x$$

for all $\varphi \in C_0^\infty(\Omega)$. We put $D^\alpha u = v_\alpha$.

Let $k \in \mathbf{Z}^+$, $\gamma \in (\mathbf{Z}^+)^n$, $p \in [1, \infty)$ or $p = \infty$. By $W_p^k(\Omega; B)$ $(W_p^{\gamma_1,\dots,\gamma_n}(\Omega; B)$, or $W_p^{(\gamma_1,\dots,\gamma_n)}(\Omega; B))$ we denote the Banach space of all (classes of) functions $u : \Omega \to B$ such that the (generalized) derivatives $D^\alpha u$ belong to $L_p(\Omega; B)$ for all $\alpha \in \Gamma_1$ $(\Gamma_2$, or $\Gamma_3)$, where $\Gamma_1 - \Gamma_3$ are given by (2.1.1)–(2.1.3). The norm is given by

$$\Big(\sum_{\alpha\in\Gamma_j} \|D^\alpha u\|^p_{L_p(\Omega;B)}\Big)^{1/p} \quad \text{if} \quad p \in [1, \infty)$$

and

$$\sum_{\alpha\in\Gamma_j} \|D^\alpha u\|_{L_\infty(\Omega;B)} \quad \text{if} \quad p = \infty\,.$$

If $k \in \mathbf{R}^+$ is not an integer, then $W_p^k(\Omega; B)$ denotes the subspace of $W_p^{[k]}(\Omega; B)$ (for $k \in \mathbf{R}$, $[k]$ denotes the greatest integer less than or equal to k) consisting of all $u \in W_p^{[k]}(\Omega; B)$ such that

$$I_\alpha(u) = \int_\Omega \int_\Omega \frac{\|D^\alpha u(x) - D^\alpha u(y)\|_B^p}{\|x - y\|^{n+p(k-[k])}} \,\mathrm{d}x\,\mathrm{d}y < \infty$$

for any $|\alpha| = [k]$. The norm is given by

$$\Big(\|u\|^p_{W_p{}^{[k]}(\Omega;B)} + \sum_{|\alpha|=[k]} I_\alpha(u)\Big)^{1/p}.$$

Instead of $W_2^k(\Omega; B)$ $(W_2^{\gamma_1,\dots,\gamma_n}(\Omega; B)$, $W_2^{(\gamma_1,\dots,\gamma_n)}(\Omega; B))$ we write $H^k(\Omega; B)$ $(H^{\gamma_1,\dots,\gamma_n}(\Omega; B)$, $H^{(\gamma_1,\dots,\gamma_n)}(\Omega; B))$.

Evidently, $H^0(\Omega;\ B) = L_2(\Omega; B)$. If $B = H$ is a Hilbert space, then $H^k(\Omega; H)$ $(H^{\gamma_1,\dots,\gamma_n}(\Omega; H)$, $H^{(\gamma_1,\dots,\gamma_n)}(\Omega; H))$ is also a Hilbert space with the inner product

$$\sum_{\alpha\in\Gamma_j} \int_\Omega \langle D^\alpha u, D^\alpha v\rangle_H \,\mathrm{d}x\,.$$

If $B = \mathbf{R}$, we write $W_p^k(\Omega)$ instead of $W_p^k(\Omega; \mathbf{R})$ and similarly in the other cases.

If $\Omega = (a, b) \subset \mathbf{R}$, we write $W_p^k(a, b; B)$ instead of $W_p^k((a, b); B)$.

Now let S be one of the spaces defined in *this* section. Then clearly $C_0^\infty(\Omega; B)$ is a subspace of S, and we define

$$\dot{S} = \operatorname{cl} C_0^\infty(\Omega; B) \quad \text{in} \quad S .$$

2.4. *Periodic functions*

Let $\omega > 0$ and let B be a Banach space. A measurable function $u : \mathbf{R} \to B$ is called *ω-periodic* if

$$u(t + \omega) = u(t)$$

for almost all $t \in \mathbf{R}$.

Let $k \in \mathbf{Z}^+$ or $k = \infty$. By $C_\omega^k(\mathbf{R}; B)$ we denote the subspace of $C^k(\mathbf{R}; B)$ of all ω-periodic functions.

We write $C_\omega(\mathbf{R}; B)$ instead of $C_\omega^0(\mathbf{R}; B)$.

Let $p \in [1, \infty)$ or $p = \infty$. By $L_{p,\omega}(\mathbf{R}; B)$ we denote the Banach space of all (classes of) ω-periodic functions $u : \mathbf{R} \to B$ for which $u \mid (0, \omega) \in L_p(0, \omega; B)$. The norm is given by $\|u \mid (0, \omega)\|_{L_p(0,\omega;B)}$.

Let $\varkappa \in \mathbf{R}^+$, $p \in [1, \infty)$ or $p = \infty$. By $W_{p,\omega}^\varkappa(\mathbf{R}; B)$ we denote the Banach space of all (classes of) ω-periodic functions $u : \mathbf{R} \to B$ such that $u \mid (-M, M) \in W_p^\varkappa(-M, M; B)$ for all $M > 0$. The norm is given by $\|u \mid (0, \omega)\|_{W_p^\varkappa(0,\omega;B)}$.

Instead of $W_{2,\omega}^\varkappa(\mathbf{R}; B)$ we write $H_\omega^\varkappa(\mathbf{R}; B)$. Evidently, $H_\omega^0(\mathbf{R}; B) = L_{2,\omega}(\mathbf{R}; B)$. If $B = H$ is a Hilbert space, then so is $H_\omega^\varkappa(\mathbf{R}; H)$.

If $B = \mathbf{R}$, we write simply $C_\omega^k(\mathbf{R})$, $L_{p,\omega}(\mathbf{R})$, $W_{p,\omega}^\varkappa(\mathbf{R})$, and $H_\omega^\varkappa(\mathbf{R})$.

THEOREM 2.4.1 *The system of functions* $\{\omega^{-1/2} e^{ij\nu t}\}_{j\in\mathbf{Z}}$ $(\nu = 2\pi/\omega)$ *forms an orthonormal base in* $H_\omega^0(\mathbf{R}; \mathbf{C})$. *Furthermore, the system of functions*

$$\{\omega^{-1/2}(1 + \nu^2 j^2 + \ldots + \nu^{2k} j^{2k})^{-1/2} e^{ij\nu t}\}_{j\in\mathbf{Z}}$$

forms an orthonormal base in $H_\omega^k(\mathbf{R}; \mathbf{C})$ *for any* $k \in \mathbf{Z}^+$.

2.5. *Representation of vector-valued functions*

The following theorems form a fundamental tool for the development of Fourier methods. Their proofs follow from the results introduced in Sections 1.4–1.6, 2.2 and 2.4. In the proofs of Theorems 2.5.1 and 2.5.2 one also uses the Dini theorem (see, for example, [11], p. 135).

THEOREM 2.5.1 *Let $\Omega \subset \mathbf{R}^n$ be a domain, $k \in \mathbf{Z}^+$, H a separable Hilbert space, and $\{e_j\}_{j \in \mathbf{J}}$ an orthonormal base in H. Let $u : \Omega \to H$ and let us write*

$$u_j(x) = \langle u(x), e_j \rangle, \quad x \in \Omega, \quad j \in \mathbf{J}. \tag{2.5.1}$$

Then $u \in C^k(\Omega; H)$ if and only if the following conditions are satisfied:

(*a*) $u_j \in C^k(\Omega; \mathbf{S})$, $\quad j \in \mathbf{J}$,
(*b*) $\sup_{x \in \Omega} \sum_{j \in \mathbf{J}} |D^\alpha u_j(x)|^2 < \infty$, $\quad |\alpha| \leqq k$,
(*c*) *the series* $\sum_{j \in \mathbf{J}} |D^\alpha u_j(x)|^2$, $|\alpha| \leqq k$, *converge uniformly on any compact subset of Ω.*

Moreover,

$$\|u\|_{C^k(\Omega; H)} = \sum_{|\alpha| \leqq k} \sup_{x \in \Omega} \big(\sum_{j \in \mathbf{J}} |D^\alpha u_j(x)|^2 \big)^{1/2}$$

for $u \in C^k(\Omega; H)$.

THEOREM 2.5.2 *Let $\Omega \subset \mathbf{R}^n$ be a domain, $k \in \mathbf{Z}^+$, $\sigma \in \mathbf{R}^+$, H a Hilbert space, and $A : \mathscr{D}(A) \subset H \to H$ a positive self-adjoint operator such that $A^{-1} \in \mathscr{L}(H, H)$ is compact. Let us denote by $\{e_j\}_{j \in \mathbf{J}}$ an orthonormal base in H of eigenvectors of A and by $\{\lambda_j\}_{j \in \mathbf{J}}$ the corresponding system of eigenvalues. Let $u : \Omega \to H$ and let us define $u_j(x)$, as in (2.5.1).*
Then $u \in C^k(\Omega; \mathscr{D}(A^\sigma))$ if and only if

(*a*) $u_j \in C^k(\Omega; \mathbf{S})$, $\quad j \in \mathbf{J}$,
(*b*) $\sup_{x \in \Omega} \sum_{j \in \mathbf{J}} \lambda_j^{2\sigma} |D^\alpha u_j(x)|^2 < \infty$, $\quad |\alpha| \leqq k$,
(*c*) *the series* $\sum_{j \in \mathbf{J}} \lambda_j^{2\sigma} |D^\alpha u_j(x)|^2$, $|\alpha| \leqq k$, *converge uniformly on any compact subset of Ω.*

Moreover,

$$\|u\|_{C^k(\Omega; \mathscr{D}(A^\sigma))} = \sum_{|\alpha| \leqq k} \sup_{x \in \Omega} \big(\sum_{j \in \mathbf{J}} \lambda_j^{2\sigma} |D^\alpha u_j(x)|^2 \big)^{1/2}$$

for $u \in C^k(\Omega; \mathscr{D}(A^\sigma))$.

THEOREM 2.5.3 (*cf.* [21], *p.* 251). *Let $\varkappa \in \mathbf{R}^+$ and let H be a Hilbert space. Let $u \in H^0_\omega(\mathbf{R}; H)$ and let us write*

$$u_j = \omega^{-1/2} \int_0^\omega e^{-ij\nu\tau} u(\tau) \, \mathrm{d}\tau, \quad j \in \mathbf{Z}, \quad (\nu = 2\pi/\omega).$$

Then $u \in H^\varkappa_\omega(\mathbf{R}; H)$ if and only if $\sum_{j \in \mathbf{Z}} |j|^{2\varkappa} \|u_j\|_H^2 < \infty$.

Moreover,

$$\|u\|_{H_\omega{}^\varkappa(\mathbf{R};H)} \sim \left(\sum_{j\in\mathbf{Z}}(1 + |j|^{2\varkappa}) \|u_j\|_H^2\right)^{1/2} .$$

THEOREM 2.5.4 *Let $\varkappa \in \mathbf{R}^+$ and let H be a separable Hilbert space with an orthonormal base $\{e_k\}_{k\in\mathbf{K}}$. Let $u \in H_\omega^0(\mathbf{R}; H)$ and let us write*

$$u_{jk} = \omega^{-1/2} \int_0^\omega \langle u(\tau), e_k\rangle\, e^{-ij\nu\tau}\, d\tau\,, \quad j \in \mathbf{Z}\,, \quad k \in \mathbf{K}\,, \quad (\nu = 2\pi/\omega)\,. \tag{2.5.2}$$

Then $u \in H_\omega^\varkappa(\mathbf{R}; H)$ if and only if

$$\sum_{j\in\mathbf{Z}} \sum_{k\in\mathbf{K}} (1 + |j|^{2\varkappa}) |u_{jk}|^2 < \infty .$$

Moreover,

$$\|u\|_{H_\omega{}^\varkappa(\mathbf{R};H)} \sim \left(\sum_{j\in\mathbf{Z}} \sum_{k\in\mathbf{K}} (1 + |j|^{2\varkappa}) |u_{jk}|^2\right)^{1/2} .$$

THEOREM 2.5.5 *Let $\varkappa \in \mathbf{R}^+$, $\sigma \in \mathbf{R}^+$, let H be a Hilbert space, and $A : \mathscr{D}(A) \subset \subset H \to H$ a positive self-adjoint operator such that $A^{-1} \in \mathscr{L}(H, H)$ is compact. Let us denote by $\{e_k\}_{k\in\mathbf{K}}$ an orthonormal base in H of eigenvectors of A and by $\{\lambda_k\}_{k\in\mathbf{K}}$ the corresponding system of eigenvalues. Let $u \in H_\omega^0(\mathbf{R}; H)$ and let us define u_{jk} as in (2.5.2).*
Then $u \in H_\omega^\varkappa(\mathbf{R}; \mathscr{D}(A^\sigma))$ if and only if

$$\sum_{j\in\mathbf{Z}} \sum_{k\in\mathbf{K}} (1 + |j|^{2\varkappa})\, \lambda_k^{2\sigma} |u_{jk}|^2 < \infty .$$

Moreover,

$$\|u\|_{H_\omega{}^\varkappa(\mathbf{R};\mathscr{D}(A^\sigma))} \sim \left(\sum_{j\in\mathbf{Z}} \sum_{k\in\mathbf{K}} (1 + |j|^{2\varkappa})\, \lambda_k^{2\sigma} |u_{jk}|^2\right)^{1/2} .$$

2.6. *Periodic functions of two arguments*

In the following chapters we shall frequently deal with real functions defined on $\mathbf{R} \times [0, \pi]$ or, more generally, on

$$Q = \mathbf{R} \times \Omega\,,$$

where Ω is a domain in $\mathbf{R}^n$. These functions have the arguments t (time variable) and x (space variable) and are *ω-periodic in t* which means that

$$u(t + \omega, x) = u(t, x)\,, \quad (t, x) \in \mathbf{R} \times \Omega\,.$$

Let $k \in \mathbf{Z}^+$, $l \in \mathbf{Z}^+$. By $C^k_\omega(Q)$ $(C^{k,l}_\omega(Q), C^{(k,l)}_\omega(Q))$ we denote the Banach space of all functions $u : Q \to \mathbf{R}$ that are ω-periodic in t and such that $D^\alpha_t D^\beta_x u \in \in C(Q)$ for any $(\alpha, \beta) \in \Delta_1$ (Δ_2, Δ_3), where

$$\Delta_1 = \{(\alpha, \beta) \in (\mathbf{Z}^+)^{n+1}, \alpha + |\beta| \leqq k\}, \tag{2.6.1}$$

$$\Delta_2 = \{(\alpha, \beta) \in (\mathbf{Z}^+)^{n+1}, \alpha = 0 \text{ and } |\beta| \leqq l \text{ or } \alpha \leqq k \text{ and } \beta = 0\}, \tag{2.6.2}$$

$$\Delta_3 = \{(\alpha, \beta) \in (\mathbf{Z}^+)^{n+1}, \alpha \leqq k, |\beta| \leqq l\}. \tag{2.6.3}$$

The norm is given by

$$\sum_{(\alpha,\beta)\in\Delta_j} \sup \{|D^\alpha_t D^\beta_x u(t, x)|; (t, x) \in Q\}.$$

Quite similarly we define the spaces $C^k_{\text{loc},\omega}(Q)$, $C^{k,l}_{\text{loc},\omega}(Q)$, and $C^{(k,l)}_{\text{loc},\omega}(Q)$.

Let us remark that in Chapter II we shall define spaces of smooth two-argument functions that are periodic not only with respect to t but also with respect to x.

Next, by $H^k_\omega(Q)$ $(H^{k,l}_\omega(Q), H^{(k,l)}_\omega(Q))$ we denote the Banach space of all (classes of) functions $u : Q \to \mathbf{R}$ that are ω-periodic in t and such that the (generalized) derivatives $D^\alpha_t D^\beta_x$ belong to $L_2((-M, M) \times \Omega)$ for all $M > 0$ and $(\alpha, \beta) \in \Delta_1(\Delta_2, \Delta_3)$ where $\Delta_1 - \Delta_3$ are given by (2.6.1)–(2.6.3). The norm is given by

$$\left(\sum_{(\alpha,\beta)\in\Delta_j} \int_{(0,\omega)\times\Omega} (D^\alpha_t D^\beta_x u(t, x))^2 \, dt \, dx\right)^{1/2}.$$

Thus, the subscript ω is reserved to indicate the property of functions that are ω-periodic with respect to t. On the other hand, a dot over the symbol of a space characterizes the behaviour of functions only with respect to x. More precisely, let S_ω be one of the spaces $H^k_\omega(Q)$, $H^{k,l}_\omega(Q)$, or $H^{(k,l)}_\omega(Q)$. We define

$$\dot{S}_\omega = \text{cl}\,\{u \in C^\infty_\omega(Q); u(t, \cdot) \in C^\infty_0(\Omega), t \in \mathbf{R}\} \quad \text{in} \quad S_\omega.$$

It can be shown that the spaces of periodic function-valued functions (Sec. 2.4) can be identified in a natural manner with the spaces of periodic functions of two arguments. For example, $C^k_\omega(\mathbf{R}; C^l(\Omega))$ can be identified with $C^{(k,l)}_\omega(Q)$, $H^k_\omega(\mathbf{R}; H^l(\Omega))$ with $H^{(k,l)}_\omega(Q)$, $H^k_\omega(\mathbf{R}; \dot{H}^l(\Omega))$ with $\dot{H}^{(k,l)}_\omega(Q)$ etc. In what follows, we use these identifications without further comment.

2.7. *Embedding theorems*

In the theory of Sobolev spaces of functions defined on a domain $\Omega \subset \mathbf{R}^n$, the geometric and regularity properties of Ω (or of its boundary) play a significant role. Therefore segment, cone and horn conditions, domains with continuous, Lipschitz continuous or C^∞-boundaries, and so on were introduced. We make no attempt to explain all these notions here, referring the reader to special monographs on the subject, for example, [1] or [4]. In what follows, whenever the theorems from this section are applied, Ω is assumed to have all the necessary properties of the above-mentioned character. Neither do we quote the precise definitions of function spaces defined on $\partial\Omega$ such as $C^k(\partial\Omega; B)$ and $L_p(\partial\Omega; B)$.

THEOREM 2.7.1 (*see* [1], *p.* 97). *Let $\Omega \subset \mathbf{R}^n$ be an open set with the cone property. Let $k \in \mathbf{N}$, $p \in [1, \infty)$. Let $kp \leqq n$. Then $W_p^k(\Omega; \mathbf{S})$ is embedded in $L_q(\Omega; \mathbf{S})$ where $p \leqq q \leqq np/(n - kp)$ if $n - kp > 0$ and $p \leqq q < \infty$ if $n - kp = 0$. Let $kp > n$. Then $W_p^k(\Omega; \mathbf{S})$ is embedded in $C^m(\Omega; \mathbf{S})$, where $m \in \mathbf{Z}^+$, $m < k - n/p$.*

Observe that the last assertion has the following meaning: if $u \in W_p^k(\Omega; \mathbf{S})$ then strictly speaking, $D^\alpha u$ (for $|\alpha| \leqq m$) is not a function defined everywhere on Ω but rather an equivalence class of such functions defined and equal to one another up to sets of measure zero; the assertion tells us that the equivalence class in question contains a continuous bounded function on Ω.

THEOREM 2.7.2 (*see* [1], *pp.* 97, 144). *Let $\Omega \subset R^n$ be an open bounded set with a Lipschitz continuous boundary. Let $k \in \mathbf{N}$, $p \in [1, \infty)$, $kp > n$. Then $W_p^k(\Omega; \mathbf{S})$ is compactly embedded in $C^m(\bar{\Omega}; \mathbf{S})$, where $m \in \mathbf{Z}^+$, $m < k - n/p$.*

THEOREM 2.7.3 (*see* [1], *p.* 144). *Let $\Omega \subset \mathbf{R}^n$ be an open bounded set with the cone property. Let $k \in \mathbf{N}$, $p \in [1, \infty)$, $kp \leqq n$. Then $W_p^k(\Omega; \mathbf{S})$ is compactly embedded in $L_q(\Omega; \mathbf{S})$, where $1 \leqq q < np/(n - kp)$.*

Next we state a very special embedding theorem for fractional order spaces of ω-periodic functions.

THEOREM 2.7.4 (*see* [1], *p.* 218, [21], *p.* 253). *Let $0 < \varkappa \leqq 1/2$. Then $H_\omega^\varkappa(\mathbf{R}; \mathbf{S})$ is embedded in $L_{q,\omega}(\mathbf{R}; \mathbf{S})$ where $1 \leqq q \leqq 2/(1 - 2\varkappa)$ if $\varkappa < 1/2$ and $1 \leqq q < \infty$ if $\varkappa = 1/2$. Let $\varkappa > 1/2$. Then $H_\omega^\varkappa(\mathbf{R}; \mathbf{S})$ is embedded in $C_\omega^m(\mathbf{R}; \mathbf{S})$, where $m \in \mathbf{Z}^+$, $m < \varkappa - 1/2$.*

The embedding theorems for anisotropic Sobolev spaces are expressed in the following assertions.

THEOREM 2.7.5 (*see* [4], *pp.* 143–145). *Let* $\gamma = (\gamma_1, \ldots, \gamma_n) \in \mathbf{N}^n$ *and let* $\Omega \subset \mathbf{R}^n$ *be an open set with the weak* γ*-horn property. Let* $\alpha = (\alpha_1, \ldots, \alpha_n) \in (\mathbf{Z}^+)^n$, $p \in [1, \infty)$.

(1) *Suppose that* $p \leqq q$,

$$\delta = \sum_{j=1}^{n} \frac{\alpha_j}{\gamma_j} + \left(\frac{1}{p} - \frac{1}{q}\right) \sum_{j=1}^{n} \frac{1}{\gamma_j} \leqq 1$$

and that for $\delta = 1$ *either* $1 < p = q < \infty$ *or* $1 < p < q < \infty$ *or* $1 = p < q = \infty$. *Then* $u \in W_p^{\gamma_1, \ldots, \gamma_n}(\Omega; \mathbf{S})$ *implies that* $D^\alpha u \in L_q(\Omega; \mathbf{S})$ *and*

$$\|D^\alpha u\|_{L_q(\Omega; \mathbf{S})} \leqq c_1 \|u\|$$

with $c_1 > 0$ *independent of* u (*the norm* $\|u\|$ *is taken in* $W_p^{\gamma_1, \ldots, \gamma_n}(\Omega; \mathbf{S})$).

(2) *Suppose that*

$$\delta = \sum_{j=1}^{n} \frac{\alpha_j}{\gamma_j} + \frac{1}{p} \sum_{j=1}^{n} \frac{1}{\gamma_j} < 1$$

or $\delta = 1$, $p = 1$. *Then* $u \in W_p^{\gamma_1, \ldots, \gamma_n}(\Omega; \mathbf{S})$ *implies that* $D^\alpha u \in C(\Omega; \mathbf{S})$ *and*

$$\sup_{x \in \Omega} |D^\alpha u(x)| \leqq c_2 \|u\|$$

with $c_2 > 0$ *independent of* u. (*Moreover, there are finitely many subsets* $\Omega_l \subset \Omega$ *such that* $\cup\, \Omega_l = \Omega$ *and* $D^\alpha u$ *is uniformly continuous on every* Ω_l.)

Note that for example $\mathbf{R}^2$ and the rectangle $\{(x_1, x_2) \in \mathbf{R}^2;\ a < x_1 < b,\ c < x_2 < d\}$ have the weak γ-horn property for any $\gamma \in \mathbf{N}^2$. On the other hand, the circle $\{(x_1, x_2) \in \mathbf{R}^2;\ x_1^2 + x_2^2 < 1\}$ has the weak γ-horn property if and only if $2^{-1}\gamma_1 \leqq \gamma_2 \leqq 2\gamma_1$ (see [4] for details). Incidentally, from the above theorem we deduce that $W_p^k(\Omega; \mathbf{S}) = W_p^{k, \ldots, k}(\Omega; \mathbf{S})$ (with equivalent norms) provided that $k \in \mathbf{N}$, $p \in (1, \infty)$ and Ω has the cone property (or, what is the same, the weak $(k, \ldots, k)$-horn property). This is not true for the analogous C-spaces.

2.8. *Traces of functions*

The following theorem gives conditions under which we may speak about values of a function from $W_p^k(\Omega; \mathbf{S})$ on the boundary $\partial\Omega$.

THEOREM 2.8.1 (*see* [26], *p.* 86). *Let* $\Omega \subset \mathbf{R}^n$ *be an open bounded set with a Lipschitz continuous boundary. Let* $k \in \mathbf{N}$, $p \in [1, \infty)$, $kp \leqq n$. *Let* $1 \leqq$

$\leqq q \leqq (n-1)p/(n-kp)$ *if* $kp < n$ *and* $1 \leqq q < \infty$ *if* $kp = n$. *Then there exists a unique operator* $T \in \mathscr{L}(W_p^k(\Omega; \mathbf{S}), L_q(\partial\Omega; \mathbf{S}))$ *such that*

$$Tu = u \mid \partial\Omega\,, \quad u \in C^\infty(\bar{\Omega}; \mathbf{S})\,.$$

If under the same assumptions $u \in W_p^k(\Omega; \mathbf{S}) \cap C(\bar{\Omega}; \mathbf{S})$, *then* $Tu = u \mid \partial\Omega$ (= the *trace of* u *on* $\partial\Omega$). When speaking about values of $u \in W_p^k(\Omega; \mathbf{S})$ (or of its derivatives) on $\partial\Omega$ we have always in mind the values Tu though we write simply u instead of Tu. T is called the *trace operator*.

Finally, we give another characterization of the spaces $\dot{W}_p^k(\Omega; \mathbf{S})$, which were introduced in Sec. 2.3.

THEOREM 2.8.2 (*see* [26], *pp.* 87, 90). *Let* $\Omega \subset \mathbf{R}^n$ *be an open bounded set with a Lipschitz continuous boundary. Then*

$$\dot{W}_p^1(\Omega; \mathbf{S}) = \{u \in W_p^1(\Omega; \mathbf{S});\ u = 0 \text{ on } \partial\Omega\}\,.$$

More generally, let $k \in \mathbf{N}$, *and let* Ω *be sufficiently regular. Then*

$$\dot{W}_p^k(\Omega; \mathbf{S}) = \left\{u \in W_p^k(\Omega; \mathbf{S});\ u = \frac{\partial u}{\partial \nu} = \ldots = \frac{\partial^{k-1} u}{\partial \nu^{k-1}} = 0 \text{ on } \partial\Omega\right\}.$$

Here $\partial^j u/\partial \nu^j$ *denotes the j-th outward normal derivative, which is defined by*

$$\frac{\partial^j u}{\partial \nu^j} = \sum_{|\alpha| = j} \frac{j!}{\alpha!} D^\alpha u\, \nu^\alpha\,,$$

where $\alpha! = \alpha_1!\, \alpha_2! \ldots \alpha_n!$, $\nu = (\nu_1, \ldots, \nu_n)$ *is the unit vector of the outward normal and* $\nu^\alpha = \nu_1^{\alpha_1} \nu_2^{\alpha_2} \ldots \nu_n^{\alpha_n}$.

Observe that $D^\alpha u \in L_p(\partial\Omega; \mathbf{S})$ for $|\alpha| \leqq k-1$ and ν exists almost everywhere on $\partial\Omega$ (by [26], Lemma 4.2, p. 88).

2.9. *The substitution operators*

In Chapters III–VII we are concerned with inhomogeneous differential equations whose right-hand sides involve an (in general non-linear) operator $F(u)$. In practice this operator F is often given in terms of a real-valued function f:

$$F(u)(x) = f(x, u(x), u_{x_1}(x), u_{x_2}(x), \ldots, u_{x_n}(x), \ldots)\,, \quad x \in \Omega \subset \mathbf{R}^n\,.$$

In this case F is said to be a *substitution* (or *Nemyckiĭ*) *operator*. We need to know when F realizes a continuous operator from a Sobolev space into another one. The following theorem helps us to answer this question.

THEOREM 2.9.1 (*see* [41], *p.* 213). *Let* $\Omega \subset \mathbf{R}^n$ *be a measurable bounded set. Let* $f : \Omega \times \mathbf{R}^l \to \mathbf{R}$ *be such that* $f(x, u)$ *is continuous in* u *for almost all*

$x \in \Omega$ and measurable in x for every $u \in \mathbf{R}^l$. Let p_k, $r \in [1, \infty)$ $(k = 1, \ldots, l)$. Then the operator

$$F(u_1, u_2, \ldots, u_l)(x) = f(x, u_1(x), u_2(x), \ldots, u_l(x))$$

maps $L_{p_1}(\Omega) \times L_{p_2}(\Omega) \times \ldots \times L_{p_l}(\Omega)$ continuously into $L_r(\Omega)$ if and only if

$$|f(x, u_1, u_2, \ldots, u_l)| \leqq a(x) + b \sum_{k=1}^{l} |u_k|^{p_k/r},$$

where $a \in L_r(\Omega)$ and $b \geqq 0$.

Furthermore, in a certain simple case the next theorem tells us whether a substitution operator is G- or F-differentiable.

THEOREM 2.9.2 (*see* [29], *p.* 16). *Let $\Omega \subset \mathbf{R}^n$ be a measurable bounded set. Let $f : \Omega \times \mathbf{R} \to \mathbf{R}$ be such that $f(x, u)$ is continuous in u for almost all $x \in \Omega$, measurable in x for every $u \in \mathbf{R}$ and $f(., 0) \in L_2(\Omega)$. Suppose further that f'_u is bounded on $\Omega \times \mathbf{R}$ and continuous in u for almost all $x \in \Omega$. Then the operator $F(u)(x) = f(x, u(x))$ maps $L_2(\Omega)$ continuously into $L_2(\Omega)$, is G-differentiable at every point $u \in L_2(\Omega)$ and*

$$(D\,F(u)\,h)(x) = f'_u(x, u(x))\,h(x).$$

The operator F is F-differentiable at a point $u_0 \in L_2(\Omega)$ if and only if $f(x, u)$ is linear in u for almost all $x \in \Omega$.

§ 3. Existence theorems for operator equations

3.1. *Theorems of a "metric" character*

Let (X_1, d_1), and (X_2, d_2) be two metric spaces. A mapping $F : \mathcal{D}(F) \subset X_1 \to X_2$ is said to be *Lipschitz continuous* if there exists a $\lambda > 0$ such that

$$d_2(F(x), F(y)) \leqq \lambda\, d_1(x, y), \quad x, y \in \mathcal{D}(F).$$

If $\lambda \in [0, 1]$ ($[0, 1)$) then F is called *non-expanding* (*contracting*).

Let F be a mapping of a set $M \subset \mathcal{D}(F)$ into itself. Then $x \in M$ is called a *fixed point of F* if $F(x) = x$. The best known and very important theorem on the existence (and uniqueness) of a fixed point of F is the following *Banach contraction principle.*

THEOREM 3.1.1 (*Banach, see* [36], *p.* 14). *Let X be a complete metric space and $F : X \to X$ a contracting mapping. Then F has a unique fixed point.*

Let us quote two other theorems of a similar character.

THEOREM 3.1.2 (*see* [5], *p.* 143). *Let X be a complete metric space, $F : X \to X$ and suppose that there exists an $n \in \mathbf{N}$ such that F^n is contracting. Then F has a unique fixed point.*

THEOREM 3.1.3 (*see Browder* VII: [15]). *Let B be a reflexive Banach space, V a continuous uniformly convex real-valued function on B, $V(0) = 0$, $V(x) > 0$ for $x \neq 0$, and $\{x; V(x) \leqq r\}$ bounded for any r. Further, let K be a non-empty bounded closed convex subset of B and suppose that a mapping $F : K \to K$ satisfies $V(F(x) - F(y)) \leqq V(x - y)$ for all $x, y \in K$. Then F has a fixed point in K.*

3.2. *Theorems of a topological character*

The theorems below have a non-trivial background in deep results on functions in finite-dimensional spaces.

THEOREM 3.2.1 (*Brouwer, see* [13], *p.* 23). *Let K be a non-empty bounded closed convex subset of $\mathbf{R}^n$ and $F : K \to K$ a continuous mapping. Then F has a fixed point.*

THEOREM 3.2.2 (*Schauder, see* [13], *p.* 43). *Let K be a non-empty bounded closed convex subset of a Banach space B and $F : K \to K$ a compact continuous mapping. Then F has a fixed point.*

COROLLARY 3.2.1 *Let K be a non-empty convex compact subset of a Banach space B and $F : K \to K$ be a continuous mapping. Then F has a fixed point.*

The following theorem is a consequence of the Tihonov theorem (see [12], p. 456) which cannot be stated here in its full generality for lack of the necessary tools.

THEOREM 3.2.3 (*Tihonov*). *Let K be a non-empty bounded closed convex subset of a separable reflexive Banach space B and $F : K \to K$ a weakly continuous mapping. Then F has a fixed point.*

The following result is a generalization of Theorems 3.1.1 and 3.2.2.

THEOREM 3.2.4 (*see* [34]). *Let K be a non-empty bounded closed convex subset of a Banach space B and $F : K \to K$ a continuous mapping with $F = F_1 + F_2$, where F_1 is contracting and F_2 is compact. Then F has a fixed point.*

3.3. *Equations with monotone operators*

Of the many results on solutions to equations with monotone (or maximal monotone) operators let us quote at least two. First we recall some definitions.

Let B be a Banach space, B^* its dual, and $F : \mathscr{D}(F) \subset B \to B^*$. Then F is called *monotone* if $\langle F(u) - F(v), u - v\rangle \geqq 0$ for all $u, v \in \mathscr{D}(F)$; F is called *maximal monotone* if it is monotone and $\langle g - F(v), u - v\rangle \geqq 0$ for all $v \in \mathscr{D}(F)$ implies that $u \in \mathscr{D}(F)$ and $F(u) = g$; F is called *hemicontinuous* if the function $t \to \langle F(u + tv), w\rangle$ is continuous on $[0, 1]$ for any $u, v \in \mathscr{D}(F)$ and $w \in B$; F is called *coercive* if $\|u\|^{-1} \langle F(u), u\rangle \to \infty$ for $\|u\| \to \infty$.

THEOREM 3.3.1 (*Browder-Minty, see* [14], *p.* 74). *Let B be a real reflexive Banach space and $F : B \to B^*$ be a hemicontinuous, monotone, coercive operator. Then $F(B) = B^*$, in other words, the equation $F(u) = g$ has a solution for any $g \in B^*$.*

THEOREM 3.3.2 (*see* [14], *p.* 78). *Let B be a real reflexive Banach space, let $\Lambda : \mathscr{D}(\Lambda) \subset B \to B^*$ be a hemicontinuous maximal monotone operator with the linear domain $\mathscr{D}(\Lambda)$ and let $F : B \to B^*$ be a hemicontinuous, monotone, coercive operator. Then the equation $\Lambda(u) + F(u) = g$ has a solution $u \in \mathscr{D}(\Lambda)$ for any $g \in B^*$.*

3.4. *Equations depending on a parameter; local theorems*

Throughout this book the following two theorems will play a fundamental role in the solution of weakly non-linear problems.

THEOREM 3.4.1 (*cf.* [11], *p.* 265). *Let (X_1, d_1) and (X_2, d_2) be two complete metric spaces, let $x_0 \in X_1$, $p_0 \in X_2$, $\alpha, \beta > 0$, $0 \leqq \lambda < 1$ and suppose that:*

(1) *$G : B(x_0, \alpha; X_1) \times B(p_0, \beta; X_2) \to X_1$ is continuous;*
(2) *$d_1(G(x_1, p), \; G(x_2, p)) \leqq \lambda\, d_1(x_1, x_2)$ for $x_1, x_2 \in B(x_0, \alpha; X_1)$ and $p \in B(p_0, \beta; X_2)$;*
(3) *$d_1(G(x_0, p), x_0) < \alpha(1 - \lambda)$ for $p \in B(p_0, \beta; X_2)$.*

Then the equation $x = G(x, p)$ has a unique solution $x = x^*(p) \in B(x_0, \alpha; X_1)$ for any $p \in B(p_0, \beta; X_2)$. Moreover, $x^* : B(p_0, \beta; X_2) \to B(x_0, \alpha; X_1)$ is continuous.

COROLLARY 3.4.1 *Let B_1 and B be two Banach spaces, $\varepsilon_0 > 0$, $\varrho > 0$, $\lambda > 0$, and suppose that:*

(1) *$F : B(0, \varrho; B_1) \times [-\varepsilon_0, \varepsilon_0] \to B$ is continuous;*
(2) *$\Lambda \in \mathscr{L}(B, B_1)$;*
(3) *$\|F(x_1, \varepsilon) - F(x_2, \varepsilon)\|_B \leqq \lambda \|x_1 - x_2\|_{B_1}$ for $x_1, x_2 \in B(0, \varrho; B_1)$ and $\varepsilon \in [-\varepsilon_0, \varepsilon_0]$.*

Then there exists an $\varepsilon^* \in (0, \varepsilon_0]$ such that the equation

$$x = \varepsilon \Lambda F(x, \varepsilon)$$

has a unique solution $x = x^*(\varepsilon) \in B(0, \varrho; B_1)$ for any $\varepsilon \in [-\varepsilon^*, \varepsilon^*]$. Moreover, $x^* \in C([-\varepsilon^*, \varepsilon^*]; B_1)$.

THEOREM 3.4.2 (*the implicit function theorem, cf.* [36], *pp.* 14–18). *Let B, B_1 and B_2 be Banach spaces, let $x_0 \in B_1$, $p_0 \in B_2$ and let U_1 and U_2 be neighbourhoods of x_0 in B_1 and p_0 in B_2, respectively. Suppose that:*

(1) *$G : U_1 \times U_2 \to B$ is continuous;*

(2) *$G(x_0, p_0) = 0$;*

(3) *the (partial) derivative $D_x G$ exists and is continuous (as a mapping of $U_1 \times U_2$ into $\mathscr{L}(B_1, B)$);*

(4) *$(D_x G(x_0, p_0))^{-1}$ exists and belongs to $\mathscr{L}(B, B_1)$.*

Then there exist neighbourhoods $V_1 \subset U_1$ and $V_2 \subset U_2$ of x_0 and p_0, respectively, such that the equation

$$G(x, p) = 0 \tag{3.4.1}$$

has a unique solution $x = x^*(p) \in V_1$ for any $p \in V_2$. Moreover, $x^* : V_2 \to V_1$ is continuous and $x^*(p_0) = x_0$.

Remark 3.4.1 (see [11], p. 270). If there exists a continuous (total) G-derivative DG, then the solution x^* of (3.4.1) has a continuous G-derivative Dx^* on $V_2' \subset V_2$, where V_2' is a neighbourhood of p_0. Moreover, $Dx^*(p) = -(D_x G(x^*(p), p))^{-1} D_p G(x^*(p), p)$ for all $p \in V_2'$.

Remark 3.4.2 In the following chapters we shall often deal with the problem $(\mathscr{P}^\varepsilon)$ consisting in the solution of one of the following two equations

$$x = \varepsilon \Lambda F(x, \varepsilon)\,, \quad \varepsilon \in [-\varepsilon_0, \varepsilon_0]\,, \quad (\varepsilon_0 > 0)\,, \tag{3.4.2}$$

$$G(x, \varepsilon) = 0\,, \quad \varepsilon \in [-\varepsilon_0, \varepsilon_0]\,. \tag{3.4.3}$$

We suppose that the assumptions under which the problem $(\mathscr{P}^\varepsilon)$ is investigated allow us to apply Corollary 3.4.1 or Theorem 3.4.2 to the equation in question. When we solve (3.4.2) or (3.4.3), the conclusion reached by means of the assertion above has the following form:

There exists an $\varepsilon^* > 0$ and a $\varrho > 0$ such that the equations (3.4.2) and (3.4.3) have a unique solution $x = x^*(\varepsilon) \in B(0, \varrho; B_1)$ or $x = x^*(\varepsilon) \in B(x_0, \varrho; B_1)$, respectively, for any $\varepsilon \in [-\varepsilon^*, \varepsilon^*]$. Moreover, $x^* \in C([-\varepsilon^*, \varepsilon^*]; B_1)$.

We express this conclusion briefly in the form:

The problem $(\mathscr{P}^{\varepsilon})$ *has a locally unique solution* $x^*(\varepsilon)$ *such that* $x^*(0) = 0$ *or, respectively,* $x^*(0) = x_0$.

3.5. *Equations depending on a parameter; global theorems*

In this section we deal briefly with the existence of a solution to a certain equation on a prescribed interval of a parameter λ. The following theorem can be proved by standard methods using, for example, Theorem 3.4.1.

THEOREM 3.5.1 *Let* B_1 *and* B_2 *be two Banach spaces and* $\{A(\lambda);\ \lambda \in [0, 1]\}$ *a family of linear operators from* B_1 *into* B_2 *that satisfy the following conditions:*

(1) *there exists a constant* $c > 0$ *such that* $\|A(\lambda)\, x\| \geqq c\|x\|$ *for* $\lambda \in [0, 1]$ *and* $x \in \mathscr{D}(A(\lambda))$;

(2) $A(0)^{-1} \in \mathscr{L}(B_2, B_1)$;

(3) *the mapping* $\lambda \to A(\lambda)\, A(0)^{-1}$ $(\lambda \in [0, 1])$ *belongs to* $C([0, 1], \mathscr{L}(B_2, B_2))$.

Then $A(1)^{-1} \in \mathscr{L}(B_2, B_1)$, *in other words, the problem* $A(1)\, x = y$ *has a solution for any* $y \in B_2$.

The following deep result can be obtained by topological methods.

THEOREM 3.5.2 (*Leray-Schauder, cf.* [23]). *Let* K *be a non-empty bounded convex open set in a Banach space* B *and let* $F : \bar{K} \times [0, 1] \to B$. *Suppose that:*

(1) F *is a compact and continuous mapping;*

(2) *there exists a unique element* $x_0 \in K$ *such that* $F(x_0, 0) = x_0$;

(3) *the F-derivative* $F'_x(x_0, 0)$ *exists and* $I - F'_x(x_0, 0)$ *has an inverse in* $\mathscr{L}(B, B)$;

(4) $F(x, \lambda) \neq x$ *for* $x \in \partial K$ *and* $\lambda \in [0, 1)$.

Then there exists an element $x_1 \in \bar{K}$ *such that* $F(x_1, 1) = x_1$.

Remark 3.5.1 Let F be a mapping from $\bar{K} \times [0, 1]$ to B such that the mapping $x \to F(x, \lambda)$ is compact and continuous for any $\lambda \in [0, 1]$, while the mapping $\lambda \to F(x, \lambda)$ is continuous uniformly in $x \in \bar{K}$. Then F satisfies the condition (1) of Theorem 3.5.2.

Chapter II

Preliminaries from the theory of differential equations

§ 1. Boundary-value and eigenvalue problems for elliptic and ordinary differential operators

1.1. *Elliptic operators*

Let $E(x, D)$ be a linear differential operator of order l, that is,

$$E(x, D) = \sum_{|\alpha| \leq l} a_\alpha(x)\, D^\alpha\,, \tag{1.1.1}$$

where the coefficients $a_\alpha(x)$ are complex-valued functions defined in an open domain Ω in $\mathbf{R}^n$. It is assumed that $\sum_{|\alpha| = l} |a_\alpha(x)| \not\equiv 0$, $x \in \Omega$. We associate with $E(x, D)$ the homogeneous polynomial in $\xi = (\xi_1, \ldots, \xi_n)$

$$E_0(x, \xi) = \sum_{|\alpha| = l} a_\alpha(x)\, \xi^\alpha \tag{1.1.2}$$

(note that $\xi^\alpha = \xi_1^{\alpha_1} \ldots \xi_n^{\alpha_n}$, $\alpha = (\alpha_1, \ldots, \alpha_n)$). The corresponding differential operator $E_0(x, D)$ is called the principal part of $E(x, D)$.

The operator $E = E(x, D)$ is said to be *elliptic at a point* x^0 if $E_0(x^0, \xi) \neq 0$ for any $\xi \neq 0$; E is *uniformly elliptic in a domain* Ω if there is a constant $C > 0$ such that

$$C^{-1}|\xi|^l \leqq |E_0(x, \xi)| \leqq C|\xi|^l\,, \quad \xi \in \mathbf{R}^n\,, \quad x \in \Omega\,, \tag{1.1.3}$$

where

$$|\xi| = \Big(\sum_{j=1}^{n} \xi_j^2\Big)^{1/2}\,;$$

E is *strongly elliptic* if there is a function $C(x)$ such that

$$\operatorname{Re}\,(C(x)\, E_0(x, \xi)) > 0\,, \quad \xi \in \mathbf{R}^n\,, \quad \xi \neq 0\,, \quad x \in \Omega\,; \tag{1.1.4}$$

E is *properly elliptic at a point* $x^0 \in \mathbf{R}^n$ if it is elliptic, $l = 2m$ and for any linearly independent vectors ξ, $\xi' \in \mathbf{R}^n$ the polynomial $E_0(x^0, \xi + \tau\xi')$ of a complex variable τ has m roots with positive imaginary parts. E is elliptic or properly elliptic in $\Omega \subset \mathbf{R}^n$ if it is elliptic or properly elliptic at each point of Ω, respectively.

The relations between the different kinds of ellipticity can be found in [14], p. 122.

The operator E is often considered in the form

$$E(x, D) = \sum_{|\alpha|,|\beta| \leqq m} (-1)^{|\alpha|} D^\alpha(a_{\alpha\beta}(x) D^\beta), \tag{1.1.5}$$

where $m \in \mathbf{N}$ and $D^\gamma a_{\alpha\beta}(x)$ exist for any α, β, $|\alpha| \leqq m$, $|\beta| \leqq m$ and for all $\gamma \leqq \alpha$.

Now let $\{L_j\}_{j=1}^q$, $(q \in \mathbf{N})$ be a system of boundary operators given by

$$L_j = \sum_{|k| \leqq m_j} b_{jk}(x) D^k, \quad j = 1, \ldots, q, \quad x \in \partial\Omega, \tag{1.1.6}$$

where $b_{jk} \in C(\partial\Omega)$, $m_j \leqq 2m - 1$. In what follows the notions of a normal system of boundary operators and of a system of boundary operators covering the operator $E(x, D)$ on $\partial\Omega$ are often used; for their definitions see [14], pp. 124–126.

Normal boundary operators that cover $E(x, D)$ are called here *admissible with respect to* $E(x, D)$ (or briefly E*-admissible*). If

$$E(x, D) = -\sum_{j,k=1}^n \frac{\partial}{\partial x_j}\left(a_{jk}(x)\frac{\partial}{\partial x_k}\right) + \sum_{j=1}^n a_j(x)\frac{\partial}{\partial x_j} + a_0(x)$$

($x \in \Omega$, a_{jk} are real-valued and differentiable at x) and if the conormal derivative $\partial/\partial n$ is defined as $\partial/\partial n = \sum_{j,k=1}^n a_{jk}(x)\, v_j(x)\, \partial/\partial x_k$, where $v_j(x)$, $j = 1, \ldots, n$ are the components of the outward normal $v(x)$ to $\partial\Omega$, then it can be shown that the *Dirichlet boundary operator* $L_1 u = u$, the *Newton boundary operator* $L_1 u = (\partial u/\partial n) + \sigma(x)\, u$ and the *combined boundary operator* $L_1 u = u$ on $\partial_1\Omega$, $L_1 u = (\partial u/\partial n) + \sigma(x)\, u$ on $\partial_2\Omega$ ($\partial_1\Omega \cup \partial_2\Omega = \partial\Omega$, $\partial_1\Omega \cap \partial_2\Omega = \emptyset$), are E-admissible, provided that $a_{jk} = a_{kj}$ and that $E(x, D)$ is uniformly elliptic in Ω. This ensures the admissibility of all boundary conditions used in Chapters III–V.

THEOREM 1.1.1 ([14], *p.* 176). *Let $\Omega \subset \mathbf{R}^n$ be a bounded domain of the class C^∞, let the operators $E(x, D)$ and L_j be given by* (1.1.5) *and* (1.1.6), *respectively, where $a_{\alpha\beta} \in C^\infty(\bar{\Omega})$ and $b_{jk} \in C^\infty(\partial\Omega)$, $E(x, D)$ is properly elliptic, and the system $\{L_j\}_{j=1}^m$ is E-admissible. If $u \in H^{2m}(\Omega)$, $L_j u = 0$ on $\partial\Omega$, $j = 1, \ldots, m$, and*

$Eu \in H^r(\Omega)$, $(r \geqq 0)$, then $u \in H^{2m+r}(\Omega)$ and there exists a constant $c_r > 0$ independent of u and such that

$$\|u\|_{2m+r} \leqq c_r(\|Eu\|_r + \|u\|_0). \tag{1.1.7}$$

(Here and in what follows $\|\cdot\|_k$ and $\langle\cdot,\cdot\rangle_k$ denote, respectively, the norm and the scalar product in $H^k(\Omega)$.)

COROLLARY 1.1.1 *Let Ω, $E(x, D)$ and L_j satisfy the assumptions of Theorem 1.1.1. We regard E as an operator in $H^0(\Omega)$ with domain $\mathscr{D}(E) = \{u \in H^{2m}(\Omega); L_j u = 0$ on $\partial\Omega$, $j = 1, \dots, m\}$. Then $\mathscr{D}(E^l) = \{u \in H^{2ml}(\Omega); L_j(E^k u) = 0$ on $\partial\Omega$, $j = 1, \dots, m$, $k = 1, \dots, l - 1\}$ for $l \in \mathbf{N}$, and there exists a constant $c_l > 0$ such that*

$$c_l^{-1}\|u\|_{2ml} \leqq \|E^l u\|_0 + \|u\|_0 \leqq c_l\|u\|_{2ml} \tag{1.1.8}$$

for $u \in \mathscr{D}(E^l)$.

The *proof* can be carried out by induction with respect to l on the basis of Theorem 1.1.1.

THEOREM 1.1.2 *(see [14], p. 211). Let Ω, $E = E(x, D)$ and L_j $(j = 1, \dots, m)$ satisfy the assumptions of Theorem 1.1.1. Suppose that E is formally self-adjoint on $\mathscr{D}_\infty = \{u \in C^\infty(\overline{\Omega}); (L_j u)(x) = 0, j = 1, \dots, m, x \in \partial\Omega\}$ in $H^0(\Omega)$. Then the operator E on $\mathscr{D}(E) = \{u \in H^{2m}(\Omega); L_j u = 0$ on $\partial\Omega$, $j = 1, \dots, m\}$ is self-adjoint in $H^0(\Omega)$.*

Before giving the example of the operator most frequently occuring in this book, the Laplace operator in one and more dimensions, we quote two well-known Lemmas.

LEMMA 1.1.1 *((Green's formula) see [16], p. 121). Let $\Omega \subset \mathbf{R}^n$ be a bounded domain with a Lipschitz continuous boundary and let $u \in W_p^1(\Omega)$, $v \in W_q^1(\Omega)$ where $1/p + 1/q \leqq (n + 1)/n$ if $1 \leqq p < n$, $1 \leqq q < n$; $q > 1$ if $p \geqq n$; and $p > 1$ if $q \geqq n$ then*

$$\int_\Omega \frac{\partial u}{\partial x_j}(x)\, v(x)\, dx = \int_{\partial\Omega} u(x)\, v(x)\, \nu_j(x)\, dS - \int_\Omega u(x) \frac{\partial v}{\partial x_j}(x)\, dx\,,$$

$j = 1, \dots, n$, where $(\nu_1(x), \dots, \nu_n(x))$ is the outward normal at $x \in \partial\Omega$.

LEMMA 1.1.2 *(Friedrichs, see [16], p. 20). Let $\Omega \subset \mathbf{R}^n$ be a bounded domain with a Lipschitz continuous boundary. Then for $u \in H^1(\Omega)$,*

$$\|u\|_1 \leqq \text{const.}\left(\int_{\partial\Omega} |u(x)|^2\, dS + \int_\Omega \sum_{j=1}^n \left|\frac{\partial u}{\partial x_j}(x)\right|^2 dx\right)^{1/2}.$$

In particular, if $u \in \dot{H}^1(\Omega)$, *then*

$$\|u\|_1 \leqq \text{const.} \left(\int_\Omega \sum_{j=1}^{n} \left| \frac{\partial u}{\partial x_j}(x) \right|^2 \mathrm{d}x \right)^{1/2}.$$

Example 1.1.1 Let $n = 1$, $\Omega = (0, \pi)$, $E = -\mathrm{d}^2/\mathrm{d}x^2$, $(L_1 u)(x) = u(x)$, $x \in \partial\Omega = \{0, \pi\}$. Then integrating by parts we find easily that E is formally self-adjoint on $\mathscr{D}(E) = \{u \in H^2(0, \pi);\ u = 0 \text{ on } \partial\Omega\}$ $(= \{u \in H^2(0, \pi);\ u(0) = = u(\pi) = 0\}$ according to I: Theorem 2.7.1 p. 24). Applying Theorem 1.1.2 we see that E is self-adjoint in $H^0(0, \pi)$. Further we have

$$\int_0^\pi [-u''(x)\, u(x)]\, \mathrm{d}x = \int_0^\pi |u'(x)|^2\, \mathrm{d}x \geqq \int_0^\pi |u(x)|^2\, \mathrm{d}x$$

for $u \in \mathscr{D}(E)$ so that E is positive definite in $H^0(\Omega)$.

The analogous results hold for the Laplace operator

$$E = -\Delta_n = -\sum_{j=1}^{n} \frac{\partial^2}{\partial x_j^2}$$

in $\Omega \subset \mathbf{R}^n$, when Ω satisfies the assumptions of Theorem 1.1.1. Here the fact that E is formally self-adjoint on $\mathscr{D}(E) = \{u \in H^2(\Omega);\ u = 0 \text{ on } \partial\Omega\}$ follows from Lemma 1.1.1 and that E is positive definite follows from Lemma 1.1.2:

$$\int_\Omega [-(\Delta_n u)(x)\, u(x)]\, \mathrm{d}x = \int_\Omega \sum_{j=1}^{n} \left(\frac{\partial u}{\partial x_j}(x) \right)^2 \mathrm{d}x \geqq c_0 \|u\|_0^2$$

for $u \in \mathscr{D}(E)$, $c_0 > 0$ a constant.

Let $E = c - \Delta_n$, $L_1 = (\partial/\partial v) + \sigma(x)$, where c is real, $\sigma \in C^\infty(\partial\Omega)$. Then E is formally self-adjoint on $\{u \in C^\infty(\bar{\Omega});\ L_1 u = 0 \text{ on } \partial\Omega\}$ (Lemma 1.1.1). Again by Theorem 1.1.2, E is self-adjoint on $\mathscr{D}(E)$ in $H^0(\Omega)$. Moreover, as we shall see in Theorem 1.3.2, we have $\langle (c - \Delta_n)\, u, u \rangle_0 \geqq (c + c_1)\, \|u\|_0^2$, $u \in \mathscr{D}(E)$ (c_1 is a constant) so that E is positive definite if $c + c_1 > 0$.

Remark 1.1.1 We note that the requirements on smoothness of $\partial\Omega$, $a_{\alpha\beta}(x)$ and $b_{jk}(x)$ in the present section can be weakened substantially (see [16], pp. 201–220).

1.2. *Boundary-value problems for ordinary differential equations*

In Chapters III, IV, and V the elliptic operator E will often reduce to $-\mathrm{d}^2/\mathrm{d}x^2$, and in Chap. VI to $\mathrm{d}^4/\mathrm{d}x^4$. Therefore, we state here several results from the theory of boundary value problems for ordinary differential equations. We introduce the necessary notation. If $A = (a_{jk})$, $j = 1, \ldots, m$, $k = 1, \ldots, n$

is an $m \times n$-matrix, then $A' = (a'_{jk})$, $j = 1, \ldots, n$, $k = 1, \ldots, m$, denotes its transpose, where $a'_{jk} = a_{kj}$. If $A = (a_{jk})$ is an $m \times n$-matrix and $B = (b_{jk})$ an $m \times p$-matrix, then (A, B) denotes the matrix $C = (c_{jk})$, where $c_{jk} = a_{jk}$ for $j = 1, \ldots, m$, $k = 1, \ldots, n$ and $c_{jk} = b_{j,k-n}$, $j = 1, \ldots, m$, $k = n + 1, \ldots, n + p$. Similarly for $\begin{pmatrix} A \\ B \end{pmatrix}$. Further, rank A denotes the maximal number of linearly independent rows or columns of A and det A denotes the determinant of a square matrix A. If y is a real-valued function m–1 times differentiable at a point $x \in \mathbf{R}$, then we write $\mathrm{J}_m\, y(x) = (y(x), y'(x), \ldots, y^{(m-1)}(x))'$ (the *jet of order m of y*).

Now let us consider the boundary value problem $(\mathscr{B})$ in the form

$$Ly \equiv \sum_{j=0}^{m} p_j(x) \frac{\mathrm{d}^{m-j} y}{\mathrm{d}x^{m-j}}(x) = g(x)\,, \quad x \in (a, b)\,, \tag{1.2.1}$$

$$(L_1 y, L_2 y, \ldots, L_l y)' \equiv M\, \mathrm{J}_m\, y(a) + N\, \mathrm{J}_m\, y(b) = \gamma\,, \tag{1.2.2}$$

where $p_j \in C^{m-j}([a, b])$, $p_0(x) \neq 0$, $(x \in [a, b])$, $g \in C([a, b])$, M and N are constant $l \times m$-matrices such that $\operatorname{rank}(M, N) = l$ $(1 \leqq l \leqq m)$ and $\gamma \in \mathbf{R}^l$ is a constant column vector. Suppose that two boundary-value problems are given by the operator L and the two systems of boundary operators $\{L_j\}_{j=1}^{l}$ and $\{L'_j\}_{j=1}^{l}$ given respectively by (1.2.2) and $(L'_1 y, L'_2 y, \ldots, L'_l y)' = M' \mathrm{J}_m\, y(a) + N' \mathrm{J}_m\, y(b)$, where M' and N' are constant $l \times m$-matrices. We say that *the boundary-value problems* in question *are equivalent* if $L_j y = 0$ if and only if $L'_j y = 0$, $j = 1, 2, \ldots, l$, for $y \in C^\infty([a, b])$. It can be easily shown that this occurs if and only if

$$M' = RM\,, \quad N' = RN \tag{1.2.2'}$$

with a non-singular matrix R.

The matrix $B(x) = (b_{jk}(x))_{j,k=1}^{m}$, where

$$b_{jk}(x) = \begin{cases} 0 \text{ for } m \geqq j > m - k + 1 \\ \displaystyle\sum_{q=j}^{m-k+1} (-1)^{q-1} \binom{q-1}{j-1} p_{m-q-k+1}^{(q-j)}(x) \text{ otherwise,} \end{cases}$$

$x \in [a, b]$, is non-singular and will be useful later (see [4], p. 61).

Example 1.2.1 Let $(Ly)(x) = (\mathrm{d}^4 y/\mathrm{d}x^4)(x)$, $x \in (a, b)$. Then an easy computation yields

$$B(x) = \begin{pmatrix} 0 & 0 & 0 & 1 \\ 0 & 0 & -1 & 0 \\ 0 & 1 & 0 & 0 \\ -1 & 0 & 0 & 0 \end{pmatrix}, \quad x \in (a, b)\,. \tag{1.2.3}$$

Example 1.2.2 Let $(Ly)(x) = -(a(x)\,y'(x))' + b(x)\,y'(x) + c(x)$, where a, b, and c are real-valued functions, and a is differentiable for $x \in (0, \pi)$. Then a simple calculation gives

$$B(x) = \begin{pmatrix} b(x) & -a(x) \\ a(x) & 0 \end{pmatrix}, \quad x \in (0, \pi).$$

It can be proved (see [4], p. 159) that there exist constant $m \times (2m - l)$-matrices P, Q and $m \times l$-matrices, P_c, Q_c satisfying the relations

$$-MB(a)^{-1}\,P + NB(b)^{-1}\,Q = 0\,, \quad \operatorname{rank}(P^{`}, Q^{`}) = 2m - l \tag{1.2.4}$$

and

$$-MB(a)^{-1}\,P_c + NB(b)^{-1}\,Q_c = I\,, \quad \operatorname{rank}(P_c^{`}, Q_c^{`}) = l\,, \tag{1.2.5}$$

respectively. The systems of boundary operators $\{M_j\}_{j=1}^{2m-l}$ and $\{M_{jc}\}_{j=1}^{l}$ given, respectively, by

$$(M_1 z, M_2 z, \ldots, M_{2m-l} z) = \mathrm{J}_m\, z(a)^{`}\, P + \mathrm{J}_m\, z(b)^{`}\, Q \tag{1.2.6}$$

and

$$(M_{1c} z, M_{2c} z, \ldots, M_{lc} z) = \mathrm{J}_m\, z(a)^{`}\, P_c + \mathrm{J}_m\, z(b)^{`}\, Q_c \tag{1.2.7}$$

are called *the systems of adjoint and complementary adjoint boundary operators* (or *forms*) if P, Q and P_c, Q_c satisfy (1.2.4) and (1.2.5), respectively.

Problem 1.2.1 Prove that the problems

$$\begin{aligned} &\ddot{y}(t) = 0\,, \quad t \in (0, \omega)\,, \quad (\omega > 0)\,, \\ &y(0) - y(\omega) = 0\,, \\ &\dot{y}(0) - \dot{y}(\omega) = 0 \end{aligned}$$

and

$$\begin{aligned} -\,&y''(x) = 0\,, \quad x \in (0, \pi)\,, \\ &y(0) \;= 0\,, \\ &y(\pi) \;= 0 \end{aligned}$$

and

$$\begin{aligned} -\,&y''(x) = 0\,, \quad x \in (0, \pi)\,, \\ &y'(0) + \sigma_0\, y(0) = 0\,, \\ &y'(\pi) + \sigma_1\, y(\pi) = 0\,, \quad (\sigma_0, \sigma_1 \in \mathbf{R} \text{ being arbitrary but fixed}) \end{aligned}$$

and

$$\begin{aligned} -\,&y''(x) = 0\,, \quad x \in (0, \pi)\,, \\ &y'(0) + \sigma\, y(0) = 0\,, \\ &y(\pi) = 0\,, \quad (\sigma \in \mathbf{R} \text{ being arbitrary but fixed}) \end{aligned}$$

are formally self-adjoint (see below) and that the matrices

$$P_c = \begin{pmatrix} 0 & -1 \\ 1 & 0 \end{pmatrix}, \quad Q_c = \begin{pmatrix} 0 & 0 \\ 0 & 0 \end{pmatrix}$$

and

$$P_c = \begin{pmatrix} 0 & 0 \\ -1 & 0 \end{pmatrix}, \quad Q_c = \begin{pmatrix} 0 & 0 \\ 0 & 1 \end{pmatrix}$$

and

$$P_c = \begin{pmatrix} 1 & 0 \\ 0 & 0 \end{pmatrix}, \quad Q_c = \begin{pmatrix} 0 & -1 \\ 0 & 0 \end{pmatrix}$$

and

$$P_c = \begin{pmatrix} 1 & 0 \\ 0 & 0 \end{pmatrix}, \quad Q_c = \begin{pmatrix} 0 & 0 \\ 0 & 1 \end{pmatrix}$$

define, respectively, their complementary adjoint boundary forms.

The following result can be verified easily.

LEMMA 1.2.1 *If P_1, Q_1 and P_{1c}, Q_{1c} are two other pairs of matrices satisfying* (1.2.4) *and* (1.2.5), *respectively, then there exists a non-singular constant* $(2m - l) \times (2m - l)$*-matrix R and a constant* $(2m - l) \times l$*-matrix S such that*

$$\begin{aligned} P_1 &= PR\,, & Q_1 &= QR \\ P_{1c} &= P_c + PS\,, & Q_{1c} &= Q_c + QS\,. \end{aligned} \tag{1.2.8}$$

Let L^* denote a formal adjoint of L (here and henceforth we use the same notation for adjoint and formal adjoint operators), that is

$$(L^* z)(x) = \sum_{j=0}^{m} (-1)^{m-j} \frac{\mathrm{d}^{m-j}}{\mathrm{d}x^{m-j}} (p_j(x)\, z(x))\,,$$

$z \in C^m([a, b])$. The boundary-value problem $(\mathscr{B}^*)$ given by

$$\begin{aligned} &(L^* z)(x) = h(x)\,, \quad x \in (a, b)\,, \quad (h \in C([a, b]))\,, \\ &\mathrm{J}_m\, z(a)'\, P + \mathrm{J}_m\, z(b)'\, Q = \delta\,, \quad (\delta \in \mathbf{R}^{2m-l}) \end{aligned} \tag{1.2.9}$$

is said to be *formally adjoint to* $(\mathscr{B})$.

The problem $(\mathscr{B})$ is called *formally self-adjoint* if it is equivalent to its adjoint problem $(\mathscr{B}^*)$. In what follows we give an existence theorem for the problem $(\mathscr{B})$, two lemmas on formal self-adjointness of $(\mathscr{B})$ and a theorem on self-adjointness of the operator L in $H^0(a, b)$.

THEOREM 1.2.1 *(see [4], p. 171). Let $p_j \in C^{m-j}([a, b])$, $j = 0, 1, \ldots, m$, and $g \in C([a, b])$. Then the boundary-value problem $(\mathscr{B})$ has a (classical) solution $y \in C^m([a, b])$ if and only if every classical solution $z(x)$ of the homogeneous adjoint boundary-value problem (that is, (1.2.9) with $h = 0$, $\delta = 0$) satisfies the relation*

$$\int_a^b z(x)\, g(x)\, \mathrm{d}x = (\mathrm{J}_m\, z(a)^{\backprime}\, P_c + \mathrm{J}_m\, z(b)^{\backprime}\, Q_c)\, \gamma\,, \tag{1.2.10}$$

where P_c and Q_c are arbitrary matrices satisfying (1.2.5).

LEMMA 1.2.2 *The boundary value problem $(\mathscr{B})$ is formally self-adjoint if and only if $l = m = 2n$, the differential operator (1.2.1) has the form*

$$(Ly)(x) = \sum_{j=0}^{n} (p_j(x)\, y^{(n-j)}(x))^{(n-j)}$$

and the matrices M and N in (1.2.2) satisfy the relation

$$-MB(a)^{-1}\, M^{\backprime} + NB(b)^{-1}\, N^{\backprime} = 0\,. \tag{1.2.11}$$

Proof. The first condition is well known (see [15], p. 14). The other follows readily from (1.2.4) and (1.2.2') with $M' = P^{\backprime}$, $N' = Q^{\backprime}$.

In particular, for $m = 2$ we have

LEMMA 1.2.3 *(see [7], p. 55). The problem $(\mathscr{B})$ with*

$$(Ly)(x) = (p(x)\, y'(x))' + q(x)\, y(x) \tag{1.2.12}$$

and $l = 2$ is formally self-adjoint if and only if

$$p(b) \det M = p(a) \det N\,.$$

THEOREM 1.2.2 *Let L be given by (1.2.1) with $l = m$, $\mathscr{D}(L) = \{y \in H^m(a, b);$ $M\, \mathrm{J}_m\, y(a) + N\, \mathrm{J}_m\, y(b) = 0\}$ and suppose that the boundary-value problem $(\mathscr{B})$ is formally self-adjoint. Then the operator L is self-adjoint in $H^0(a, b)$.*

Proof. Clearly, L is formally self-adjoint on $\mathscr{D}(L)$ in $H^0(a, b)$. Let L^* be the adjoint of L in $H^0(a, b)$ with the domain $\mathscr{D}(L^*)$. By a standard method it can be shown that $\mathscr{R}(L)$ is closed and $\mathscr{N}(L)$ is a finite-dimensional subspace of $C^m([a, b])$ (see [10], § 20). Now $\mathscr{N}(L)^\perp \cap C([a, b]) \subseteq \mathscr{R}(L)$ from Theorem 1.2.1 and $\mathscr{R}(L) = \mathscr{N}(L^*)^\perp$ from I: Theorem 1.5.1. Further, $\mathscr{N}(L^*)$ is closed and $C([a, b])$ is dense in $H^0(a, b)$, so that

$$\mathscr{N}(L)^\perp \cap C([a, b]) \subseteq \mathscr{N}(L^*)^\perp \tag{1.2.13}$$

implies that $\mathscr{N}(L^*) \subseteq \mathscr{N}(L)$. For from (1.2.13) we have $\mathscr{N}(L^*) \subseteq (\mathscr{N}(L)^\perp \cap$ $\cap\, C([a, b]))^\perp$ and if $f \in \mathscr{N}(L)^\perp$, $f_n \in C([a, b])$, $f_n = g_n + h_n$, $g_n \in \mathscr{N}(L)^\perp$,

$h_n \in \mathcal{N}(L) \subset C([a, b])$, $f_n \to f$ in $H^0(a, b)$, then $g_n \in \mathcal{N}(L)^\perp \cap C([a, b])$ and $g_n \to f$. As L^* is an extension of L we have $\mathcal{N}(L) \subseteq \mathcal{N}(L^*)$ and of course $\mathcal{R}(L) \subseteq \mathcal{R}(L^*)$. Thus $\mathcal{N}(L) = \mathcal{N}(L^*)$. Since $\mathcal{R}(L^*) \subseteq \mathcal{N}(L^{**})^\perp \subseteq \mathcal{N}(L)^\perp = \mathcal{R}(L)$, we have also $\mathcal{R}(L) = \mathcal{R}(L^*)$. Let $z \in \mathcal{D}(L^*)$. Then $L^*z = g \in \mathcal{R}(L)$ that is, there is a $y \in \mathcal{D}(L) \subset \mathcal{D}(L^*)$ such that $Ly = g$. This means that $y - z \in \mathcal{N}(L^*) = \mathcal{N}(L) \subseteq \mathcal{D}(L)$ and $z = (z - y) + y \in \mathcal{D}(L)$.

1.3. *Eigenvalue problems for elliptic operators*

When applying the Fourier method (see II: Sec. 3.4, III: §§ 1, 2, IV: § 1, V: § 6 and VI: § 1), it is useful to know conditions under which the operator E given by (1.1.5) possesses a system of eigenfunctions $\{v_k\}_{k=1}^\infty$ that forms an orthonormal base in $H^0(\Omega)$. In this context let us mention a general theorem and introduce some of its applications.

THEOREM 1.3.1 *Let Ω, E, and L_j satisfy the assumptions of Theorem 1.1.2. Then E is self-adjoint in $H^0(\Omega)$ and possesses in $H^0(\Omega)$ a complete orthonormal system of eigenfunctions with the corresponding system of eigenvalues $\{\lambda_k\}_{k=1}^\infty$, $(|\lambda_1| \leqq |\lambda_2| \leqq \ldots)$. Each eigenvalue is of finite multiplicity and $\lim_{k\to\infty} |\lambda_k| = \infty$.*

The *proof* is based on the fact that E has a compact resolvent $(\lambda - E)^{-1} : H^0(\Omega) \to H^0(\Omega)$ for some λ (Theorem 1.1.1, I: Theorem 2.7.3). Applying I: Theorem 1.6.3 the assertion is obtained easily.

THEOREM 1.3.2 *Let $\Omega \subset \mathbf{R}^n$ be bounded and of class C^∞,*

$$(Ev)(x) = -\sum_{j,k=1}^{n} \frac{\partial}{\partial x_j}\left(a_{jk}(x)\frac{\partial v}{\partial x_k}(x)\right) + a_0(x)\,v(x),$$

where $a_{jk} \in C^\infty(\bar{\Omega})$, $a_0 \in C^\infty(\bar{\Omega})$,

$$\sum_{j,k=1}^{n} a_{jk}(x)\,\xi_j\xi_k \geqq c_1 \sum_{j=1}^{n} \xi_j^2,\ (\xi_1, \ldots, \xi_n) \in \mathbf{R}^n$$

and let $(L_1u)(x) = u(x)$, $x \in \partial\Omega$ or $(L_1u)(x) = (\partial u/\partial n)(x) + \sigma(x)\,u(x)$, where $\partial u/\partial n$ is the conormal derivative defined in Sec. 1.1, $\sigma \in C^\infty(\partial\Omega)$, $x \in \partial\Omega$.
Then E with $\mathcal{D}(E) = \{u \in H^2(\Omega);\ L_1u = 0 \text{ on } \partial\Omega\}$ is self-adjoint in $H^0(\Omega)$, bounded below, and possesses a complete orthonormal system of eigenfunctions $\{v_k\}_{k=1}^\infty$ in $H^0(\Omega)$ for its system of eigenvalues $\{\lambda_k\}_{k=1}^\infty$ $(\lambda_1 \leqq \lambda_2 \leqq \ldots)$. Each eigenvalue is of finite multiplicity and $\lim_{k\to\infty} \lambda_k = \infty$.

Furthermore, $\mathscr{D}(E_\lambda^{1/2}) \equiv \{v \in H^0(\Omega); \sum_{k=1}^{\infty} |\lambda_k| \, |\langle v, v_k \rangle_0|^2 < \infty\}$ *is equal to* $\dot{H}^1(\Omega)$ *if* $L_1 u = u$ *and to* $H^1(\Omega)$ *if* $L_1 u = \partial u/\partial n + \sigma(x)\, u$, *where* $F_\lambda = E + \lambda$ *with* λ *sufficiently large.*

Proof. Let $v \in \mathscr{D}(E)$ and $a_1 \leqq a_0(x) \leqq a_2$, $x \in \Omega$. Then by Lemma 1.1.1,

$$\langle Ev, v \rangle_0 = -\int_{\partial\Omega} \frac{\partial v}{\partial n}(x)\, v(x)\, dS +$$

$$+ \int_\Omega \sum_{j,k=1}^{n} a_{jk}(x) \frac{\partial v}{\partial x_j}(x) \frac{\partial v}{\partial x_k}(x)\, dx + \int_\Omega a_0(x)\, v(x)^2\, dx \geqq$$

$$\geqq -\int_{\partial\Omega} \frac{\partial v}{\partial n}(x)\, v(x)\, dS + c_1 \int_\Omega \sum_{j=1}^{n} \left| \frac{\partial v}{\partial x_j}(x) \right|^2 dx + a_1 \|v\|_0^2 . \tag{1.3.1}$$

If $L_1 u = u$, then (1.3.1) and Lemma 1.1.2 yield $\langle Ev, v \rangle_0 \geqq c_2 \|v\|_1^2 - c_3 \|v\|_0^2$ and $\langle Ev, v \rangle_0 \leqq c_4 \|v\|_1^2$, where c_2, c_3 and c_4 are positive constants independent of v.

If $L_1 u = \partial u/\partial n + \sigma(x)\, u$ and $\sigma(x) \geqq \sigma_0$, then (1.3.1) yields

$$\langle Ev, v \rangle_0 \geqq \sigma_0 \int_{\partial\Omega} v(x)^2\, dS + c_1 \int_\Omega \sum_{j=1}^{n} \left| \frac{\partial v}{\partial x_j}(x) \right|^2 dx + a_1 \|v\|_0^2 .$$

Using Lemma 1.1.2 we obtain

$$\langle Ev, v \rangle_0 \geqq -c_5 \int_{\partial\Omega} v(x)^2\, dS + c_6 \|v\|_1^2 + a_1 \|v\|_0^2$$

with some $c_5 > 0$ and $c_6 > 0$. By the Lions lemma (see [16], p. 108); [14], Theorem 16.1) the inequality $\|v\|_{H^{1/2}(\Omega)} \leqq \varepsilon \|v\|_1 + c_7 \|v\|_0$ holds for an arbitrarily small $\varepsilon > 0$ so that owing to $\|v\|_{H^0(\partial\Omega)} \leqq \text{const.} \|v\|_{H^{1/2}(\Omega)}$, (see [14], p. 47) we finally conclude that $\langle Ev, v \rangle_0 \geqq c_8 \|v\|_1^2 - c_9 \|v\|_0^2$, where $c_8 > 0$ and $c_9 > 0$. As

$$\int_{\partial\Omega} v(x)^2\, dS \leqq \text{const.} \|v\|_1^2$$

(see I: Theorem 2.8.1), (1.3.1) gives $\langle Ev, v \rangle_0 \geqq \text{const.} \|v\|_1^2$.

We have proved that in the both cases E is bounded below, i.e. there exists a constant λ such that E_λ is positive and the graph norm of $E_\lambda^{1/2}$ is equivalent to that of $H^1(\Omega)$ on $\mathscr{D}(E)$. Thus the relations $\mathscr{D}(E_\lambda^{1/2}) = \dot{H}^1(\Omega)$ or $\mathscr{D}(E_\lambda^{1/2}) = H^1(\Omega)$ follow from the fact that $\mathscr{D}(E_\lambda^{1/2})$ is precisely the closure of $\mathscr{D}(E)$ in the norm of $\mathscr{D}(E_\lambda^{1/2})$, that is $\|\cdot\|_1$. The other assertions follow immediately from Theorem 1.3.1, the assumptions of which are satisfied, as can be shown with help of Lemma 1.1.1.

Remark 1.3.1 The requirements on smoothness of Ω, $a_{jk}(x)$ and $\sigma(x)$ can be weakened (see Remark 1.1.1).

Throughout the rest of this section we denote by $\{v_k\}_{k=1}^{\infty}$ an orthonormal base of eigenfunctions of E in $H^0(\Omega)$ with $\mathscr{D}(E) = \{u \in H^{2m}(\Omega);\ L_j u = 0$ on $\partial\Omega,\ j = 1, \ldots, m\}$.

The following two theorems can be derived easily from I: Theorems 2.5.1, 2.5.5, I: Sec. 2.6 and from Theorems 1.3.1 and 1.3.2.

THEOREM 1.3.3 (1) *Let Ω and E satisfy the assumptions of Theorem* 1.3.1, *and let $I \subset \mathbf{R}$ be a compact interval. If $u : I \to H^0(\Omega)$, $u(t) = \sum_{k=1}^{\infty} u_k(t)\, v_k$, then $u \in C^m(I; \mathscr{D}(E^\beta))$, $(m \in \mathbf{Z}^+, \beta \in \mathbf{R}^+)$ if and only if $u_k \in C^m(I)$, $k \in \mathbf{N}$ and the series $\sum_{k=1}^{\infty} |\lambda_k|^{2\beta}\, |u_k^{(l)}(t)|^2$, $l = 0, 1, \ldots, m$, converge uniformly in I. The relation $\|u\|_{(m,\beta)} = (\max\{\sum_{k=1}^{\infty} (1 + |\lambda_k|)^{2\beta}\, |u_k^{(l)}(t)|^2;\ t \in I,\ l = 0, 1, \ldots, m\})^{1/2}$, $u \in C^m(I;$ $\mathscr{D}(E^\beta))$ defines a norm, which is equivalent to that of $C^m(I; \mathscr{D}(E^\beta))$.*

(2) *Let Ω, E and $\sigma(x)$ satisfy the assumptions of Theorem* 1.3.2 *and let $I \subset \mathbf{R}$ be a compact interval. If $L_1 u = u$ or $L_1 u = \partial u/\partial n + \sigma(x)\, u$, respectively, then for all $m \in \mathbf{Z}^+$*

$$C^m(I; \mathscr{D}(E^{1/2})) = C^m(I; \dot{H}^1(\Omega)),$$
$$C^m(I; \mathscr{D}(E)) = C^m(I; H^2(\Omega) \cap \dot{H}^1(\Omega))$$

or

$$C^m(I; \mathscr{D}(E^{1/2})) = C^m(I; H^1(\Omega)),$$
$$C^m(I; \mathscr{D}(E)) = C^m(I; \{u \in H^2(\Omega);\ \partial u/\partial n + \sigma(x)\, u = 0 \quad \text{on} \quad \partial\Omega\}).$$

respectively.

THEOREM 1.3.4 (1) *Let Ω and E satisfy the assumptions of Theorem* 1.3.1. *If $u \in H^0_\omega(\mathbf{R}; H^0(\Omega))$,*

$$u(t) = \sum_{j=-\infty}^{\infty} \sum_{k=1}^{\infty} u_{jk}\, e^{ij\nu t} v_k \quad \left(\nu = \frac{2\pi}{\omega}\right),$$

then $u \in H^\alpha_\omega(\mathbf{R}; \mathscr{D}(E^\beta))$ $(\alpha, \beta \in \mathbf{R}^+)$ if and only if

$$\|u\|_{\alpha,\beta} = \left(\sum_{j=-\infty}^{\infty} \sum_{k=1}^{\infty} (1 + |j|)^{2\alpha}\, (1 + |\lambda_k|)^{2\beta}\, |u_{jk}|^2\right)^{1/2} < \infty\,.$$

The norm in $H^\alpha_\omega(\mathbf{R}; \mathscr{D}(E^\beta))$ is equivalent to $\|\cdot\|_{\alpha,\beta}$.

(2) *Let* Ω *and* E *satisfy the assumptions of Theorem* 1.3.2. *If* $L_1 u = u$, *then the spaces* $H^m_\omega(\mathbf{R}; H^0(\Omega))$, $H^m_\omega(\mathbf{R}; \mathscr{D}(E^{1/2}))$ *and* $H^m_\omega(\mathbf{R}; \mathscr{D}(E))$ *can be identified with* $H^{m,0}_\omega(\mathbf{R} \times \Omega)$, $\dot{H}^{(m,1)}_\omega(\mathbf{R} \times \Omega)$, *and* $H^{(m,2)}_\omega(\mathbf{R} \times \Omega) \cap \dot{H}^{(m,1)}(\mathbf{R} \times \Omega)$, *respectively*.

Remark 1.3.2 Note that the norms in spaces of the types $\bigcap_{j=1}^{p} H^{\alpha_j}_\omega(\mathbf{R}; \mathscr{D}(E^{\beta_j}))$ and $\bigcap_{j=1}^{p} C^{m_j}_\omega(\mathbf{R}; \mathscr{D}(E^{k_j}))$ are defined by the maximum or the sum of the norms in the spaces that are included in the intersection. In the following chapters we use frequently the spaces $H^1_\omega(\mathbf{R}; H^0(\Omega)) \cap H^0_\omega(\mathbf{R}; \mathscr{D}(E))$ and $H^2_\omega(\mathbf{R}; H^0(\Omega)) \cap \cap H^0_\omega(\mathbf{R}; \mathscr{D}(E))$, where E satisfies the assumptions of Theorem 1.3.2. Their norms are equivalent to those of $H^{1,2}_\omega(\mathbf{R} \times \Omega)$ and $H^2_\omega(\mathbf{R} \times \Omega)$, respectively (see Theorem 1.1.1).

Remark 1.3.3 If $E = -\Delta_n$ on $\mathscr{D}(E) = H^2(\Omega) \cap \dot{H}^1(\Omega)$, then we use the notation $\mathscr{D}(E^\beta) = {}^0H^{2\beta}(\Omega)$, $(\beta \geqq 0)$. By Theorem 1.3.2 ${}^0H^1(\Omega) = \dot{H}^1(\Omega)$.

In particular, if $\Omega = (0, \pi)$ and $E = -\mathrm{d}^2/\mathrm{d}x^2$ then ${}^0H^r(0, \pi) = \{u \in \in H^r(0, \pi);\ u^{(2l)}(0) = u^{(2l)}(\pi) = 0,\ l = 0, 1, \ldots, (r/2) - 1$ if r is even, $l = = 0, 1, \ldots, [r/2]$ if r is odd$\}$ for all $r \in \mathbf{N}$, by Corollary 1.1.1.
Let us also remark that $H^{2\beta}_\omega(\mathbf{R}) = \mathscr{D}(E^\beta)$, where now $E = -\mathrm{d}^2/\mathrm{d}t^2$ with $\mathscr{D}(E) = H^2_\omega(\mathbf{R})$.

Example 1.3.1 It is easy to check that the problem given by

$$\begin{aligned} -v''(x) &= \lambda\, v(x)\,, \quad x \in (0, \pi)\,, \\ v(0) &= 0\,, \\ v(\pi) &= 0 \end{aligned} \tag{1.3.2}$$

has the eigenvalues $\lambda_k = k^2$ and the corresponding eigenfunctions $v_k(x) = = (2/\pi)^{1/2} \sin kx$, $x \in [0, \pi]$. By Theorem 1.3.2, the system $\{v_k\}_{k=1}^{\infty}$ is an orthonormal base in $H^0(0, \pi)$.

Example 1.3.2 The problem given by

$$\begin{aligned} -v''(x) &= \lambda\, v(x)\,, \quad x \in (0, \pi)\,, \\ v'(0) + \sigma_0\, v(0) &= 0\,, \\ v'(\pi) + \sigma_1\, v(\pi) &= 0 \end{aligned} \tag{1.3.3}$$

is more complicated, and we shall treat it in more detail. We find easily that the positive eigenvalues λ_k satisfy the equation

$$(\lambda + \sigma_0\sigma_1) \sin(\pi\lambda^{1/2}) + (\sigma_0 - \sigma_1)\, \lambda^{1/2} \cos(\pi\lambda^{1/2}) = 0\,, \tag{1.3.4}$$

so that if $\sigma_0 = \sigma_1$, then $\lambda_k = k^2$ and $v_k(x) = \cos kx - (\sigma_0/k) \sin kx, x \in [0, \pi]$. If $\sigma_1 \neq \sigma_0$, then there exists an $m \in \mathbf{Z}$ such that

$$\lambda_{k+m}^{1/2} = k + \frac{1}{k} \frac{\sigma_1 - \sigma_0}{\pi} + 0\left(\frac{1}{k^3}\right), \tag{1.3.5}$$

and

$$v_k(x) = \cos \lambda_k^{1/2} x - \frac{\sigma_0}{\lambda_k^{1/2}} \sin \lambda_k^{1/2} x, \quad x \in [0, \pi], \tag{1.3.6}$$

are the eigenfunctions corresponding to λ_k, $(k = 1, 2, \ldots)$. Further, it can be verified that $\lambda = 0$ is an eigenvalue with the eigenfunction $v(x) = 1 - \sigma_0 x$ if and only if $\varrho(\sigma_0, \sigma_1) \equiv \sigma_1 - \sigma_0 - \pi\sigma_0\sigma_1 = 0$. Finally, since the negative eigenvalues $\lambda = -\mu^2$ are given by the equation

$$(\mu^2 - \sigma_0\sigma_1) \sinh \mu\pi + \mu(\sigma_1 - \sigma_0) \cosh \mu\pi = 0, \tag{1.3.7}$$

we see easily that for $\sigma_1 = \sigma_0, \lambda = -\sigma_0^2$ is an eigenvalue with the eigenfunction $v(x) = \cosh \sigma_0 x - \sinh \sigma_0 x$. For $\sigma_1 \neq \sigma_0$, transforming (1.3.7) with help of the formula

$$Ar \operatorname{Tanh} x = \frac{1}{2} \log \frac{1 + x}{1 - x}$$

to the form

$$2\pi\mu = \log \frac{(\mu + \sigma_0)(\mu - \sigma_1)}{(\mu - \sigma_0)(\mu + \sigma_1)} \equiv q(\mu)$$

and taking into account that

$$q'(\mu) = 2 \frac{(\sigma_1 - \sigma_0)(\mu^2 + \sigma_0\sigma_1)}{(\mu^2 - \sigma_0)(\mu^2 - \sigma_1)},$$

we find that the problem (1.3.3) has no negative eigenvalue if

(a) $\sigma_0 \leqq 0 \leqq \sigma_1, \quad \sigma_0 \neq \sigma_1,$

or

(b) $\sigma_0\sigma_1 > 0, \quad \sigma_0 < \sigma_1, \quad \varrho(\sigma_0, \sigma_1) \leqq 0;$

has exactly one negative eigenvalue if

(a) $\sigma_0\sigma_1 > 0, \quad \sigma_0 < \sigma_1, \quad \varrho(\sigma_0, \sigma_1) > 0$

or

(b) $\sigma_0\sigma_1 > 0, \quad \sigma_1 < \sigma_0$

or

(c) $\sigma_1 \leqq 0 \leqq \sigma_0, \quad \sigma_0 \neq \sigma_1, \quad \varrho(\sigma_0, \sigma_1) \leqq 0;$

and has two (which is the maximal possible number of negative eigenvalues) if $\sigma_1 < 0 < \sigma_0$ and $\varrho(\sigma_0, \sigma_1) > 0$. In the last two cases any eigenfunction $v(x)$ corresponding to an eigenvalue $\lambda < 0$ is of the form

$$\cosh(-\lambda)^{1/2} x - \frac{\sigma_0}{(-\lambda)^{1/2}} \sinh(-\lambda)^{1/2} x . \tag{1.3.8}$$

By Theorem 1.3.2 the system of all eigenfunctions $\{v_k(x)\}_{k=n}^{\infty}$, $(n \in \{-2, -1, 0, 1\}$ in accordance with the non-positive eigenvalues) of the problem (1.3.3) is a complete orthogonal base in each of these cases. This base becomes orthonormal if every eigenfunction $v_k(x)$ is multiplied by

$$\left(\int_0^\pi |v_k(x)|^2 \, dx\right)^{-1/2} = 0(1), \quad k = n, n+1, n+2, \ldots .$$

Example 1.3.3 By the same arguments as in Example 1.3.2 it can be checked readily that the problem given by

$$\begin{aligned} -v''(x) &= \lambda\, v(x), \quad x \in (0, \pi), \\ v'(0) + \sigma_0\, v(0) &= 0, \\ v(\pi) &= 0 \end{aligned} \tag{1.3.9}$$

has positive eigenvalues $\lambda_k = (k + \frac{1}{2})^2$ and the corresponding eigenfunctions $v_k(x) = \cos(k + \frac{1}{2})\, x$ if $\sigma_0 = 0$, whereas in the case $\sigma_0 \neq 0$ there exists an $m \in \mathbf{Z}$ such that

$$\lambda_{k+m}^{1/2} = k + \frac{1}{2} + \frac{\sigma_0}{\pi(k + \frac{1}{2})} + 0\left(\frac{1}{k^3}\right), \tag{1.3.10}$$

$$v_k(x) = \cos\lambda_k^{1/2} x - \frac{\sigma_0}{\lambda_k^{1/2}} \sin \lambda_k^{1/2} x, \quad x \in [0, \pi]$$

for $k = 1, 2, \ldots$.

Further, $\lambda_0 = 0$ is an eigenvalue of the problem (1.3.9) with the eigenfunction $v_0(x) = 1 - x/\pi$ if and only if $\sigma_0\pi = 1$. Finally, the problem (1.3.9) has no negative eigenvalue if $\sigma_0\pi \leqq 1$ and has exactly one (which is the root of the equation $\sinh(-\lambda)^{1/2}\pi = (-\lambda)^{1/2}\sigma_0^{-1}$) with the eigenfunction

$$v_{-1}(x) = \cosh(-\lambda_{-1})^{1/2} x - \frac{\sigma_0}{(-\lambda_{-1})^{1/2}} \sinh(-\lambda_{-1})^{1/2} x \quad \text{if} \quad \sigma_0\pi > 1 .$$

Again by Theorem 1.3.2, the system $\{v_k\}_{k=n}^{\infty}$ $(n \in \{-1, 0, 1\}$ in accordance with the non-positive eigenvalues) is a complete orthogonal base in $H^0(0, \pi)$. This base becomes orthonormal if every $v_k(x)$ is multiplied by

$$\left(\int_0^\pi |v_k(x)|^2 \, dx\right)^{-1/2} = 0(1), \quad k = n, n+1, n+2, \ldots .$$

§ 2. The wave and the telegraph equations

2.1. *The Cauchy problem for the wave equation in* $\mathbf{R}$

Let us consider the Cauchy problem

$$u_{tt}(t, x) - u_{xx}(t, x) = g(t, x), \quad t \in \mathbf{R}, \quad x \in \mathbf{R}, \tag{2.1.1}$$

$$u(0, x) = \varphi(x), \quad u_t(0, x) = \psi(x), \quad x \in \mathbf{R}. \tag{2.1.2}$$

The following result is standard (see [2], pp. 2–4, 17–18).

LEMMA 2.1.1 *Let* $\varphi \in C^2_{\text{loc}}(\mathbf{R})$, $\psi \in C^1_{\text{loc}}(\mathbf{R})$, $g \in C^{\alpha, 1-\alpha}_{\text{loc}}(\mathbf{R}^2)$, *where* $\alpha = 0$ *or* $\alpha = 1$.
Then the Cauchy problem (2.1.1), (2.1.2) *has a unique solution* $u^* \in C^2_{\text{loc}}(\mathbf{R}^2)$, *given by the formula*

$$u^*(t, x) = \Phi(\varphi)(t, x) + \Psi(\psi)(t, x) + G(g)(t, x), \quad t \in \mathbf{R}, \quad x \in \mathbf{R}, \tag{2.1.3}$$

where

$$\Phi(\varphi)(t, x) = \tfrac{1}{2}[\varphi(x + t) + \varphi(x - t)], \tag{2.1.4}$$

$$\Psi(\psi)(t, x) = \frac{1}{2}\int_{x-t}^{x+t} \psi(\xi)\, d\xi,$$

$$G(g) = \frac{1}{2}\int_0^t \int_{x-t+\tau}^{x+t-\tau} g(\tau, \xi)\, d\xi\, d\tau, \quad t \in \mathbf{R}, \quad x \in \mathbf{R}.$$

Quite similarly we can prove the next lemma.

LEMMA 2.1.2 *Let* $a \in C^2_{\text{loc}}(\mathbf{R})$, $b \in C^1_{\text{loc}}(\mathbf{R})$, $g \in C^{\alpha, 1-\alpha}_{\text{loc}}(\mathbf{R})$, *where* $\alpha = 0$ or $\alpha = 1$.
Then the problem (2.1.1),

$$u(t, 0) = a(t), \quad u_x(t, 0) = b(t), \quad t \in \mathbf{R}, \tag{2.1.5}$$

has a unique solution $u^* \in C^2_{\text{loc}}(\mathbf{R}^2)$. *This solution is defined by*

$$u^*(t, x) = \tfrac{1}{2}[a(t + x) + a(t - x)] + \frac{1}{2}\int_{t-x}^{t+x} b(\tau)\, d\tau -$$

$$- \frac{1}{2}\int_0^x \int_{t-x+\xi}^{t+x-\xi} g(\tau, \xi)\, d\tau\, d\xi, \quad t \in \mathbf{R}, \quad x \in \mathbf{R}.$$

2.2. *The initial boundary-value problem with the Dirichlet boundary conditions*

First let us consider the linear initial boundary-value problem

$$u_{tt}(t, x) - u_{xx}(t, x) = g(t, x), \quad (t, x) \in Q = (0, \infty) \times (0, \pi) \tag{2.2.1}$$

$$u(t, 0) = 0, \quad u(t, \pi) = 0, \quad t \in \mathbf{R}^+ \tag{2.2.2}$$

$$u(0, x) = \varphi(x), \quad u_t(0, x) = \psi(x), \quad x \in [0, \pi]. \tag{2.2.3}$$

We look for a classical solution that is a function $u \in C^2_{\text{loc}}(\bar{Q})$ satisfying (2.2.1)–(2.2.3). Let $I \subseteq \mathbf{R}$ be a closed interval, $r \in \mathbf{Z}^+$ and $\alpha \in \{0, 1, \ldots, r\}$. Then we write

$$\dot{C}^{(\alpha, r-\alpha)}_{\text{loc}}(I \times [0, \pi]) = \{u \in C^{(\alpha, r-\alpha)}_{\text{loc}}(I \times [0, \pi])\,;$$

$$u(t, 0) = u(t, \pi) = 0, \quad t \in I\},$$

$$\dot{C}^{(\alpha, r-\alpha)}(I \times [0, \pi]) = \dot{C}^{(\alpha, r-\alpha)}_{\text{loc}}(I \times [0, \pi]) \cap C^{(\alpha, r-\alpha)}(I \times [0, \pi]),$$

$$^0C^{(\alpha, r-\alpha)}(I \times [0, \pi]) = \Big\{u \in C^{(\alpha, r-\alpha)}(I \times [0, \pi])\,;$$

$$\frac{\partial^{2l} u}{\partial x^{2l}}(t, 0) = \frac{\partial^{2l} u}{\partial x^{2l}}(t, \pi) = 0, \quad t \in I, \quad l = 0, 1, \ldots, \left[\frac{r-\alpha}{2}\right]\Big\},$$

$$^0C^r([0, \pi]) = \Big\{u \in C^r([0, \pi])\,;$$

$$u^{(2l)}(0) = u^{(2l)}(\pi) = 0, \quad l = 0, 1, \ldots, \left[\frac{r}{2}\right]\Big\}.$$

If B is a space of real or complex valued functions of two real variables t and x, then B_{ω_1, ω_2} $(\omega_1 > 0, \omega_2 > 0)$ denotes the space of all functions from B that are ω_1-periodic in t and ω_2-periodic in x. If $\omega_1 = 0$, then the functions are ω_2-periodic in x only. On the other hand, the subset of functions from B that are only ω_1-periodic in t is denoted simply by B_{ω_1}. The symbol B_{ω_1, ω_2} equipped with a subscript (o) or (e) denotes the subspace of B_{ω_1, ω_2} of all functions odd or even respectively, in the variable indicated by the position of the subscript in question. Thus, $C_{\omega_1, \omega_2(o)}(\mathbf{R}^2)$ denotes the subspace of $C(\mathbf{R}^2)$ of functions $u(t, x)$ that are ω_1-periodic in t and ω_2-periodic and odd in x. An analogous notation is used when B is a space of functions of several variables.

Let

$$g \in \dot{C}_{\text{loc}}^{(\alpha,1-\alpha)}(\bar{Q}), \quad \text{where} \quad \alpha = 0 \quad \text{or} \quad \alpha = 1, \tag{2.2.4}$$

$$\varphi \in {}^0C^2([0, \pi]), \quad \psi \in {}^0C^1([0, \pi]). \tag{2.2.5}$$

We define the extended functions g_e, φ_e, ψ_e by the relations

$$\begin{aligned}
g_e(t, x) &= g(t, x), \quad (t, x) \in \bar{Q}, \\
\varphi_e(x) &= \varphi(x), \quad \psi_e(x) = \psi(x), \quad x \in [0, \pi], \\
g_e(t, x) &= -g_e(t, -x) = g_e(t, x + 2\pi), \\
\varphi_e(x) &= -\varphi_e(-x) = \varphi_e(x + 2\pi), \\
\psi_e(x) &= -\psi_e(-x) = \psi_e(x + 2\pi), \quad t \in \mathbf{R}^+, \quad x \in \mathbf{R}.
\end{aligned}$$

It can be verified easily that on their whole domain of definition the extended functions are as smooth as the original ones. If no confusion can arise, then we denote the extended functions by the same symbols as the original ones.

We put

$$\sigma(x) = \tfrac{1}{2}\left[\varphi_e(x) + \int_0^x \psi_e(\xi)\, \mathrm{d}\xi + c\right], \tag{2.2.6}$$

where c is a constant. Clearly, $\sigma \in C_{2\pi}^2(\mathbf{R})$. On the other hand, if $\sigma \in C_{2\pi}^2(\mathbf{R})$ is given, then the functions $\varphi_e \in C_{2\pi(o)}^2(\mathbf{R})$ and $\psi_e \in C_{2\pi(o)}^1(\mathbf{R})$ are uniquely determined by $\varphi_e(x) = \sigma(x) - \sigma(-x)$ and $\psi_e(x) = \sigma'(x) - \sigma'(-x)$, $x \in \mathbf{R}$.

THEOREM 2.2.1 *Suppose that the assumptions (2.2.4) and (2.2.5) are satisfied. Then there exists a unique classical solution of the problem (2.2.1)–(2.2.3) given by the formula*

$$u^*(t, x) = \sigma(x + t) - \sigma(-x + t) + \frac{1}{2}\int_0^t \int_{x-t+\tau}^{x+t-\tau} g_e(\tau, \xi)\, \mathrm{d}\xi\, \mathrm{d}\tau, \quad (t, x) \in \bar{Q}. \tag{2.2.7}$$

Proof. The uniqueness of the solution follows from the energy identity

$$\int_0^\pi \int_0^t \frac{\mathrm{d}}{\mathrm{d}t}\left[(u_t(\tau, \xi))^2 + (u_x(\tau, \xi))^2\right] \mathrm{d}\tau\, \mathrm{d}\xi = 0, \quad t \in \mathbf{R}^+ \tag{2.2.8}$$

which holds for any solution u of (2.2.1)–(2.2.3) with $\varphi = 0$, $\psi = 0$, $g = 0$, where (2.2.8) is obtained after integrating by parts the identity

$$\int_0^\pi \int_0^t \left[u_{tt}(\tau, \xi) - u_{xx}(\tau, \xi)\right] u_t(\tau, \xi)\, \mathrm{d}\tau\, \mathrm{d}\xi = 0.$$

By virtue of (2.2.4) and (2.2.5), the fact that $u^*(t, x)$ given by (2.2.7) is a classical solution of (2.2.1)–(2.2.3) is easily verified by substituting the formula in (2.2.1)–(2.2.3).

Evidently, (2.2.7) defines u^* not only in the half-strip $\bar{Q}$, but in the whole half plane $Q_e = \mathbf{R}^+ \times \mathbf{R}$, while

$$u^*(t, x) = -u^*(t, -x) = u^*(t, x + 2\pi), \quad (t, x) \in Q_e .$$

This leads us to a reformulation of the original problem. The new "extended" problem is defined by the following relations:

$$u_{tt}(t, x) - u_{xx}(t, x) = g_e(t, x), \quad t \in (0, \infty), \quad x \in \mathbf{R}, \tag{2.2.1'}$$

$$u(t, x) = -u(t, -x) = u(t, x + 2\pi), \tag{2.2.2'}$$

$$u(0, x) = \varphi_e(x), \quad u_t(0, x) = \psi_e(x), \quad (t, x) \in Q_e . \tag{2.2.3'}$$

In what follows we prefer this form of the problem.

Remark 2.2.1 A more complicated but advantageous way (suitable also for non-homogeneous problems) of extending the functions φ, ψ and g is used in [19].

Remark 2.2.2 For $T > 0$, let $Q_T = (0, T) \times (0, \pi)$ and consider the problem (2.2.1')–(2.2.3') for $(t, x) \in Q_{T,e} = (0, T) \times \mathbf{R}$ and $u \in \mathscr{U} = C^2_{0,2\pi(o)}(\bar{Q}_{T,e})$ $(\|\cdot\|_{\mathscr{U}} = \|\cdot\|_{C^2(\bar{Q}_T)})$. Writing $Lu \equiv (u_{tt} - u_{xx}, u(0, x), u_t(0, x)) = (g_e, \varphi_e, \psi_e)$ if and only if $u \in \mathscr{U}$ is a solution to (2.2.1')–(2.2.3'), we obtain the following corollary.

COROLLARY 2.2.1 *We have* $L^{-1} \in \mathscr{L}(C^{(\alpha,1-\alpha)}_{0,2\pi(o)}(\bar{Q}_{T,e}) \times C^2_{2\pi(o)}(\mathbf{R}) \times C^1_{2\pi(o)}(\mathbf{R}), \mathscr{U})$ *for* $\alpha = 0$ or $\alpha = 1$, *that is there is a constant* $M > 0$ *such that*

$$\|L^{-1}(g_e, \varphi_e, \psi_e)\|_{\mathscr{U}} \leqq M(\|g_e\|_{C^{(\alpha,1-\alpha)}{}_{0,2\pi(o)}(Q_{T,e})} + \|\varphi\|_{C^2{}_{2\pi(o)}(\mathbf{R})} +$$

$$+ \|\psi\|_{C^1{}_{2\pi(o)}(\mathbf{R})}) \quad \textit{for all} \quad g \in C^{(\alpha,1-\alpha)}_{0,2\pi(o)}(\bar{Q}_{T,e}), \quad \varphi \in C^2_{2\pi(o)}(\mathbf{R}),$$

$\psi \in C^1_{2\pi(o)}(\mathbf{R})$, *where* $\alpha = 0$ or $\alpha = 1$.

Now, we consider the problem

$$u_{tt}(t, x) - u_{xx}(t, x) = F(u)(t, x), \quad (t, x) \in Q, \tag{2.2.9}$$

(2.2.2) and (2.2.3), where

$$F(u) \text{ maps } C^2_{\text{loc}}(\bar{Q}) \cap {}^0C^{(0,2)}_{\text{loc}}(\bar{Q}) \quad \text{into} \quad \dot{C}^{(\alpha,1-\alpha)}_{\text{loc}}(\bar{Q}) \tag{2.2.10}$$

with $\alpha = 0$ or $\alpha = 1$,

and where φ and ψ satisfy (2.2.5). We extend φ and ψ as above and extend F to an operator F_e in such a way that for each $u \in C^2_{\text{loc},0,2\pi(o)}(\bar{Q}_e)$ it is defined by

$$F_e(u)(t, x) = F(u \mid \bar{Q})(t, x), \quad (t, x) \in \bar{Q}, \tag{2.2.11}$$

$$F_e(u)(t, x) = -F_e(u)(t, -x) = F_e(u)(t, x + 2\pi), \quad (t, x) \in \bar{Q}_e.$$

Clearly, $F_e(u)$ maps $C^2_{\text{loc},0,2\pi(o)}(\bar{Q}_e)$ into $C^{(\alpha,1-\alpha)}_{\text{loc},0,2\pi(o)}(\bar{Q}_e)$. Suppose that the "extended" problem is given by

$$u_{tt}(t, x) - u_{xx}(t, x) = F_e(u)(t, x), \quad (t, x) \in Q_e, \tag{2.2.12}$$

(2.2.2') and (2.2.3'). It is easy to prove the following theorem.

THEOREM 2.2.2 *Suppose that the assumptions (2.2.5) and (2.2.10) are satisfied. Then every classical solution $u(t, x)$ of the problem (2.2.12), (2.2.2'), and (2.2.3') satisfies the equation*

$$u(t, x) = \sigma(x + t) - \sigma(-x + t) +$$

$$+ \frac{1}{2}\int_0^t \int_{x-t+\tau}^{x+t-\tau} F_e(u)(\tau, \xi)\, d\xi\, d\tau, \quad (t, x) \in \bar{Q}_e, \tag{2.2.13}$$

σ being given by (2.2.6). Conversely, if there is a solution $u^* \in C^2_{\text{loc},0,2\pi(o)}(\bar{Q}_e)$ of (2.2.13), then u^* satisfies (2.2.12), (2.2.2'), (2.2.3').

2.3. *The telegraph equation*

In this section we establish a formula for a solution of the Cauchy problem $(\mathcal{I})$, given by

$$u_{tt}(t, x) - u_{xx}(t, x) + c\, u(t, x) = g(t, x), \quad t > 0, \quad x \in \mathbf{R}, \tag{2.3.1}$$

$$u(0, x) = \varphi(x), \quad u_t(0, x) = \psi(x), \quad x \in \mathbf{R}. \tag{2.3.2}$$

Remark 2.3.1 A more general problem given by

$$u_{tt}(t, x) - u_{xx}(t, x) + a\, u_t(t, x) + b\, u_x(t, x) + c\, u(t, x) =$$

$$= g(t, x), \quad t > 0, \quad x \in \mathbf{R}$$

and (2.3.2) can be transformed to the above problem (2.3.1), (2.3.2) by the substitution

$$v(t, x) = u(t, x) \exp \tfrac{1}{2}(at - bx).$$

LEMMA 2.3.1 *Let* $\varphi \in C^2_{\text{loc}}(\mathbf{R})$, $\psi \in C^1_{\text{loc}}(\mathbf{R})$, $g \in C^{(\alpha,1-\alpha)}_{\text{loc}}(\mathbf{R}^+ \times \mathbf{R})$, $\alpha = 0$ *or* $\alpha = 1$.
Then the problem $(\mathscr{I})$ *has a unique classical solution*

$$u^* \in C^2_{\text{loc}}((0, \infty) \times \mathbf{R}) \cap C^{1,0}_{\text{loc}}(\mathbf{R}^+ \times \mathbf{R}) .$$

This solution is given by (2.1.3), *where*

$$\Phi(\varphi)(t, x) = \tfrac{1}{2}[\varphi(x + t) + \varphi(x - t)] + $$
$$+ \frac{1}{2}\int_{x-t}^{x+t} \frac{\partial J_0}{\partial t} (c^{1/2}[t^2 - (\xi - x)^2]^{1/2})\, \varphi(\xi)\, d\xi ,$$

$$\Psi(\psi)(t, x) = \frac{1}{2}\int_{x-t}^{x+t} J_0(c^{1/2}[t^2 - (\xi - x)^2]^{1/2})\, \psi(\xi)\, d\xi ,$$

$$G(g)(t, x) = \frac{1}{2}\int_0^t \int_{x-t+\tau}^{x+t-\tau} J_0(c^{1/2}[(\tau - t)^2 - (\xi - x)^2]^{1/2}) \,.$$
$$. \, g(\tau, \xi)\, d\xi\, d\tau , \quad t \in \mathbf{R}^+ , \quad x \in \mathbf{R} , \tag{2.3.3}$$

J_0 *being the usual Bessel function* (*see* [21], *p.* 16).
For the *proof*, see [5], vol. II, pp. 692–695.

Now let us consider the initial boundary-value problem given by the equation

$$u_{tt}(t, x) + 2a\, u_t(t, x) - u_{xx}(t, x) + c\, u(t, x) = g(t, x) , \tag{2.3.4}$$
$$(t, x) \in Q = \mathbf{R} \times (0, \pi) , \quad (a \neq 0,\ c \in \mathbf{R} \text{ are constants}) ,$$

the Dirichlet boundary condition

$$u(t, 0) = u(t, \pi) = 0 , \quad t \in \mathbf{R} , \tag{2.3.5}$$

and the initial conditions

$$u(0, x) = \varphi(x) , \quad u_t(0, x) = \psi(x) , \quad x \in [0, \pi] . \tag{2.3.6}$$

Suppose that $\varphi \in {}^0C^2([0, \pi])$, $\psi \in {}^0C^1([0, \pi])$ and $g \in \dot{C}^{(\alpha,1-\alpha)}_{\text{loc}}(\mathbf{R} \times [0, \pi])$, where $\alpha = 0$ or $\alpha = 1$. Then the functions φ, ψ, and g can be extended in x to odd and 2π-periodic functions φ_e, ψ_e and g_e, respectively, which retain the smoothness properties of the original functions. Treating the extended problem given by

$$u_{tt}(t, x) + 2a\, u_t(t, x) - u_{xx}(t, x) + c\, u(t, x) = g_e(t, x) , \quad (t, x) \in \mathbf{R}^2 , \tag{2.3.4'}$$

(2.2.2′) and (2.2.3′) and proceeding just as in Sec. 2.2, we obtain the next theorem.

THEOREM 2.3.1 *Let* $\varphi \in {}^0C^2([0, \pi]), \psi \in {}^0C^1([0, \pi])$ and $g \in \dot{C}_{\text{loc}}^{(\alpha, 1-\alpha)}(\mathbf{R} \times [0, \pi])$ *where* $\alpha = 0$ *or* $\alpha = 1$.
Then the problem given by (2.3.4′), (2.2.2′) and (2.2.3′) has a unique solution $u^* \in C_{\text{loc}}^2(\mathbf{R}^2)$. *This solution is* 2π*-periodic and odd in* x *and is given by (2.1.3) with* Φ, Ψ and G *as in (2.3.3), where we write* $c - a^2$ *instead of* c *and* e^{-at} $\varphi_e(x)$, $e^{-at}(a\varphi_e(x) + \psi_e(x))$, *and* $e^{-a(t-\tau)}\, g_e(\tau,\xi)$ *instead of* $\varphi(x)$, $\psi(x)_e$ *and* $g(\tau,\xi)$ *respectively.*

2.4. *The Fourier method*

The Fourier method is another useful device for solving the initial boundary-value problem (2.2.1)–(2.2.3) or even more general initial boundary-value problems for hyperbolic (as well as parabolic) equations.

Let us consider the problems $(\mathscr{IB}_1)$ and $(\mathscr{IB}_2)$ given, respectively, by the equation

$$u_{tt}(t, x) + E\, u(t, x) + c\, u(t, x) = g(t, x)\,,$$

$$(t, x) \in Q = [0, T] \times \Omega\,, \quad (T > 0)\,, \tag{2.4.1}$$

and the boundary condition

$$u(t, x) = 0\,, \quad (t, x) \in [0, T] \times \partial\Omega \tag{2.4.2}$$

or

$$\partial u/\partial n\, (t, x) + \sigma(x)\, u(t, x) = 0\,, \quad (t, x) \in [0, T] \times \partial\Omega \tag{2.4.2′}$$

and the initial conditions

$$u(0, x) = \varphi(x)\,, \quad u_t(0, x) = \psi(x)\,, \quad x \in \Omega\,, \tag{2.4.3}$$

where $\Omega \subset \mathbf{R}^n$ is a bounded domain, $c \in \mathbf{R}$,

$$E\, v(x) = -\sum_{j,k=1}^{n} (a_{jk}(x)\, v_{x_k}(x))_{x_j} + a_0(x)\, v(x)\,, \quad x \in \Omega\,.$$

We keep the notation as well as the assumptions on a_{jk}, a_0, σ and $\partial\Omega$ in Theorem 1.3.2. Without loss of generality we may suppose that E is positive definite.

The problems $(\mathscr{IB}_1)$ and $(\mathscr{IB}_2)$ can be regarded as an abstract problem $(\mathscr{IB})$:

$$u_{tt}(t) + E\, u(t) + c\, u(t) = g(t)\,, \quad t \in [0, T]\,, \tag{2.4.1′}$$

$$u(0) = \varphi\,, \quad u_t(0) = \psi\,, \tag{2.4.3′}$$

$(E : \mathscr{D}(E) \subset H^0(\Omega) \to H^0(\Omega)$, where either $\mathscr{D}(E) = \{v \in H^2(\Omega);\ v(x) = 0$ on $\partial\Omega\} = H^2(\Omega) \cap \dot{H}^1(\Omega)$ or $\mathscr{D}(E) = \{v \in H^2(\Omega);\ (\partial v/\partial n)(x) + \sigma(x)\, v(x) = 0$ on $\partial\Omega\})$.

By a solution of $(\mathscr{I}\mathscr{B})$ we mean a function

$$u \in \mathscr{U} = C([0, T]; \mathscr{D}(E)) \cap C^1([0, T]; \mathscr{D}(E^{1/2})) \cap C^2([0, T]; H^0(\Omega))$$

satisfying (2.4.1′) and (2.4.3′).

The norm on $\mathscr{U}$ is introduced by

$$\|u\|_{\mathscr{U}} = \max\{\max_{t\in[0,T]} \|E\, u(t)\|_0, \max_{t\in[0,T]} \|E^{1/2}\, u_t(t)\|_0, \max_{t\in[0,T]} \|u_{tt}(t)\|_0\}\,.$$

Henceforth, we write $u(t, x)$ instead of $u(t)$.

We look for a solution in the form of a Fourier series

$$u(t, x) = \sum_{k=1}^{\infty} u_k(t)\, v_k(x)\,, \tag{2.4.4}$$

where $\{v_k\}_{k=1}^{\infty}$ is an orthonormal base in $H^0(\Omega)$ of eigenfunctions of the operator E, and

$$u_k(t) = \langle u(t, \cdot), v_k\rangle_0\,.$$

By Theorem 1.3.3, a function $u : [0, T] \to H^0(\Omega)$ belongs to $\mathscr{U}$ if and only if $u_k \in C^2([0, T])$ for all $k \in \mathbf{N}$ and the series

$$\sum_{k=1}^{\infty} \lambda_k^2\, u_k^2(t)\,, \tag{2.4.5}$$

$$\sum_{k=1}^{\infty} \lambda_k\, \dot{u}_k^2(t)\,, \tag{2.4.6}$$

$$\sum_{k=1}^{\infty} \ddot{u}_k^2(t) \tag{2.4.7}$$

are uniformly convergent in $t \in [0, T]$, where the λ_k are the eigenvalues of E corresponding to v_k.

Now

$$\|u\|_{\mathscr{U}} = \max\{\max_{t\in[0,T]} (\sum_{k=1}^{\infty} \lambda_k^2\, u_k^2(t))^{1/2}\,, \tag{2.4.8}$$

$$\max_{t\in[0,T]} (\sum_{k=1}^{\infty} \lambda_k\, \dot{u}_k^2(t))^{1/2}\,, \quad \max_{t\in[0,T]} (\sum_{k=1}^{\infty} \ddot{u}_k^2(t))^{1/2}\}\,.$$

As we shall see later, it is natural to assume that $\varphi \in \mathscr{D}(E)$, $\psi \in \mathscr{D}(E^{1/2})$, $g \in C([0, T]; \mathscr{D}(E^{1/2}))$, that is (see again Theorem 1.3.3),

$$\varphi(x) = \sum_{k=1}^{\infty} \varphi_k v_k(x), \quad \varphi_k = \langle \varphi, v_k \rangle_0, \tag{2.4.9}$$

$$\psi(x) = \sum_{k=1}^{\infty} \psi_k v_k(x), \quad \psi_k = \langle \psi, v_k \rangle_0, \tag{2.4.10}$$

$$g(t, x) = \sum_{k=1}^{\infty} g_k(t) v_k(x), \quad g_k(t) = \langle g(t, \cdot), v_k \rangle_0, \tag{2.4.11}$$

where

$$\sum_{k=1}^{\infty} \lambda_k^2 \varphi_k^2 \text{ is convergent}, \tag{2.4.12}$$

$$\sum_{k=1}^{\infty} \lambda_k \psi_k^2 \text{ is convergent}, \tag{2.4.13}$$

$$\sum_{k=1}^{\infty} \lambda_k g_k^2(t) \text{ is uniformly convergent on } [0, T]. \tag{2.4.14}$$

Obviously,

$$\|\varphi\|_{\mathscr{D}(E)} = \Big(\sum_{k=1}^{\infty} \lambda_k^2 \varphi_k^2\Big)^{1/2}, \tag{2.4.15}$$

$$\|\psi\|_{\mathscr{D}(E^{1/2})} = \Big(\sum_{k=1}^{\infty} \lambda_k \psi_k^2\Big)^{1/2}, \tag{2.4.16}$$

$$\|g\|_{C([0,T];\mathscr{D}(E^{1/2}))} = \max_{t \in [0,T]} \Big(\sum_{k=1}^{\infty} \lambda_k g_k^2(t)\Big)^{1/2}. \tag{2.4.17}$$

Substituting (2.4.4), (2.4.9)–(2.4.11) in (2.4.1′), (2.4.3′), using

$$E v_k(x) = \lambda_k v_k(x), \quad k \in \mathbf{N},$$

and comparing the corresponding coefficient of $v_k(x)$, we obtain for u_k the following initial value problem:

$$\ddot{u}_k(t) + (\lambda_k + c) u_k(t) = g_k(t), \quad t \in (0, T), \tag{2.4.18}$$

$$u_k(0) = \varphi_k, \quad \dot{u}_k(0) = \psi_k. \tag{2.4.19}$$

Its solution is

$$u_k(t) = \varphi_k \cos(\lambda_k + c)^{1/2} t + \psi_k(\lambda_k + c)^{-1/2} \sin(\lambda_k + c)^{1/2} t +$$

$$+ (\lambda_k + c)^{-1/2} \int_0^t \sin((\lambda_k + c)^{1/2}(t - \tau)) g_k(\tau) \, d\tau \tag{2.4.20}$$

if $\lambda_k + c > 0$ for $k \geqq k^+$, say, $k^+ \in \mathbf{N}$;

$$u_k(t) = \varphi_k + \psi_k t + \int_0^t (t - \tau)\, g_k(\tau)\, d\tau \tag{2.4.21}$$

if $\lambda_k + c = 0$;

$$u_k(t) = \varphi_k \cosh(-\lambda_k - c)^{1/2}\, t + \psi_k(-\lambda_k - c)^{-1/2} \sinh(-\lambda_k - c)^{1/2}\, t + \\ + (-\lambda_k - c)^{-1/2} \int_0^t \sinh(-\lambda_k - c)^{1/2} (t - \tau)\, g_k(\tau)\, d\tau \tag{2.4.22}$$

if $\lambda_k + c < 0$.

It is easy to see that $u_k \in C^2([0, T])$ for all $k \in \mathbf{N}$. Evidently, if $\varphi = \psi = g = 0$ then $u = 0$ which implies the uniqueness of the solution to the problem $(\mathcal{IB})$.

Let us prove the uniform convergence of the series (2.4.5)–(2.4.7). Owing to the Schwarz inequality, (2.4.20) implies for $k \geqq k^+$ that

$$u_k^2(t) \leqq 3\left(\varphi_k^2 + (\lambda_k + c)^{-1}\psi_k^2 + (\lambda_k + c)^{-1}\, T \int_0^T g_k^2(\tau)\, d\tau\right).$$

There are constants $c_1, c_2 > 0$ such that

$$c_1 \lambda_k \leqq \lambda_k + c \leqq c_2 \lambda_k \quad \text{for all} \quad k \geqq k^+ .$$

By virtue of (2.4.14),

$$\sum_{k=1}^{\infty} \lambda_k \int_0^T g_k^2(\tau)\, d\tau = \int_0^T \sum_{k=1}^{\infty} \lambda_k\, g_k^2(\tau)\, d\tau < +\infty . \tag{2.4.23}$$

For $k \geqq k^+$,

$$\lambda_k^2\, u_k^2(t) \leqq \text{const.} \left(\lambda_k^2 \varphi_k^2 + \lambda_k \psi_k^2 + \lambda_k \int_0^T g_k^2(\tau)\, d\tau\right), \quad t \in [0, T],$$

and (2.4.12), (2.4.13) and (2.4.23) yield the uniform convergence of the series (2.4.5).

Further,

$$\dot{u}_k(t) = -(\lambda_k + c)^{1/2}\, \varphi_k \sin(\lambda_k + c)^{1/2}\, t + \psi_k \cos(\lambda_k + c)^{1/2}\, t + \\ + \int_0^t \cos((\lambda_k + c)^{1/2} (t - \tau))\, g_k(\tau)\, d\tau, \quad k \geqq k^+ \tag{2.4.24}$$

and just as above we obtain the uniform convergence of the series (2.4.6). The uniform convergence of (2.4.7) follows from the relation $\ddot{u}_k(t) = g_k(t) - (\lambda_k + c)\, u_k(t)$. Hence, $u \in \mathcal{U}$.

As regards (2.4.8), (2.4.15)–(2.4.17), it is easily seen that there is a constant $M > 0$ such that

$$\|u\|_{\mathscr{U}} \leqq M(\|\varphi\|_{\mathscr{D}(E)} + \|\psi\|_{\mathscr{D}(E^{1/2})} + \|g\|_{C([0,T];\mathscr{D}(E^{1/2}))}) . \tag{2.4.25}$$

We have now proved the following theorem.

THEOREM 2.4.1 *Let*

(a) $g \in C([0, T]; \mathscr{D}(E^{1/2}))$,

(b) $\varphi \in \mathscr{D}(E)$, $\psi \in \mathscr{D}(E^{1/2})$.

Then the problem $(\mathscr{I}\mathscr{B})$ *has a unique solution*

$$u = S(\varphi, \psi, g) \in \mathscr{U}$$

and

$$S \in \mathscr{L}(\mathscr{D}(E) \times \mathscr{D}(E^{1/2}) \times C([0, T]; \mathscr{D}(E^{1/2})), \mathscr{U}) .$$

This solution is given by (2.4.4), *where the* u_k *are determined by* (2.4.20)–(2.4.22).

Remark 2.4.1 We may suppose that $g \in C^1([0, T]; H^0(\Omega))$ instead of $g \in$ $\in C([0, T]; \mathscr{D}(E^{1/2}))$. To obtain an estimate of u_k (or $\dot{u}_k$) we need only integrate by parts.

Remark 2.4.2 The norms $\|\cdot\|_{\mathscr{D}(E)}$ and $\|\cdot\|_{\mathscr{D}(E^{1/2})}$ are equivalent to $\|\cdot\|_2$ and $\|\cdot\|_1$, respectively, and $\mathscr{U}$ is a subspace of

$$C([0, T]; H^2(\Omega)) \cap C^1([0, T]; H^1(\Omega)) \cap C^2([0, T]; H^0(\Omega)) ,$$

(see Theorem 1.3.2).
The series (2.4.4) can be differentiated term-by-term with respect to $t, x_1, \ldots, x_n$ once or twice, and the resulting series are convergent in the norm of $C([0,T];$ $H^0(\Omega))$.

Remark 2.4.3 The same method can be applied if more regular (in particular classical, that is, $u \in C([0, T]; \mathscr{D}(E^{([n/2]+3)/2})) \cap C^1([0, T]; \mathscr{D}(E^{[n/2]/2+1})) \cap$ $\cap C^2([0, T]; \mathscr{D}(E^{([n/2]+1)/2})))$ solutions are looked for. However, in this case some additional conditions on the behaviour of φ, ψ, and g on the boundary $\partial\Omega$ must be imposed (for the precise formulation see [11] pp. 101 and 111; see also Corollary 1.1.1).

Remark 2.4.4 This method fails if the boundary condition (2.4.2) or (2.4.2′) is not homogeneous. In this case it is necessary to transform the problem to one with a homogeneous boundary condition for a new unknown v by putting $u = u_0 + v$ where u_0 is a function satisfying the inhomogeneous boundary condition.

A similar theorem can be also established for a more general telegraph equation. Suppose that the problem $(\mathscr{I}\mathscr{B})$ consists of finding a $u \in \mathscr{U}$ satisfying

$$u_{tt}(t, x) + 2a\, u_t(t, x) + E\, u(t, x) + c\, u(t, x) = g(t, x)\,,$$
$$(t, x) \in [0, T] \times \Omega\,, \quad a \neq 0 \tag{2.4.26}$$

(2.4.2) or (2.4.2'), and (2.4.3).

The assumptions on a_{jk}, a_0, σ, and $\partial\Omega$ as well as on φ, ψ, and g remain unchanged.

Looking again for a solution u in the form (2.4.4) we obtain u_k as the solution of the Cauchy problem

$$\ddot{u}_k(t) + 2a\, \dot{u}_k(t) + (\lambda_k + c)\, u_k(t) = g_k(t)\,, \quad t \in (0, T)\,, \tag{2.4.27}$$

$$u_k(0) = \varphi_k\,, \quad \dot{u}_k(0) = \psi_k\,, \tag{2.4.28}$$

that is

$$u_k(t) = \varphi_k\, e^{-at} \cos (\lambda_k + c - a^2)^{1/2}\, t +$$
$$+ (a\varphi_k + \psi_k)\, e^{-at} (\lambda_k + c - a^2)^{-1/2} \sin (\lambda_k + c - a^2)^{1/2}\, t +$$
$$+ (\lambda_k + c - a^2)^{-1/2} \int_0^t e^{-a(t-\tau)} \sin ((\lambda_k + c - a^2)^{1/2}\, (t - \tau))\, g_k(\tau)\, \mathrm{d}\tau \tag{2.4.29}$$

if $\lambda_k + c - a^2 > 0$,

$$u_k(t) = ((a\varphi_k + \psi_k)\, t + \varphi_k)\, e^{-at} + \int_0^t e^{-a(t-\tau)}\, (t - \tau)\, g_k(\tau)\, \mathrm{d}\tau \tag{2.4.30}$$

if $\lambda_k + c - a^2 = 0$,

$$u_k(t) = \varphi_k\, e^{-at} \cosh (a^2 - \lambda_k - c)^{1/2}\, t +$$
$$+ (a\varphi_k + \psi_k)\, e^{-at}(a^2 - \lambda_k - c)^{-1/2} \sinh (a^2 - \lambda_k - c)^{1/2}\, t +$$
$$+ (a^2 - \lambda_k - c)^{-1/2} \int_0^t e^{-a(t-\tau)} \sinh ((a^2 - \lambda_k - c)^{1/2}\, (t - \tau))\, g_k(\tau)\, \mathrm{d}\tau \tag{2.4.31}$$

if $\lambda_k + c - a^2 < 0$.

By a similar argument we obtain the following theorem.

THEOREM 2.4.2 *Let*

$$(a)\ g \in C([0, T]; \mathscr{D}(E^{1/2})),$$
$$(b)\ \varphi \in \mathscr{D}(E), \quad \psi \in \mathscr{D}(E^{1/2}).$$

Then the problem $(\mathscr{I}\mathscr{B})$ *given by* (2.4.26), (2.4.2) *or* (2.4.2′) *and* (2.4.3) *has a unique solution*

$$u = S(\varphi, \psi, g) \in \mathscr{U}$$

with $S \in \mathscr{L}(\mathscr{D}(E) \times \mathscr{D}(E^{1/2}) \times C([0, T]; \mathscr{D}(E^{1/2})), \mathscr{U})$.
This solution is given by (2.4.4), *where the* u_k *are determined by* (2.4.29)–(2.4.31).

§ 3. The heat equation

3.1. *The Cauchy problem*

By the Cauchy problem for the heat equation we mean the problem $(\mathscr{I}_n)$ given by

$$u_t(t, x) - \Delta_n u(t, x) + c\, u(t, x) = g(t, x),$$
$$(t, x) \in (0, T) \times \mathbf{R}^n, \quad (T > 0), \tag{3.1.1}$$
$$u(0, x) = \varphi(x), \quad x \in \mathbf{R}^n. \tag{3.1.2}$$

By the substitution $v(t, x) = e^{ct} u(t, x)$ the problem $(\mathscr{I}_n)$ transforms to the problem

$$v_t(t, x) - \Delta_n v(t, x) = e^{ct} g(t, x), \quad (t, x) \in (0, T) \times \mathbf{R}^n,$$
$$v(0, x) = \varphi(x), \quad x \in \mathbf{R}^n.$$

Applying the results in [6], pp. 25 and 29, to the transformed problem we obtain the following theorem.

THEOREM 3.1.1 *Let* $\varphi \in C_{\text{loc}}(\mathbf{R}^n)$ *and let* $g \in C_{\text{loc}}((0, T] \times \mathbf{R}^n)$ *be locally Hölder continuous in* $x \in \mathbf{R}^n$ *uniformly with respect to* t. *Furthermore, let* K *and* $\alpha < (4T)^{-1}$ *be constants such that*

$$|\varphi(x)| \leqq K\, e^{\alpha|x|^2}, \quad x \in \mathbf{R}^n,$$
$$|g(t, x)| \leqq K\, e^{\alpha|x|^2}, \quad t \in (0, T], \quad x \in \mathbf{R}^n, \quad (|x| = (\sum_{j=1}^{n} x_j^2)^{1/2}).$$

Then the problem $(\mathscr{I}_n)$ *has a unique classical solution* $u^* \in C^{1,2}_{\mathrm{loc}}((0, T] \times \times \mathbf{R}^n) \cap C_{\mathrm{loc}}([0, T] \times \mathbf{R}^n)$. *This solution is given by*

$$u^*(t, x) = (4\pi t)^{-n/2} e^{-ct} \int_{\mathbf{R}^n} \exp\left(-\frac{|x - \xi|^2}{4t}\right) \varphi(\xi) \,\mathrm{d}\xi +$$

$$+ (4\pi)^{-n/2} \int_0^t \int_{\mathbf{R}^n} (t - \tau)^{-n/2} e^{-c(t-\tau)} \exp\left(-\frac{|x - \xi|^2}{4(t - \tau)}\right) g(\tau, \xi) \,\mathrm{d}\xi \,\mathrm{d}\tau\,,$$

$$t \in (0, T]\,, \quad x \in \mathbf{R}^n\,, \quad u^*(0, x) = \varphi(x)\,, \quad x \in \mathbf{R}^n \tag{3.1.3}$$

and satisfies $|u(t, x)| \leqq M\, e^{-ct}\, e^{\beta|x|^2}$ *for some constants* M, β *and all* $t \in [0, T]$, $x \in \mathbf{R}^n$.

3.2. *The initial boundary-value problems*

Suppose that $(\mathscr{I}\mathscr{B})$ is given by

$$u_t(t, x) - u_{xx}(t, x) + c\, u(t, x) = g(t, x)\,,$$

$$(t, x) \in Q = (0, T) \times (0, \pi)\,, (T > 0)\,, \tag{3.2.1}$$

$$u(t, 0) = h_0(t)\,, \quad u(t, \pi) = h_1(t)\,, \quad t \in [0, T]\,, \tag{3.2.2}$$

$$u(0, x) = \varphi(x)\,, \quad x \in [0, \pi]\,. \tag{3.2.3}$$

We put

$$\Theta(t, \eta) = 1 + 2 \sum_{n=1}^{\infty} e^{-\pi^2 n^2 t} \cos 2\pi n\eta = (\pi t)^{-1/2} \sum_{n=-\infty}^{\infty} e^{-(\eta+n)^2/t}\,, \tag{3.2.4}$$

$$\Gamma(t, \eta, \xi) = \frac{1}{2\pi}\left[\Theta\left(\frac{t}{\pi^2}, \frac{\eta - \xi}{2\pi}\right) - \Theta\left(\frac{t}{\pi^2}, \frac{\eta + \xi}{2\pi}\right)\right]. \tag{3.2.5}$$

THEOREM 3.2.1 *Let*

(a) $g \in C^{0,1}(\overline{Q})$,
(b) $h_0, h_1 \in C^1([0, T])$,
(c) $\varphi \in C^2([0, \pi])$.

Suppose that the following compatibility conditions hold:

$$\varphi(0) = h_0(0)\,, \quad \varphi(\pi) = h_1(0)\,,$$

$$h_0'(0) - \varphi''(0) + c\,\varphi(0) = g(0, 0),\ h_1'(0) - \varphi''(\pi) + c\,\varphi(\pi) = g(0, \pi)$$

Then the problem $(\mathscr{I}\mathscr{B})$ *has a unique solution* $u^* \in C^{1,2}(\bar{Q})$. *This is given by*

$$u^*(t, x) \equiv S(\varphi, g, h_0, h_1)(t, x) = e^{-ct} \int_0^\pi \Gamma(t, x, \xi)\, \varphi(\xi)\, \mathrm{d}\xi +$$

$$+ e^{-ct} \int_0^t \int_0^\pi e^{c\tau}\, \Gamma(t - \tau, x, \xi)\, g(\tau, \xi)\, \mathrm{d}\xi\, \mathrm{d}\tau -$$

$$- \frac{1}{\pi} e^{-ct} \int_0^t e^{c\tau}\, h_0(\tau) \frac{\partial}{\partial x}\, \Theta\left(\frac{t - \tau}{\pi^2}, \frac{x}{2\pi}\right) \mathrm{d}\tau +$$

$$+ \frac{1}{\pi} e^{-ct} \int_0^t e^{c\tau}\, h_1(\tau) \frac{\partial}{\partial x}\, \Theta\left(\frac{t - \tau}{\pi^2}, \frac{\pi - x}{2\pi}\right) \mathrm{d}\tau \qquad (3.2.6)$$

and

$$S \in \mathscr{L}(C^2([0, \pi]) \times C^{0,1}(\bar{Q}) \times C^1([0, T]) \times C^1([0, T]),\ C^{1,2}(\bar{Q}))\,.$$

Proof. The formula (3.2.6) is derived in [2] pp. 104–107. The fact that $u^* \in C^{1,2}(\bar{Q})$ can be verified by an easy but rather lengthy calculation (see [22]).

Remark 3.2.1 An existence theorem for the corresponding weakly non-linear problem can easily be established with the help of I: Theorem 3.1.1.

Remark 3.2.2 A similar formula holds when the Dirichlet boundary conditions (3.2.2) are replaced by the Neumann boundary conditions

$$u_x(t, 0) = h_0(t)\,,$$

$$u_x(t, \pi) = h_1(t)\,, \quad t \in [0, T]\,.$$

3.3. *The Fourier method*

The problem (3.2.1)–(3.2.3) with $h_0 = h_1 = 0$ as well as the more general problem given by

$$u_t(t, x) + Eu\,(t, x) + c\,u(t, x) = g(t, x)\,, \quad t \in (0, T)\,, \quad x \in \Omega\,, \qquad (3.3.1)$$

$$u(0, x) = \varphi(x)\,, \quad x \in \Omega \qquad (3.3.2)$$

and (2.4.2) can be treated by the Fourier method. Here Ω, E, and c satisfy the same assumptions as in Sec. 2.4. Again, the solution is looked for in the form (2.4.4), where, $\{v_k\}_{k=1}^\infty$ is the system of eigenfunctions of the operator E in $H^0(\Omega)$ with $\mathscr{D}(E) = H^2(\Omega) \cap \dot{H}^1(\Omega)$. We search for the solution in the space $\mathscr{U} = C^1([0, T]; H^0(\Omega)) \cap C^0([0, T]; H^2(\Omega) \cap \dot{H}^1(\Omega))$. The norm on $\mathscr{U}$ is defined by

$$\|u\|_{\mathscr{U}} = \max\{\max_{t\in[0,T]} \|E\, u(t)\|_{H^0(\Omega)}, \quad \max_{t\in[0,T]} \|u_t(t)\|_{H^0(\Omega)}\}.$$

We can state the following theorem.

THEOREM 3.3.1 *Let $\varphi \in H^2(\Omega) \cap \dot{H}^1(\Omega)$ and $g \in C([0, T]; {}^0H^{2\alpha}(\Omega))$ for some $\alpha > 0$. Then there exists a unique solution $u \in \mathscr{U}$ to the problem* (3.3.1), (3.3.2). *This is given by* (2.4.4), *where $u_k(t)$ are defined by* (3.3.4). *Moreover,*

$$\|u\|_{\mathscr{U}} \leqq \text{const.}\left(\|\varphi\|_{H^2(\Omega)} + \|g\|_{C([0,T];{}^0H^{2\alpha}(\Omega))}\right).$$

Proof. Substituting the series (2.4.4) in (3.3.1)–(3.3.2), and proceeding just as in Sec. 2.4, we find that $u_k(t)$ must satisfy the relations

$$\dot{u}_k(t) + (\lambda_k + c)\, u_k(t) = g_k(t), \quad t \in (0, T),$$
$$u_k(0) = \varphi_k, \tag{3.3.3}$$

where

$$g_k(t) = \int_\Omega g(t, x)\, v_k(x)\, dx,$$

$$\varphi_k = \int_\Omega \varphi(x)\, v_k(x)\, dx, \quad k = 1, 2, \dots .$$

Evidently we have

$$u_k(t) = e^{-(\lambda_k + c)t}\, \varphi_k + \int_0^t e^{-(\lambda_k+c)(t-\tau)}\, g_k(\tau)\, d\tau. \tag{3.3.4}$$

According to Theorem 1.3.3, p. 42, the function u belongs to $\mathscr{U}$ if and only if it is of the form (2.4.4), where

$$\sum_{k=1}^{\infty} (\dot{u}_k^2(t) + \lambda_k^2\, u_k^2(t)) \quad \text{converges uniformly for} \quad t \in [0, T]$$

$$\text{and} \quad u_k \in C^1([0, T]). \tag{3.3.5}$$

Clearly,

$$\|u\|_{\mathscr{U}} = \max\{\max_{t\in[0,T]} (\sum_{k=1}^{\infty} \lambda_k^2\, u_k^2(t))^{1/2}, \quad \max_{t\in[0,T]} (\sum_{k=1}^{\infty} \dot{u}_k^2(t))^{1/2}\}. \tag{3.3.6}$$

Using Theorem 1.3.3, we find that $\sum_{k=1}^{\infty} \lambda_k^2 \varphi_k^2$ is convergent and $\sum_{k=1}^{\infty} \lambda_k^{2\alpha} g_k^2(t)$ is uniformly convergent for $t \in [0, T]$. Because of (3.3.4), (3.3.2) and (3.3.3) to establish (3.3.5) it suffices to show that the series

$$\sum_{k=1}^{\infty} \lambda_k^2 \left(\int_0^t e^{-\lambda_k(t-\tau)} g_k(\tau) \, d\tau \right)^2$$

is uniformly convergent for $t \in [0, T]$.
Using I: Theorem 1.4.2 (c), p. 12 (the Parseval identity) we obtain

$$\left(\sum_{k=1}^{n} \lambda_k^2 \left(\int_0^t e^{-\lambda_k(t-\tau)} g_k(\tau) \, d\tau \right)^2 \right)^{1/2} =$$

$$= \left\| \int_0^t \sum_{k=1}^{n} \lambda_k^{1-\alpha} e^{-\lambda_k(t-\tau)} \lambda_k^{\alpha} g_k(\tau) v_k \, d\tau \right\|_0 \leqq$$

$$\leqq \int_0^t \left\| \sum_{k=1}^{n} \lambda_k^{1-\alpha} e^{-\lambda_k(t-\tau)} \lambda_k^{\alpha} g_k(\tau) v_k \right\|_0 d\tau =$$

$$= \int_0^t \left(\sum_{k=1}^{n} \lambda_k^{2(1-\alpha)} e^{-2\lambda_k(t-\tau)} \lambda_k^{2\alpha} g_k^2(\tau) \right)^{1/2} d\tau \leqq$$

$$\leqq \int_0^t \left(\max_{\lambda \in [0,\infty)} \lambda^{2(1-\alpha)} e^{-2\lambda(t-\tau)} \right)^{1/2} \left(\sum_{k=1}^{n} \lambda_k^{2\alpha} g_k^2(\tau) \right)^{1/2} d\tau \leqq$$

$$\leqq \int_0^t \frac{\text{const.}}{(t-\tau)^{1-\alpha}} \sup_{t \in [0,T]} \left(\sum_{k=1}^{\infty} \lambda_k^{2\alpha} g_k^2(t) \right)^{1/2} d\tau \leqq$$

$$\leqq \text{const.} \, \|g\|_{C([0,T];\, {}^0H^{2\alpha}(\Omega))} ,$$

and our assertion follows immediately.

Chapter III

The heat equation

§ 1. The (t, s)-Fourier method

1.1. *The linear case*

Let us investigate conditions under which there exists a solution to the problem $(\mathscr{P}_\omega)$ given by

$$u_t(t, x) - u_{xx}(t, x) + c\, u(t, x) = g(t, x)\,, \quad (t, x) \in Q = \mathbf{R} \times (0, \pi)\,, \tag{1.1.1}$$

$$u(t, 0) = u(t, \pi) = 0\,, \quad t \in \mathbf{R}\,, \tag{1.1.2}$$

$$u(t + \omega, x) - u(t, x) = 0\,, \quad (t, x) \in Q\,, \tag{1.1.3}$$

where $c \in \mathbf{R}$ and g is a function ω-periodic in t.

By a solution we mean a function

$$u \in \mathscr{U} = H^{1,2}_\omega(Q) \cap \dot{H}^1_\omega(Q)$$

that satisfies (1.1.1) almost everywhere in Q. Note that

$$\dot{H}^1_\omega(Q) = \mathrm{cl}\,\{u \in C^\infty_\omega(\bar{Q});\, u(t, 0) = u(t, \pi) = 0,\, t \in \mathbf{R}\} \quad \text{in} \quad H^1_\omega(Q)\,.$$

The space $\mathscr{U}$ is equipped with the norm of $H^{1,2}_\omega(Q)$, that is,

$$\|u\|_{1,2} = \left(\int_0^\omega \int_0^\pi [(u(t, x))^2 + (u_t(t, x))^2 + (u_x(t, x))^2 + \right.$$

$$\left. + (u_{xx}(t, x))^2]\, \mathrm{d}x\, \mathrm{d}t\right)^{1/2}.$$

By II: Example 1.3.1, I: Theorem 2.4.1 and I: Theorem 2.2.1 the system $\{(2/\pi\omega)^{1/2}\, e^{ij\nu t} \sin kx\}_{j\in\mathbf{Z},k\in\mathbf{N}}$,

where

$$\nu = \frac{2\pi}{\omega},$$

forms an orthonormal base in $H^0_\omega(Q)$. Hence, every function $u \in H^0_\omega(Q)$ can be written as a Fourier series

$$u(t, x) = \left(\frac{2}{\pi\omega}\right)^{1/2} \sum_{j,k} u_{jk}\, e^{ij\nu t} \sin kx \tag{1.1.4}$$

(here and in what follows we use the convention that $\sum_{j,k} = \sum_{j=-\infty}^{\infty} \sum_{k=1}^{\infty}$), where

$$u_{jk} = \left(\frac{2}{\pi\omega}\right)^{1/2} \int_0^\omega \int_0^\pi u(\tau, \xi)\, e^{-ij\nu\tau} \sin k\xi \, d\xi \, d\tau \,.$$

By II: Theorem 1.3.4, $u \in H^0_\omega(Q)$ belongs to $\mathscr{U}$ if and only if the series

$$\sum_{j,k} (j^2 + k^4)\, |u_{jk}|^2 \tag{1.1.5}$$

converges. The square root of the series (1.1.5) defines a norm $\|\cdot\|_{\mathscr{U}}$ on $\mathscr{U}$, which is equivalent to the norm $\|\cdot\|_{1,2}$.

Suppose that $g \in H^0_\omega(Q)$. Then,

$$g(t, x) = \left(\frac{2}{\pi\omega}\right)^{1/2} \sum_{j,k} g_{jk}\, e^{ij\nu t} \sin kx \,, \tag{1.1.6}$$

where

$$g_{jk} = \left(\frac{2}{\pi\omega}\right)^{1/2} \int_0^\omega \int_0^\pi g(\tau, \xi)\, e^{-ij\nu\tau} \sin k\xi \, d\xi \, d\tau$$

and

$$\|g\|_0^2 = \sum_{j,k} |g_{jk}|^2 < \infty \,.$$

Note that all functions are supposed to be real-valued so that the partial sums of a series of the form (1.1.6) defined as

$$\sum_{j=-m}^{m} \sum_{k=1}^{n} g_{jk}\, e^{ij\nu t} \sin kx \,, \quad m, n \in \mathbf{N}$$

are also real (because $\bar{g}_{jk} = g_{-j,k}$, $j \in \mathbf{Z}$, $k \in \mathbf{N}$).

Substituting the series (1.1.4), (1.1.6) in (1.1.1) and taking into account that two Fourier series are equal if and only if the corresponding Fourier coefficients are equal, we obtain for the u_{jk} a system of algebraic equations

$$(ij\nu + k^2 + c)\, u_{jk} = g_{jk}\,, \quad j \in \mathbf{Z}\,, \quad k \in \mathbf{N}\,. \tag{1.1.7}$$

We have to distinguish between two cases

(i) $c \neq -k^2$, $k \in \mathbf{N}$;
(ii) $c = -k_0^2$ for some $k_0 \in \mathbf{N}$.

In the case (i) we find from (1.1.7) that

$$u_{jk} = \frac{g_{jk}}{ijv + k^2 + c}, \quad j \in \mathbf{Z}, \quad k \in \mathbf{N}; \tag{1.1.8}$$

when substituted in (1.1.5), this yields

$$\|u\|_{\mathscr{U}}^2 = \sum_{j,k} (j^2 + k^4) |u_{jk}|^2 = \sum_{j,k} \frac{j^2 + k^4}{v^2 j^2 + (k^2 + c)^2} |g_{jk}|^2 =$$

$$= \sum_{j,k} \frac{j^2 + k^4}{v^2 j^2 + (1 + ck^{-2})^2 k^4} |g_{jk}|^2 \leqq$$

$$\leqq \left(\min_{k \in \mathbf{N}} [v^2, (1 + ck^{-2})^2]\right)^{-1} \sum_{j,k} |g_{jk}|^2 = C \|g\|_0^2,$$

where C is a constant independent of g.

Before stating the result let us define an operator L by

$$Lu = u_t - u_{xx} + cu$$

on $\mathscr{D}(L) = \mathscr{U}$. We have proved that L has the inverse L^{-1} and

$$\|L^{-1} g\|_{\mathscr{U}} \leqq \text{const.} \|g\|_0, \quad g \in H_\omega^0(Q),$$

which means that $L^{-1}: H_\omega^0(Q) \to \mathscr{U}$ is continuous.

THEOREM 1.1.1 (*The non-critical case*). *Let $g \in H_\omega^0(Q)$ and $c \neq -k^2$ for all $k \in \mathbf{N}$.*
Then the problem $(\mathscr{P}_\omega)$ has a unique solution $u \in \mathscr{U}$,

$$u(t, x) = (L^{-1} g)(t, x) =$$

$$= \left(\frac{2}{\pi\omega}\right)^{1/2} \sum_{j,k} \frac{g_{jk}}{ijv + k^2 + c} e^{ijvt} \sin kx \quad (v = 2\pi/\omega) \tag{1.1.9}$$

and $L^{-1} \in \mathscr{L}(H_\omega^0(Q), \mathscr{U})$.

In the case (ii) the left-hand side of (1.1.7) vanishes for $(j, k) = (0, k_0)$. Hence $(\mathscr{P}_\omega)$ has a solution only if $g_{0k_0} = 0$, that is,

$$\int_0^\omega \int_0^\pi g(\tau, \xi) \sin k_0 \xi \, d\xi \, d\tau = 0. \tag{1.1.10}$$

If (1.1.10) holds, then the solution of (1.1.7) for $(j, k) = (0, k_0)$ (satisfying $u_{0k_0} = \bar{u}_{0k_0}$) is an arbitrary $d \in \mathbf{R}$ and

$$u(t, x) = d \sin k_0 x\,, \quad d \in \mathbf{R} \tag{1.1.11}$$

is the solution of $(\mathscr{P}_\omega)$ with $g \equiv 0$ (the homogeneous problem).
For $(j, k) \neq (0, k_0)$ the u_{jk} are determined by (1.1.8) and we can estimate as above. For the sake of simplicity of our further formulations let us introduce a certain notation. Let $\mathscr{V}$ be the subspace of functions $u \in H^0_\omega(Q)$ of the form (1.1.11) and $\mathscr{W}$ its orthogonal complement in $H^0_\omega(Q)$, that is, the (closed) subspace of functions $u \in H^0_\omega(Q)$ of the form

$$u(t, x) = \left(\frac{2}{\pi\omega}\right)^{1/2} \sum_{\substack{j,k \\ (j,k) \neq (0,k_0)}} u_{jk}\, e^{ij\nu t} \sin kx$$

(or equivalently the subspace of functions $u \in H^0_\omega(Q)$ for which

$$\int_0^\omega \int_0^\pi u(\tau, \xi) \sin k_0 \xi \, d\xi \, d\tau = 0)\,.$$

Let $g \in \mathscr{W}$. Then

$$u(t, x) = v(t, x) + w(t, x)\,,$$

where $v \in \mathscr{U} \cap \mathscr{V}$ is arbitrary, $w \in \mathscr{U} \cap \mathscr{W}$ is uniquely determined and $\|w\|_{\mathscr{U}} \leqq C\|g\|_0$ with $C > 0$ independent of g. (The solution is a sum of the general solution to the corresponding homogeneous problem and a particular solution to the inhomogeneous one.) Since the null-space of L, that is $\mathscr{U} \cap \mathscr{V}$, is non-trivial ($\dim \mathscr{U} \cap \mathscr{V} = 1$), an inverse to L does not exist. But if L is regarded as restricted to $\mathscr{U} \cap \mathscr{W}$ an inverse (denoted again by L^{-1} and sometimes called a *right inverse to* L) exists on $\mathscr{W}$ and

$$\|L^{-1} g\|_{\mathscr{U}} \leqq \text{const.}\, \|g\|_0\,, \quad g \in \mathscr{W}\,.$$

We have proved the following result.

THEOREM 1.1.2 (*The critical case*). *Let $g \in H^0_\omega(Q)$ and let $k_0 \in \mathbf{N}$ be such that $c = -k_0^2$.*
Then the problem $(\mathscr{P}_\omega)$ has a solution $u \in \mathscr{U}$ if and only if (1.1.10) holds, that is, $g \in \mathscr{W}$. If this condition is satisfied, then the solutions can be written in the form

$$u(t, x) = d \sin k_0 x + w(t, x)\,,$$

where $d \in \mathbf{R}$ is arbitrary, $w \in \mathscr{U} \cap \mathscr{W}$,

$$w(t, x) = (L^{-1}g)(t, x) =$$

$$= \left(\frac{2}{\pi\omega}\right)^{1/2} \sum_{\substack{j,k \\ (j,k) \neq (0,k_0)}} \frac{g_{jk}}{ijv + k^2 + c} e^{ijvt} \sin kx \quad (v = 2\pi/\omega) \tag{1.1.12}$$

and $L^{-1} \in \mathscr{L}(\mathscr{W}, \mathscr{U} \cap \mathscr{W})$.

Remark 1.1.1 By I: Theorem 2.7.5 we find that $u \in \mathscr{U}$ implies that $u \in C_\omega(Q)$ and $u_x \in L_{6,\omega}(Q)$. (The fact that $u \in C_\omega(\overline{Q})$ can be deduced simply as follows: using the Cauchy inequality and the inequality $a^2 + b^2 \geqq a^\beta b^{2-\beta}$, $a \geqq 0$, $b \geqq 0$, $0 \leqq \beta \leqq 2$ we find that $\sum_{j,k} |u_{jk}| \leqq \text{const.} \|u\|_{\mathscr{U}}$, which yields the required result.)

Remark 1.1.2 Sometimes we are interested in a solution that is smoother, for example, in $C_\omega^{1,2}(Q)$, than that in $\mathscr{U}$ we have just found. We can use the expansions (1.1.4) and (1.1.6) and immediately establish that

$$\sum_{j,k} (j^4 + k^8) |u_{jk}|^2 \leqq \text{const.} \sum_{j,k} (j^2 + k^4) |g_{jk}|^2 .$$

(We treat for simplicity the non-critical case.) This means by II: Theorem 1.3.4 that $u \in H_\omega^{2,4}(Q) \cap H_\omega^0(\mathbf{R}; {}^0H^4(0, \pi))$ (that is, $u \in H_\omega^{2,4}(Q)$ and $u(t, 0) = u(t, \pi) = u_{xx}(t, 0) = u_{xx}(t, \pi) = 0$, $t \in \mathbf{R}$) provided that $g \in H_\omega^{1,2}(Q) \cap \dot{H}_\omega^1(Q)$. By I: Theorem 2.7.5 we get $u \in C_\omega^{1,2}(Q)$. Thus, we have obtained the classical solution, but only under the assumption that $g(t, 0) = g(t, \pi) = 0$, $t \in \mathbf{R}$. It turns out that this restriction can be removed by means of another regularization procedure:

Let $g \in H_\omega^{1,0}(Q)$, $c \neq -k^2$, $k \in \mathbf{N}$,

then

$$\sum_{j,k} (1 + j^2)(j^2 + k^4) |u_{jk}|^2 \leqq \text{const.} \sum_{j,k} (1 + j^2) |g_{jk}|^2 ,$$

which implies that $u_t \in H_\omega^{1,2}(Q) \cap \dot{H}_\omega^1(Q)$. In particular, $u_{tt}, u_{txx} \in H_\omega^0(Q)$. We write

$$u_{xx}(t, x) = -g(t, x) + u_t(t, x) + c\, u(t, x)$$

and suppose that $g \in H_\omega^{1,2}(Q)$, so that the right-hand side has the second derivative with respect to x, which belongs to $H_\omega^0(Q)$. Hence, $u_{xxxx} \in H_\omega^0(Q)$, and therefore, $u \in H_\omega^{2,4}(Q) \cap \dot{H}_\omega^1(Q)$. Clearly, the operator L defined now on $H_\omega^{2,4}(Q) \cap \dot{H}_\omega^1(Q)$ has a continuous inverse on $H_\omega^{1,2}(Q)$. A solution from the same space can be obtained also in the critical case, provided that $g \in H_\omega^{1,2}(Q) \cap \mathscr{W}$.

Remark 1.1.3 The space $\mathscr{U} = H_{\omega}^{1,2}(Q) \cap \dot{H}_{\omega}^{1}(Q)$ can be written equivalently as

$$H_{\omega}^{0}(\mathbf{R}; \mathscr{D}(E)) \cap H_{\omega}^{1,0}(Q),$$

where E is the operator in $H^{0}(\Omega)$ defined by $\mathscr{D}(E) = \{v \in H^{2}(0, \pi); v(0) = = v(\pi) = 0\}$ (that is, $\mathscr{D}(E) = H^{2}(0, \pi) \cap \dot{H}^{1}(0, \pi)$) and $Ev = -v_{xx}$.

A similar procedure can be carried out for the problem $(\mathscr{P}_{\omega}^{1})$ given by

$$u_t(t, x) - u_{xx}(t, x) + c' u(t, x) = g(t, x), \quad (t, x) \in Q, \tag{1.1.1'}$$

the Newton boundary conditions

$$\begin{aligned} &u_x(t, 0) + \sigma_0 u(t, 0) = 0, \\ &u_x(t, \pi) + \sigma_1 u(t, \pi) = 0, \quad t \in \mathbf{R} \end{aligned} \tag{1.1.13}$$

and the periodicity condition (1.1.3). Here $c', \sigma_0, \sigma_1 \in \mathbf{R}$.

Let us denote by E' the operator defined by

$$\mathscr{D}(E') = \{v \in H^{2}(0, \pi); v_x(0) + \sigma_0 v(0) = 0, v_x(\pi) + \sigma_1 v(\pi) = 0\}$$

and $E'v = -v_{xx}$. The eigenvalue problem $E'v = \lambda' v$ is studied in II: Example 1.3.2. It is shown that there exists a countable set of eigenvalues $\{\lambda_k'\}_{k=1}^{\infty}$, λ_k' being simple, real and positive with at most two exceptions, $\lambda_1' < \ldots < \lambda_k' < \ldots$ with ∞ as the only limit point, and that the corresponding system of eigenfunctions $\{v_k\}_{k=1}^{\infty} \subset C^{\infty}([0, \pi])$ forms an orthonormal base in $H^{0}(0, \pi)$. To remove possible non-positive eigenvalues we choose a real number c_0 so that an operator E defined by $\mathscr{D}(E) = \mathscr{D}(E')$ and $E = E' + c_0 I$ (I is the identity mapping) is already positive definite in $H^{0}(0, \pi)$, that is, for some $\gamma > 0$ we have

$$\langle Ev, v \rangle_{H^{0}(0,\pi)} \geqq \gamma \|v\|_{H^{0}(0,\pi)}^{2}, \quad v \in \mathscr{D}(E).$$

(We put $c_0 > -\lambda_1'$ if $\lambda_1' \leqq 0$, and $c_0 = 0$ otherwise.) The eigenvalues of E are $\lambda_k = \lambda_k' + c_0 > 0$, $k \in \mathbf{N}$. The system of eigenfunctions remains unchanged. Now, setting $c' - c_0 = c$, we can rewrite (1.1.1') in the form

$$u_t(t, x) + E\, u(t, x) + c\, u(t, x) = g(t, x). \tag{1.1.1''}$$

By a solution to $(\mathscr{P}_{\omega}^{1})$ we mean a function

$$u \in \mathscr{U}_1 = H_{\omega}^{0}(\mathbf{R}; \mathscr{D}(E)) \cap H_{\omega}^{1,0}(Q)$$

satisfying (1.1.1'') almost everywhere in Q. The space $\mathscr{U}_1$ is again equipped with the norm of $H_{\omega}^{1,2}(Q)$. By II: Theorem 1.3.4 a function $u \in H_{\omega}^{0}(Q)$ written as

$$u(t, x) = \omega^{-1/2} \sum_{j,k} u_{jk}\, e^{ij\nu t}\, v_k(x), \quad \nu = 2\pi/\omega,$$

$$\left(\text{that is, } u_{jk} = \omega^{-1/2} \int_0^\omega \int_0^\pi u(\tau, \xi)\, e^{-ij\nu\tau}\, v_k(\xi)\, d\xi\, d\tau\right)$$

belongs to $\mathscr{U}_1$ if and only if

$$\sum_{j,k} (j^2 + \lambda_k^2)\, |u_{jk}|^2 < \infty . \tag{1.1.14}$$

(We recall that according to I: Theorem 2.2.1, the system

$$\{\omega^{-1/2}\, e^{ij\nu t}\, v_k(x)\}_{j\in\mathbf{Z}, k\in\mathbf{N}}$$

forms an orthonormal base in $H^0_\omega(Q)$.) The square root of the series (1.1.14) defines a norm $\|\cdot\|_{\mathscr{U}_1}$ on $\mathscr{U}_1$ which is equivalent to $\|\cdot\|_{1,2}$.

Suppose that $g \in H^0_\omega(Q)$, hence

$$g(t, x) = \omega^{-1/2} \sum_{j,k} g_{jk}\, e^{ij\nu t}\, v_k(x)$$

$$\left(\text{that is, } g_{jk} = \omega^{-1/2} \int_0^\omega \int_0^\pi g(\tau, \xi)\, e^{-ij\nu\tau}\, v_k(\xi)\, d\xi\, d\tau\right)$$

with

$$\|g\|_0^2 = \sum_{j,k} |g_{jk}|^2 .$$

We can proceed just as in the preceding problem with the only difference that now we obtain for u_{jk} the system

$$(ij\nu + \lambda_k + c)\, u_{jk} = g_{jk} ,$$

and hence the cases

(i) $c \neq -\lambda_k , \quad k \in \mathbf{N}$;
(ii) $c = -\lambda_{k_0}$ for some $k_0 \in \mathbf{N}$

have to be distinguished. We denote by L_1 the operator defined by $\mathscr{D}(L_1) = \mathscr{U}_1$ and $L_1 u = u_t + Eu + cu$. If $c = -\lambda_{k_0}$ for some $k_0 \in \mathbf{N}$, then we denote by $\mathscr{V}_1$ the subspace of functions of $H^0_\omega(Q)$ of the form

$$u(t, x) = d\, v_{k_0}(x) , \quad d \in \mathbf{R} ,$$

and let $\mathscr{W}_1 = \mathscr{V}_1^\perp$ in $H^0_\omega(Q)$.

THEOREM 1.1.3 (*The non-critical case*). *Let* $g \in H^0_\omega(Q)$ *and* $c \neq -\lambda_k$ *for all* $k \in \mathbf{N}$.
Then the problem $(\mathscr{P}^1_\omega)$ *has a unique solution* $u = L_1^{-1} g \in \mathscr{U}_1$ *and* $L_1^{-1} \in \mathscr{L}(H^0_\omega(Q), \mathscr{U}_1)$.

THEOREM 1.1.4 (*The critical case*). *Let $g \in H^0_\omega(Q)$ and let $k_0 \in \mathbf{N}$ be such that $c = -\lambda_{k_0}$.*
Then the problem $(\mathscr{P}^1_\omega)$ has a solution if and only if $g \in \mathscr{W}_1$. If this condition is satisfied, then the solutions can be written in the form

$$u(t, x) = d\, v_{k_0}(x) + w(t, x),$$

where $d \in \mathbf{R}$ is arbitrary, $w = L_1^{-1} g \in \mathscr{U}_1 \cap \mathscr{W}_1$ and $L_1^{-1} \in \mathscr{L}(\mathscr{W}_1, \mathscr{U}_1 \cap \mathscr{W}_1)$.

Remark 1.1.4 Remarks quite similar to Remarks 1.1.1 and 1.1.2 can also be made for the problem $(\mathscr{P}^1_\omega)$.

Remark 1.1.5 The solution of the problem $(\mathscr{P}_\omega)$ satisfies the boundary condition (1.1.2) in the classical sense (see Remark 1.1.1). On the other hand, the solution u of the problem $(\mathscr{P}^1_\omega)$ satisfies the boundary conditions (1.1.13) only in the generalized sense: there exists a sequence of functions $u_n \in C^\infty_\omega(\bar{Q})$, $n \in \mathbf{N}$, satisfying (1.1.13) and

$$\|u_n - u\|_{1,2} \to 0 \quad \text{for} \quad n \to \infty .$$

Remark 1.1.6 If the boundary conditions (1.1.2) and (1.1.13) are replaced respectively by the in homogeneous ones

$$u(t, 0) = h_0(t),$$
$$u(t, \pi) = h_1(t), \quad t \in \mathbf{R}, \tag{1.1.2'}$$

and

$$u_x(t, 0) + \sigma_0\, u(t, 0) = h_0(t),$$
$$u_x(t, \pi) + \sigma_1\, u(t, \pi) = h_1(t), \quad t \in \mathbf{R}, \tag{1.1.13'}$$

where h_0 and h_1 are (sufficiently smooth) ω-periodic functions, then by putting (for example)

$$z(t, x) = u(t, x) + \pi^{-1}(x - \pi)\, h_0(t) - \pi^{-1} x\, h_1(t),$$

and

$$z(t, x) = u(t, x) - 2^{-1}(h_0(t) - h_1(t)) \sin x -$$
$$- 4^{-1}(h_0(t) + h_1(t)) \sin 2x,$$

respectively, the problem is reduced to one for z with homogeneous boundary conditions of the type (1.1.2) and (1.1.13), respectively, with the right-hand side of the form

$$g(t, x) + \pi^{-1}(x - \pi)\,(h_0'(t) + c\, h_0(t)) - \pi^{-1} x (h_1'(t) + c\, h_1(t)),$$

and

$$g(t, x) - 2^{-1}[h_0'(t) - h_1'(t) + (1 + c)(h_0(t) - h_1(t))] \sin x -$$
$$- 4^{-1}[h_0'(t) + h_1'(t) + (4 + c)(h_0(t) + h_1(t))] \sin 2x .$$

For instance, the assumptions of Theorems 1.1.1 and 1.1.3 are satisfied if $g \in H_\omega^0(Q)$ and $h_0, h_1 \in H_\omega^1(\mathbf{R})$.

Remark 1.1.7 Consider the slightly more general problems defined by

$$v_t(t, x) - v_{xx}(t, x) + 2b\, v_x(t, x) + c_1\, v(t, x) = g_1(t, x), \quad (t, x) \in Q,$$
$$v(t, 0) = v(t, \pi) = 0, \quad t \in \mathbf{R},$$

or

$$v_x(t, 0) + \sigma_{01}\, v(t, 0) = 0, \quad v_x(t, \pi) + \sigma_{11}\, v(t, \pi) = 0, \quad t \in \mathbf{R},$$

and by

$$v(t + \omega, x) - v(t, x) = 0, \quad (t, x) \in Q,$$

$(b, c_1, \sigma_{01}, \sigma_{11} \in \mathbf{R})$.

If we put $v(t, x) = e^{bx}\, u(t, x)$,

these problems turn into the problems $(\mathscr{P}_\omega)$ and $(\mathscr{P}_\omega^1)$, respectively, with $c = c_1 + b^2$, $g(t, x) = e^{-bx} g_1(t, x)$, and $\sigma_j = \sigma_{j1} + b$, $j = 0, 1$.

Remark 1.1.8 The problem given by

$$v_\tau(\tau, \xi) - a^2 v_{\xi\xi}(\tau, \xi) + c_1\, v(\tau, \xi) = g_1(\tau, \xi), \ (\tau, \xi) \in \mathbf{R} \times (0, l), \ (l > 0),$$
$$v(\tau, 0) = v(\tau, l) = 0, \quad \tau \in \mathbf{R},$$
$$v(\tau + \omega_1, \xi) - v(\tau, \xi) = 0, \quad (\tau, \xi) \in \mathbf{R} \times (0, l)$$

$(a \in \mathbf{R} \setminus \{0\}$, $c_1 \in \mathbf{R}$, g_1 is ω_1-periodic in $\tau)$ goes over into the problem $(\mathscr{P}_\omega)$ by putting

$$x = \pi l^{-1} \xi, \quad t = a^2 \pi^2 l^{-2} \tau \quad (\omega = a^2 \pi^2 l^{-2} \omega_1)$$

and $u(t, x) = v(\tau, \xi)$, $g(t, x) = a^{-2} \pi^{-2} l^2\, g_1(\tau, \xi)$. An analogous remark holds for $(\mathscr{P}_\omega^1)$.

Problem 1.1.1 Derive conditions under which there exists a solution to the problem given by (1.1.1), (1.1.3) and by the combined boundary conditions

$$\begin{aligned} u_x(t, 0) + \sigma\, u(t, 0) &= 0, \\ u(t, \pi) &= 0, \end{aligned} \qquad t \in \mathbf{R}, \quad (\sigma \in \mathbf{R}).$$

(See II: Example 1.3.3.)

1.2. *The weakly non-linear case*

The results of the preceding section can be used to study the problem $(\mathscr{P}^{\varepsilon}_{\omega})$ defined by

$$u_t(t, x) - u_{xx}(t, x) + c\,u(t, x) = \varepsilon\, F(u, \varepsilon)\,(t, x)\,,$$
$$(t, x) \in Q\,, \quad \varepsilon \in [-\varepsilon_0, \varepsilon_0]\,, \quad (\varepsilon_0 > 0) \tag{1.2.1}$$

and by (1.1.2), (1.1.3).

We suppose here that the operator F is a continuous mapping from $\mathscr{U} \times [-\varepsilon_0, \varepsilon_0]$ into $H^0_\omega(Q)$. The notation is the same as in the previous section.

A function $u \in C([-\varepsilon^*, \varepsilon^*]; \mathscr{U})$ $(\varepsilon^* \in (0, \varepsilon_0])$ is called a solution of $(\mathscr{P}^{\varepsilon}_{\omega})$ if $u(\varepsilon)$ for every $\varepsilon \in [-\varepsilon^*, \varepsilon^*]$ satisfies (1.2.1) almost everywhere in Q.

First let us deal with the case

$$c \neq -k^2\,, \quad k \in \mathbf{N}\,.$$

We suppose that

the operator $F(u, \varepsilon)$ *maps for some* $\varrho > 0$

$$B(0, \varrho; \mathscr{U}) \times [-\varepsilon_0, \varepsilon_0] \ \textit{ into } H^0_\omega(Q) \textit{ continuously} \tag{1.2.2}$$

and

there exists a $\lambda > 0$ *such that*

$$\|F(u_1, \varepsilon) - F(u_2, \varepsilon)\|_0 \leqq \lambda \|u_1 - u_2\|_{\mathscr{U}} \tag{1.2.3}$$

for all $u_1, u_2 \in B(0, \varrho; \mathscr{U})$ *and* $\varepsilon \in [-\varepsilon_0, \varepsilon_0]$.

According to Theorem 1.1.1, u is a solution of $(\mathscr{P}^{\varepsilon}_{\omega})$ if and only if u solves the equation

$$u = \varepsilon L^{-1}\, F(u, \varepsilon)\,,$$

where L^{-1} is given by (1.1.9) and $u \in C([-\varepsilon^*, \varepsilon^*]; \mathscr{U})$. Since

$$L^{-1} \in \mathscr{L}(H^0_\omega(Q), \mathscr{U})\,, \tag{1.2.4}$$

we can apply I: Corollary 3.4.1 with $B = H^0_\omega(Q)$, $B_1 = \mathscr{U}$, $\Lambda = L^{-1}$ and obtain the following theorem.

THEOREM 1.2.1 (The non-critical case). *Let* $c \neq -k^2$ *for all* $k \in \mathbf{N}$. *Suppose that* (1.2.2) *and* (1.2.3) *hold.*

Then the problem $(\mathscr{P}^{\varepsilon}_{\omega})$ *has a locally unique solution* $u^*(\varepsilon) \in \mathscr{U}$ *such that* $u^*(0) = 0$.

Remark 1.2.1 The phrase "*a problem has a locally unique solution* u^* *such that...*" is used in the sense explained in I: Sec. 3.4 (p. 31).

Remark 1.2.2 The pair of conditions (1.2.2), (1.2.3) is equivalent to the following:

$$\begin{aligned}&\textit{the operator } F(u, \varepsilon) \textit{ maps for some } \varrho > 0\\ &B(0, \varrho; \mathscr{U}) \times [-\varepsilon_0, \varepsilon_0] \textit{ into } H^0_\omega(Q) \qquad (1.2.2')\\ &\textit{and is continuous in } \varepsilon \textit{ for any } u \in B(0, \varrho; \mathscr{U})\end{aligned}$$

and (1.2.3).

Remark 1.2.3 Let

$$F(u, \varepsilon)(t, x) = f(t, x, u(t, x), u_t(t, x), u_x(t, x), u_{xx}(t, x), \varepsilon).$$

The hypotheses of Theorem 1.2.1 are satisfied if $f = f(t, x, u_0, u_1, u_2, u_3, \varepsilon)$ is continuous on $M_1 = \mathbf{R} \times [0, \pi] \times [-\varrho, \varrho] \times \mathbf{R}^3 \times [-\varepsilon_0, \varepsilon_0]$ (for some $\varrho > 0$), ω-periodic in t, and Lipschitz continuous in u_0, u_1, u_2, u_3 uniformly with respect to t, x and ε, that is, there exists a $\lambda > 0$ such that

$$\begin{aligned}&|f(t, x, u_0, u_1, u_2, u_3, \varepsilon) - f(t, x, v_0, v_1, v_2, v_3, \varepsilon)| \leqq\\ &\leqq \lambda(|u_0 - v_0| + |u_1 - v_1| + |u_2 - v_2| + |u_3 - v_3|),\\ &(t, x, u_0, u_1, u_2, u_3, \varepsilon), \quad (t, x, v_0, v_1, v_2, v_3, \varepsilon) \in M_1.\end{aligned}$$

If these conditions are satisfied, then from Remark 1.1.1 we conclude that when u lies in a ball in $\mathscr{U}$,

$$|f(t, x, u, u_t, u_x, u_{xx}, \varepsilon)| \leqq \text{const.}\,(1 + |u_t| + |u_x| + |u_{xx}|).$$

The restriction on the growth of the function f in u_t, u_x, and u_{xx} may sometimes be inconvenient. The situation becomes easier to deal with if f is sufficiently smooth and we look for a smoother solution, for example $u \in H^{2,4}_\omega(Q) \cap \dot{H}^1_\omega(Q)$ whose derivatives occurring in f are continuous and bounded on Q. Hence, let f be ω-periodic in t, continuous on $M_2 = \mathbf{R} \times [0, \pi] \times [-\varrho, \varrho]^4 \times [-\varepsilon_0, \varepsilon_0]$ together with its partial derivatives

$$D_t^\alpha D_x^\beta D_{u_0}^{\gamma_0} D_{u_1}^{\gamma_1} D_{u_2}^{\gamma_2} D_{u_3}^{\gamma_3} f, \quad \alpha + \beta + \sum_{j=0}^{3} \gamma_j \leqq 2, \quad 2\alpha + \beta \leqq 2,$$

these being Lipschitz continuous in u_0, u_1, u_2, u_3, uniformly with respect to t, x and ε. Using the embedding theorem I: Theorem 2.7.5 (namely

$$\begin{aligned}&\max(\|u\|_{C_\omega^{1,2}(Q)}, \|u_{tx}\|_{H_\omega^0(Q)}, \|u_{txx}\|_{H_\omega^0(Q)},\\ &\|u_{xxx}\|_{L_{4,\omega}(Q)}) \leqq \text{const.}\, \|u\|_{H_\omega^{2,4}(Q)} \quad \text{for all} \quad u \in H^{2,4}_\omega(Q))\end{aligned}$$

we find that F maps $B(0, \tilde{\rho}; H_\omega^{2,4}(Q) \cap \dot{H}_\omega^1(Q)) \times [-\varepsilon_0, \varepsilon_0]$ into $H_\omega^{1,2}(Q)$, is Lipschitz continuous in u and continuous in ε. Since

$$L^{-1} \in \mathscr{L}(H_\omega^{1,2}(Q), \quad H_\omega^{2,4}(Q) \cap \dot{H}_\omega^1(Q))$$

by Remark 1.1.2, the existence of a solution $u \in C([-\varepsilon^*, \varepsilon^*]; H_\omega^{2,4}(Q) \cap \cap \dot{H}_\omega^1(Q))$ to $(\mathscr{P}_\omega^\varepsilon)$ follows like the existence of a solution $u \in \mathscr{U}$ in the theorem just proved.

Let us now turn to the case when there is a $k_0 \in \mathbf{N}$ such that

$$c = -k_0^2 .$$

By Theorem 1.1.2 u is a solution of $(\mathscr{P}_\omega^\varepsilon)$ if and only if

$$u(\varepsilon)(t, x) = d(\varepsilon) \sin k_0 x + w(\varepsilon)(t, x) ,$$

where (w, d) is a solution of the system

$$G_1(w, d, \varepsilon) \equiv w(t, x) - \varepsilon L^{-1} F(d \sin k_0 x + w(t, x), \varepsilon)(t, x) = 0 ,$$

$$G_2(w, d, \varepsilon) \equiv \int_0^\omega \int_0^\pi F(d \sin k_0 x + w(t, x), \varepsilon)(\tau, \xi) \sin k_0 \xi \, d\xi \, d\tau = 0 \tag{1.2.5}$$

with L^{-1} given by (1.1.12), and $d \in C([-\varepsilon^*, \varepsilon^*])$, $w \in C([-\varepsilon^*, \varepsilon^*]; \mathscr{U} \cap \mathscr{W})$. Passing to the limit $\varepsilon \to 0$ and taking into account the continuity of F we obtain the next lemma.

LEMMA 1.2.1 *The problem* $(\mathscr{P}_\omega^\varepsilon)$ *with* $c = -k_0^2$ *for some* $k_0 \in \mathbf{N}$ *has a solution only if the bifurcation equation*

$$\int_0^\omega \int_0^\pi F(d_0 \sin k_0 x, 0)(\tau, \xi) \sin k_0 \xi \, d\xi \, d\tau = 0 \tag{1.2.6}$$

has a solution $d_0 = d_0^* \in \mathbf{R}$.

Some sufficient conditions of solubility are given by the following theorem.

THEOREM 1.2.2 (*The critical case*). *Let* $c = -k_0^2$ *for some* $k_0 \in \mathbf{N}$. *Suppose that the following assumptions hold*:

(*a*) *the equation* (1.2.6) *has a solution* $d_0 = d_0^* \in \mathbf{R}$;

(*b*) *the operator* $F(u, \varepsilon)$ *and its G-derivative* $F'_u(u, \varepsilon)$ *map* $B(d_0^* \sin k_0 x, \varrho; \mathscr{U}) \times [-\varepsilon_0, \varepsilon_0]$ *for some* $\varrho > 0$ *continuously into* $H_\omega^0(Q)$ *and into* $\mathscr{L}(\mathscr{U}, H_\omega^0(Q))$, *respectively*;

(*c*) $\displaystyle\int_0^\omega \int_0^\pi F'_u(d_0^* \sin k_0 x, 0)(\sin k_0 x)(\tau, \xi) \sin k_0 \xi \, d\xi \, d\tau \neq 0 .$

Then the problem $(\mathscr{P}_\omega^\varepsilon)$ *has a locally unique solution* $u^*(\varepsilon) \in \mathscr{U}$ *such that*

$$u^*(0)(t, x) = d_0^* \sin k_0 x .$$

Proof. To prove the existence of a solution $w^*(\varepsilon)$, $d^*(\varepsilon)$ to the system (1.2.5) we apply the implicit function theorem I: Theorem 3.4.2 (p. 30) setting $B = B_1 = (\mathscr{U} \cap \mathscr{W}) \times \mathbf{R}$, $B_2 = \mathbf{R}$, $x = (w, d)$, $p = \varepsilon$, $x_0 = (0, d_0^*)$, $p_0 = 0$, $G = (G_1, G_2)$. Its second assumption is satisfied by virtue of (a). The first is satisfied due to (b) and

$$L^{-1} \in \mathscr{L}(\mathscr{W}, \mathscr{U} \cap \mathscr{W}) . \tag{1.2.7}$$

The third assumption is satisfied by virtue of (b) with due regard to (1.2.7) and I: Theorems 1.8.5, 1.8.6 on the differentiation of a composite function and of a linear mapping. Note that, in particular,

$$G'_{(w,d)}(0, d_0^*, 0)(\bar{w}, \bar{d}) =$$

$$= \left(\bar{w}, \int_0^\omega \int_0^\pi F'_u(d_0^* \sin k_0 x, 0)(\bar{d} \sin k_0 x + \bar{w}(t, x))(\tau, \xi) \sin k_0 \xi \, d\xi \, d\tau\right) .$$

Finally, the fourth assumption is satisfied by virtue of (c). In fact, the equation

$$G'_{(w,d)}(0, d_0^*, 0)(\bar{w}, \bar{d}) = (z, \zeta)$$

has a unique solution $\bar{w}^* = z$,

$$\bar{d}^* = \left(\int_0^\omega \int_0^\pi F'_u(d_0^* \sin k_0 x, 0)(\sin k_0 x)(\tau, \xi) \sin k_0 \xi \, d\xi \, d\tau\right)^{-1}$$

$$\left(\zeta - \int_0^\omega \int_0^\pi F'_u(d_0^* \sin k_0 x, 0)(z)(\tau, \xi) \sin k_0 \xi \, d\xi \, d\tau\right)$$

for any $(z, \zeta) \in (\mathscr{U} \cap \mathscr{W}) \times \mathbf{R}$. It remains to use I: Theorem 1.2.2. Hence, all the assumptions of I: Theorem 3.4.2 are verified and by applying this theorem we obtain the required result immediately. The proof is now complete.

Remark 1.2.4 Let

$$F(u, \varepsilon)(t, x) = f(t, x, u(t, x), u_t(t, x), u_x(t, x), u_{xx}(t, x), \varepsilon) .$$

By Remark 1.1.1, F maps $B(d_0^* \sin k_0 x, \tilde{\rho}; \mathscr{U}) \times [-\varepsilon_0, \varepsilon_0]$ into $H_\omega^0(Q)$ if f is continuous on $M_3 = \mathbf{R} \times [0, \pi] \times [-|d_0^*| - \varrho, |d_0^*| + \varrho] \times \mathbf{R}^3 \times [-\varepsilon_0, \varepsilon_0]$ (for $k_0 = 1$ it suffices to take $M_3 = \mathbf{R} \times [0, \pi] \times [-\varrho, d_0^* + \varrho] \times \mathbf{R}^3 \times [-\varepsilon_0, \varepsilon_0]$ (if $d_0^* > 0$) or $M_3 = \mathbf{R} \times [0, \pi] \times [d_0^* - \varrho, \varrho] \times \mathbf{R}^3 \times [-\varepsilon_0, \varepsilon_0]$ (if $d_0^* < 0$)), ω-periodic in t and $|f(t, x, u_0, u_1, u_2, u_3, \varepsilon)| \leqq$

$\leqq$ const. $(1 + |u_1| + |u_2|^3 + |u_3|)$ for $(t, x, u_0, u_1, u_2, u_3, \varepsilon) \in M_3$. It turns out that F has a continuous G-derivative F'_u only if f is linear in u_t and u_{xx}.

Let us remove these restrictions by looking for a smoother solution belonging, for example, to $H^{2,4}_\omega(Q) \cap \dot{H}^1_\omega(Q)$. Suppose that f together with its partial derivatives

$$D_t^\alpha D_x^\beta D_{u_0}^{\gamma_0} D_{u_1}^{\gamma_1} D_{u_2}^{\gamma_2} D_{u_3}^{\gamma_3} f\,, \quad \text{where} \quad \alpha + \beta + \sum_{j=0}^{3} \gamma_j \leqq 3\,,\ 2\alpha + \beta \leqq 2\,,$$

is continuous on $M_4 = \mathbf{R} \times [0, \pi] \times [-|d_0^*| - \varrho, |d_0^*| + \varrho] \times [-\varrho, \varrho] \times \times [-k_0|d_0^*| - \varrho, k_0|d_0^*| + \varrho] \times [-k_0^2|d_0^*| - \varrho, k_0^2|d_0^*| + \varrho] \times [-\varepsilon_0, \varepsilon_0]$ and that f is ω-periodic in t. Then the assumption (b) of Theorem 1.2.2 (modified for $\mathscr{U} = H^{2,4}_\omega(Q) \cap \dot{H}^1_\omega(Q)$ and for the space $H^{1,2}_\omega(Q)$ of the right-hand sides) is satisfied and since

$$L^{-1} \in \mathscr{L}(H^{1,2}_\omega(Q) \cap \mathscr{W}\,, \quad H^{2,4}_\omega(Q) \cap \dot{H}^1_\omega(Q) \cap \mathscr{W})$$

(see Remark 1.1.2), the corresponding existence theorem can be asserted.

Example 1.2.1 Let $k_0 \in \mathbf{N}$, $c = -k_0^2$, $\omega = 2\pi$ and $F(u, \varepsilon)(t, x) = \sin k_0 x\,.\,\cos^2 k_0 t + \gamma u^2 \sin k_0 x$, $(t, x) \in Q$, $\gamma \neq 0$. Then the bifurcation equation (1.2.6) is $2 + 3\gamma d_0^2 = 0$ and the condition (c) of Theorem 1.2.2 is $\frac{3}{2}\pi^2 \gamma d_0^* \neq 0$. Thus, if $\gamma < 0$, then there are two 2π-periodic solutions of the given problem such that $u^*(0)(t, x) = (-2/3\gamma)^{1/2} \sin k_0 x$ and $u^*(0)(t, x) = -(-2/3\gamma)^{1/2} \sin k_0 x$, respectively. On the other hand, if $\gamma > 0$, then (for $|\varepsilon|$ small) our problem has no solution.

Remark 1.2.5 The more general problem $(\hat{\mathscr{P}}^\varepsilon_\omega)$ given by

$$u_t(t, x) - u_{xx}(t, x) + c\,u(t, x) = g(t, x) + \varepsilon\,F_1(u, \varepsilon)(t, x),\ (t, x) \in Q$$

and by (1.1.2), (1.1.3) can be investigated in the following way. If the corresponding linear problem $(\hat{\mathscr{P}}_\omega)$ has no solution, then the given problem has (for $|\varepsilon|$ small) no solution either. On the other hand, if $(\hat{\mathscr{P}}_\omega)$ has a solution, we take an arbitrary but fixed solution, say u_0. Setting $u = u_0 + v$, we obtain for v the problem $(\mathscr{P}^\varepsilon_\omega)$ with $F(v, \varepsilon) = F_1(u_0 + v, \varepsilon)$.

Remark 1.2.6 The Schauder fixed point theorem (see I: Theorem 3.2.2) can be also applied to the solution of non-linear problems corresponding to $(\mathscr{P}_\omega)$. See ŠŤASTNOVÁ and FUČÍK [83].

The weakly non-linear problem $(\mathscr{P}^{1,\varepsilon}_\omega)$ given by (1.2.1), (1.1.13) and by (1.1.3) can be treated in the same way. Restatements and proofs of Theorems 1.2.1 and 1.2.2 hold verbatim if $(\mathscr{P}^\varepsilon_\omega)$, $\mathscr{U}$, k^2 and $\sin k_0 x$ are replaced by $(\mathscr{P}^{1,\varepsilon}_\omega)$, $\mathscr{U}_1$, λ_k and v_{k_0}, respectively.

1.3. *The n-dimensional case*

The (t, s)-Fourier method can also be used with advantage for solving problems given on $\mathbf{R} \times \Omega$, Ω being a bounded domain in $\mathbf{R}^n$.

Let us investigate the problem $(\mathscr{P}_\omega)$ defined by

$$u_t(t, x) - \sum_{j,k=1}^{n} (a_{jk}(x)\, u_{x_k}(t, x))_{x_j} + a_0(x)\, u(t, x) = g(t, x)\,,$$

$$(t, x) \in Q = \mathbf{R} \times \Omega\,, \tag{1.3.1}$$

$$u(t, x) = 0\,, \quad (t, x) \in \mathbf{R} \times \partial\Omega \tag{1.3.2}$$

or

$$\frac{\partial u}{\partial n}(t, x) + \sigma(x)\, u(t, x) = 0\,, \quad (t, x) \in \mathbf{R} \times \partial\Omega\,, \tag{1.3.2'}$$

where $\partial/\partial n$ denotes the conormal derivative, that is,

$$\frac{\partial}{\partial n} = \sum_{j,k=1}^{n} a_{jk}(x) \cos\left(\nu(x), x_j\right) \frac{\partial}{\partial x_k}\,,$$

$\nu(x)$ is the outward normal to $\partial\Omega$ at x and $(\nu(x), x_j)$ is the angle between ν and the j-th coordinate axis, and by

$$u(t + \omega, x) - u(t, x) = 0\,, \quad (t, x) \in Q\,. \tag{1.3.3}$$

We make the following assumptions:

(1) $\Omega \subset \mathbf{R}^n$ *is an open bounded domain of class* C^∞;

(2) $a_0, a_{jk} \in C^\infty(\Omega)$, $j, k = 1, \ldots, n$,

$a_{jk}(x) = a_{kj}(x)$, $x \in \Omega$, $j, k = 1, \ldots, n$;

(3) *there exists an* $m > 0$ *such that*

$$\sum_{j,k=1}^{n} a_{jk}(x)\, \xi_j \xi_k \geqq m \sum_{j=1}^{n} \xi_j^2\,, \quad x \in \bar{\Omega}\,, \quad (\xi_1, \ldots, \xi_n) \in \mathbf{R}^n\,;$$

(4) $\sigma \in C^\infty(\partial\Omega)$.

We know by II: Theorem 1.3.2 (p. 40) that there is a constant $c \in \mathbf{R}$ such that the (uniformly strongly elliptic) operator

$$E\, v(x) = -\sum_{j,k=1}^{n} (a_{jk}(x)\, v_{x_k}(x))_{x_j} + (a_0(x) - c)\, v(x)\,, \quad x \in \Omega$$

with the domain

$$\mathscr{D}(E) = \{v \in H^2(\Omega);\ v(x) = 0,\ x \in \partial\Omega \text{ (in the sense of traces)}\}$$
$$= H^2(\Omega) \cap \dot{H}^1(\Omega)$$

or,

$$\mathscr{D}(E) = \left\{v \in H^2(\Omega);\ \frac{\partial v}{\partial n}(x) + \sigma(x)\, v(x) = 0,\ x \in \partial\Omega \right.$$
$$\left. \text{(in the sense of traces)}\right\}$$

is positive definite in $H^0(\Omega)$, self-adjoint, and has a countable set of eigenvalues with ∞ as the only limit point, all eigenvalues are positive, each of finite multiplicity. We arrange them, with due regard to their multiplicity, in ascending order:

$$\lambda_1 < \lambda_2 \leqq \ldots \leqq \lambda_k \leqq \ldots .$$

We can choose the corresponding system $\{v_k\}_{k=1}^{\infty}$ of eigenfunctions to form an orthonormal base in $H^0(\Omega)$.

Note that by virtue of the embedding theorem I: Theorem 2.8.1 the derivative $\partial v/\partial n$ has a meaning in some $L_q(\partial\Omega)$ and that $\mathscr{D}(E)$ in both cases is closed in $H^2(\Omega)$. For $v \in \mathscr{D}(E)$ we define the graph norm (see I: Theorem 1.5.2) by

$$\|v\|_{\mathscr{D}(E)} = \|Ev\|_{H^0(\Omega)} .$$

By II: Theorem 1.1.1 we find that the norms $\|\cdot\|_{\mathscr{D}(E)}$ and $\|\cdot\|_{H^2(\Omega)}$ are equivalent on $\mathscr{D}(E)$.

Moreover, for $v \in \mathscr{D}(E)$ we have $v = \sum_{k=1}^{\infty} \langle v, v_k\rangle_{H^0(\Omega)}\, v_k$ and

$$\|v\|^2_{\mathscr{D}(E)} = \sum_{k=1}^{\infty} \lambda_k^2 \langle v, v_k\rangle^2_{H^0(\Omega)} .$$

By a solution to the problem $(\mathscr{P}_\omega)$ we mean a function

$$u \in \mathscr{U} = H^0_\omega(\mathbf{R}; \mathscr{D}(E)) \cap H^{1,0}_\omega(Q)$$

for which

$$u_t(t, x) + E\, u(t, x) + c\, u(t, x) = g(t, x) \tag{1.3.1'}$$

almost everywhere in Q. (Note that $\mathscr{U} = H^{1,2}_\omega(Q) \cap \dot{H}^1_\omega(Q)$ or, $\mathscr{U} = \mathrm{cl}\,\{u \in C^\infty_\omega(\bar{Q}), (\partial u/\partial n)(t, x) + \sigma(x)\,u(t, x) = 0, (t, x) \in \mathbf{R} \times \partial\Omega\}$ in $H^{1,2}_\omega(Q)$.) The space $\mathscr{U}$ is equipped with the norm of $H^{1,2}_\omega(Q)$, that is,

$$\|u\|_{1,2} = \left(\int_0^\omega \int_\Omega [(u_t(t, x))^2 + \sum_{|\beta| \leqq 2} (D^\beta_x u(t, x))^2]\, dx\, dt\right)^{1/2}.$$

A function $u \in H^0_\omega(Q)$,

$$u(t, x) = \omega^{-1/2} \sum_{j,k} u_{jk}\, e^{ij\nu t}\, v_k(x)\,, \quad (\nu = 2\pi/\omega)\,, \tag{1.3.4}$$

belongs to $\mathscr{U}$ if and only if

$$\sum_{j,k} (j^2 + \lambda_k^2)\, |u_{jk}|^2 < \infty$$

and the square root of this series defines a norm $\|\cdot\|_{\mathscr{U}}$ on $\mathscr{U}$, which is equivalent to the norm $\|\cdot\|_{1,2}$. (See II: Theorem 1.3.4 and II: Remark 1.3.2.)

Suppose that $g \in H^0_\omega(Q)$, that is,

$$g(t, x) = \omega^{-1/2} \sum_{j,k} g_{jk}\, e^{ij\nu t}\, v_k(x) \tag{1.3.5}$$

and

$$\|g\|_0^2 = \sum_{j,k} |g_{jk}|^2 < \infty\,.$$

Substituting (1.3.4), (1.3.5) in (1.3.1′) we obtain

$$(ij\nu + \lambda_k + c)\, u_{jk} = g_{jk}\,, \quad (j, k) \in \mathbf{Z} \times \mathbf{N}$$

We must distinguish the following two cases:

(i) $-c \neq \lambda_k\,, \quad k \in \mathbf{N}\,;$

(ii) $-c$ is an eigenvalue of multiplicity r (for the sake of definiteness we write $\lambda_{k_0 - 1} < -c = \lambda_{k_0} = \lambda_{k_0+1} = \ldots = \lambda_{k_0+r-1} < \lambda_{k_0+r}$ $(\lambda_0 \equiv 0)$).

In the case (i) we find

$$u_{jk} = (ij\nu + \lambda_k + c)^{-1}\, g_{jk}\,, \quad (j, k) \in \mathbf{Z} \times \mathbf{N}$$

which yields

$$\|u\|^2_{\mathscr{U}} = \sum_{j,k} (j^2 + \lambda_k^2)\, |u_{jk}|^2 =$$

$$= \sum_{j,k} (j^2 + \lambda_k^2)\, (\nu^2 j^2 + (\lambda_k + c)^2)^{-1}\, |g_{jk}|^2 \leqq$$

$$\leqq \text{const.} \sum_{j,k} |g_{jk}|^2 = \text{const.}\, \|g\|_0^2\,.$$

In the case (ii), necessary conditions for the existence of a solution are

$$g_{0,k_0} = g_{0,k_0+1} = \dots = g_{0,k_0+r-1} = 0\,. \tag{1.3.6}$$

The homogeneous problem for $g \equiv 0$ has in the case (ii) an r-parameter family of solutions

$$u(t, x) = d_0\, v_{k_0}(x) + \dots + d_{r-1}\, v_{k_0+r-1}(x)\,, \quad d_0, \dots, d_{r-1} \in \mathbf{R}\,.$$

Let $S_1 = \{(0, k);\, k = k_0, \dots, k_0 + r - 1\}$ and $S_2 = (\mathbf{Z} \times \mathbf{N}) \setminus S_1$. Let $\mathscr{V} = \operatorname{lin}\{v_{k_0}, \dots, v_{k_0+r-1}\}$, $\mathscr{W} = \mathscr{V}^{\perp}$ in $H^0_{\omega}(Q)$. Assuming that $g \in \mathscr{W}$ and estimating the particular solution of $(\mathscr{P}_{\omega})$, which is given by

$$w(t, x) = \omega^{-1/2} \sum_{(j,k)\in S_2} (ij\nu + \lambda_k + c)^{-1}\, g_{jk}\, e^{ij\nu t} v_k(x)\,, \tag{1.3.7}$$

as above, we obtain $\|w\|_{\mathscr{U}} \leqq \text{const.}\, \|g\|_0$.

We define an operator L by $\mathscr{D}(L) = \mathscr{U}$ and $Lu = u_t + Eu + cu$ and summarize the preceding results in the following two theorems.

THEOREM 1.3.1 (*The non-critical case*). *Let $g \in H^0_{\omega}(Q)$ and $c \neq -\lambda_k$ for all $k \in \mathbf{N}$.*
Then the problem $(\mathscr{P}_{\omega})$ has a unique solution $u \in \mathscr{U}$,

$$u(t, x) = (L^{-1}g)(t, x) = \omega^{-1/2} \sum_{j,k} (ij\nu + \lambda_k + c)^{-1}\, g_{jk}\, e^{ij\nu t} v_k(x)$$

and $L^{-1} \in \mathscr{L}(H^0_{\omega}(Q), \mathscr{U})$.

THEOREM 1.3.2 (*The critical case*). *Let $g \in H^0_{\omega}(Q)$ and let $k_0 \in \mathbf{N}$ be such that*

$$\lambda_{k_0-1} < -c = \lambda_{k_0} = \lambda_{k_0+1} = \dots = \lambda_{k_0+r-1} < \lambda_{k_0+r}\,. \tag{1.3.8}$$

Then the problem $(\mathscr{P}_{\omega})$ has a solution if and only if (1.3.6) holds, that is, $g \in \mathscr{W}$. If this condition is satisfied, then the solutions can be written in the form

$$u(t, x) = \sum_{l=0}^{r-1} d_l\, v_{k_0+l}(x) + w(t, x)\,,$$

where $d_0, \dots, d_{r-1} \in \mathbf{R}$ are arbitrary, $w = L^{-1}g \in \mathscr{U} \cap \mathscr{W}$ is given by (1.3.7) and $L^{-1} \in \mathscr{L}(\mathscr{W}, \mathscr{U} \cap \mathscr{W})$.

Remark 1.3.1 Using I: Theorem 2.7.5 we can establish some additional properties of a function $u \in \mathscr{U}$. (The domain $(0, \omega) \times \Omega$ satisfies the assumption of that theorem.) We have

$$u \in L_{q,\omega}(Q)\,, \quad q \in [1, \infty) \quad \text{if} \quad n = 2\,,$$

$$q = 2(n + 2)/(n - 2) \quad \text{if} \quad n > 2$$

and

$$u_{x_j} \in L_{q,\omega}(Q), \quad q = 2(n+2)/n \quad \text{if} \quad n \geqq 2 \quad (j = 1, \dots, n).$$

Remark 1.3.2 The same theorem of Chap. I tells us when the solution u is classical. For example, if $u \in H^{s,2s}_{\omega}(Q)$, where $s = [(n+6)/4] + 1$, then $u \in C^{1,2}_{\omega}(Q)$.

Let $g \in H^0_{\omega}(Q)$ and let u be the solution of $(\mathscr{P}_{\omega})$. (Assume, for simplicity, $c \neq -\lambda_k$, $k \in \mathbf{N}$.) Let us write

$$u(t, x) = \omega^{-1/2} \sum_{j=-\infty}^{\infty} u_j(x)\, e^{ij\nu t},$$

$$g(t, x) = \omega^{-1/2} \sum_{j=-\infty}^{\infty} g_j(x)\, e^{ij\nu t},$$

where $u_j(x)$ and $g_j(x)$ are the Fourier coefficients respectively of u and g relative to the system $\{\omega^{-1/2} e^{ij\nu t}\}_{j \in \mathbf{Z}}$ (cf. Sec. 2.1). We prove the following implication. If $s \in \mathbf{N}$ and $g \in H^{s-1,2(s-1)}_{\omega}(Q)$, i.e.

$$\|g\|^2_{s-1,2(s-1)} \sim \sum_{j=-\infty}^{\infty} (j^{2(s-1)} \|g_j\|^2_{H^0(\Omega)} + \|g_j\|^2_{H^{2(s-1)}(\Omega)}) < \infty,$$

then

(a) $j^{2s} \|u_j\|^2_{H^0(\Omega)} \leqq \text{const.}\, j^{2(s-1)} \|g_j\|^2_{H^0(\Omega)}, \quad j \in \mathbf{Z}$

(b) $\|u_j\|^2_{H^{2s}(\Omega)} \leqq \text{const.}\, (\|g_j\|^2_{H^{2(s-1)}(\Omega)} + (1 + j^{2(s-1)}) \|g_j\|^2_{H^0(\Omega)}), \quad j \in \mathbf{Z}$.

Hence, $u \in H^{s,2s}_{\omega}(Q) \cap \dot{H}^1_{\omega}(Q)$ and $\|u\|_{s,2s} \leqq \text{const.}\ \|g\|_{s-1,2(s-1)}$. (a) is a consequence of the fact (see p. 79) that $g \in H^{s-1,0}_{\omega}(Q)$ implies

$$\sum_{j=-\infty}^{\infty} (1 + j^{2(s-1)}) (j^2 \|u_j\|^2_{H^0(\Omega)} + \|u_j\|^2_{H^2(\Omega)}) \leqq$$

$$\leqq \text{const.} \sum_{j=-\infty}^{\infty} (1 + j^{2(s-1)}) \|g_j\|^2_{H^0(\Omega)} < \infty.$$

To prove (b) we proceed by induction. For $s = 1$, the result follows from the estimates above. Suppose that (b) is valid for $s = r$ and let $g \in H^{r,2r}_{\omega}(Q)$. The function u_j $(j \in \mathbf{Z})$ is determined by $u_j \in \mathscr{D}(E)$ and

$$E\, u_j(x) = g_j(x) - (ij\nu + c)\, u_j(x),$$

where the right-hand side belongs to $H^{2r}(\Omega)$. Since E is strongly elliptic, we obtain by II: Theorem 1.1.1 that $u_j \in H^{2(r+1)}(\Omega)$ and

$$\|u_j\|^2_{H^{2(r+1)}(\Omega)} \leqq C(\|g_j\|^2_{H^{2r}(\Omega)} + (1 + j^2) \|u_j\|^2_{H^{2r}(\Omega)})$$

with $C > 0$ independent of j and g. By I: Theorem 2.7.5 we know that $g_t \in H_\omega^{0,2(r-1)}(Q)$ and $\|g_t\|_{0,2(r-1)} \leqq \text{const.} \|g\|_{r,2r}$. Hence it is easy to see that (b) holds for $s = r + 1$, and this completes the proof.

Remark 1.3.3 Note that the requirements on smoothness of the functions a_{jk}, a and σ can be weakened considerably. (For instance, in Theorems 1.3.1 and 1.3.2, it suffices to suppose that $a_{jk} \in C^1(\bar{\Omega})$, $a_0 \in C(\bar{\Omega})$, $\sigma \in C^1(\partial\Omega)$.) An analogous remark holds for Ω.

Let us proceed to the investigation of the corresponding weakly non-linear problem $(\mathscr{P}_\omega^\varepsilon)$ given by

$$u_t(t, x) + E\, u(t, x) + c\, u(t, x) =$$

$$= \varepsilon F(u, \varepsilon)(t, x), \quad (t, x) \in Q, \quad \varepsilon \in [-\varepsilon_0, \varepsilon_0], \quad (\varepsilon_0 > 0) \tag{1.3.9}$$

(1.3.2) or (1.3.2′), and by (1.3.3).

A function $u \in C([-\varepsilon^*, \varepsilon^*]; \mathscr{U})$ $(\varepsilon^* \in (0, \varepsilon_0])$ is called a solution of $(\mathscr{P}_\omega^\varepsilon)$ if $u(\varepsilon)$ for every $\varepsilon \in [-\varepsilon^*, \varepsilon^*]$ satisfies (1.3.9) almost everywhere in Q.

THEOREM 1.3.3 (*The non-critical case*). *Let $c \neq -\lambda_k$ for all $k \in \mathbf{N}$. Suppose the following assumptions are satisfied*:

(*a*) *the operator $F(u, \varepsilon)$ maps $B(0, \varrho; \mathscr{U}) \times [-\varepsilon_0, \varepsilon_0]$ for some $\varrho > 0$ continuously into $H_\omega^0(Q)$*;

(*b*) *there exists a $\lambda > 0$ such that*

$$\|F(u_1, \varepsilon) - F(u_2, \varepsilon)\|_0 \leqq \lambda \|u_1 - u_2\|_{\mathscr{U}}$$

for all $u_1, u_2 \in B(0, \varrho; \mathscr{U})$ and $\varepsilon \in [-\varepsilon_0, \varepsilon_0]$.
Then the problem $(\mathscr{P}_\omega^\varepsilon)$ has a locally unique solution $u^*(\varepsilon) \in \mathscr{U}$ such that $u^*(0) = 0$.

The *proof* is formally the same as that of Theorem 1.2.1 and we leave it to the reader.

THEOREM 1.3.4 (*The critical case*). *Let $k_0 \in \mathbf{N}$ be such that* (1.3.8) *holds. Suppose that the following assumptions are satisfied*:

(*a*) *the system of equations*

$$\int_0^\omega \int_\Omega F\Big(\sum_{l=0}^{r-1} d_l\, v_{k_0+l}, 0\Big)(\tau, \xi)\, v_{k_0+m}(\xi)\, d\xi\, d\tau = 0, \qquad m = 0, \ldots, r-1$$

has a solution $\boldsymbol{d}_0 = \boldsymbol{d}_0^* = (d_{0,0}^*, \ldots, d_{0,r-1}^*) \in \mathbf{R}^r$;

(b) *the operator* $F(u, \varepsilon)$ *and its G-derivative* $F'_u(u, \varepsilon)$ *map*

$$B\left(\sum_{l=0}^{r-1} d_{0,l}^{*} v_{k_0+l}, \varrho; \mathscr{U}\right) \times [-\varepsilon_0, \varepsilon_0]$$

for some $\varrho > 0$ *continuously into* $H_\omega^0(Q)$ *and into* $\mathscr{L}(\mathscr{U}, H_\omega^0(Q))$, *respectively;*

(c) *the determinant*

$$\det\left[\int_0^\omega \int_\Omega F'_u\left(\sum_{l=0}^{r-1} d_{0,l}^{*} v_{k_0+l}, 0\right)(v_{k_0+j})(\tau, \xi)\, v_{k_0+m}(\xi)\, d\xi\, d\tau\right]_{j,m=0}^{r-1} \neq 0 .$$

Then the problem $(\mathscr{P}_\omega^\varepsilon)$ *has a locally unique solution* $u^*(\varepsilon) \in \mathscr{U}$ *such that*

$$u^*(0) = \sum_{l=0}^{r-1} d_{0,l}^{*} v_{k_0+l} .$$

Proof. By Theorem 1.3.2 u is a solution to $(\mathscr{P}_\omega^\varepsilon)$ if $u = \sum_{l=0}^{r-1} d_l v_{k_0+l} + w$, where $\boldsymbol{d} = (d_0, \ldots, d_{r-1})$, w is a solution of the system

$$G_1(w, \boldsymbol{d}, \varepsilon) \equiv w(t, x) - \varepsilon L^{-1} F\left(\sum_{l=0}^{r-1} d_l v_{k_0+l} + w, \varepsilon\right)(t, x) = 0 ,$$

$$G_2(w, \boldsymbol{d}, \varepsilon) \equiv \left(\int_0^\omega \int_\Omega F\left(\sum_{l=0}^{r-1} d_l v_{k_0+l} + w, \varepsilon\right)(\tau, \xi)\, v_{k_0}(\xi)\, d\xi\, d\tau , \ldots\right.$$

$$\left.\ldots, \int_0^\omega \int_\Omega F\left(\sum_{l=0}^{r-1} d_l v_{k_0+l} + w, \varepsilon\right)(\tau, \xi)\, v_{k_0+r-1}(\xi)\, d\xi\, d\tau\right) = 0 ,$$

that is continuous in ε.

Put $B = B_1 = (\mathscr{U} \times \mathscr{W}) \times \mathbf{R}^r$, $B_2 = \mathbf{R}$, $x = (w, \boldsymbol{d})$, $p = \varepsilon$, $x_0 = (0, \boldsymbol{d}_0^*)$, $p_0 = 0$, $G = (G_1, G_2)$. Let us show that all the assumptions of I: Theorem 3.4.2 are satisfied. The second assumption is satisfied by virtue of (a), the first and third follow from (b). Further, we have

$$G'_{(w,d)}(0, \boldsymbol{d}_0^*, 0)(\bar{w}, \bar{\boldsymbol{d}}) = \left(\bar{w},\right.$$

$$\int_0^\omega \int_\Omega F'_u\left(\sum_{l=0}^{r-1} d_{0,l}^{*} v_{k_0+l}, 0\right)\left(\sum_{l=0}^{r-1} \bar{d}_l v_{k_0+l} + \bar{w}\right)(\tau, \xi)\, v_{k_0}(\xi)\, d\xi\, d\tau , \ldots$$

$$\left.\ldots, \int_0^\omega \int_\Omega F'_u\left(\sum_{l=0}^{r-1} d_{0,l}^{*} v_{k_0+l}, 0\right)\left(\sum_{l=0}^{r-1} \bar{d}_l v_{k_0+l} + \bar{w}\right)(\tau, \xi)\, v_{k_0+r-1}(\xi)\, d\xi\, d\tau\right),$$

and the fourth assumption follows from (c) by using I: Theorem 1.2.2. Applying the mentioned implicit function theorem we find that there exists a unique $(w^*(\varepsilon), \boldsymbol{d}^*(\varepsilon)) \in B_1$ such that

$$G_1(w^*(\varepsilon), \boldsymbol{d}^*(\varepsilon), \varepsilon) = 0\,, \quad G_2(w^*(\varepsilon), \boldsymbol{d}^*(\varepsilon), \varepsilon) = 0$$

for all sufficiently small $|\varepsilon|$, that w^*, and $\boldsymbol{d}^*$ are continuous in ε and $w^*(\varepsilon) \to 0$, $\boldsymbol{d}^*(\varepsilon) \to \boldsymbol{d}_0^*$ as $\varepsilon \to 0$. Writing

$$u^*(\varepsilon) = \sum_{l=0}^{r-1} d_l^*(\varepsilon)\, v_{k_0+l} + w^*(\varepsilon)$$

we complete the proof.

Problem 1.3.1 Let F be a substitution operator represented by a function f and $n = 2, 3$. Find conditions on f under which it satisfies the assumptions of the above theorems modified for $H_\omega^{3,6}(Q) \cap \dot{H}_\omega^1(Q)$ and $H_\omega^{2,4}(Q)$.

§ 2. The *t*-Fourier and *s*-Fourier methods

2.1. *The t-Fourier method*: $0 \leqq x \leqq \pi$

Let us apply the t-Fourier method to the problem $(\mathscr{P}_\omega)$ given by (1.1.1)–(1.1.3).

Now a solution is looked for in the form of a Fourier series with respect to $\{\omega^{-1/2}\, e^{2\pi i j t/\omega}\}_{j \in \mathbf{Z}}$, which is an orthonormal base in $H_\omega^0(\mathbf{R})$. On the other hand, it is not necessary to work in a Hilbert space with respect to the variable x (however, see Remark 2.1.4 for that case).

By a solution to $(\mathscr{P}_\omega)$ we mean a function u from the space

$$\mathscr{U} = C(I; H_\omega^1(\mathbf{R})) \cap C^1(I; H_\omega^{1/2}(\mathbf{R})) \cap C^2(I; H_\omega^0(\mathbf{R}))\,, \quad I = [0, \pi] \tag{2.1.1}$$

satisfying (1.1.1) for all $x \in (0, \pi)$ and (1.1.2), all in the sense of $H_\omega^0(\mathbf{R})$ (almost everywhere in $\mathbf{R}$).

(Actually, $(\mathscr{P}_\omega)$ is considered as a problem

$$-u_{xx}(x) + c\, u(x) + D_t\, u(x) = g(x)\,, \quad x \in (0, \pi)\,,$$

$$u(0) = u(\pi) = 0$$

in $H_\omega^0(\mathbf{R})$ for an abstract function $u : I \to H_\omega^0(\mathbf{R})$.)

Let us recall that

$$H_\omega^\alpha(\mathbf{R}) = \{u \in H_\omega^0(\mathbf{R});\ u(t) = \omega^{-1/2} \sum_{j=-\infty}^{\infty} u_j\, e^{ij\nu t},$$

$$\nu = 2\pi/\omega\,, \quad \sum_{j=-\infty}^{\infty} |j|^{2\alpha} |u_j|^2 < \infty\} \quad \text{for} \quad \alpha \geqq 0\,,$$

and the norm is given by

$$\|u\|_{H_\omega{}^\alpha(\mathbf{R})} = \Big(\sum_{j=-\infty}^{\infty} (1 + |j|^{2\alpha})\, |u_j|^2\Big)^{1/2}$$

(see I: Theorem 2.5.3).

The space $\mathscr{U}$ is equipped with the norm

$$\|u\| = \max\{\sup_{x\in I} \|u(\cdot, x)\|_{H_\omega{}^1(\mathbf{R})}\,,$$

$$\sup_{x\in I} \|u_x(\cdot, x)\|_{H_\omega{}^{1/2}(\mathbf{R})}\,, \quad \sup_{x\in I} \|u_{xx}(\cdot, x)\|_{H_\omega{}^0(\mathbf{R})}\}\,. \tag{2.1.2}$$

A function $u : I \to H_\omega^0(\mathbf{R})$ can be expanded for each $x \in I$ in a Fourier series

$$u(t, x) = \omega^{-1/2} \sum_{j=-\infty}^{\infty} u_j(x)\, e^{ij\nu t} \Big(= \lim_{m\to\infty} \omega^{-1/2} \sum_{j=-m}^{m} u_j(x)\, e^{ij\nu t}\Big), \tag{2.1.3}$$

where

$$u_j(x) = \omega^{-1/2} \int_0^\omega u(\tau, x)\, e^{-ij\nu\tau}\, d\tau\,, \quad \nu = 2\pi/\omega\,, (\bar{u}_j = u_{-j})\,.$$

Applying I: Theorem 2.5.2 (p. 21) and II: Remark 1.3.3 (p. 43) we find that $u \in \mathscr{U}$ if and only if $u_j \in C^2(I;\, \mathbf{C})$ for all $j \in \mathbf{Z}$ and the series

$$\sum_{j=-\infty}^{\infty} j^2 |u_j(x)|^2\,, \quad \sum_{j=-\infty}^{\infty} |j|\, |u_j'(x)|^2\,, \quad \sum_{j=-\infty}^{\infty} |u_j''(x)|^2 \tag{2.1.4}$$

converge uniformly on I. The norm (2.1.2) is equivalent to

$$\|u\|_{\mathscr{U}} = \max\{\sup_{x\in I} \Big(\sum_{j=-\infty}^{\infty} (1 + j^2)\, |u_j(x)|^2\Big)^{1/2}\,,$$

$$\sup_{x\in I} \Big(\sum_{j=-\infty}^{\infty} (1 + |j|)\, |u_j'(x)|^2\Big)^{1/2}\,, \quad \sup_{x\in I} \Big(\sum_{j=-\infty}^{\infty} |u_j''(x)|^2\Big)^{1/2}\}\,. \tag{2.1.5}$$

We suppose that $g \in \mathscr{G} = C(I;\, H_\omega^\alpha(\mathbf{R}))$ for some $\alpha > 0$ and write

$$g(t, x) = \omega^{-1/2} \sum_{j=-\infty}^{\infty} g_j(x)\, e^{ij\nu t}\,, \tag{2.1.6}$$

where

$$g_j(x) = \omega^{-1/2} \int_0^\omega g(\tau, x)\, e^{-ij\nu\tau}\, d\tau\,, \quad (\bar{g}_j = g_{-j})\,.$$

By I: Theorem 2.5.2 a function $g : I \to H^0_\omega(\mathbf{R})$ belongs to $\mathcal{G}$ if and only if $g_j \in C(I; \mathbf{C})$, $j \in \mathbf{Z}$, and

$$\sum_{j=-\infty}^{\infty} |j|^{2\alpha}\, |g_j(x)|^2 \quad \text{converges uniformly on } I\,. \tag{2.1.7}$$

The norm on $\mathcal{G}$ is defined by

$$\|g\|_{\mathcal{G}} = \sup_{x\in I} \Big(\sum_{j=-\infty}^{\infty} (1 + |j|^{2\alpha})\, |g_j(x)|^2 \Big)^{1/2}\,. \tag{2.1.8}$$

Substituting (2.1.3) and (2.1.6) in (1.1.1) and comparing the corresponding Fourier coefficients, we obtain

$$u_j''(x) - (c + i\nu j)\, u_j(x) = -g_j(x)\,, \quad x \in (0, \pi)\,, \quad j \in \mathbf{Z}\,. \tag{2.1.9}$$

To ensure that (1.1.2) holds, we must require

$$u_j(0) = u_j(\pi) = 0\,, \quad j \in \mathbf{Z}\,. \tag{2.1.10}$$

The solution of the boundary-value problem (2.1.9), (2.1.10) is

$$u_j(x) = \frac{1}{\varkappa_j} \Big(\int_0^x \frac{\sinh \varkappa_j \xi}{\sinh \varkappa_j \pi} \sinh \varkappa_j(\pi - x)\, g_j(\xi)\, d\xi\, + $$

$$+ \int_x^\pi \frac{\sinh \varkappa_j x}{\sinh \varkappa_j \pi} \sinh \varkappa_j(\pi - \xi)\, g_j(\xi)\, d\xi \Big)\,, \quad x \in I\,, \tag{2.1.11}$$

where

$$\varkappa_j = (c + i\nu j)^{1/2} = \mu_j + i\nu_j\,,$$

$$\mu_j = 2^{-1/2}(c + (c^2 + \nu^2 j^2)^{1/2})^{1/2}\,,$$

$$\nu_j = 2^{-1/2}(-c + (c^2 + \nu^2 j^2)^{1/2})^{1/2} \operatorname{sgn} j$$

($\operatorname{sgn} j = +1$ for $j \geqq 0$, $\operatorname{sgn} j = -1$ for $j < 0$) if either $j \neq 0$ (and c is arbitrary) or $j = 0$ and $c \neq 0, -k^2$, $k \in \mathbf{N}$. If $j = 0$, $c = 0$, then

$$u_0(x) = \pi^{-1} x \int_0^\pi (\pi - \xi)\, g_0(\xi)\, d\xi - \int_0^x (x - \xi)\, g_0(\xi)\, d\xi\,, \quad x \in I\,. \tag{2.1.12}$$

A solution of (2.1.9), (2.1.10) for $j = 0$ and $c = -k_0^2$ for some $k_0 \in \mathbf{N}$ exists if and only if

$$\int_0^\pi g_0(\xi) \sin k_0\xi \, d\xi = 0 \tag{2.1.13}$$

(the right-hand side of (2.1.9) is orthogonal to any solution of the homogeneous adjoint problem). If this condition is satisfied then

$$u_0(x) = \delta \sin k_0 x - k_0^{-1} \int_0^x \sin k_0(x - \xi)\, g_0(\xi) \, d\xi \,, \quad x \in I \,, \tag{2.1.14}$$

where $\delta \in \mathbf{R}$ is arbitrary. In all cases $u_j \in C^2(I; \mathbf{C})$.

For $x, y \in \mathbf{R}$ we have

$$|\sinh (x + iy)|^2 = 2^{-1}(\cosh 2x - \cos 2y)$$

and consequently there is a constant $C_1 > 0$ such that

$$\left|\frac{\sinh \varkappa_j\xi \sinh \varkappa_j(\pi - x)}{\sinh \varkappa_j\pi}\right| \leqq$$

$$\leqq \frac{1}{2}\left(\frac{(e^{2\mu_j\xi} + e^{-2\mu_j\xi} + 2)\,(e^{2\mu_j(\pi-x)} + e^{-2\mu_j(\pi-x)} + 2)}{e^{2\mu_j\pi} + e^{-2\mu_j\pi} - 2}\right)^{1/2} \leqq$$

$$\leqq \frac{1}{2}\,\frac{(e^{\mu_j\xi} + e^{-\mu_j\xi})\,(e^{\mu_j(\pi-x)} + e^{-\mu_j(\pi-x)})}{e^{\mu_j\pi} - e^{-\mu_j\pi}} \leqq$$

$$\leqq C_1\, e^{-\mu_j(x-\xi)}\,, \quad \xi\,, x \in I\,, \quad j \in \mathbf{Z} \setminus \{0\}$$

Further, there is $C_2 > 0$ such that

$$C_2|j|^{1/2} \leqq \mu_j \leqq C_2^{-1}|j|^{1/2}\,, \quad j \in \mathbf{Z} \setminus \{0\}$$

and for any real β there is $C_3 > 0$ such that

$$y^\beta\, e^{-y} \leqq C_3\,, \quad y \in \mathbf{R}^+\,.$$

Thus, we get from (2.1.11) with help of the Schwarz inequality

$$|u_j(x)|^2 \leqq \frac{C_4}{\mu_j^2 + \nu_j^2} \int_0^\pi e^{-\mu_j|x-\xi|} \, d\xi \int_0^\pi e^{-\mu_j|x-\xi|}\, |g_j(\xi)|^2 \, d\xi \leqq$$

$$\leqq \frac{C_5}{|j|^{3/2}} \int_0^\pi e^{-C_2|j|^{1/2}|x-\xi|} |g_j(\xi)|^2 \, d\xi \leqq$$

$$\leqq \frac{C_6}{|j|^{2-2\alpha}} \int_0^\pi \frac{1}{|x - \xi|^{1-4\alpha}}\, |g_j(\xi)|^2 \, d\xi\,,$$

where $C_4 - C_6$ are some positive constants independent of $x \in I$ and $j \in \mathbf{Z} \setminus \{0\}$. According to (2.1.7) we can write

$$\sum_{j \in \mathbf{Z} \setminus \{0\}} j^2 |u_j(x)|^2 \leqq$$

$$\leqq C_6 \sup_{\xi \in I} \sum_{j \in \mathbf{Z} \setminus \{0\}} |j|^{2\alpha} |g_j(\xi)|^2 \int_0^\pi \frac{1}{|x - \xi|^{1 - 4\alpha}} \, \mathrm{d}\xi$$

and immediately obtain the uniform convergence of the first series in (2.1.4). By differentiating the formula (2.1.11) the uniform convergence of the other series in (2.1.4) can be established similarly.

Let us distinguish two cases:

(i) $c \neq -k^2 , \quad k \in \mathbf{N}$,

(ii) $c = -k_0^2$ for some $k_0 \in \mathbf{N}$.

In the case (i) there exists, evidently, a $C_7 > 0$ independent of g such that

$$\|u\|_{\mathscr{U}} \leqq C_7 \|g\|_{\mathscr{G}}.$$

In the case (ii), by (2.1.14), the homogeneous problem $(\mathscr{P}_\omega)$ has a one-parameter family of solutions $u(t, x) = d \sin k_0 x$, $d \in \mathbf{R}$. In the inhomogeneous case, suppose that the necessary solubility condition (2.1.13) holds. Then the general solution of $(\mathscr{P}_\omega)$ can be written in the form $u(t, x) = d \sin k_0 x + w(t, x)$, where $d \in \mathbf{R}$ is arbitrary and w is a particular solution of $(\mathscr{P}_\omega)$, which is uniquely determined by the requirement to belong to the space $\mathscr{U} \cap \mathscr{W}$,

$$\mathscr{W} = \Big\{ w \in C(I; H^0_\omega(\mathbf{R})); \; w(t, x) = \omega^{-1/2} \sum_{j=-\infty}^{\infty} w_j(x) \, e^{ijvt} ,$$

$$\int_0^\pi w_0(\xi) \sin k_0 \xi \, \mathrm{d}\xi = 0$$

$$\left(\text{or, equivalently, } \int_0^\omega \int_0^\pi w(\tau, \xi) \sin k_0 \xi \, \mathrm{d}\xi \, \mathrm{d}\tau = 0 \right) \Big\} .$$

that is,

$$w(t, x) = \omega^{-1/2} \sum_{j=-\infty}^{\infty} u_j(x) \, e^{ijvt} , \tag{2.1.15}$$

where u_j, $j \in \mathbf{Z} \setminus \{0\}$, are given by (2.1.11) and

$$u_0(x) = (k_0\pi)^{-1} \int_0^\pi (\pi - \xi) \cos k_0\xi g_0(\xi) \, d\xi \sin k_0 x$$

$$- k_0^{-1} \int_0^x \sin k_0(x - \xi) \, g_0(\xi) \, d\xi \, .$$

Moreover, $\|w\|_{\mathscr{U}} \leqq C_8 \|g\|_{\mathscr{G}}$ with $C_8 > 0$ independent of g.

We define the operator L by $\mathscr{D}(L) = \mathscr{U}$ and $Lu = u_t - u_{xx} + cu$ and summarize our results.

THEOREM 2.1.1 (*The non-critical case*). *Let* $g \in \mathscr{G} = C(I; H_\omega^\alpha(\mathbf{R}))$ *for some* $\alpha > 0$. *Let* $c \neq -k^2$ *for all* $k \in \mathbf{N}$.
Then the problem $(\mathscr{P}_\omega)$ *has a unique solution* $u \in \mathscr{U}$,

$$u(t, x) = (L^{-1}g)(t, x) = \omega^{-1/2} \sum_{j=-\infty}^{\infty} u_j(x) \, e^{ij\nu t} \quad (\nu = 2\pi/\omega) \, ,$$

where the u_j *are determined by* (2.1.11) *and* (2.1.12) *respectively and* $L^{-1} \in \mathscr{L}(\mathscr{G}, \mathscr{U})$.

THEOREM 2.1.2 (*The critical case*). *Let* $g \in \mathscr{G}$. *Let* $c = -k_0^2$ *for some* $k_0 \in \mathbf{N}$. *Then the problem* $(\mathscr{P}_\omega)$ *has a solution if and only if* $g \in \mathscr{G} \cap \mathscr{W}$. *If this condition is satisfied, then the solutions have the form*

$$u(t, x) = d \sin k_0 x + w(t, x) \, ,$$

where $d \in \mathbf{R}$ *is arbitrary,* $w = L^{-1}g \in \mathscr{U} \cap \mathscr{W}$ *is given by* (2.1.15) *and* $L^{-1} \in \mathscr{L}(\mathscr{G} \cap \mathscr{W}, \mathscr{U} \cap \mathscr{W})$.

To specify the smoothness properties of functions from $\mathscr{U}$ and $\mathscr{G}$ we need I: Theorem 2.7.4 and the following lemma.

LEMMA 2.1.1 *Let* $\alpha > \frac{1}{2}$ *and* $u \in C(I; H_\omega^\alpha(\mathbf{R}))$. *Then* $u \in C_\omega(\bar{Q})$.

Proof. A function $u \in C(I; H_\omega^\alpha(\mathbf{R}))$ can be identified with a function $u = u(t, x)$ defined on $\mathbf{R} \times I$. By I: Theorem 2.7.4, $u(t, x)$ is a continuous function in t for all $x \in I$. By the same theorem

$$\sup_{t \in \mathbf{R}} |u(t, x + \xi) - u(t, x)| \leqq \text{const.} \, \|u(., x + \xi) - u(., x)\|_{H_\omega^\alpha(\mathbf{R})} \, ,$$

which shows that $u(t, x)$ is continuous in x, uniformly with respect to t. The assertion now follows immediately.

Remark 2.1.1 By the preceding lemma and by I: Theorem 2.7.4 we obtain that $u \in \mathscr{U}$ implies that $u \in C_\omega(\bar{Q})$ and $u_x \in C(I; L_{q,\omega}(\mathbf{R}))$, $q \in [1, \infty)$. Actually, by the proof of Theorem 2.1.1 it is seen that the solution belongs to $C(I; H_\omega^{1+\delta}(\mathbf{R})) \cap C^1(I; H_\omega^{1/2+\delta}(\mathbf{R})) \cap C^2(I; H_\omega^\delta(\mathbf{R}))$ where $0 < \delta < \alpha$, so that $u \in C_\omega^{0,1}(\bar{Q})$. If $g \in C(I; H_\omega^1(\mathbf{R}))$ (or $g \in C(I; H_\omega^1(\mathbf{R})) \cap \mathscr{W}$), then the solution belongs to $\mathscr{U}_1 = C(I; H_\omega^{2-\delta}(\mathbf{R})) \cap C^1(I; H_\omega^{3/2-\delta}(\mathbf{R})) \cap C^2(I; H_\omega^{1-\delta}(\mathbf{R})) \subset$ $\subset C_\omega^{1,2}(\bar{Q})$ for an arbitrary $0 < \delta < 1/2$.

Problem 2.1.1 Prove that Theorem 2.1.1 is no longer true if $g \in C(I; H_\omega^0(\mathbf{R}))$. (Cf. Theorem 2.3.3.)

Let us treat the corresponding weakly non-linear problem $(\mathscr{P}_\omega^\varepsilon)$, which is given by

$$u_t(t, x) - u_{xx}(t, x) + c\, u(t, x) = \varepsilon\, F(u, \varepsilon)\,(t, x)\,,$$

$$(t, x) \in \mathbf{R} \times (0, \pi)\,, \quad \varepsilon \in [-\varepsilon_0, \varepsilon_0]\,, \quad (\varepsilon_0 > 0)$$

and (1.1.2), (1.1.3).

A function $u \in C([-\varepsilon^*, \varepsilon^*]; \mathscr{U})$ $(\varepsilon^* \in (0, \varepsilon_0])$ is said to be a solution if $u(\varepsilon)$ for every $\varepsilon \in [-\varepsilon^*, \varepsilon^*]$ satisfies the above equation for all $x \in (0, \pi)$ and (1.1.2), both in the sense of $H_\omega^0(\mathbf{R})$.

Using the same approach as in Sec. 1.2 we obtain the following two theorems.

THEOREM 2.1.3 (*The non-critical case*). *Let* $c \neq -k^2$ *for all* $k \in \mathbf{N}$. *Let the following assumptions be satisfied*:

(*a*) *for some* $\varrho > 0$ *the operator* $F(u, \varepsilon)$ *maps* $B(0, \varrho; \mathscr{U}) \times [-\varepsilon_0, \varepsilon_0]$ *continuously into* $\mathscr{G}$;

(*b*) *there is a* $\lambda > 0$ *such that*

$$\|F(u_1, \varepsilon) - F(u_2, \varepsilon)\|_{\mathscr{G}} \leqq \lambda \|u_1 - u_2\|_{\mathscr{U}}$$

for all $u_1, u_2 \in B(0, \varrho; \mathscr{U})$ *and* $\varepsilon \in [-\varepsilon_0, \varepsilon_0]$.
Then the problem $(\mathscr{P}_\omega^\varepsilon)$ *has a locally unique solution* $u^*(\varepsilon) \in \mathscr{U}$ *such that* $u^*(0) = 0$.

THEOREM 2.1.4 (*The critical case*). *Let* $c = -k_0^2$ *for some* $k_0 \in \mathbf{N}$. *Let the following assumptions be satisfied*:

(*a*) *the equation*

$$\int_0^\omega \int_0^\pi F(d_0 \sin k_0 x, 0)\,(\tau, \xi) \sin k_0 \xi \; d\xi \; d\tau = 0$$

has a solution $d_0 = d_0^* \in \mathbf{R}$;

(*b*) *the operator $F(u, \varepsilon)$ and its G-derivative $F'_u(u, \varepsilon)$ map $B(d_0^* \sin k_0 x, \varrho; \mathscr{U}) \times [-\varepsilon_0, \varepsilon_0]$ continuously into $\mathscr{G}$ and $\mathscr{L}(\mathscr{U}, \mathscr{G})$, respectively;*

$$(c) \int_0^\omega \int_0^\pi F'_u(d_0^* \sin k_0 x, 0)(\sin k_0 x)(\tau, \xi) \sin k_0 \xi \, d\xi \, d\tau \neq 0 .$$

Then the problem $(\mathscr{P}_\omega^\varepsilon)$ has a locally unique solution $u^*(\varepsilon) \in \mathscr{U}$ such that $u^*(0) = d_0^* \sin k_0 x$.

Remark 2.1.2 Let

$$F(u, \varepsilon)(t, x) = f(t, x, u(t, x), u_x(t, x), \varepsilon)$$

and let $f = f(t, x, u_0, u_1, \varepsilon)$ be ω-periodic in t, let f and its partial derivatives f_t, f_{u_0}, f_{u_1} be continuous on $\mathbf{R} \times [0, \pi] \times [-\varrho, \varrho]^2 \times [-\varepsilon_0, \varepsilon_0]$ and Lipschitz continuous in u_0 and u_1, uniformly with respect to t, x, ε. Let $c \neq -k^2$, $k \in \mathbf{N}$. Then all the assumptions of Theorem 2.1.3 modified for $\mathscr{U} = C_\omega^{1,2}(\bar{Q})$ and $\mathscr{G} = C_\omega^{1,0}(\bar{Q})$ (cf. Remark 2.1.1) are satisfied. Similarly we can formulate the assumptions on f in the critical case.

Remark 2.1.3 KOPÁČKOVÁ [49] applies her result obtained for a more general problem (see VII: Sec. 4.9) to $(\mathscr{P}_\omega)$. Wanting to avoid fractional derivatives, she requires that $g \in C(I; H_\omega^2(\mathbf{R}))$ and obtains $u \in C(I; H_\omega^2(\mathbf{R})) \cap C^2(I; H_\omega^1(\mathbf{R}))$.

ŠMULEV in [79] proves with the aid of the t-Fourier method the existence of an ω-periodic solution of the problem

$$u_t - \sum_{j,k=1}^{n} a_{jk}(x) u_{x_j x_k} - \sum_{j=1}^{n} b_j(x) u_{x_j} - c(x) u = g(t, x), \ (t, x) \in \mathbf{R} \times \Omega ,$$

$$u(t, x) = h(t, x), \quad (t, x) \in \mathbf{R} \times \partial\Omega ,$$

where $\Omega \subset \mathbf{R}^n$ is a bounded domain, $-c(x) \geqq c_0 > 0$ and the assumptions (1)–(3) in Sec. 1.3 are satisfied. The functions g and h are assumed to be ω-periodic in t and to have continuous derivatives with respect to t up to the fourth order. The last assumption makes it impossible to carry over this result to an (even weakly) non-linear problem.

MINASJAN [63] deals with classical solutions to a boundary value problem for the heat equation in three dimensional space in the cylindrically symmetrical case.

Remark 2.1.4 The t-Fourier method can also be applied to solve the n-dimensional problems studied in Sec. 1.3. Using I: Theorem 2.5.3 and deriving some a priori estimates of u_j and Eu_j in the norm of $H^0(\Omega; \mathbf{C})$, we can again arrive at Theorems 1.3.1 and 1.3.2.

2.2. The t-Fourier method: $-\infty < x < \infty$

The t-Fourier method enables us to investigate the problem $(\mathscr{P}_\omega)$ given by the equations (1.1.1), (1.1.3) for $(t, x) \in \mathbf{R}^2$.

By a solution we mean a function u from the space $\mathscr{U}$ given by (2.1.1) with $I = \mathbf{R}$ satisfying the equation (1.1.1) for all $x \in \mathbf{R}$ in the sense of $H^0_\omega(\mathbf{R})$. Note that the boundary condition (1.1.2) is now replaced by boundedness in x and that this property is included in the definition of $\mathscr{U}$.

By I: Theorem 2.5.2 (p. 21) a function $u : \mathbf{R} \to H^0_\omega(\mathbf{R})$ written in the form (2.1.3) belongs to $\mathscr{U}$ if and only if $u_j \in C^2(\mathbf{R}; \mathbf{C})$, $j \in \mathbf{Z}$, the series (2.1.4) converge locally uniformly on $\mathbf{R}$ and their sums are bounded functions on $\mathbf{R}$. The space $\mathscr{U}$ is normed by (2.1.2) and this norm is equivalent to (2.1.5) (where, of course, I is replaced by $\mathbf{R}$).

We assume that $g \in C(\mathbf{R}; H^\alpha_\omega(\mathbf{R}))$ for some $\alpha > 0$. Writing g in the form (2.1.6) we conclude by I: Theorem 2.5.2 that $g_j \in C(\mathbf{R}; \mathbf{C})$, $j \in \mathbf{Z}$, the series (2.1.7) converges locally uniformly on $\mathbf{R}$ and its sum is bounded on $\mathbf{R}$. The norm of g is given by (2.1.8) (where $I = \mathbf{R}$).

Proceeding as in the previous section we find that the coefficients u_j have to satisfy the equation (2.1.9) for $x \in \mathbf{R}$ and to belong to $C^2(\mathbf{R}; \mathbf{C})$.

We have

$$\mu_j = \operatorname{Re} \varkappa_j = \operatorname{Re}(c + i\nu j)^{1/2} = 2^{-1/2}(c + (c^2 + \nu^2 j^2)^{1/2})^{1/2} > 0$$

provided that either $j = 0$, $c > 0$ or $j \neq 0$. In these cases the general solution to (2.1.9) can be written as

$$u_j(x) = e^{\varkappa_j x}\left(a_j - (2\varkappa_j)^{-1}\int_0^x e^{-\varkappa_j \xi}\, g_j(\xi)\, \mathrm{d}\xi\right) +$$

$$+ e^{-\varkappa_j x}\left(b_j + (2\varkappa_j)^{-1}\int_0^x e^{\varkappa_j \xi}\, g_j(\xi)\, \mathrm{d}\xi\right) \quad (a_j, b_j \in \mathbf{C})$$

and the only solution in $C^2(\mathbf{R}; \mathbf{C})$ is

$$u_j(x) = (2\varkappa_j)^{-1}\left(\int_x^\infty e^{\varkappa_j(x-\xi)} g_j(\xi)\, \mathrm{d}\xi + \int_{-\infty}^x e^{-\varkappa_j(x-\xi)} g_j(\xi)\, \mathrm{d}\xi\right) =$$

$$= (2\varkappa_j)^{-1}\int_{-\infty}^\infty e^{-\varkappa_j|x-\xi|} g_j(\xi)\, \mathrm{d}\xi\,.$$

If $c = 0$, then

$$u_0(x) = a_0 + b_0 x + \int_0^x (\xi - x)\, g_0(\xi)\, \mathrm{d}\xi$$

and $u_0 \in C^2(\mathbf{R}; \mathbf{C})$ if and only if the constant b_0 can be chosen in such a way that

$$\left| b_0 x + \int_0^x (\xi - x)\, g_0(\xi)\, \mathrm{d}\xi \right| \leqq \text{const.}\,, \quad \left| \int_0^x g_0(\xi)\, \mathrm{d}\xi \right| \leqq \text{const.}\,, \qquad x \in \mathbf{R}\,.$$

Finally, if $c < 0$, then

$$u_0(x) = a_0 \cos\left[(-c)^{1/2} x\right] + b_0 \sin\left[(-c)^{1/2} x\right] +$$

$$+ (-c)^{-1/2} \int_0^x \sin\left[(-c)^{1/2} (\xi - x)\right] g_0(\xi)\, \mathrm{d}\xi$$

and $u_0 \in C^2(\mathbf{R}; \mathbf{C})$ if and only if

$$\left| \int_0^x \sin\left[(-c)^{1/2} \xi\right] g_0(\xi)\, \mathrm{d}\xi \right| \leqq \text{const.}\,, \quad \left| \int_0^x \cos\left[(-c)^{1/2} \xi\right] g_0(\xi)\, \mathrm{d}\xi \right| \leqq$$

$$\leqq \text{const.}\,, \quad x \in \mathbf{R}\,.$$

However, we leave the case $c \leqq 0$ out of our further consideration since the conditions on g_0 do not represent a suitable starting point for our methods used in the weakly non-linear case.

If $c > 0$ we can write similarly as in the previous section

$$|u_j(x)|^2 \leqq \frac{\text{const.}}{\mu_j^2 + \nu_j^2} \int_{-\infty}^{\infty} e^{-\mu_j|x-\xi|}\, \mathrm{d}\xi \int_{-\infty}^{\infty} e^{-\mu_j|x-\xi|} |g_j(\xi)|^2\, \mathrm{d}\xi \leqq$$

$$\leqq \frac{\text{const.}}{|j|^{3/2}} \int_{-\infty}^{\infty} e^{-C_2|j|^{1/2}|x-\xi|} |g_j(\xi)|^2\, \mathrm{d}\xi \leqq$$

$$\leqq \frac{\text{const.}}{|j|^{3/2}} \left(\int_{|x-\xi| \leqq 1} \frac{1}{\left(|j|^{1/2}|x - \xi|\right)^{1-4\alpha}} |g_j(\xi)|^2\, \mathrm{d}\xi + \right.$$

$$\left. + \int_{|x-\xi|>1} \frac{1}{|j|(x - \xi)^2} |g_j(\xi)|^2\, \mathrm{d}\xi \right), \quad j \neq 0\,.$$

Hence by the Lebesgue monotone convergence theorem and by the properties of g

$$\sum_{j \neq 0} j^2 |u_j(x)|^2 \leqq$$

$$\leqq \text{const.} \sup_{\xi \in \mathbf{R}} \sum_{j \neq 0} |j|^{2\alpha} |g_j(\xi)|^2 \left(\int_0^1 \frac{\mathrm{d}\xi}{\xi^{1-4\alpha}} + \int_1^{\infty} \frac{\mathrm{d}\xi}{\xi^2} \right)$$

and we obtain that $\sum_{j=-\infty}^{\infty} j^2|u_j(x)|^2$ converges locally uniformly on $\mathbf{R}$. Evidently, $\sup_{x\in\mathbf{R}} \sum_{j=-\infty}^{\infty} j^2|u_j(x)|^2 \leqq \text{const.} \sup_{x\in\mathbf{R}} \sum_{j=-\infty}^{\infty} |j|^{2\alpha} |g_j(x)|^2$. Similarly for the two other series in (2.1.4). So there exists a constant $C > 0$ such that $\|u\|_{\mathscr{U}} \leqq \leqq C\|g\|_{C(\mathbf{R},H_\omega{}^\alpha(\mathbf{R}))}$.

Defining the operator L by $\mathscr{D}(L) = \mathscr{U}$ and $Lu = u_t - u_{xx} + cu$ we can state the following theorem.

THEOREM 2.2.1 (*The non-critical case*). *Let $g \in C(\mathbf{R};\ H^\alpha_\omega(\mathbf{R}))$ for some $\alpha > 0$ and $c > 0$.*
Then the problem $(\mathscr{P}_\omega)$ has a unique solution $u \in \mathscr{U}$, $u = L^{-1}g$ and $L^{-1} \in \in \mathscr{L}(C(\mathbf{R};\ H^\alpha_\omega(\mathbf{R})), \mathscr{U})$.

There is another way to solve the preceding problem, namely to use the Fourier transform for the solution of the equation (2.1.9), where $x \in \mathbf{R}$. Of course, in this case it is convenient to look for a solution to $(\mathscr{P}_\omega)$ in the space

$$\mathscr{U} = H^0(\mathbf{R};\ H^1_\omega(\mathbf{R})) \cap H^2(\mathbf{R};\ H^0_\omega(\mathbf{R}))\,.$$

Applying the Fourier transform to (2.1.9) we obtain

$$(ijv + \xi^2 + c)\,\hat{u}_j(\xi) = \hat{g}_j(\xi)\,,$$

where

$$\hat{u}_j(\xi) = (2\pi)^{-1/2}\int_{-\infty}^{\infty} e^{-ix\xi}\,u_j(x)\,\mathrm{d}x\,,$$

$$\hat{g}_j(\xi) = (2\pi)^{-1/2}\int_{-\infty}^{\infty} e^{-ix\xi}\,g_j(x)\,\mathrm{d}x\,.$$

It can be shown (see, for example, II: [14]) that $u \in H^0_\omega(\mathbf{R}^2)$ belongs to $\mathscr{U}$ if and only if $\hat{u}_j \in H^0(\mathbf{R};\ \mathbf{C})$, $j \in \mathbf{Z}$,

$$\int_{-\infty}^{\infty} \sum_{j=-\infty}^{\infty} (1 + j^2 + \xi^4)\,|\hat{u}_j(\xi)|^2\,\mathrm{d}\xi < \infty\,. \tag{2.2.1}$$

Since

$$|\hat{u}_j(\xi)|^2 = \frac{|\hat{g}_j(\xi)|^2}{v^2j^2 + (\xi^2 + c)^2}$$

it suffices to require that $c > 0$ and

$$\int_{-\infty}^{\infty} \sum_{j=-\infty}^{\infty} |\hat{g}_j(\xi)|^2\,\mathrm{d}\xi < \infty\,, \tag{2.2.2}$$

that is, $g \in H^0(\mathbf{R}; H^0_\omega(\mathbf{R}))$. The norms on $\mathscr{U}$ ($= H^{1,2}_\omega(\mathbf{R}^2)$ in the more usual notation) and $H^0(\mathbf{R}; H^0_\omega(\mathbf{R}))$ ($= H^0_\omega(\mathbf{R}^2)$) are equivalent to the square roots of the expressions in (2.2.1) and (2.2.2), respectively. So we have the following result.

THEOREM 2.2.2 (*The non-critical case*). *Let* $g \in H^0_\omega(\mathbf{R}^2)$ *and* $c > 0$. *Then the problem* $(\mathscr{P}_\omega)$ *has a unique solution* $u = L^{-1}g \in H^{1,2}_\omega(\mathbf{R}^2)$, *where* $L^{-1} \in \in \mathscr{L}(H^0_\omega(\mathbf{R}^2), H^{1,2}_\omega(\mathbf{R}^2))$.

Problem 2.2.1 Use the Fourier transform to investigate more general problems (for example for the equation $u_t(t, x) - \Delta\, u(t, x) + c\, u(t, x) = g(t, x)$) for all $x \in \mathbf{R}^n$.

Problem 2.2.2 Derive counterparts to Theorems 2.2.1 and 2.2.2 for the weakly non-linear problem.

2.3. *The s-Fourier method*

We apply the result (as well as the notation) of II: Sec. 3.3 to the problem $(\mathscr{P}_\omega)$ given by (1.1.1)–(1.1.3).

(Now $(\mathscr{P}_\omega)$ is regarded as an abstract problem in $H^0(0, \pi)$:

$$u_t(t) - D_x^2\, u(t) + c\, u(t) = g(t)\,, \quad t \in \mathbf{R}\,,$$

$$u(t) \in \mathscr{D}(-D_x^2) = H^2(0, \pi) \cap \dot{H}^1(0, \pi)\,, \quad t \in \mathbf{R}\,,$$

$$u(t + \omega) - u(t) = 0\,, \quad t \in \mathbf{R}$$

for an abstract function $u : \mathbf{R} \to H^0(0, \pi)$.)

By a solution to $(\mathscr{P}_\omega)$ we mean a function

$$u \in \mathscr{U} = C_\omega(\mathbf{R}; H^2(0, \pi) \cap \dot{H}^1(0, \pi)) \cap C^1_\omega(\mathbf{R}; H^0(0, \pi))$$

satisfying (1.1.1) for all $t \in \mathbf{R}$ in the sense of $H^0(0, \pi)$.

The norm on $\mathscr{U}$ is given by

$$\|u\|_{\mathscr{U}} = \max \{\max_{t \in \mathbf{R}} \|u(t, .)\|_{H^2(0,\pi)}\,, \quad \max_{t \in \mathbf{R}} \|u_t(t, .)\|_{H^0(0,\pi)}\}$$

and when II: (3.3.6) is taken into account

$$\|u\|_{\mathscr{U}} = \max \{ \max_{t \in [0,\omega]} (\sum_{k=1}^{\infty} k^4\, u_k^2(t))^{1/2}\,, \quad \max_{t \in [0,\omega]} (\sum_{k=1}^{\infty} \dot{u}_k^2(t))^{1/2}\}\,,$$

where

$$u_k(t) = \left(\frac{2}{\pi}\right)^{1/2} \int_0^\pi u(t, \xi) \sin k\xi \,\mathrm{d}\xi\,.$$

The corresponding initial boundary-value problem $(\mathcal{IB})$ is given by the equation (1.1.1) with the boundary conditions (1.1.2) and the initial condition

$$u(0, x) = \varphi(x), \quad x \in (0, \pi).$$

According to II: Theorem 3.3.1 (p. 61), the problem $(\mathcal{IB})$ has a unique solution

$$u = S(\varphi, g) \in C([0, \omega]; H^2(0, \pi) \cap \dot{H}^1(0, \pi)) \cap C^1([0, \omega]; H^0(0, \pi))$$

provided that $g \in C([0, \omega]; {}^0H^{2\alpha}(0, \pi))$ for some $\alpha > 0$ and $\varphi \in H^2(0, \pi) \cap \cap \dot{H}^1(0, \pi)$. (Note that the space ${}^0H^{\beta}(0, \pi)$ $(\beta \in \mathbf{R}^+)$ can be described as

$$\left\{u \in H^0(0, \pi); u(x) = \left(\frac{2}{\pi}\right)^{1/2} \sum_{k=1}^{\infty} u_k \sin kx, \sum_{k=1}^{\infty} k^{2\beta} u_k^2 < \infty\right\}\right).$$

Thus, assuming that $g \in \mathcal{G} = C_\omega(\mathbf{R}; {}^0H^{2\alpha}(0, \pi))$, u is ω-periodic in t if and only if there exists a solution $\varphi \in H^2(0, \pi) \cap \dot{H}^1(0, \pi)$ of the equation

$$T\varphi = \varphi, \tag{2.3.1}$$

where T is the so-called *translation operator* defined by

$$(T\varphi)(x) = S(\varphi, g)(\omega, x). \tag{2.3.2}$$

For since g is ω-periodic in t, by (2.3.1), and (2.3.2) the function $v(t, x) = u(t + \omega, x)$ solves (1.1.1)–(1.1.2) with the same right-hand side and the same initial condition and by virtue of the uniqueness of the solution to $(\mathcal{IB})$, $v(t, x) = u(t, x)$ in Q. (Henceforth the same argument will be used frequently without further comment.)

The equation (2.3.1) is equivalent to the infinite system for the Fourier coefficients

$$u_k(\omega) = \varphi_k, \quad k \in \mathbf{N}.$$

By II: (3.3.4),

$$u_k(t) = e^{-(k^2+c)t}\varphi_k + \int_0^t e^{-(k^2+c)(t-\tau)} g_k(\tau)\, d\tau,$$

and therefore the φ_k have to satisfy the relations

$$\varphi_k(1 - e^{-(k^2+c)\omega}) = \int_0^\omega e^{-(k^2+c)(\omega-\tau)} g_k(\tau)\, d\tau, \quad k \in \mathbf{N}. \tag{2.3.3}$$

We must distinguish two cases:

(i) $c \neq -k^2, \quad k \in \mathbf{N}$;

(ii) $c = -k_0^2$ for some $k_0 \in \mathbf{N}$.

In the first case the φ_k are uniquely determined for all $k \in \mathbf{N}$. Put

$$\varphi(x) = \left(\frac{2}{\pi}\right)^{1/2} \sum_{k=1}^{\infty} \varphi_k \sin kx \, .$$

In II: Sec. 3.3 we have shown that

$$\sum_{k=1}^{\infty} k^4 \left(\int_0^{\omega} e^{-(k^2+c)(\omega-\tau)} \, g_k(\tau) \, \mathrm{d}\tau \right)^2 \leqq \text{const.} \, \|g\|_{\mathscr{G}}^2 \, ,$$

hence by II: Theorem 1.3.3, $\varphi \in {}^0H^2(0, \pi) = H^2(0, \pi) \cap \dot{H}^1(0, \pi)$ and $\|\varphi\|_{H^2(0,\pi)} \leqq \text{const.} \, \|g\|_{\mathscr{G}}$. By II: Theorem 3.3.1 the function φ determines a unique solution u of the problem $(\mathscr{I}\mathscr{B})$, which is simultaneously a solution to $(\mathscr{P}_{\omega})$ and $\|u\|_{\mathscr{U}} \leqq \text{const.} \, \|g\|_{\mathscr{G}}$.

Explicitly,

$$u_k(t) = \left(1 - e^{-(k^2+c)\omega}\right)^{-1} \int_0^{\omega} e^{-(k^2+c)(\omega+t-\tau)} \, g_k(\tau) \, \mathrm{d}\tau \, +$$

$$+ \int_0^t e^{-(k^2+c)(t-\tau)} \, g_k(\tau) \, \mathrm{d}\tau \, , \quad k \in \mathbf{N} \, . \tag{2.3.4}$$

The coefficients can be expressed in another way for k with $k^2 + c > 0$. The ω-periodicity of u_k and g_k yields

$$u_k(t) = \lim_{n \to \infty, n \in \mathbf{N}} u_k(t + n\omega) =$$

$$= \lim_{n \to \infty} \int_0^{t+n\omega} e^{-(k^2+c)(t+n\omega-\tau)} \, g_k(\tau) \, \mathrm{d}\tau =$$

$$= \lim_{n \to \infty} \int_0^{t+n\omega} e^{-(k^2+c)\sigma} \, g_k(t - \sigma) \, \mathrm{d}\sigma \, ,$$

thus,

$$u_k(t) = \int_0^{\infty} e^{-(k^2+c)\tau} \, g_k(t - \tau) \, \mathrm{d}\tau \, , \quad k \in \mathbf{N} \, , \quad k^2 + c > 0 \, . \tag{2.3.5}$$

In the case (ii) the system (2.3.3) has a solution if and only if

$$\int_0^{\omega} g_{k_0}(\tau) \, \mathrm{d}\tau = 0 \, .$$

Suppose that this condition is satisfied. Then φ_{k_0} is arbitrary, while the other φ_k are uniquely determined. We choose

$$\varphi_{k_0} = \omega^{-1} \int_0^{\omega} \tau \, g_{k_0}(\tau) \, \mathrm{d}\tau \, .$$

Putting

$$\varphi(x) = \left(\frac{2}{\pi}\right)^{1/2} \sum_{k=1}^{\infty} \varphi_k \sin kx ,$$

we can verify that $\varphi \in H^2(0, \pi) \cap \dot{H}^1(0, \pi)$ and that $\|\varphi\|_{H^2(0,\pi)} \leqq \text{const.}\ \|g\|_{\mathscr{G}}$. According to II: Theorem 3.3.1, this function φ determines a solution w of the problem $(\mathscr{P}_\omega)$ and $\|w\|_{\mathscr{U}} \leqq \text{const.}\ \|g\|_{\mathscr{G}}$. Since

$$\int_0^\omega \int_0^\tau g_{k_0}(\sigma)\, d\sigma\, d\tau = \int_0^\omega (\omega - \tau)\, g_{k_0}(\tau)\, d\tau ,$$

this solution is the only one with the property that

$$\int_0^\omega w_{k_0}(\tau)\, d\tau = 0 .$$

It is evident that the homogeneous problem $(\mathscr{P}_\omega)$ has the general solution $v(t, x) = d \sin k_0 x$, $d \in \mathbf{R}$, consequently the general solution to $(\mathscr{P}_\omega)$ is $u(t, x) = v(t, x) + w(t, x)$. To simplify further statements we write

$$\mathscr{W} = \left\{ w \in C_\omega(\mathbf{R};\, H^0(0, \pi));\ w(t, x) = \left(\frac{2}{\pi}\right)^{1/2} \sum_{k=1}^{\infty} w_k(t) \sin kx ,\right.$$

$$\left. \int_0^\omega w_{k_0}(\tau)\, d\tau = 0 \ \left(\text{or, equivalently,} \int_0^\omega \int_0^\pi w(\tau, \xi) \sin k_0 \xi\, d\xi\, d\tau = 0\right)\right\} .$$

Denoting by L the operator defined by $\mathscr{D}(L) = \mathscr{U}$, $Lu = u_t - u_{xx} + cu$ and summarizing our results we obtain the following two theorems.

THEOREM 2.3.1 (*The non-critical case*). *Let* $g \in \mathscr{G} = C_\omega(\mathbf{R};\, {}^0H^{2\alpha}(0, \pi))$ *for some* $\alpha > 0$. *Let* $c \neq -k^2$ *for all* $k \in \mathbf{N}$.
Then the problem $(\mathscr{P}_\omega)$ *has a unique solution* $u \in \mathscr{U}$,

$$u(t, x) = (L^{-1}g)(t, x) = \left(\frac{2}{\pi}\right)^{1/2} \sum_{k=1}^{\infty} u_k(t) \sin kx ,$$

where the u_k *are determined by* (2.3.4) (*or* (2.3.5) *for* k, $k^2 + c > 0$) *and* $L^{-1} \in \mathscr{L}(\mathscr{G}, \mathscr{U})$.

THEOREM 2.3.2 (*The critical case*). *Let* $g \in \mathscr{G}$, *and* $c = -k_0^2$ *for some* $k_0 \in \mathbf{N}$. *Then the problem* $(\mathscr{P}_\omega)$ *has a solution* $u \in \mathscr{U}$ *if and only if* $g \in \mathscr{G} \cap \mathscr{W}$. *If this condition is satisfied then the solution can be written in the form*

$$u(t, x) = d \sin k_0 x + w(t, x) ,$$

where $d \in \mathbf{R}$ is arbitrary,

$$w(t, x) = (L^{-1}g)(t, x) = \left(\frac{2}{\pi}\right)^{1/2} \sum_{k=1}^{\infty} u_k(t) \sin kx ,$$

the u_k are determined by (2.3.4) for $k \neq k_0$ (or (2.3.5) for $k > k_0$),

$$u_{k_0}(t) = \int_0^t g_{k_0}(\tau)\, d\tau + \omega^{-1} \int_0^\omega \tau\, g_{k_0}(\tau)\, d\tau$$

and $L^{-1} \in \mathscr{L}(\mathscr{G} \cap \mathscr{W}, \mathscr{U} \cap \mathscr{W})$.

Problem 2.3.1 Prove that $u \in \mathscr{U}$ implies that $u \in C_\omega^{0,1}(\mathbf{R} \times [0, \pi])$.

Problem 2.3.2 Formulate and prove existence theorems for the corresponding weakly non-linear problem.

Problem 2.3.3 Investigate the n-dimensional problem (1.3.1)–(1.3.3) by means of the s-Fourier method.

It is natural to ask whether Theorem 2.3.1 is perhaps true even for $\alpha = 0$ (cf. HAVLOVÁ [19]). The following theorem shows that this is not the case.

THEOREM 2.3.3 *Let $c = 0$, $\omega = 2$.* Define

$$g_k(t) = 0 , \quad t \in \mathbf{R} , \quad k \neq 2^m , \quad m \in \mathbf{N} ,$$

$$g_{2^m}(t) = \begin{cases} m^{-1/2}(2^{2m+3}t - 3) , & t \in (3 \,.\, 2^{-(2m+3)}, 2^{-(2m+1)}) , \\ m^{-1/2} , & t \in [2^{-(2m+1)}, 2^{-2m}] , \\ m^{-1/2}(-2^{2m+1}t + 3) , & t \in (2^{-2m}, 3 \,.\, 2^{-(2m+1)}) , \\ 0 \text{ otherwise in } [0, 1] , \end{cases}$$

extended as an even and 2-periodic function on $\mathbf{R}$, $m \in \mathbf{N}$.

Then the function

$$g(t, x) = \sum_{k=1}^{\infty} g_k(t) \sin kx$$

belongs to $C_2(\mathbf{R}; H^0(0, \pi))$, but the problem $(\mathscr{P}_\omega)$ has no solution in the space $\mathscr{U}$.

Proof. The first assertion is obvious. Suppose that there is a solution $u \in \mathscr{U}$. Necessarily

$$u(t, x) = \sum_{k=1}^{\infty} \int_0^\infty e^{-k^2\tau} g_k(t - \tau)\, d\tau \sin kx . \tag{2.3.6}$$

But then

$$\sup_{t\in[-1,1]} \sum_{k=1}^{\infty} k^4 u_k^2(t) \geqq \sum_{k=1}^{\infty} k^4 u_k^2(0) =$$

$$= \sum_{m=1}^{\infty} \left(2^{2m} \int_0^{\infty} \exp(-2^{2m}\tau)\, g_{2^m}(-\tau)\, \mathrm{d}\tau\right)^2 =$$

$$= \sum_{m=1}^{\infty} \left(\int_0^{\infty} e^{-\sigma} g_{2^m}(-2^{-2m}\sigma)\, \mathrm{d}\sigma\right)^2 \geqq \sum_{m=1}^{\infty} \left(\int_{1/2}^{1} e^{-\sigma} g_{2^m}(2^{-2m}\sigma)\, \mathrm{d}\sigma\right)^2 =$$

$$= \sum_{m=1}^{\infty} \frac{1}{m} \left(\int_{1/2}^{1} e^{-\sigma}\, \mathrm{d}\sigma\right)^2 = \infty$$

which is a contradiction.

Concerning our construction, we note that a function g for which there is no solution $u \in \mathscr{U}$ of $(\mathscr{P}_\omega)$ must satisfy the conditions

$$\sup_{t\in[-1,1]} \sum_{k=1}^{\infty} g_k^2(t) < \infty \quad \text{and} \quad \sum_{k=1}^{\infty} \left(\sup_{t\in[-1,1]} |g_k(t)|\right)^2 = \infty\,.$$

For according to (2.3.6)

$$|u_k(t)| \leqq k^{-2} \sup_{t\in[-1,1]} |g_k(t)|\,, \quad t \in \mathbf{R}\,, \quad k \in \mathbf{N}\,,$$

and if $\sum_{k=1}^{\infty} \left(\sup_{t\in[-1,1]} |g_k(t)|\right)^2 < \infty$, then $\sum_{k=1}^{\infty} k^4 u_k^2(t)$ (and similarly $\sum_{k=1}^{\infty} \dot{u}_k^2(t)$) is uniformly convergent.

Remark 2.3.1 The s-Fourier method is also used by Karimov [22]–[30]. A system of weakly non-linear heat equations is studied by Mitrjakov [64]–[65] with the same shortcomings that appear in the papers of Karimov.

Remark 2.3.2 The spectral resolution method described in VII: [62] can be regarded as a natural generalization of the s-Fourier method. This makes it possible to deal with equations with "elliptic parts" represented by operators whose spectra do not necessarily reduce to their point parts.

Remark 2.3.3 Comparing the results established by the (t, s)-, t-, and s-Fourier methods for the problem given by (1.1.1)–(1.1.3), the reader sees at once the advantage the (t, s)-Fourier method has in the case of a weakly non-linear problem whose perturbation operator F is represented by a function f; namely, f may include besides u and u_x, also u_t and u_{xx}, which is not possible in the two other methods. Nevertheless, if the character of the perturbed problem does not require f to involve u_t or u_{xx}, then the t- and s-Fourier methods are more suitable in the sense that weaker smoothness conditions are to be imposed on f to obtain a classical solution. The t-Fourier method has another advantage, namely, that it allows to tackle problems with inhomogeneous boundary conditions directly and to deal with problems given on the whole x-space.

Remark 2.3.4 The choice of spaces used in the methods investigated above affects influence upon the functional analytical properties of the operator L. So, for instance, the operator $L : H^{1,2}_{\omega}(Q) \cap \dot{H}^1_{\omega}(Q) \subset H^0_{\omega}(Q) \to H^0_{\omega}(Q)$ defined in Sec. 1.1 and 1.3 has a closed range in $H^0_{\omega}(Q)$. On the other hand, by Theorem 2.3.3 the operator $L : C_{\omega}(\mathbf{R}; H^2(0, \pi) \cap \dot{H}^1(0, \pi)) \cap C^1_{\omega}(\mathbf{R}; H^0(0, \pi)) \subset$ $\subset C_{\omega}(\mathbf{R}; H^0(0, \pi)) \to C_{\omega}(\mathbf{R}; H^0(0, \pi))$ defined in Sec. 2.3 and by Problem 2.1.1 the operator $L : C(I; H^1_{\omega}(\mathbf{R})) \cap C^1(I; H^{1/2}_{\omega}(\mathbf{R})) \cap C^2(I; H^0_{\omega}(\mathbf{R})) \subset$ $\subset C(I; H^0_{\omega}(\mathbf{R})) \to C(I; H^0_{\omega}(\mathbf{R}))$ defined in Sec. 2.1, 2.2 have not closed ranges in $C_{\omega}(\mathbf{R}; H^0(0, \pi))$, and $C(I; H^0_{\omega}(\mathbf{R}))$, respectively.

§ 3. The Poincaré method

3.1. *The linear case*

Let us investigate the problem $(\mathscr{P}_{\omega})$ given by

$$u_t(t, x) - u_{xx}(t, x) + c\, u(t, x) = g(t, x)\,, \quad (t, x) \in Q = \mathbf{R} \times (0, \pi)\,, \tag{3.1.1}$$

$$u(t, 0) = h_0(t)\,, \quad u(t, \pi) = h_1(t)\,, \quad t \in \mathbf{R}\,, \tag{3.1.2}$$

$$u(t + \omega, x) - u(t, x) = 0\,, \quad (t, x) \in Q\,, \tag{3.1.3}$$

where $c \in \mathbf{R}$, $g \in \mathscr{G} = C^{0,1}_{\omega}(\bar{Q})$, h_0, $h_1 \in \mathscr{H} = C^1_{\omega}(\mathbf{R})$.

By a solution we mean a function

$$u \in \mathscr{U} = C^{1,2}_{\omega}(\bar{Q})$$

satisfying (3.1.1) and (3.1.2).

The corresponding initial boundary-value problem $(\mathscr{I}\mathscr{B})$ is given by (3.1.1), (3.1.2), and by

$$u(0, x) = \varphi(x)\,, \quad x \in [0, \pi]\,, \tag{3.1.4}$$

where $\varphi \in C^2([0, \pi])$ and the compatibility conditions

$$\varphi(0) = h_0(0)\,, \quad \varphi(\pi) = h_1(0)\,,$$

$$h_0'(0) - \varphi''(0) + c\,\varphi(0) = g(0, 0),\ h_1'(\pi) - \varphi''(\pi) + c\,\varphi(\pi) = g(0, \pi) \tag{3.1.5}$$

are fulfilled. By II: Theorem 3.2.1, the problem $(\mathcal{I}\mathcal{B})$ has a unique solution $u = S(\varphi, g, h_0, h_1) \in C^{1,2}([0, \omega] \times [0, \pi])$ and

$$\|u\|_{C^{1,2}([0,\omega]\times[0,\pi])} \leqq C_1(\|\varphi\|_{C^2([0,\pi])} + \|g\|_{\mathcal{G}} + \|h_0\|_{\mathcal{H}} + \|h_1\|_{\mathcal{H}}) \tag{3.1.6}$$

with $C_1 > 0$ independent of φ, g, h_0 and h_1.

Hence, a solution to $(\mathcal{P}_\omega)$ exists if and only if there is a function $\varphi \in C^2([0, \pi])$ satisfying (3.1.5) that is a fixed point of the translation operator T defined by

$$(T\varphi)(x) = S(\varphi, g, h_0, h_1)(\omega, x) .$$

When we put

$$(K\varphi)(x) = e^{-c\omega} \int_0^\pi \Gamma(\omega, x, \xi)\, \varphi(\xi)\, d\xi ,$$

$$\gamma(x) = \int_0^\omega e^{-c(\omega-\tau)} \int_0^\pi \Gamma(\omega - \tau, x, \xi)\, g(\tau, \xi)\, d\xi\, d\tau -$$

$$- \frac{1}{\pi} \int_0^\omega e^{-c(\omega-\tau)}\, h_0(\tau) \frac{\partial}{\partial x}\, \Theta\left(\frac{\omega - \tau}{\pi^2}, \frac{x}{2\pi}\right) d\tau +$$

$$+ \frac{1}{\pi} \int_0^\omega e^{-c(\omega-\tau)}\, h_1(\tau) \frac{\partial}{\partial x}\, \Theta\left(\frac{\omega - \tau}{\pi^2}, \frac{\pi - x}{2\pi}\right) d\tau$$

(for the definitions of Θ and Γ see II: (3.2.4)–(3.2.5)) we have to solve the equation

$$\varphi - K\varphi = \gamma . \tag{3.1.7}$$

This is a Fredholm integral equation of the second kind whose kernel is symmetric and belongs to $C^\infty([0, \pi] \times [0, \pi])$, while its right-hand side belongs to $C^2([0, \pi])$. Note that according to these smoothness properties it suffices to look for $\varphi \in H^0(0, \pi)$. By Fredholm's third theorem (see, for example, I: [32]), a solution exists if and only if the right-hand side is orthogonal to every solution ψ of the homogeneous equation

$$\psi - K\psi = 0 . \tag{3.1.8}$$

We look for a solution of (3.1.8) in the form of a Fourier series

$$\psi(x) = \sum_{k=1}^{\infty} \psi_k \sin kx . \tag{3.1.9}$$

Observe that any H^0–solution of the form (3.1.9) belongs to $C^\infty([0, \pi])$, because

$$\Gamma(\omega, x, \xi) = \frac{2}{\pi} \sum_{n=1}^{\infty} e^{-n^2\omega} \sin nx \sin n\xi .$$

Substituting (3.1.9) in (3.1.8), we find that

$$\psi_k(1 - e^{-(c+k^2)\omega}) = 0\,, \quad k \in \mathbf{N}\,.$$

We have to distinguish two cases:

(i) $c \neq -k^2\,, \quad k \in \mathbf{N}\,;$

(ii) $c = -k_0^2$ for some $k_0 \in \mathbf{N}\,.$

In the first case the only solution of (3.1.8) is the trivial one, hence, the equation (3.1.7) has a unique solution φ for any right-hand side. In addition, by I: Theorem 1.2.2 (p. 10), there exists a $C_2 > 0$ such that $\|\varphi\|_0 \leqq C_2\|\gamma\|_0$ ($\|\cdot\|_0$ stands for the norm in $H^0(0, \pi)$). In accordance with the properties of $K\varphi$ and γ we obtain

$$\varphi \in C^2([0, \pi])\,, \quad (3.1.5) \text{ holds} \tag{3.1.10}$$

and

$$\|\varphi\|_{C^2([0,\pi])} \leqq C_3(\|g\|_{\mathscr{G}} + \|h_0\|_{\mathscr{H}} + \|h_1\|_{\mathscr{H}}) \tag{3.1.11}$$

with $C_3 > 0$ independent of g, h_0, h_1.

We define the operator $L : \mathscr{U} \to \mathscr{G} \times \mathscr{H} \times \mathscr{H}$ by $Lu = (u_t - u_{xx} + cu, u(t, 0), u(t, \pi))$. By (3.1.6) and (3.1.11) the following theorem holds.

THEOREM 3.1.1 (*The non-critical case*). *Let $g \in \mathscr{G}$, h_0, $h_1 \in \mathscr{H}$, and $c \neq -k^2$ for all $k \in \mathbf{N}$.*
Then the problem $(\mathscr{P}_\omega)$ has a unique solution $u = L^{-1}(g, h_0, h_1) \in \mathscr{U}$ and $L^{-1} \in \mathscr{L}(\mathscr{G} \times \mathscr{H} \times \mathscr{H}, \mathscr{U})$.

Now we turn to the case (ii) when $c = -k_0^2$. Then (3.1.8) has precisely one linearly independent solution $\psi(x) = \sin k_0 x$. The solubility condition for (3.1.7) is

$$\int_0^\pi \gamma(\xi) \sin k_0\xi \, d\xi = 0\,. \tag{3.1.12}$$

After some rearrangement (3.1.12) can be rewritten in the form

$$\int_0^\omega \int_0^\pi g(\tau, \xi) \sin k_0\xi \, d\xi \, d\tau + $$

$$+ k_0 \int_0^\omega (h_0(\tau) + (-1)^{k_0+1} h_1(\tau)) \, d\tau = 0\,. \tag{3.1.12'}$$

If this condition is satisfied, then by I: Theorem 1.6.4 (p. 14) and I: Theorem 1.2.2 (p. 10), all solutions of (3.1.7) are given by $\delta \sin k_0 x + \varphi_p(x)$, where $\delta \in \mathbf{R}$ is arbitrary and φ_p is a particular solution of (3.1.7) which is uniquely determined in the subspace of $H^0(0, \pi)$ of functions orthogonal to $\sin k_0 x$ and $\|\varphi_p\|_0 \leqq C_4 \|\gamma\|_0$ with $C_4 > 0$ independent of γ. We can choose φ so that (3.1.10), and (3.1.11) hold and then the general solution of $(\mathscr{P}_\omega)$ is given by

$$u(t, x) = d \sin k_0 x + w(t, x), \quad d \in \mathbf{R}, \tag{3.1.13}$$

where w is uniquely determined in the space $\mathscr{U} \cap \mathscr{W}$,

$$\mathscr{W} = \left\{ w \in H^0_\omega(Q); \int_0^\omega \int_0^\pi w(\tau, \xi) \sin k_0 \xi \, d\xi \, d\tau = 0 \right\}$$

and

$$\|w\|_{\mathscr{U}} \leqq C_5(\|g\|_{\mathscr{G}} + \|h_0\|_{\mathscr{H}} + \|h_1\|_{\mathscr{H}}),$$

with C_5 independent of g, h_0, h_1.

THEOREM 3.1.2 (*The critical case*). *Let $g \in \mathscr{G}, h_0, h_1 \in \mathscr{H}$, and $c = -k_0^2$ for some $k_0 \in \mathbf{N}$.*
Then the problem $(\mathscr{P}_\omega)$ has a solution $u \in \mathscr{U}$ if and only if (3.1.12') holds. If this condition is satisfied, then the solutions can be written in the form (3.1.13), where $w = L^{-1}(g, h_0, h_1)$ and $L^{-1} \in \mathscr{L}(\{(g, h_0, h_1) \in \mathscr{G} \times \mathscr{H} \times \mathscr{H}; g, h_0, h_1$ satisfy (3.1.12')$\}, \mathscr{U} \cap \mathscr{W})$.

Remark 3.1.1 LAUEROVÁ [57] deals with the problem given by

$$u_t(t, x) - a(t)\, u_{xx}(t, x) + c(t)\, u(t, x) = g(t, x), \quad (t, x) \in \mathbf{R} \times (0, \pi),$$

(3.1.2), (3.1.3) and finds again conditions for its solubility.

3.2. *The weakly non-linear case*

Let us consider the problem $(\mathscr{P}^\varepsilon_\omega)$ given by

$$u_t(t, x) - u_{xx}(t, x) + c\, u(t, x) = \varepsilon F(u, \varepsilon)(t, x),$$

$$(t, x) \in Q, \quad \varepsilon \in [-\varepsilon_0, \varepsilon_0], \quad (\varepsilon_0 > 0), \tag{3.2.1}$$

$$u(t, 0) = \varepsilon X_0(u, \varepsilon)(t), \quad u(t, \pi) = \varepsilon X_1(u, \varepsilon)(t), \quad t \in \mathbf{R} \tag{3.2.2}$$

and by (3.1.3), where F, X_0, X_1 are operators whose properties are stated in the theorems below.

A function $u \in C([-\varepsilon^*, \varepsilon^*]; \mathscr{U})$ $(\varepsilon^* \in (0, \varepsilon_0])$ is said to be a solution of $(\mathscr{P}^\varepsilon_\omega)$ if $u(\varepsilon)$ for every $\varepsilon \in [-\varepsilon^*, \varepsilon^*]$ satisfies (3.2.1) and (3.2.2).

THEOREM 3.2.1 (*The non-critical case*). *Let $c \neq -k^2$ for all $k \in \mathbf{N}$ and suppose that the following assumptions are satisfied*:

(*a*) *the operator $F(u, \varepsilon)$ maps $B(0, \varrho; \mathscr{U}) \times [-\varepsilon_0, \varepsilon_0]$ for some $\varrho > 0$ continuously into $\mathscr{G}$*;

(*b*) *the operators* $X_j(u, \varepsilon)$, $j = 0, 1$, *map* $B(0, \varrho; \mathscr{U}) \times [-\varepsilon_0, \varepsilon_0]$ *continuously into* $\mathscr{H}$;

(*c*) *there exists a* $\lambda > 0$ *such that*

$$\|F(u_1, \varepsilon) - F(u_2, \varepsilon)\|_{\mathscr{G}} \leqq \lambda \|u_1 - u_2\|_{\mathscr{U}},$$

$$\|X_j(u_1, \varepsilon) - X_j(u_2, \varepsilon)\|_{\mathscr{H}} \leqq \lambda \|u_1 - u_2\|_{\mathscr{U}}, \quad j = 0, 1$$

for all $u_1, u_2 \in B(0, \varrho; \mathscr{U})$ *and* $\varepsilon \in [-\varepsilon_0, \varepsilon_0]$.
Then the problem $(\mathscr{P}^{\varepsilon}_{\omega})$ *has a locally unique solution* $u^*(\varepsilon) \in \mathscr{U}$ *such that* $u^*(0) = 0$.

Proof. According to Theorem 3.1.1 the equation

$$u = \varepsilon L^{-1}(F(u, \varepsilon), \quad X_0(u, \varepsilon), \quad X_1(u, \varepsilon))$$

has to be solved in $\mathscr{U}$. We can apply I: Corollary 3.4.1 just as in Sec. 1.2. Therefore we omit the details.

THEOREM 3.2.2 (*The critical case*). *Let* $c = -k_0^2$ *for some* $k_0 \in \mathbf{N}$ *and suppose that the following assumptions are satisfied*:

(*a*) *the equation*

$$\int_0^{\omega} \int_0^{\pi} F(d_0 \sin k_0 x, 0)(\tau, \xi) \sin k_0 \xi \, d\xi \, d\tau +$$

$$+ k_0 \int_0^{\omega} (X_0(d_0 \sin k_0 x, 0)(\tau) + (-1)^{k_0+1} X_1(d_0 \sin k_0 x, 0)(\tau)) \, d\tau = 0 \tag{3.2.3}$$

has a solution $d_0 = d_0^* \in \mathbf{R}$;

(*b*) *the operator* $F(u, \varepsilon)$ *and its G-derivative* $F'_u(u, \varepsilon)$ *map* $B(d_0^* \sin k_0 x, \varrho; \mathscr{U}) \times [-\varepsilon_0, \varepsilon_0]$ *for some* $\varrho > 0$ *continuously into* $\mathscr{G}$ *and* $\mathscr{L}(\mathscr{U}, \mathscr{G})$, *respectively*;

(*c*) *the operators* $X_j(u, \varepsilon)$, $j = 0, 1$, *and their G-derivatives* $X'_{ju}(u, \varepsilon)$ *map* $B(d_0^* \sin k_0 x, \varrho; \mathscr{U}) \times [-\varepsilon_0, \varepsilon_0]$ *continuously into* $\mathscr{H}$ *and* $\mathscr{L}(\mathscr{U}, \mathscr{H})$ *respectively*;

$$(d) \int_0^{\omega} \int_0^{\pi} F'_u(d_0^* \sin k_0 x, 0)(\sin k_0 x)(\tau, \xi) \sin k_0 \xi \, d\xi \, d\tau +$$

$$+ k_0 \int_0^{\omega} (X'_{0u}(d_0^* \sin k_0 x, 0)(\sin k_0 x)(\tau) +$$

$$+ (-1)^{k_0+1} X'_{1u}(d_0^* \sin k_0 x, 0)(\sin k_0 x)(\tau)) \, d\tau \neq 0 .$$

Then the problem $(\mathscr{P}^{\varepsilon}_{\omega})$ *has a locally unique solution* $u^*(\varepsilon) \in \mathscr{U}$ *such that* $u^*(0)(t, x) = d_0^* \sin k_0 x$.

Proof. By Theorem 3.1.2 the problem $(\mathscr{P}^{\varepsilon}_{\omega})$ is equivalent to the following system for $(w, d) \in (\mathscr{U} \cap \mathscr{W}) \times \mathbf{R}$:

$$G_1(w, d, \varepsilon) \equiv w(t, x) - \varepsilon L^{-1}(F(d \sin k_0 x + w(t, x), \varepsilon),$$

$$X_0(d \sin k_0 x + w(t, x), \varepsilon), \quad X_1(d \sin k_0 x + w(t, x), \varepsilon)) = 0,$$

$$G_2(w, d, \varepsilon) \equiv \int_0^{\omega} \int_0^{\pi} F(d \sin k_0 x + w(t, x), \varepsilon)(\tau, \xi) \sin k_0 \xi \, d\xi \, d\tau +$$

$$+ k_0 \int_0^{\omega} (X_0(d \sin k_0 x + w(t, x), \varepsilon)(\tau) +$$

$$+ (-1)^{k_0 + 1} X_1(d \sin k_0 x + w(t, x), \varepsilon)(\tau)) \, d\tau = 0.$$

All assumptions of I: Theorem 3.4.2 can be verified as in the proof of Theorem 1.2.2. Applying the implicit function theorem we see that $d^*(\varepsilon)$ and $w^*(\varepsilon)$ are continuous in $\varepsilon \in [-\varepsilon^*, \varepsilon^*]$. When we set $u^*(\varepsilon) = d^*(\varepsilon) \sin k_0 x + w^*(\varepsilon)$, we see that u^* is the solution of $(\mathscr{P}^{\varepsilon}_{\omega})$.

Remark 3.2.1 Let

$$F(u, \varepsilon)(t, x) = f(t, x, u(t, x), u_x(t, x), \varepsilon),$$

$$X_j(u, \varepsilon)(t) = \chi_j(t, u(t, 0), u(t, \pi), \varepsilon), \quad j = 1, 2,$$

and let $f = f(t, x, u_0, u_1, \varepsilon)$, $\chi_j = \chi_j(t, v_0, v_1, \varepsilon)$ be ω-periodic in t.

The assumptions of Theorem 3.2.1 are satisfied if f and its partial derivatives f_x, f_{u_0}, f_{u_1} are continuous on $\mathbf{R} \times [0, \pi] \times [-\varrho, \varrho]^2 \times [-\varepsilon_0, \varepsilon_0]$ (ϱ a positive number) and Lipschitz continuous in u_0 and u_1 uniformly with respect to t, x, and ε, while χ_j and their partial derivatives $\chi_{jt}, \chi_{jv_0}, \chi_{jv_1}$ are continuous on $\mathbf{R} \times [-\varrho, \varrho]^2 \times [-\varepsilon_0, \varepsilon_0]$ and Lipschitz continuous in v_0 and v_1, uniformly with respect to t and ε.

The assumptions (b) and (c) of Theorem 3.2.2 are satisfied if f together with its partial derivatives

$$D_x^{\beta} D_{u_0}^{\gamma_0} D_{u_1}^{\gamma_1} f, \quad \beta + \gamma_0 + \gamma_1 \leqq 2, \quad \beta \leqq 1$$

is continuous on $\mathbf{R} \times [0, \pi] \times [-|d_0^*| - \varrho, \ |d_0^*| + \varrho] \times [-k_0|d_0^*| - \varrho, k_0|d_0^*| + \varrho] \times [-\varepsilon_0, \varepsilon_0]$, and the χ_j together with their partial derivatives

$$D_t^{\alpha} D_{v_0}^{\gamma_0} D_{v_1}^{\gamma_1} \chi_j, \quad \alpha + \gamma_0 + \gamma_1 \leqq 2, \quad \alpha \leqq 1,$$

are continuous on $\mathbf{R} \times [-\varrho, \varrho]^2 \times [-\varepsilon_0, \varepsilon_0]$.

Remark 3.2.2 ŠŤASTNOVÁ and VEJVODA [84] study the problem $(\mathscr{P}_\omega)$ by the same method assuming that h_0 and h_1 belong to $C_\omega(\mathbf{R})$. They obtain a solution in the space $\{u \in C_\omega(Q) \cap C^{1,2}_{\mathrm{loc},\omega}(Q)$;

$$\sup_{(t,x)\in Q} |x(\pi - x)\, u_x(t, x)| < \infty\,, \quad \sup_{(t,x)\in Q} |x^2(\pi - x)^2\, u_{xx}(t, x)| < \infty\}\,.$$

Remark 3.2.3 BYKOV and GORŠKOV [8] investigate in detail the case when the bifurcation equation (3.2.3) has a root, but the expression in (d) vanishes at this root.

Problem 3.2.1 Using II: (3.1.3) investigate ω-periodic solutions to the equation (3.1.1) where x ranges over the whole space $\mathbf{R}$.

Hint: Suppose that $c > 0$ and use I: Theorem 3.1.1.

§ 4. Supplements and comments on the linear and weakly non-linear heat equation

4.1. *The adjoint problem method*

In this section we investigate a one-dimensional problem, slightly more general than that discussed in the previous paragraphs, with general boundary conditions (together with some special cases). We show how necessary conditions for the solubility of corresponding inhomogeneous problems can be derived in an elementary manner, namely, by making use of a formally adjoint problem. This approach is simply called *the adjoint problem method*. (To get necessary and sufficient conditions we had to use I: Theorem 1.5.1 (or the analogous theorem for Banach spaces, see e.g. I: [39], Sec. 4.6). But it would be necessary to fix spaces and domains of definition, to investigate the closedness of ranges (cf. Remark 2.3.4) and to determine adjoint operators. Unfortunately, the lack of room doesn't allow us to take this way.)

Suppose that the problem $(\mathscr{P}_\omega)$ is given by

$$(Lu)(t, x) = u_t(t, x) - (a(t, x)\, u_x(t, x))_x + b(t, x)\, u_x(t, \boldsymbol{x}) +$$

$$+ c(t, x)\, u(t, x) = g(t, x)\,, \quad (t, x) \in Q = \mathbf{R} \times (0, \pi)\,, \tag{4.1.1}$$

$$u(0, x) - u(\omega, x) = 0\,, \quad x \in [0, \pi]\,, \tag{4.1.2}$$

$$M(t) \begin{pmatrix} u(t, 0) \\ u_x(t, 0) \end{pmatrix} + N(t) \begin{pmatrix} u(t, \pi) \\ u_x(t, \pi) \end{pmatrix} = \begin{pmatrix} h_0(t) \\ h_1(t) \end{pmatrix}, \quad t \in \mathbf{R}\,, \tag{4.1.3}$$

where $a, b \in C^{0,1}_{\omega}(\mathbf{R} \times [0, \pi])$, $a(t, x) > 0$, $(t, x) \in Q$, $c \in C_{\omega}(\bar{Q})$, $g \in C_{\omega}(\bar{Q})$, $h_0, h_1 \in C_{\omega}(\mathbf{R})$, M, N are 2×2-matrices whose elements belong to $C_{\omega}(\mathbf{R})$, and

$$\operatorname{rank}(M, N) = 2, \quad t \in \mathbf{R}.$$

((M, N) denotes the 2×4-matrix formed by M and N.)

For any sufficiently smooth u and v we have

$$\int_0^{\omega} \int_0^{\pi} vLu \, d\xi \, d\tau = \int_0^{\omega} \int_0^{\pi} u(-v_t - (av_x)_x - (bv)_x + cv) \, d\xi \, d\tau +$$

$$+ \int_0^{\pi} (v(\omega, \xi) \, u(\omega, \xi) - v(0, \xi) \, u(0, \xi)) \, d\xi +$$

$$+ \int_0^{\omega} (-a(\tau, \pi) \, v(\tau, \pi) \, u_x(\tau, \pi) + a(\tau, \pi) \, u(\tau, \pi) \, v_x(\tau, \pi) +$$

$$+ b(\tau, \pi) \, u(\tau, \pi) \, v(\tau, \pi) + a(\tau, 0) \, v(\tau, 0) \, u_x(\tau, 0) -$$

$$- a(\tau, 0) \, u(\tau, 0) \, v_x(\tau, 0) - b(\tau, 0) \, u(\tau, 0) \, v(\tau, 0)) \, d\tau =$$

$$= \int_0^{\omega} \int_0^{\pi} uL^*v \, d\xi \, d\tau + \int_0^{\pi} u(0, \xi) \, (-v(0, \xi) + v(\omega, \xi)) \, d\xi +$$

$$+ \int_0^{\omega} (v(\tau, 0), v_x(\tau, 0), v(\tau, \pi), v_x(\tau, \pi)) \, A(\tau) \, .$$

$$. \, (u(\tau, 0), u_x(\tau, 0), u(\tau, \pi), u_x(\tau, \pi))' \, d\tau \, ,$$

where

$$A(t) = \begin{pmatrix} -B(t, 0) & 0 \\ 0 & B(t, \pi) \end{pmatrix}, \quad B(t, x) = \begin{pmatrix} b(t, x) & -a(t, x) \\ a(t, x) & 0 \end{pmatrix}, \quad (t, x) \in \bar{Q}.$$

Taking this and II: Sec. 1.2 into account, we define the formally adjoint problem $(\mathscr{P}^*_{\omega})$ to $(\mathscr{P}_{\omega})$ by

$$(L^*v)(t, x) = -v_t(t, x) - (a(t, x) \, v_x(t, x))_x - (b(t, x) \, v(t, x))_x +$$

$$+ c(t, x) \, v(t, x) = 0, \quad (t, x) \in Q, \tag{4.1.4}$$

$$v(0, x) - v(\omega, x) = 0, \quad x \in [0, \pi], \tag{4.1.5}$$

$$(v(t, 0), v_x(t, 0)) \, P(t) + (v(t, \pi), v_x(t, \pi)) \, Q(t) = 0, \quad t \in \mathbf{R}, \tag{4.1.6}$$

where the 2×2-matrices P and Q are determined by

$$-M(t) \, B^{-1}(t, 0) \, P(t) + N(t) \, B^{-1}(t, \pi) \, Q(t) = 0, \quad t \in \mathbf{R},$$

$$\operatorname{rank}(P', Q') = 2. \tag{4.1.7}$$

Observe that

$$B^{-1}(t, x) = a^{-2}(t, x)\begin{pmatrix} 0 & a(t, x) \\ -a(t, x) & b(t, x) \end{pmatrix}, \quad (t, x) \in Q .$$

The 2×2-matrices P_c and Q_c defining the complementary adjoint (spatial) boundary forms

$$(v(t, 0), v_x(t, 0))\, P_c(t) + (v(t, \pi), v_x(t, \pi))\, Q_c(t)$$

are determined by

$$\begin{aligned} &-M(t)\, B^{-1}(t, 0)\, P_c(t) + N(t)\, B^{-1}(t, \pi)\, Q_c(t) = I\,, \quad t \in \mathbf{R}\,, \\ &\operatorname{rank}\,(P_c', Q_c') = 2\,. \end{aligned} \tag{4.1.8}$$

Note that the matrices P, Q, P_c, and Q_c are chosen so that

$$A = \begin{pmatrix} P_c & P \\ Q_c & Q \end{pmatrix}\begin{pmatrix} M & N \\ M_c & N_c \end{pmatrix},$$

where M_c and N_c are suitable 2×2-matrices such that

$$\operatorname{rank}\begin{pmatrix} M & N \\ M_c & N_c \end{pmatrix} = 4\,.$$

Now, if in the formula for

$$\int_0^\omega \int_0^\pi vLu \, d\xi \, d\tau$$

obtained above u is a solution of $(\mathscr{P}_\omega)$, while v is a solution of the formally adjoint problem $(\mathscr{P}_\omega^*)$, we obtain the following theorem:

THEOREM 4.1.1 *The problem* $(\mathscr{P}_\omega)$ *has a solution only if*

$$\begin{aligned} &\int_0^\omega \int_0^\pi v(\tau, \xi)\, g(\tau, \xi)\, d\xi\, d\tau - \int_0^\omega ((v(\tau, 0), v_x(\tau, 0))\, P_c(\tau) + \\ &+ (v(\tau, \pi), v_x(\tau, \pi))\, Q_c(\tau))\, (h_0(\tau), h_1(\tau))'\, d\tau = 0 \end{aligned} \tag{4.1.9}$$

for every solution v *of* $(\mathscr{P}_\omega^*)$.

Now let us turn to some special cases of the problem $(\mathscr{P}_\omega)$. We shall investigate the problems $(\mathscr{P}_\omega^j)$, $j = 1, 2, 3, 4$ given by

$$u_t(t, x) - u_{xx}(t, x) + c\, u(t, x) = g(t, x)\,, \tag{4.1.10}$$

$$u(0, x) - u(\omega, x) = 0 \tag{4.1.11}$$

for $j = 1$

$$u(t, 0) = h_0(t)\,, \quad u(t, \pi) = h_1(t)\,, \tag{4.1.12}$$

for $j = 2$,

$$u_x(t, 0) + \sigma_0\, u(t, 0) = h_0(t)\,, \quad (\sigma_0, \sigma_1 \in \mathbf{R})\,,$$
$$u_x(t, \pi) + \sigma_1\, u(t, \pi) = h_1(t)\,, \tag{4.1.13}$$

for $j = 3$,

$$u_x(t, 0) + \sigma\, u(t, 0) = h_0(t)\,, \quad (\sigma \in \mathbf{R})\,,$$
$$u(t, \pi) = h_1(t)\,, \tag{4.1.14}$$

for $j = 4$

$$u(t, 0) - u(t, \pi) = 0\,,$$
$$u_x(t, 0) - u_x(t, \pi) = 0\,. \tag{4.1.15}$$

We find easily that the formally adjoint problems $(\mathscr{P}_\omega^{j*})$ are defined by

$$-v_t(t, x) - v_{xx}(t, x) + c\, v(t, x) = 0\,, \tag{4.1.16}$$

$$v(0, x) - v(\omega, x) = 0 \tag{4.1.17}$$

and $(j = 1)$

$$v(t, 0) = v(t, \pi) = 0\,, \tag{4.1.18}$$

$(j = 2)$

$$v_x(t, 0) + \sigma_0\, v(t, 0) = 0\,,$$
$$v_x(t, \pi) + \sigma_1\, v(t, \pi) = 0\,, \tag{4.1.19}$$

$(j = 3)$

$$v_x(t, 0) + \sigma\, v(t, 0) = 0\,,$$
$$v(t, \pi) = 0\,, \tag{4.1.20}$$

and $(j = 4)$

$$v(t, 0) - v(t, \pi) = 0\,,$$
$$v_x(t, 0) - v_x(t, \pi) = 0 \tag{4.1.21}$$

and the complementary adjoint forms are

$$(-v_x(t, 0), v_x(t, \pi))\,, \tag{4.1.22}$$

respectively,

$$(v(t, 0), -v(t, \pi)), \tag{4.1.23}$$

respectively,

$$(v(t, 0), v_x(t, \pi)), \tag{4.1.24}$$

respectively,

$$(-v_x(t, 0), v(t, 0)). \tag{4.1.25}$$

The reader can find readily that the problems $(\mathscr{P}_\omega^{j*})$ have "in general" only trivial solutions. Hence, the relation (4.1.9) imposes no restriction on g, h_0, and h_1. The situation in "critical" cases is described by the following lemmas the proofs of which follow immediately from Theorem 4.1.1, from the fact that the problems $(\mathscr{P}_\omega^{j*})$ have the same solutions as the homogeneous problems $(\mathscr{P}_\omega^j)$ which depend only on x, and from II: Examples 1.3.1–1.3.3.

LEMMA 4.1.1 *The problem* $(\mathscr{P}_\omega^1)$ *with* $c = -k_0^2$ *for some* $k_0 \in \mathbf{N}$ *has a solution only if*

$$\int_0^\omega \int_0^\pi g(\tau, \xi) \sin k_0\xi \, d\xi \, d\tau + k_0 \int_0^\omega (h_0(\tau) + (-1)^{k_0+1} h_1(\tau)) \, d\tau = 0 .$$

LEMMA 4.1.2 *The problem* $(\mathscr{P}_\omega^2)$ *with* $c = -\lambda_{k_0}$, *where* λ_{k_0} $(k_0 \in \mathbf{N} \cup \{0, -1, -2\})$ *is an eigenvalue of the problem* II: (1.3.3) (*see* II: *Example* 1.3.2, *p.* 43), *has a solution only if*

$$\int_0^\omega \int_0^\pi g(\tau, \xi) \, v_{k_0}(\xi) \, d\xi \, d\tau + \int_0^\omega (-h_0(\tau) \, v_{k_0}(0) + h_1(\tau) \, v_{k_0}(\pi)) \, d\tau = 0 ,$$

where v_{k_0} *is an eigenfunction corresponding to* λ_{k_0}.

LEMMA 4.1.3 *The problem* $(\mathscr{P}_\omega^3)$ *with* $c = -\lambda_{k_0}$, *where* λ_{k_0} $(k_0 \in \mathbf{N} \cup \{0, -1\})$ *is an eigenvalue of the problem* II: (1.3.9) (*see* II: *Example* 1.3.3, *p.* 45), *has a solution only if*

$$\int_0^\omega \int_0^\pi g(\tau, \xi) \, v_{k_0}(\xi) \, d\xi \, d\tau - \int_0^\omega (h_0(\tau) \, v_{k_0}(0) + h_1(\tau) \, v'_{k_0}(\pi)) \, d\tau = 0 ,$$

where v_{k_0} *is an eigenfunction corresponding to* λ_{k_0}.

LEMMA 4.1.4 *The problem* $(\mathscr{P}_\omega^4)$ *with* $c = -4k_0^2$ *for some* $k_0 \in \mathbf{N}$ *has a solution only if*

$$\int_0^\omega \int_0^\pi g(\tau, \xi) \cos 2k_0\xi \, d\xi \, d\tau = \int_0^\omega \int_0^\pi g(\tau, \xi) \sin 2k_0\xi \, d\xi \, d\tau = 0 .$$

Problem 4.1.1 Show (for example by means of the t-Fourier method) that the above conditions are also sufficient (g, h_0, and h_1 being sufficiently smooth).

4.2. *Comments on other results*

RABINOWITZ [76] states that the problem

$$u_t - u_{xx} = \varepsilon f(t, x, u, u_t, u_x, u_{tt}, u_{tx}, u_{xx}), \quad (t, x) \in \mathbf{R} \times (0, \pi),$$
$$u(t, 0) = u(t, \pi) = 0, \quad t \in \mathbf{R},$$
$$u(t + \omega, x) - u(t, x) = 0, \quad (t, x) \in \mathbf{R} \times (0, \pi)$$

has a smooth solution, provided that f is sufficiently smooth and $|\varepsilon|$ sufficiently small. This result is incidental to a similar assertion for the telegraph equation (see IV: Sec. 4.3).

FARLOW [12] proves the existence of a unique classical ω-periodic solution to the problem

$$u_t(t, x) - \sum_{j,k=1}^{n} a_{jk}(t, x)\, u_{x_k x_j}(t, x) -$$
$$- \sum_{j=1}^{n} b_j(t, x)\, u_{x_j}(t, x) - c(t, x)\, u(t, x) = g(t, x), \quad (t, x) \in \mathbf{R} \times \Omega, \quad (4.2.1)$$
$$\frac{\partial u}{\partial n}(t, x) + \sigma(t, x)\, u(t, x) = h(t, x), \quad (t, x) \in \mathbf{R} \times \partial\Omega,$$

where $\Omega \subset \mathbf{R}^n$ is a bounded domain,

$$\sum_{j,k=1}^{n} a_{jk}(t, x)\, \xi_j \xi_k \geqq m \sum_{j=1}^{n} \xi_j^2 \quad \text{for some} \quad m > 0,$$

and $-c(t, x) \geqq 0$. The technique is the same as in [80]. In [13], the result is extended to the corresponding weakly non-linear problem.

KONO [48], deals with the ω-periodic solutions of the problem given by (4.2.1) and $u(t, x) = h(t, x)$, $(t, x) \in \mathbf{R} \times \partial\Omega$, where $\Omega \subseteq \mathbf{R}^n$ is unbounded (in particular $\Omega = \mathbf{R}^n$) and $-c(t, x) \geqq \text{const.} > 0$. The proof is based on the fact that the classical ω-periodic solutions u_N on $\mathbf{R} \times (\Omega \cap B(0, N; \mathbf{R}^n))$, which are assumed to exist, converge as $N \to \infty$ to a unique classical ω-periodic solution.

§ 5. Comments on strongly non-linear parabolic equations

In what follows we assume tacitly that all functions in equations and boundary conditions are ω-periodic in t and that the operator

$$u \to \sum_{j,k=1}^{n} a_{jk}\, u_{x_k x_j} \quad \text{or} \quad u \to \sum_{j,k=1}^{n} (a_{jk}\, u_{x_k})_{x_j}$$

is elliptic, that is, $\sum_{j,k=1}^{n} a_{jk}(t, x, u)\, \xi_j \xi_k \geqq m(\varrho) \sum_{j=1}^{n} \xi_j^2$ for $|u| \leqq \varrho$, where m is a positive non-increasing function. We do not quote the assumptions on the smoothness of these functions and of a domain $\Omega \subset \mathbf{R}^n$ which is supposed to be bounded.

The solutions are frequently looked for in the space which we denote here by $C_\omega^{2+\alpha}(\bar{Q})$, $Q = \mathbf{R} \times \Omega$, and which is defined as follows. Given a function $u \in C_\omega(\bar{Q})$, we introduce the following norms:

$$\|u\|_\alpha = \sup \{|u(t, x)|; (t, x) \in Q\} +$$
$$+ \sup \{|u(t, x) - u(t', x')| \, (|t - t'| + \sum_{j=1}^{n} (x_j - x_j')^2)^{-\alpha/2} ;$$
$$(t, x), (t', x') \in Q, (t, x) \neq (t', x')\} ,$$

$$\|u\|_{1+\alpha} = \|u\|_\alpha + \sum_{j=1}^{n} \|u_{x_j}\|_\alpha ,$$

$$\|u\|_{2+\alpha} = \|u\|_{1+\alpha} + \sum_{j=1}^{n} \|u_{x_j}\|_{1+\alpha} + \|u_t\|_\alpha ,$$

where $0 < \alpha < 1$. We say that $u \in C_\omega^s(\bar{Q})$ if $\|u\|_s < \infty$ $(s = \alpha, 1 + \alpha, 2 + \alpha)$.

5.1. *Results of Prodi, Vaghi and Bange. (Differential inequality techniques)* Apparently, the first papers on strongly non-linear periodic parabolic problems are [73] and [74] by PRODI, who investigates ω-periodic solutions of the equation

$$u_t = u_{xx} + f(t, x, u, u_x) , \quad (t, x) \in \mathbf{R} \times (0, l) \tag{5.1.1}$$

with the boundary conditions

$$u(t, 0) = u(t, l) = 0 , \quad t \in \mathbf{R} . \tag{5.1.2}$$

Setting $\bar{f}(x, u, u_x) = \max_{t \in \mathbf{R}} f(t, x, u, u_x)$ and $\underline{f}(x, u, u_x) = \min_{t \in \mathbf{R}} f(t, x, u, u_x)$ he assumes that there are solutions u_1 and u_2 of the equations $u_1'' + \bar{f}(x, u_1, u_1') = 0$ and $u_2'' + \underline{f}(x, u_2, u_2') = 0$ such that $u_1(0) > 0$, $u_1(l) > 0$, $u_2(0) < 0$,

$u_2(l) < 0$ and $u_2(x) < u_1(x)$ for $x \in [0, l]$. Assuming that $|f(t, x, u, p)| \leqq c_M(|p|)$, $t \in \mathbf{R}$, $x \in [0, l]$, $p \in \mathbf{R}$, $|u| \leqq M$ (M a positive number), where c_M is a positive non-decreasing function and $c_M(|p|) = o(p^2)$ as $p \to \infty$, he proves the existence of a classical ω-periodic solution to the problem (5.1.1), (5.1.2). If f does not depend on u_x he shows that there is an ω-periodic solution of (5.1.1), (5.1.2), provided that $f(t, x, u)\, u^{-1} < K < \pi^2 l^{-2}$ for sufficiently large $|u|$.

In [75] similar results are proved for the problem given by (5.1.1) and

$$u_x(t, 0) = \varphi_1(t, u(t, 0)), \quad u_x(t, l) = \varphi_2(t, u(t, l)), \quad t \in \mathbf{R}.$$

Following Prodi [74], Vaghi proves in [86] and [88] the existence and uniqueness of a bounded, periodic or almost periodic solution to the problem (5.1.1), (5.1.2) with $f = f(t, x, u, p) = o(p^2)$ as $p \to \infty$. In [87], she studies the same questions for the equation $u_t - u_{xx} = f(t, x, u, u_x) + \varphi(t, x, u)\, u_x^2$.

The differential inequality techniques in [74] are also used by Bange in [3] and [4] for the study of the existence of ω-periodic solutions to (5.1.1) (in [3] f does not depend on u_x) with inhomogeneous Dirichlet conditions.

5.2. *Results of Kolesov, Klimov, Amann, Tsai, Deuel and Hess. (Use of upper and lower solutions)*

In [47] (preceded by preliminary communications [41] and [42]), Kolesov deals with the problem of the existence, uniqueness, and stability of a classical ω-periodic solution to the equation

$$Lu \equiv u_t - \sum_{j,k=1}^{n} a_{jk}(t, x, u)\, u_{x_k x_j} = f(t, x, u, \operatorname{grad} u), \quad (t, x) \in \mathbf{R} \times \Omega; \tag{5.2.1}$$

where

$$\operatorname{grad} u = (u_{x_1}, \ldots, u_{x_n}),$$

with the boundary condition

$$u(t, x) = 0, \quad (t, x) \in \mathbf{R} \times \partial\Omega. \tag{5.2.2}$$

Under the hypotheses

(a) $u\, f(t, x, u, 0) \leqq c_1 + c_2 u^2$, $(t, x, u) \in \mathbf{R} \times \bar{\Omega} \times \mathbf{R}$;

(b) $|f(t, x, u, p)| \leqq d_1(\varrho) + d_2(\varrho)\, |p|^2$, $(t, x, u, p) \in \mathbf{R} \times \bar{\Omega} \times [-\varrho, \varrho] \times \mathbf{R}^n$, $\varrho > 0$;

(c) there exist functions v and w (the so-called *lower solution* and *upper solution*, respectively) that satisfy

$$Lv \leqq f(t, x, v, \operatorname{grad} v), \quad Lw \geqq f(t, x, w, \operatorname{grad} w), \quad (t, x) \in [0, \omega] \times \bar{\Omega},$$

$$v(t, x) \leqq 0, \quad w(t, x) \geqq 0, \quad (t, x) \in [0, \omega] \times \partial\Omega,$$

$$v(0, x) \leqq v(\omega, x), \quad w(0, x) \geqq w(\omega, x), \quad w(0, x) \geqq v(0, x), \quad x \in \bar{\Omega},$$

the author proves that there exists an ω-periodic solution u to (5.2.1), (5.2.2), for which, moreover, $v(t, x) \leqq u(t, x) \leqq w(t, x)$ for $(t, x) \in \mathbf{R} \times \bar{\Omega}$. (This implies as an interesting corollary the following assertion: If for some $\varrho_1, \varrho_2 \geqq 0$ and for all $t \in \mathbf{R}$ and $x \in \Omega$ the inequalities $f(t, x, -\varrho_1, 0) \geqq 0, f(t, x, \varrho_2, 0) \leqq 0$ hold, then the problem (5.2.1), (5.2.2) has at least one ω-periodic solution u for which $-\varrho_1 \leqq u(t, x) \leqq \varrho_2$.) Some modifications of the above theorem are stated for the case when the a_{jk} are independent of u and f is linear in u and grad u. In the same paper the uniqueness and asymptotic stability are also studied. In another part of the paper the existence of positive ω-periodic solutions is dealt with for a more special type of the equation (5.2.1). The author uses the Poincaré method and the complete continuity of the translation (or "quasi-translation") operator.

The papers [44] and [46] of the same author are two almost identical preliminary communications (without proofs) on the existence of a stable ω-periodic solution to the problem (5.2.1), (5.2.2) (where the a_{jk} do not depend on u); they are based on a general theorem concerning the stability of a fixed point of a completely continuous non-linear operator.

KLIMOV [32] is interested in ω-periodic solutions to (5.2.1), (5.2.2) with the a_{jk} depending on grad u.

AMANN [1] deals with ω-periodic solutions to the problem

$$Lu \equiv u_t + Eu = f(t, x, u, \operatorname{grad} u) \quad \text{on} \quad \mathbf{R} \times \Omega, \tag{5.2.3}$$

$$Bu \equiv b_0 u + \delta \frac{\partial u}{\partial \beta} = 0 \quad \text{on} \quad \mathbf{R} \times \partial\Omega,$$

where

$$Eu = -\sum_{j,k=1}^{n} a_{jk}(t, x)\, u_{x_k x_j} + \sum_{j=1}^{n} b_j(t, x)\, u_{x_j} + c(t, x)\, u, \tag{5.2.4}$$

either $\delta = 0$ and $b_0 = 1$, or $\delta = 1$ and b_0 is a non-negative function on $\partial\Omega$ ($\partial/\partial\beta$ means the derivative in a non-tangential direction). The function f satisfies the following growth condition: there exists a non-negative function c such that

$$|f(t, x, u, p)| \leqq c(\varrho)(1 + |p|^2)$$

for every $\varrho \geqq 0$ and $(t, x, u, p) \in \mathbf{R} \times \bar{\Omega} \times [-\varrho, \varrho] \times \mathbf{R}^n$.

Using fixed-point theorems in appropriate ordered Banach spaces and the compactness of the translation operator he proves the existence of at least one solution $u \in C^{2+\alpha}_{\omega}(\mathbf{R} \times \bar{\Omega})$ provided that there exist a lower solution v and an upper solution w that are B-related (a lower solution v and an upper solution w are said to be B related if there exists a function u with $Bu = 0$, such that $v(0, .) \leqq u \leqq w(0, .)$). This hypothesis is fulfilled for example if there exist non-negative functions a and b such that

$$f(t, x, u, p) \leqq a(t, x) u + b(t, x), \quad (t, x, u, p) \in \mathbf{R} \times \bar{\Omega} \times \mathbf{R}^{+} \times \mathbf{R}^{n},$$

$$f(t, x, u, p) \geqq a(t, x) u - b(t, x), \quad (t, x, -u, p) \in \mathbf{R} \times \bar{\Omega} \times \mathbf{R}^{+} \times \mathbf{R}^{n},$$

$c \geqq 0$ (if $\delta = 1$, $b_0 = 0$, then $c > 0$), the coefficients of $E - a$ are independent of t and the smallest eigenvalue of the problem $(E - a)\, v = \lambda v$, $Bu = 0$ is positive.

The author further shows that there exist at least three periodic solutions if there are two such pairs of lower and upper solutions which are such that the initial value of the upper solution of one pair is not greater than the initial value of the lower solution of the other pair.

The corresponding system of equations is also investigated (with more restrictive conditions on f).

Many results of other authors are generalized by TSAI [85], where the equation (5.2.3) with inhomogeneous Dirichlet condition is studied. In addition, much attention is paid to the corresponding systems of equations.

The techniques of upper and lower solutions are used also by DEUEL [10] to study of the Dirichlet and Neumann problems for the equation

$$u_t - \sum_{j=1}^{n} \frac{\partial}{\partial x_j} A_j(t, x, u, \operatorname{grad} u) + f(t, x, u, \operatorname{grad} u) =$$

$$= g(t, x), \quad (t, x) \in \mathbf{R} \times \Omega,$$

where the functions A_j satisfy conditions of Leray-Lions type (see II: [15], p. 322) and

$$|f(t, x, u, p)| \leqq k(t, x) + c|p|^{q-1},$$

$$(t, x, u, p) \in \mathbf{R} \times \Omega \times [-\varrho_0, \varrho_0] \times \mathbf{R}^{n}, \quad (q \geqq 2, k \in L_{q/(q-1)}).$$

The weak ω-periodic solutions are looked for in L_q-spaces. The corresponding variational inequality is also treated.

The results for the Dirichlet problem are extended by DEUEL and HESS [11] to the case when

$$|f(t, x, u, p)| \leqq k(t, x) + c|p|^{q-\delta}$$

with some $\delta > 0$.

5.3. *Results of Šmulev, Fife, Kusano, Kružkov, Gaines and Walter. (A priori estimates techniques)*

Šmulev in [80] first treats the equation

$$Lu \equiv u_t + Eu = g \quad \text{on} \quad \mathbf{R} \times \Omega \tag{5.3.1}$$

with the boundary condition

$$u(t, x) = h(t, x), \quad (t, x) \in \mathbf{R} \times \partial\Omega \quad (h \in C_\omega^{2+\alpha}(\mathbf{R} \times \bar{\Omega})) . \tag{5.3.2}$$

The operator E is given by (5.2.4). The author proves the following theorem (by continuation with respect to the parameter (see I: Theorem 3.5.1)).

Theorem 5.3.1 *Let a_{jk} b_j, c, g be Hölder continuous, $c(t, x) \geqq c_0 > 0$. Then the problem (5.3.1), (5.3.2) has a unique solution $u \in C_\omega^{2+\alpha}(\mathbf{R} \times \bar{\Omega})$ and* $\|u\|_{2+\alpha} \leqq \text{const.} (\|g\|_\alpha + \|h\|_{2+\alpha})$.

Further, the author proves the existence of a solution $u \in C_\omega^{2+\alpha}(\mathbf{R} \times \bar{\Omega})$ to the equation

$$u_t - \sum_{j,k=1}^{n} (a_{jk}(t, x, u)\, u_{x_k})_{x_j} = b(t, x, u, \operatorname{grad} u) \tag{5.3.3}$$

with the boundary condition (5.3.2) under the main assumptions

$$u\, b(t, x, u, p)|_{p=0} \leqq K_1 - K_2 u^2 \quad \text{for} \quad u \in \mathbf{R}, |b| + |\operatorname{grad} b| +$$

$$+ |b_u| \leqq K_1(|p|^2 + 1), \quad |\operatorname{grad}_p b| \leqq K_2(|p| + 1) \quad \text{for} \quad |u| \leqq M, p \in \mathbf{R}.$$

In this case the author applies the Leray-Schauder theorem (see I: Theorem 3.5.2) connecting the given equation with the simple heat equation:

$$v_t = \sum_{j,k=1}^{n} (\lambda\, a_{jk}(t, x, u)\, v_{x_k} + (1 - \lambda)\, \delta_{jk} v_{x_k})_{x_j} +$$

$$+ \lambda f(t, x, u, \operatorname{grad} u), \quad \lambda \in [0, 1] . \tag{5.3.4}$$

To verify the hypotheses of the Leray-Schauder theorem the author derives a number of a priori estimates for (5.3.4).

In [81] similar results are proved for the equations (5.3.1) and (5.2.1) with the Newton boundary condition.

Fife in [14] and [15] deals with the ω-periodic solutions to the problem

$$u_t = \sum_{j,k=1}^{n} (a_{jk}(t, x, u)\, u_{x_k})_{x_j} + \sum_{j=1}^{n} b_j(t, x, u, \operatorname{grad} u)\, u_{x_j} +$$

$$+ c(t, x, u, \operatorname{grad} u)\, u , \quad (t, x) \in \mathbf{R} \times \Omega ,$$

$$u = h \quad \text{on} \quad \mathbf{R} \times \partial\Omega$$

and obtains a result similar to that of ŠMULEV [80]. The proof is based on a preceding general discussion of the existence of a unique bounded global solution.

The results of ŠMULEV and FIFE are generalized by KUSANO in [54] and [55] to systems of parabolic equations.

KRUŽKOV proves in [53] the existence of a solution $u \in C_\omega^{2+\alpha}(\mathbf{R} \times [-l, l])$ to the inhomogeneous Dirichlet problem for the equation

$$u_t - a(t, x, u, u_x)\, u_{xx} = b(t, x, u, u_x)\,. \tag{5.3.5}$$

The following assumptions are made: $a = a(t, x, u, p) \geqq a_0 > 0$, $|b| \leqq \leqq aKp^2 + H$ for $|u| \leqq M$, $p \in \mathbf{R}$, $u\, b(t, x, u, 0) < 0$ for $|u| \geqq M$.

Further, he studies the equation

$$u_t = a(t, x, u, u_x, u_{xx})$$

under the assumptions: $a_r(t, x, u, p, r) \geqq a_0 > 0$, $a(t, x, u, \xi, -K\xi^2) \leqq H$, $-a(t, x, u, \xi, K\xi^2) \leqq H$ for $|u| \leqq M$, $\xi \in \mathbf{R}$, $|a_x| + a \leqq a_r K_0(r^2 + 1)$, $|a_t| \leqq \leqq K_0\, a_r(|r|^{2+\varepsilon} + 1)$, $|a_u| \leqq a_r K_0(|r|^{1+\varepsilon} + 1)$ for $|u| \leqq M$, $|p| \leqq M_1$, $r \in \mathbf{R}$, $(\varepsilon < 1)$, $u\, a(t, x, u, 0, 0) < 0$ for $|u| \geqq M$.

GAINES and WALTER [16] look for solutions $u \in C_\omega^{2+\alpha}(\mathbf{R} \times [0, 1])$ to the Dirichlet problem for (5.3.5) under the main assumptions: $a = a(t, x, u, p) > 0$ on $\mathbf{R} \times [0, 1] \times \mathbf{R}^2$,

$$\operatorname{sgn} u \,.\, \frac{b(t, x, u, p)}{a(t, x, u, p)} \leqq g(u, p) \text{ with } g \text{ appropriately determined}$$

$$|b|\,,\ \ |b_x| \leqq a\, K_1 p^2\,,\ \ |b_u|\,,\ \ |a_x + b_p| \leqq a\, K_1 |p|\,,\ \ ap^2 \geqq K_2\,,$$

$$|a_u| \leqq a\, K_1\,,\ \ |a_p| \leqq a\, K_3 \ \text{ for }\ |u| \leqq M\,,\ \ |p| \geqq M_1\,.$$

Their result is derived with help of some a priori estimates, the theory of differential inequalities, and the Leray-Schauder theorem (see I: Theorem 3.5.2).

5.4. *Results of Šmulev, Mal'cev, Walter and Knolle. (Methods of discretizations)*

The paper [78] by ŠMULEV is devoted to the study of ω-periodic solutions to the equation

$$u_t = a(t, x, u)\, u_{xx} + b(t, x, u, u_x)\,, \quad (t, x) \in \mathbf{R} \times (0, l) \tag{5.4.1}$$

with various boundary conditions (with a generalization to n-dimensional problems) via the difference-differential (Rothe) method. The proofs are only briefly sketched.

MAL'CEV in [58] and [61] investigates the equation (5.1.1) with the respective boundary conditions

$$u_x(t, 0) - \psi_0(t, u(t, 0)) = 0\,,$$
$$u_x(t, l) + \psi_1(t, u(t, l)) = 0\,, \quad t \in \mathbf{R}\,,$$

and

$$u_t(t, 0) = \psi_1(t, u\,(t, 0)\,,\; u_x(t, 0))\,,$$
$$u_t(t, l) = \psi_2(t, u(t, l)\,,\; u_x(t, l))\,, \quad t \in \mathbf{R}\,,$$

replacing the given problem by a system of difference-differential equations in which the variable t is discretized. In [62] the latter problem is solved by discretizing now with respect to the variable x.

WALTER in [89] proves the existence of a solution $u \in C^{2+\alpha}_{\omega}(\mathbf{R} \times [0, l])$ to the Dirichlet problem for the equation

$$u_t - u_{xx} = f(t, x, u)\,, \quad (t, x) \in \mathbf{R} \times (0, l) \tag{5.4.2}$$

under the only main assumption $f_u(t, x, u) < K < \pi^2\, l^{-2}$ on $\mathbf{R} \times [0, l] \times \mathbf{R}$. Discretizing the spatial variable he replaces (5.4.2) by a system of ordinary differential equations from whose periodic solutions an approximation to the required solution is determined by means of linear interpolation.

The method of discretization with respect to the variable x is also used by KNOLLE [33] in a study of the problem given by (5.4.1) and

$$u_x(t, 0) = -\Phi(\alpha(t)\,,\; u(t, 0))\,,\; t \in \mathbf{R}\,,$$
$$u_x(t, l) = \Phi(\beta(t)\,,\; u(t, l))\,, \quad t \in \mathbf{R}\,.$$

5.5. *Results of Palmieri, Nakao, Nanbu, Biroli and Zecca. (Problems with special type of non-linearity)*

In [71] PALMIERI deals with the existence of ω-periodic solutions to the problem

$$u_t = \sum_{j,k=1}^{n} (a_{jk}(x)\, u_{x_k})_{x_j} - c(x)\, u + g(t, x) - \beta(u)\,, \quad (t, x) \in \mathbf{R} \times \Omega\,, \tag{5.5.1}$$

$$u = 0 \quad \text{on} \quad \mathbf{R} \times \partial\Omega\,, \tag{5.5.2}$$

where β is a non-decreasing function defined on $(a, b) \subset \mathbf{R}$, with $\beta(a+) = -\infty$ if $a > -\infty$ and $\beta(b-) = \infty$ if $b < \infty$. Using an appropriate definition of a solution (according to the character of β) she bases her proof on the contractibility of the translation operator.

NAKAO [68] studies bounded, periodic and almost periodic solutions of the problem (5.5.1), (5.5.2) without any monotonicity assumptions on β. NAKAO and NANBU in [69] and [70] deal with the solutions from the same classes for the equation

$$u_t + Eu + \beta(t, x, u) + g(t, x) = 0\,, \quad (t, x) \in \mathbf{R} \times \Omega\,,$$

with Dirichlet or Newton boundary conditions, where E is an elliptic operator of the form (5.2.4), for the Dirichlet problem for the parabolic equation with an elliptic operator of order $2m$. Again, the monotonicity of β is not required, but as in [68], g is supposed to be small.

BIROLI in [5] proves the existence of generalized periodic solutions to the Dirichlet problem for the equation

$$u_t - \Delta u + \beta(u) = g(t, x)\,, \quad (t, x) \in \mathbf{R} \times \Omega\,,$$

where β is defined on $\mathbf{R}$ and $\beta(\xi)\, \xi \geqq 0$.

ZECCA in [90] obtains generalized solutions that are 2π-periodic in both variables to the "many-valued" equation

$$u_t - u_{xx} - cu \in F(t, x, u)\,, \quad (t, x) \in \mathbf{R}^2$$

and applies this result to the equation

$$u_t - u_{xx} - cu = f(t, x, u) + g(t, x)\, \varphi(v)$$

with a parameter v and $c \neq k^2$, $k \in \mathbf{N}$.

5.6. *Results of Brézis and Nirenberg, Šťastnová and Fučík. (Critical cases)*

BRÉZIS and NIRENBERG [6] deal with the Dirichlet problem for the equation

$$u_t + Eu - cu + \beta(t, x, u) = g(t, x)\,, \quad (t, x) \in \mathbf{R} \times \Omega\,,$$

where E is a positive definite self-adjoint elliptic operator of the form (5.2.4) and c is an eigenvalue λ_k of E. To take advantage of the property of the first eigenvalue λ_1 (which is simple, with the corresponding eigenfunction v_1 positive), they investigate the cases $c = \lambda_1$ and $c = \lambda_k$ $(k > 1)$ separately. The existence of a generalized ω-periodic solution is established under the following assumptions:

(I) $c = \lambda_1$ and β (continuous in u) satisfies

$$u\,\beta(t, x, u) \geqq -\delta(t, x)\,|u| - d(t, x)\,, \quad (\delta \in L_2, d \in L_1) \tag{5.6.1}$$

and

$$\int_0^\omega \int_\Omega \beta_+ v_1 \,\mathrm{d}x\,\mathrm{d}t > \int_0^\omega \int_\Omega g v_1 \,\mathrm{d}x\,\mathrm{d}t > \int_0^\omega \int_\Omega \beta_- v_1 \,\mathrm{d}x\,\mathrm{d}t$$

where

$$\beta_+(t, x) = \liminf_{u \to \infty} \beta(t, x, u)\,, \quad \beta_-(t, x) = \limsup_{u \to -\infty} \beta(t, x, u)\,;$$

(II) $c = \lambda_k$ for some $k > 1$ (more precisely, let (1.3.8) holds with $k_0 \in \mathbf{N} \setminus \{1\}$), the function β satisfies (5.6.1),

$$|\beta(t, x, u)| \leqq \gamma |u| + b(t, x)\,, \quad (\gamma > 0, b \in L_2)\,, \tag{5.6.2}$$

$$\int_0^\omega \int_\Omega \beta_+ v^+ \,\mathrm{d}x\,\mathrm{d}t - \int_0^\omega \int_\Omega \beta_- v^- \,\mathrm{d}x\,\mathrm{d}t > \int_0^\omega \int_\Omega g v \,\mathrm{d}x\,\mathrm{d}t$$

(here

$$v^+ = \max\{v, 0\}\,, \quad v^- = \max\{-v, 0\})$$

for $v = v_{k_0 + l}$, $l = 0, \ldots, r - 1$ (see Sec. 1.3), and finally, one of the conditions (a) or (b) is satisfied:

(a) (5.6.2) holds for every $\gamma > 0$ (with $b = b_\gamma$),

(b) $\beta_- \leqq \beta_+$ and (5.6.2) holds with $\gamma < \alpha$, where $\alpha = \lambda_{k_0} - \lambda_{k_0 - 1}$.

Regularity results are also stated.

Šťastnová and Fučík [83] study ω-periodic solutions to the Dirichlet problem for the equations

$$u_t - u_{xx} - cu + \beta(u) = g(t, x)\,, \quad (t, x) \in \mathbf{R} \times (0, \pi)$$

and

$$u_t - u_{xx} - \mu u^+ + \nu u^- + \beta(u) = g(t, x)\,, \quad (t, x) \in \mathbf{R} \times (0, \pi)$$

where $c, \mu, \nu \in \mathbf{R}$, with various assumptions of β which is supposed to be continuous and bounded (or growing at most linearly). The critical cases $c = k^2$ ($k \in \mathbf{N}$) are also discussed. In particular, in [82] the same authors treat the case $c = 1$.

5.7. *Results of Klimov, Krasnosel'skiĭ and Sobolevskiĭ. (Eigenvalue problem)*

KLIMOV in [31] finds conditions ensuring that the problem

$$u_t - \sum_{j,k=1}^{n} a_{jk}(t, x, u, \operatorname{grad} u)\, u_{x_k x_j} +$$

$$+ \sum_{j=1}^{n} b_j(t, x, u, \operatorname{grad} u)\, u_{x_j} + c(t, x, u, \operatorname{grad} u)\, u =$$

$$= \lambda f(t, x, u, \operatorname{grad} u, \lambda)\,, \quad (t, x) \in \mathbf{R} \times \Omega\,, \tag{5.7.1}$$

$$u = 0 \quad \text{on} \quad \mathbf{R} \times \partial\Omega \tag{5.7.2}$$

has a positive ω-periodic solution for at least one λ or for $\lambda \in (0, \lambda_0)$.

KRASNOSEL'SKIĬ and SOBOLEVSKIĬ in [52] study the existence of ω-periodic solutions to the parabolic equation

$$Eu \equiv \sum_{j,k=1}^{n} (a_{jk}(x)\, u_{x_k})_{x_j} = u_t - f(t, x, u, \lambda)\,, \quad (t, x) \in \mathbf{R} \times \Omega\,, \tag{5.7.3}$$

with the boundary condition (5.7.2), where $f(t, x, 0, \lambda) \equiv 0$. Assuming $a(x, \lambda) = f_u(t, x, 0, \lambda)$ the authors assert that the problem (5.7.3), (5.7.2) has at least one bifurcation point, provided that $\lim_{\lambda \to -\infty} a(x, \lambda) = -\infty$ and $\lim_{\lambda \to \infty} a(x, \lambda) = \infty$. If λ_0 is a bifurcation point, then clearly 0 is an eigenvalue of the operator E_{λ_0}, $E_\lambda u = Eu + a(x, \lambda)\, u$. The authors show that the converse is also true under some additional hypotheses.

5.8. *Results of other authors*

BROWDER [7] applies a general theorem on monotone operators to a periodic problem for the non-linear equation of a parabolic type

$$u_t + \sum_{|\alpha| \leqq m} D_x^\alpha A_\alpha(t, x, u, \ldots, D_x^m u) = g \quad \text{on} \quad \mathbf{R} \times \Omega\,.$$

SEIDMAN [77] studies the existence and stability of a periodic solution to the problem

$$u_t = \sum_{k=1}^{n} (\gamma(|\operatorname{grad} u|)\, u_{x_k})_{x_k} \quad \text{on} \quad \mathbf{R} \times \Omega\,, \quad u = \varphi \quad \text{on} \quad \mathbf{R} \times \partial\Omega\,,$$

where, $r\,\gamma(r) = G(r)$, G is positive, continuous and non-decreasing on $(0, \infty)$, $G(r) \geqq M_1(1 + r^\alpha)$, $r > 0$, $G(r) \geqq M_2 r^\alpha$, $r \geqq r_0$ $(\alpha > 0)$. He makes use of a theorem of Browder related to I: Theorem 3.3.1.

KRASNOSEL'SKIĬ [51] derives some properties of translation operators and indicates how to apply his abstract results to the problem (5.2.1), (5.2.2).

CHERKAS in [9] investigates classical periodic (as well as bounded and almost periodic) solutions of the equation

$$u_t - \sum_{j,k=1}^{n} a_{jk}(x)\, u_{x_k x_j} + \sum_{j=1}^{n} b_j(x)\, u_{x_j} + c(x)\, u = f(t, x, u)\,, \tag{5.8.1}$$

where

$$f(t, x, u) = \sum_{j=0}^{\infty} d_j(t, x)\, u^j(t, x) \quad \text{and} \quad (t, x) \in \mathbf{R} \times \mathbf{R}^n\,.$$

The author makes use of the theory of semigroups like TAAM in VII: [65] and [66].

KALLINA [21] proves under rather complicated hypotheses the existence of ω-periodic solutions to the equation

$$u_{xx} = a(t, x, u, \varepsilon)\, u_t + b(t, x, u)\, u_x + c(t, x, u)\, u + g(t, x)$$

on $\mathbf{R} \times (0, l)$ with the boundary conditions $u(t, 0) = h_0(t)$, $u(t, l) = h_1(t)$.

MONARI in [66] proves the existence of an ω-periodic solution $u \in C_\omega(\mathbf{R}; \dot{H}^1(\Omega)) \cap H^1(\mathbf{R}; H^0(\Omega))$ to the problem

$$u_t - \sum_{j,k=1}^{n} (a_{jk}(x)\, u_{x_k})_{x_j} + c(x)\, u + \beta(\|u(t - \tau, .)\|_{H^0(\Omega)})\, u = g(t, x)\,,$$

$(t, x) \in \mathbf{R} \times \Omega$, $u = 0$ on $\mathbf{R} \times \partial\Omega$, where β is a continuous non-negative function on $\mathbf{R}^+$. In [67] he obtains a similar result for a more general problem.

5.9. *Comments on papers of an applied character*

GOR'KOV [17], KOLESOV [43], [45] and PENDJUR [72] deal with periodic solutions to the problem given by the equation (5.8.1) with various types of boundary conditions when the right-hand side f is either $f_1(x)$ or $f_2(x)$ in accordance with a prescribed law of switching over. This problem describes mathematically some oscillating regimes arising, for example, in relay regulation of temperature or in some electrochemical processes.

KOČINA [34] looks for positive ω-periodic solutions to the Burgers equation $u_t + uu_x - u_{xx} = 0$ on $\mathbf{R} \times \mathbf{R}^+$ with the boundary conditions $u(t, 0) = h(t)$, $\lim_{x \to \infty} u(t, x) = u_\infty \leqq 0$. Putting $u = -2v_x v^{-1}$ she carries over the above equation to the linear one $v_t - v_{xx} = 0$. In [36]–[40] she studies the existence of periodic solutions to the autonomous problem $u_t - a^2 u_{xx} = 0$ on $\mathbf{R} \times \mathbf{R}^+$, $\lim_{x \to \infty} u(t, x) = 0$, $u_x(t, 0) = \chi(u(t, 0), u_t(t, 0))$, where χ is a many-valued function of a prescribed (serpent-like) form. The period depends on the solution found. Finally, in [35] she treats an analogous problem for

$$u_t + g(u)\, u_x = \varphi(u)\,, \quad u_x(t, 0) = \chi(u(t, 0), u_t(t, 0))\,.$$

KOPELL and HOWARD in [50] and [20] treat the parabolic system $u_t - \Delta u = f(u)$ and look for a solution in the form of plane waves.

§ 6. Comments on the Navier-Stokes equations and related problems

6.1. *The non-autonomous Navier-Stokes equations*

Many authors have investigated the existence of solutions to the Navier-Stokes equations

$$\frac{\partial u}{\partial t} - \nu \Delta u + (u, \nabla)\, u + \frac{1}{\varrho} \nabla p = g\,, \tag{6.1.1}$$

where

$$\nabla = \left(\frac{\partial}{\partial x_1}, \ldots, \frac{\partial}{\partial x_n}\right)$$

so that

$$(u, \nabla)\, u = \left(\sum_{j=1}^{n} \frac{\partial u_1}{\partial x_j} u_j, \ldots, \sum_{j=1}^{n} \frac{\partial u_n}{\partial x_j} u_j\right),$$

with the conditions

$$\operatorname{div} u = 0 \tag{6.1.2}$$

and with the boundary conditions

$$u \big| \partial\Omega = \psi \tag{6.1.3}$$

or in particular

$$u \big| \partial\Omega = 0\,, \tag{6.1.4}$$

which describe the motion of an incompressible fluid (u being the velocity and p the pressure of the fluid).

The system (6.1.1), (6.1.2) is often transformed to a parabolic system of the form

$$\frac{\partial u}{\partial t} + Lu + N(u) = h\,, \tag{6.1.5}$$

where $Lu = -\nu P_\sigma \Delta u, N(u) = P_\sigma((u, \nabla) u)$; P_σ is the projection onto the so called space of solenoidal functions H_σ (that is roughly speaking, onto the space of vector-valued functions u such that $\operatorname{div} u = 0$). The system (6.1.5) is in many respects similar to equations investigated in the preceding paragraphs. The operator $-P_\sigma \Delta$ is called the *Stokes operator* in what follows.

Serrin in [118] investigates the equations (6.1.1), (6.1.2) with the boundary conditions (6.1.3) in a bounded domain $\Omega = \Omega(t) \subset \mathbf{R}^3$ provided that $\Omega(t + \omega) = \Omega(t)$, $g(t + \omega) = g(t)$. The author assumes that the initial boundary-value problem is soluble and that there is at least one exponentially stable solution. The exponential stability is guaranteed by conditions imposed on the value of the Reynolds number of the flow. He proves the existence, uniqueness, and exponential stability of a strong periodic solution.

Using the Brouwer fixed-point theorem (see I: Theorem 3.2.1) Judovič [100] derives sufficient conditions for the existence of a generalized periodic solution of the equation

$$\dot{x} + Ax + Kx = g ,$$

where x is an element of a separable Hilbert space, A is a linear symmetric positive densely defined operator, and K is a non-linear non-negative operator. This theorem is applied to the problem (6.1.1), (6.1.2), (6.1.3) in a bounded domain $\Omega \subset \mathbf{R}^2$ or $\Omega \subset \mathbf{R}^3$.

Prodi [112] considers the problem (6.1.1), (6.1.2), (6.1.4) with $\Omega \subset R^2$ bounded. He derives first some properties of a weak solution to the Cauchy problem and then, using the Poincaré method and the Tihonov fixed-point theorem (see I: Theorem 3.2.3), he proves the existence of a periodic solution.

Making use of the Faedo-Galerkin method, Prouse in [113] (and for more general boundary conditions in [114]) proves the existence of weak periodic solutions to (6.1.1), (6.1.2), (6.1.4) in a bounded domain $\Omega \subset \mathbf{R}^n$. The same method is used by Lions in [105].

Zaretti [124] treats the system (6.1.1), (6.1.2) in a bounded domain $\Omega \subset \mathbf{R}^2$ with the boundary $\partial\Omega = \Gamma_1 \cup \Gamma_2 \cup \Gamma_3$, where

$$\Gamma_1 = \{(x_1, x_2); x_1 = 0, x_2 \in (0, k)\} ,$$

$$\Gamma_2 = \{(x_1, x_2); x_1 = l, x_2 \in (0, k)\} .$$

The boundary conditions are

$$\tfrac{1}{2}|u(t, x)|^2 + p(t, x) = \alpha_i(t, x) \, (x \in \Gamma_i, 0 \leqq t \leqq T, i = 1, 2)$$

$$p(t, x) = \beta(t, x) \, u(t, x) \, \nu(x) \, |u(t, x) \, \nu(x)| \, (x \in \Gamma_3, 0 \leqq t \leqq T)$$

$$|u(t, x)| = |u(t, x) \, \nu(x)| \, , \, (x \in \Gamma, \, 0 \leqq t \leqq T) \, ,$$

where the α_i and β are given functions $\beta \geqq 0$, and ν is the outward normal. By a method similar to that of PROUSE [113], the author proves the existence of at least one periodic solution.

MORIMOTO [109] looks for a periodic solution of (6.1.1), (6.1.2), (6.1.3) in the domain $\Omega = \Omega(t) \subset \mathbf{R}^m$ $(m = 2, 3)$, writing $u = \sum_k u_k(t)\, v_k(x)$, where v_k are eigenfunctions of the Stokes operator.

Some authors consider a more general (the so called reproductive) property of the Navier-Stokes equations. We say that the problem (6.1.1), (6.1.2), (6.1.3) is reproductive in t_1 if there exists an initial value u_0 such that the corresponding solution u satisfies the condition $u(t_1) = u_0$ (we call this solution reproductive). If this problem is reproductive in every t_1, we say that it is reproductive. Clearly, if the function g, the boundary conditions, and the domain Ω are time-periodic with the same period and if the corresponding problem is reproductive, then there exists a periodic solution. KANIEL and SHINBROT in [103] investigate the reproductive property of the problem (6.1.1), (6.1.2), (6.1.4), with g small enough in a bounded domain $\Omega \subset \mathbf{R}^3$. (The method can also be modified for the case $\Omega \subset \mathbf{R}^2$.) A simpler proof of the reproductivity of (6.1.1), (6.1.2), (6.1.4) and of the existence and uniqueness of a strong reproductive solution is presented in the paper [119] by SHINBROT. If g is, moreover, time-periodic, then the existence and the exponential stability of a periodic solution is proved.

TAKESHITA [122] examines the reproductive property of (6.1.1), (6.1.2), (6.1.4) in a bounded domain $\Omega \subset \mathbf{R}^2$. His work differs from SHINBROT [119] in that he does not require g to be small enough to prove the reproductive property of the given problem (but the condition is kept in proving the uniqueness). The proof is based on the Schauder fixed-point theorem (see I: Theorem 3.2.2).

TON [123] proves the existence of a weak periodic solution to (6.1.1), (6.1.2), (6.1.4) in a bounded domain $\Omega \subset \mathbf{R}^n$, using results of BROWDER VII: [12] and of LIONS VII: [46].

6.2. *The autonomous Navier-Stokes equations*

Periodic solutions of the autonomous Navier-Stokes equations are investigated mostly in connection with the bifurcation of a certain basic solution, as the Reynolds number R changes.

The existence of time-periodic solutions of the Navier-Stokes equations

$$\frac{\partial u}{\partial t} - \frac{1}{R} \Delta u + (u, \nabla)\, u + \nabla p = 0 \tag{6.2.1}$$

with the conditions (6.1.2) and (6.1.3) is studied by SATTINGER [116]. He considers the bifurcation of time-periodic solutions from a steady solution $\tilde{u}$, when R crosses a certain critical number R_c.

The equations for the perturbation of the flow are

$$\frac{\partial v}{\partial t} - \frac{1}{R} \Delta v + (\tilde{u}, \nabla) v + (v, \nabla) \tilde{u} + (v, \nabla) v + \nabla p = 0 \tag{6.2.2}$$

and (6.1.2), (6.1.4).

Applying the projection P_σ to (6.2.2) one obtains the "modified" Navier-Stokes equations in the form

$$\frac{\partial v}{\partial t} + Lv + N(v) = 0\,, \tag{6.2.3}$$

where

$$Lv = P_\sigma \left[-\frac{1}{R} \Delta v + (\tilde{u}, \nabla) v + (v, \nabla) \tilde{u} \right], \quad N(v) = P_\sigma(v, \nabla) v\,.$$

Periodic solutions of (6.2.3) are looked for in the form of a power series in a certain parameter μ. The author assumes that Ω is a bounded domain and the operator L has a simple eigenvalue. Then he proves that there exists a one-parameter family of time-periodic solutions of (6.2.1), (6.1.2), (6.1.3) with period $T(R)$ such that $T(R) \to 2\pi/\omega_c$ and the amplitude of the oscillations tends to zero as $R \to R_c$.

Similar problems are studied by JOSEPH and SATTINGER in [99]. The authors consider the problem (6.2.1), (6.1.2), (6.1.3), where the equation (6.2.1) is in an inhomogeneous form with the right-hand side $g(x)$, representing a stationary body force. Similar results are contained in the book [117] by SATTINGER.

IOOSS [93] (and in a preliminary form in [92]) considers the equation of the type

$$\frac{\partial u}{\partial t} + L_\lambda u - M(u, u) = 0\,, \tag{6.2.4}$$

which involves the equation for the perturbation of the steady flow of a viscous incompressible fluid. The same author [94], [95] deals with bifurcation of an nT-periodic solution from a T-periodic one. Several of the above results are quoted in detail in the book [108] by MARSDEN and MC CRACKEN.

JUDOVIČ in [101] investigates the motion of an incompressible homogeneous fluid in a bounded domain $\Omega \subset \mathbf{R}^3$ and proves the existence of a periodic solution. (The results are generalized to a class of ordinary differential equations in a Banach space, involving as a special case equations of parabolic type, equations of convection, equations of magnetohydrodynamics, etc.) In [102] he applies the theory developed in [101] to a class of ordinary differential equations in a Hilbert space of the type

$$-\frac{du}{dt} + Au - \lambda\, Bu = K(u, \lambda) \tag{6.2.5}$$

(where A is a self-adjoint, positive definite and coercive operator, A^{-1} is completely continuous, B is a certain linear and K a certain non-linear operator). Sufficient conditions are indicated for the appearance of auto-oscillations if the parameter λ crosses its critical value.

6.3. *Related problems*

Several authors investigate the existence of periodic solutions of problems related to those described by the Navier-Stokes equations.

Šmulev [121] constructs a periodic solution of the boundary-value problem

$$\nu \frac{\partial^2 u}{\partial x^2} - u \frac{\partial u}{\partial x} - \frac{\partial u}{\partial t} = 0\,,$$

$$u(0, t) = \psi(t)\,, \quad u(x, t) \xrightarrow{(x \to \infty)} a\,.$$

Ton [123] deals with the equations

$$\frac{\partial u}{\partial t} - \Delta u + (u, \nabla)\, u + \tfrac{1}{2}(\operatorname{div} u)\, u + \nabla p = g$$

$$\operatorname{div} u = -\varepsilon p$$

in a bounded domain $\Omega \subset \mathbf{R}^n$. He proves the existence of a weak periodic solution satisfying the condition (6.1.4), and its convergence (for $\varepsilon \to 0$) to the solution of (6.1.1), (6.1.2), (6.1.4).

The existence of a weak periodic solution to the equations

$$-\frac{\partial u}{\partial t} - \nu \sum_{i=1}^{n} \frac{\partial}{\partial x_i}\left(|\nabla u|^{q-2} \frac{\partial u}{\partial x_i}\right) + (u, \nabla)\, u + \operatorname{grad} p = g\,,$$

(6.1.2) and (6.1.4) is investigated by Lions [105].

Markman [106] treats the time-periodic convection in a viscous incompressible fluid with distributed sources of heat, filling up a vessel with rigid walls.

In [107] the same author investigates the existence of a time-periodic convective flow in a layer of a viscous incompressible fluid with free boundaries.

Silaev [120] proves the existence and uniqueness of a periodic solution to the system of the Prandtl equations of a boundary layer for a non-stationary axially symmetric flow of an incompressible fluid with suction.

Gordeev [91] investigates a system that describes tidal phenomena of an ocean.

Lions [104] using the Faedo-Galerkin method proves the existence of a weak periodic solution to the inequality

$$\int_0^T [\langle \varphi', \varphi - u \rangle + a(u, \varphi - u) + b(u, u, \varphi - u) - \langle f, \varphi - u \rangle] \, dt \geqq 0 ,$$

where

$$a(u, v) = \nu \sum_{i,k=1}^{2} \int_\Omega \frac{\partial u_k}{\partial x_i} \frac{\partial v_k}{\partial x_i} \, dx ,$$

$$b(u, v, w) = \sum_{i,k=1}^{2} \int_\Omega u_k \frac{\partial v_i}{\partial x_k} w_i \, dx .$$

A somewhat more general inequality of the Bingham flow is studied by Salvi in [115]. Their results were generalized by Naumann in [110] and [111].

Chapter IV

The telegraph equation

§ 1. The Fourier methods

1.1. *The (t, s)-Fourier method; the linear case*

We shall investigate the problem $(\mathscr{P}_\omega)$ given by

$$u_{tt}(t, x) - u_{xx}(t, x) + a\, u_t(t, x) + c\, u(t, x) = g(t, x)\,,$$

$$(t, x) \in Q = \mathbf{R} \times (0, \pi)\,, \tag{1.1.1}$$

$$u(t, 0) = u(t, \pi) = 0\,, \quad t \in \mathbf{R}\,, \tag{1.1.2}$$

$$u(t + \omega, x) - u(t, x) = 0\,, \quad (t, x) \in Q\,; \tag{1.1.3}$$

where $a, c \in \mathbf{R}$, $a \neq 0$ (throughout this chapter) and g is ω-periodic in t.

By a solution we mean a function

$$u \in \mathscr{U} = H^2_\omega(Q) \cap \dot{H}^1_\omega(Q)$$

satisfying (1.1.1) almost everywhere in Q.

Suppose that the space $\mathscr{U}$ is equipped with the norm of $H^2_\omega(Q)$, that is,

$$\|u\|_2 = \left(\sum_{\alpha+\beta \leqq 2} \int_0^\omega \int_0^\pi (D_t^\alpha D_x^\beta\, u(t, x))^2\, \mathrm{d}x\, \mathrm{d}t\right)^{1/2}\,.$$

By II: Theorem 1.3.4 (p. 42) a function $u \in H^0_\omega(Q)$,

$$u(t, x) = \left(\frac{2}{\pi\omega}\right)^{1/2} \sum_{j,k} u_{jk}\, e^{ijvt} \sin kx\,, \tag{1.1.4}$$

$$u_{jk} = \left(\frac{2}{\pi\omega}\right)^{1/2} \int_0^\omega \int_0^\pi u(\tau, \xi)\, e^{-ijv\tau} \sin k\xi\, \mathrm{d}\xi\, \mathrm{d}\tau\,, \quad v = 2\pi/\omega\,,$$

belongs to $\mathscr{U}$ if and only if

$$\sum_{j,k}(j^4 + k^4)\,|u_{jk}|^2 < \infty \tag{1.1.5}$$

(the summation is over $j \in \mathbf{Z}$, $k \subset \mathbf{N}$).
The square root of the series in (1.1.5) defines a norm $\|\cdot\|_{\mathscr{U}}$ on $\mathscr{U}$, which is equivalent to the norm $\|\cdot\|_2$.

Let

$$g \in \mathscr{G} = H^{\alpha}_{\omega}(\mathbf{R}; {}^0H^{1-\alpha}(0, \pi)) \quad \textit{for a fixed} \quad \alpha \in [0, 1]\,.$$

(For the definition of the spaces ${}^0H^{\beta}(0, \pi)$ see p. 43. In particular, g may belong to $H^{1,0}_{\omega}(Q)$ or to $\dot{H}^{0,1}_{\omega}(Q) = \operatorname{cl}\{u \in C^{\infty}_{\omega}(\overline{Q});\, u(t, 0) = u(t, \pi) = 0, t \in \mathbf{R}\}$ in $H^{0,1}_{\omega}(Q)$.)

By II: Theorem 1.3.4, a function $g \in H^0_{\omega}(Q)$,

$$g(t, x) = \left(\frac{2}{\pi\omega}\right)^{1/2} \sum_{j,k} g_{jk}\, e^{ij\nu t} \sin kx\,, \tag{1.1.6}$$

$$g_{jk} = \left(\frac{2}{\pi\omega}\right)^{1/2} \int_0^{\omega}\int_0^{\pi} g(\tau, \xi)\, e^{-ij\nu\tau} \sin k\xi \,\mathrm{d}\xi\,\mathrm{d}\tau\,,$$

belongs to $\mathscr{G}$ if and only if

$$\sum_{j,k}(1 + |j|^{2\alpha})\, k^{2(1-\alpha)}\, |g_{jk}|^2 < \infty\,. \tag{1.1.7}$$

The square root of this series defines in $H^{\alpha}_{\omega}(\mathbf{R}; {}^0H^{1-\alpha}(0, \pi))$ an equivalent norm $\|\cdot\|_{\mathscr{G}}$.

Substituting (1.1.4) and (1.1.6) in (1.1.1) we get

$$(-\nu^2 j^2 + k^2 + c + a\nu j i)\, u_{jk} = g_{jk}\,, \quad j \in \mathbf{Z}\,, \quad k \in \mathbf{N}\,.$$

If $j \neq 0$ or $c \neq -k^2$ for all $k \in \mathbf{N}$, then

$$|u_{jk}|^2 = \frac{|g_{jk}|^2}{(-\nu^2 j^2 + k^2 + c)^2 + a^2\nu^2 j^2}\,. \tag{1.1.8}$$

If $c = -k_0^2$ for some $k_0 \in \mathbf{N}$, then the problem $(\mathscr{P}_{\omega})$ has a solution only if $g_{0k_0} = 0$, that is,

$$\int_0^{\omega}\int_0^{\pi} g(\tau, \xi) \sin k_0\xi \,\mathrm{d}\xi\,\mathrm{d}\tau = 0\,, \tag{1.1.9}$$

and in this case the homogeneous problem for $g \equiv 0$ has a one-parameter family of solutions

$$u(t, x) = d \sin k_0 x\,, \quad d \in \mathbf{R}\,.$$

To establish an easy test of the convergence and an estimate of the series (1.1.5) with $|u_{jk}|^2$ defined by (1.1.8), let us prove the following lemma.

LEMMA 1.1.1 *Let* $\{\lambda_k\}_{k=1}^{\infty}$ *be a non-decreasing sequence of positive numbers,* $\mu, \nu, c \in \mathbf{R}$, $\mu \neq 0$, $\nu \neq 0$. *Then*

(1) *there exists a constant* $M > 0$ *such that for any* $s \in [0, 1]$ *we have*

$$a_{jk} \equiv \frac{j^4 + \lambda_k^2}{(-\nu^2 j^2 + \lambda_k + c)^2 + \mu^2 j^2} \leqq M|j|^{2s} \lambda_k^{1-s}, \quad (j, k) \in (\mathbf{Z} \setminus \{0\}) \times \mathbf{N};$$

(2) *if* $\sup_{k \in \mathbf{N}} \lambda_k \neq -c$, *the set* $\{a_{0k};\ k \in \mathbf{N},\ \lambda_k + c \neq 0\}$ *is bounded.*

Proof. Let $\lambda = \sup_{k \in \mathbf{N}} \lambda_k$. If $-c \geqq \lambda$, then obviously $\{a_{jk};\ j \in \mathbf{Z} \setminus \{0\},\ k \in \mathbf{N}\}$ is bounded. Let us suppose that $-c < \lambda$. Then there exists a $k^+ \in \mathbf{N}$ such that

$$\lambda_k + c \leqq 0 \quad \text{for} \quad k = 1, \ldots, k^+ - 1$$

and

$$\lambda_k + c > 0 \quad \text{for all} \quad k \geqq k^+ .$$

We define a decomposition of the set $(\mathbf{Z} \setminus \{0\}) \times \mathbf{N}$ and estimate a_{jk} on each part separately. Let

$$\Sigma_0 = \{(j, k) \in \mathbf{Z} \times \mathbf{N};\ j \neq 0,\ k = 1, \ldots, k^+ - 1\} .$$

Further, choose $\varepsilon \in (0, 1)$ and put

$$\Sigma_1 = \{(j, k) \in \mathbf{Z} \times \mathbf{N};\ k \geqq k^+, \text{ there exists a } \Theta \in (0, 1 - \varepsilon] \text{ such that } \nu|j| = \Theta(\lambda_k + c)^{1/2}\} ,$$

$$\Sigma_2 = \{(j, k) \in \mathbf{Z} \times \mathbf{N};\ k \geqq k^+, \text{ there exists a } \Theta \in (1 - \varepsilon, 1 + \varepsilon) \text{ such that } \nu|j| = \Theta(\lambda_k + c)^{1/2}\} ,$$

$$\Sigma_3 = \{(j, k) \in \mathbf{Z} \times \mathbf{N};\ k \geqq k^+, \text{ there exists a } \Theta \in [1 + \varepsilon, \infty) \text{ such that } \nu|j| = \Theta(\lambda_k + c)^{1/2}\} .$$

Clearly,

$$a_{jk} \leqq C, \quad (j, k) \in \Sigma_0$$

(C stands for various immaterial constants, independent of j, k and s). Taking into account that

$$c_1 \lambda_k^2 \leqq (\lambda_k + c)^2 \leqq c_2 \lambda_k^2, \quad k \geqq k^+ \tag{1.1.10}$$

with some $c_1, c_2 > 0$ independent of k, we find for $(j, k) \in \Sigma_1$

$$a_{jk} = \frac{j^4 + \lambda_k^2}{(-\nu|j| + (\lambda_k + c)^{1/2})^2 (\nu|j| + (\lambda_k + c)^{1/2})^2 + \mu^2 j^2} \leqq$$

$$\leqq C \frac{\Theta^4(\lambda_k + c)^2 + \lambda_k^2}{\varepsilon^2(\Theta + 1)^2 (\lambda_k + c)^2} \leqq C\,,$$

for $(j, k) \in \Sigma_3$

$$a_{jk} \leqq C \frac{\Theta^4(\lambda_k + c)^2 + \lambda_k^2}{\varepsilon^2(1 + \varepsilon)^{-2}\, \Theta^2(\Theta + 1)^2 (\lambda_k + c)^2} \leqq C \frac{1 + \Theta^{-4}}{(1 + \Theta^{-1})^2} \leqq C\,,$$

whereas for $(j, k) \in \Sigma_2$ we have

$$a_{jk} \leqq C(j^2 + \lambda_k^2 j^{-2}) \leqq Cj^2 \leqq C|j|^{2s} \lambda_k^{1-s}, \quad (s \in [0, 1])\,.$$

Combining these estimates we see that the first assertion holds. The other one is evident.

To simplify the formulation of the following theorems, let us introduce a certain notation. We define an operator L by $\mathscr{D}(L) = \mathscr{U}$ and $Lu = u_{tt} - u_{xx} + au_t + cu$. Further, if $c = -k_0^2$ for some $k_0 \in \mathbf{N}$, we denote by $\mathscr{V}$ the subspace of $H_\omega^0(Q)$ of functions of the form $u(t, x) = d \sin k_0 x$, $d \in \mathbf{R}$ and by $\mathscr{W}$ its orthogonal complement in $H_\omega^0(Q)$.

Applying Lemma 1.1.1 with $\lambda_k = k^2$, $\mu = a\nu$, $s = \alpha$, we obtain:

(i) if $c \neq -k^2$, $k \in \mathbf{N}$, then $u \in \mathscr{U}$ and $\|u\|_{\mathscr{U}} \leqq C\|g\|_{\mathscr{G}}$;

(ii) if $c = -k_0^2$ for some $k_0 \in \mathbf{N}$, $g \in \mathscr{G} \cap \mathscr{W}$, then $u = d \sin k_0 x + w$, $w \in \mathscr{U} \cap \mathscr{W}$ and $\|w\|_{\mathscr{U}} \leqq C\|g\|_{\mathscr{G}}$.

In both cases C is a constant independent of g.

Thus, we have proved two theorems.

THEOREM 1.1.1 (*The non-critical case*). *Let* $g \in \mathscr{G} = H_\omega^\alpha(\mathbf{R}; {}^0H^{1-\alpha}(0, \pi))$ *for some* $\alpha \in [0, 1]$. *Let* $a \neq 0$ *and* $c \neq -k^2$ *for all* $k \in \mathbf{N}$.
Then the problem $(\mathscr{P}_\omega)$ *has a unique solution* $u \in \mathscr{U}$,

$$u(t, x) = (L^{-1}g)(t, x) =$$

$$= \left(\frac{2}{\pi\omega}\right)^{1/2} \sum_{j,k} \frac{g_{jk}}{-\nu^2 j^2 + k^2 + c + a\nu j i} e^{ij\nu t} \sin kx \quad (\nu = 2\pi/\omega) \quad (1.1.11)$$

and

$$L^{-1} \in \mathscr{L}(\mathscr{G}, \mathscr{U})\,.$$

THEOREM 1.1.2 (*The critical case*). *Let* $g \in \mathscr{G}$. *Let* $a \neq 0$ *and there exist* $k_0 \in \mathbf{N}$ *such that* $c = -k_0^2$.
Then the problem $(\mathscr{P}_\omega)$ *has a solution* $u \in \mathscr{U}$ *if and only if* (1.1.9) *holds, that is,* $g \in \mathscr{G} \cap \mathscr{W}$. *If this condition is satisfied, the solutions can be written in the form*

$$u(t, x) = d \sin k_0 x + w(t, x)\,,$$

where $d \in \mathbf{R}$ *is arbitrary and* $w \in \mathscr{U} \cap \mathscr{W}$,

$$w(t, x) = (L^{-1} g)(t, x) =$$

$$= \left(\frac{2}{\pi\omega}\right)^{1/2} \sum_{\substack{j,k \\ (j,k) \neq (0,k_0)}} \frac{g_{jk}}{-v^2 j^2 + k^2 + c + avji} e^{ijvt} \sin kx \quad (v = 2\pi/\omega) \tag{1.1.12}$$

and

$$L^{-1} \in \mathscr{L}(\mathscr{G} \cap \mathscr{W}, \mathscr{U} \cap \mathscr{W})\,.$$

Remark 1.1.1 According to the embedding Theorems 2.7.1 and 2.7.2 of Chap. I, we see that $u \in \mathscr{U}$ implies that $u \in C_\omega(\bar{Q})$ and $u_t, u_x \in L_{q,\omega}(Q)$, $q \in$ $\in [1, +\infty)$.

Remark 1.1.2 If $g \in H_\omega^{2,0}(Q)$ (or $g \in H_\omega^{2,0}(Q) \cap \mathscr{W}$ in the critical case), we obtain immediately from Lemma 1.1.1 that $u_t \in H_\omega^2(Q) \cap \dot{H}_\omega^1(Q)$. We write

$$u_{xx}(t, x) = -g(t, x) + u_{tt}(t, x) + a\, u_t(t, x) + c\, u(t, x)\,.$$

Here the right-hand side is of class H^1 in x, provided that $g \in H_\omega^{2,1}(Q)$ (or $g \in H_\omega^{2,1}(Q) \cap \mathscr{W}$), so that $u_{xxx} \in H_\omega^0(Q)$. Hence, $u \in H_\omega^3(Q) \cap \dot{H}_\omega^1(Q)$ and by the embedding Theorem 2.7.2 of Chap. I we obtain $u \in C_\omega^1(\bar{Q})$. Clearly, $\|u\|_{H_\omega{}^3(Q)} \leqq \text{const.}\ \|g\|_{H_\omega{}^{2,1}(Q)}$. Similarly we can obtain the classical solution (namely, if $g \in H_\omega^{3,2}(Q)$ (or $g \in H_\omega^{3,2}(Q) \cap \mathscr{W}$), then $u \in H_\omega^4(Q) \cap \dot{H}_\omega^1(Q)$ which implies that $u \in C_\omega^2(\bar{Q})$).

Remark 1.1.3 Let $\omega = 2\pi$, $c = 0$, $\alpha \in (0, 1]$ and $\eta \in (0, \alpha]$. Then there exists a function $g \in H_\omega^{\alpha-\eta}(\mathbf{R}; {}^0H^{1-\alpha}(0, \pi))$ such that the problem $(\mathscr{P}_{2\pi})$ has no solution in $\mathscr{U}$. For let us define the function g by (1.1.6), where

$$g_{jk} = 0\,, \quad |j| \neq k\,; \quad g_{-k,k} = g_{kk} = k^{-3/2}\,, \quad k \in \mathbf{N}\,.$$

Evidently, $\sum_{j,k} |j|^{2(\alpha-\eta)}\, k^{2(1-\alpha)}\, |g_{jk}|^2 < \infty$, hence $g \in H_\omega^{\alpha-\eta}(\mathbf{R}; {}^0H^{1-\alpha}(0, \pi))$. According to (1.1.8), we have

$$u_{jk} = 0\,, \quad |j| \neq k\,; \quad |u_{-k,k}|^2 = |u_{kk}|^2 = a^{-2}\, k^{-5}\,,$$

and the assertion follows readily. This example shows that the assumption on g in Theorem 1.1.1 cannot be weakened.

Remark 1.1.4 A slightly more general problem given by

$$v_{tt}(t, x) + a\, v_t(t, x) - v_{xx}(t, x) + 2b\, v_x(t, x) + c\, v(t, x) =$$

$$= g_1(t, x), \quad (t, x) \in Q,$$

$$v(t, 0) = h_0(t),$$

$$v(t, \pi) = h_1(t), \quad t \in \mathbf{R},$$

and by (1.1.3), where a, b, $c \in \mathbf{R}$, $a \neq 0$, g_1, h_0, h_1 are ω-periodic in t, can be treated, after certain transformations (see III: Remark 1.1.6 and III: Remark 1.1.7) like $(\mathscr{P}_\omega)$.

To end this section we state two existence theorems concerning an n-dimensional analogue to (1.1.1). The proofs go essentially along the same lines as those of the preceding theorems, and we omit them.

Let the problem be given by

$$u_{tt}(t, x) + a\, u_t(t, x) + E\, u(t, x) + c\, u(t, x) = g(t, x), \quad (t, x) \in \mathbf{R} \times \Omega, \tag{1.1.13}$$

$$u(t, x) = 0, \quad (t, x) \in \mathbf{R} \times \partial\Omega \tag{1.1.14}$$

or

$$\frac{\partial u}{\partial n}(t, x) + \sigma(x)\, u(t, x) = 0, \quad (t, x) \in \mathbf{R} \times \partial\Omega \tag{1.1.14'}$$

and by

$$u(t + \omega, x) - u(t, x) = 0, \quad (t, x) \in \mathbf{R} \times \Omega \tag{1.1.15}$$

where $a, c \in \mathbf{R}$, $a \neq 0$, g is ω-periodic in t, $\Omega \subset \mathbf{R}^n$ is a bounded domain, $\partial/\partial n$ is the conormal derivative,

$$E\, v(x) = -\sum_{j,k=1}^{n} (a_{jk}(x)\, v_{x_k}(x))_{x_j} + a_0(x)\, v(x), \quad x \in \Omega. \tag{1.1.16}$$

Suppose that the assumptions (1)–(4) in III: Sec. 1.3 (p. 77) hold. Without loss of generality we may assume that the operator E is positive definite in $H^0(\Omega)$. The symbols λ_k, v_k ($k \in \mathbf{N}$), and $\mathscr{D}(E^\beta)$ have the same meaning as in III: Sec. 1.3.

By a solution of the above problem we mean a function

$$u \in H^0_\omega(\mathbf{R}; \mathscr{D}(E)) \cap H^{2,0}_\omega(\mathbf{R} \times \Omega)$$

satisfying (1.1.13) almost everywhere in $\mathbf{R} \times \Omega$.

THEOREM 1.1.3 (*The non-critical case*). *Let* $g \in H^\alpha_\omega(\mathbf{R}; \mathscr{D}(E^{(1-\alpha)/2}))$ *for some* $\alpha \in [0, 1]$. *Let* $a \neq 0$ *and* $c \neq -\lambda_k$ *for all* $k \in \mathbf{N}$.
Then the problem (1.1.13)–(1.1.15) *has a unique solution* u *and* $g \to u$ *is a continuous linear mapping of* $H^\alpha_\omega(\mathbf{R}; \mathscr{D}(E^{(1-\alpha)/2}))$ *into* $H^0_\omega(\mathbf{R}; \mathscr{D}(E)) \cap$ $\cap\, H^{2,0}_\omega(\mathbf{R} \times \Omega)$.

THEOREM 1.1.4 (*The critical case*). *Let* $g \in H^\alpha_\omega(\mathbf{R}; \mathscr{D}(E^{(1-\alpha)/2}))$ *for some* $\alpha \in [0, 1]$. *Let* $a \neq 0$ *and suppose that there is a* $k_0 \in \mathbf{N}$ *such that*

$$\lambda_{k_0-1} < -c = \lambda_{k_0} = \lambda_{k_0+1} = \ldots = \lambda_{k_0+r-1} < \lambda_{k_0+r}\,. \tag{1.1.17}$$

Then the problem (1.1.13)–(1.1.15) *has a solution if and only if*

$$\int_0^\omega \int_\Omega g(\tau, \xi)\, v_{k_0+l}(\xi)\, \mathrm{d}\xi\, \mathrm{d}\tau = 0\,, \quad l = 0\,, \ldots, r-1\,.$$

If these conditions are satisfied, then the solutions can be written in the form

$$u(t, x) = \sum_{l=0}^{r-1} d_l\, v_{k_0+l}(x) + w(t, x)\,,$$

where $d_0, \ldots, d_{r-1} \in \mathbf{R}$ *are arbitrary,* w *belongs to*

$$H^0_\omega(\mathbf{R}; \mathscr{D}(E)) \cap H^{2,0}_\omega(\mathbf{R} \times \Omega) \cap$$

$$\cap \left\{w; \int_0^\omega \int_\Omega w(\tau, \xi)\, v_{k_0+l}(\xi)\, \mathrm{d}\xi\, \mathrm{d}\tau = 0\,, \quad l = 0, \ldots, r-1\right\}, \tag{1.1.18}$$

and $g \to w$ *is a continuous linear mapping of*

$$H^\alpha_\omega(\mathbf{R}; \mathscr{D}(E^{(1-\alpha)/2})) \cap \left\{g; \int_0^\omega \int_\Omega g(\tau, \xi)\, v_{k_0+l}(\xi)\, \mathrm{d}\xi\, \mathrm{d}\tau = 0\,,\right.$$

$$\left. l = 0, \ldots, r-1\right\}$$

into the space (1.1.18).

Remark 1.1.5 Using II: Theorem 1.3.4 we obtain that the solution necessarily lies in $H^1_\omega(\mathbf{R}; \mathscr{D}(E^{1/2}))$.

Remark 1.1.6 By I: Theorems 2.7.1 and 2.7.2, a solution u belongs to $C_\omega(\mathbf{R} \times \times \bar{\Omega})$ if $n < 3$,

$$u \in L_{q,\omega}(\mathbf{R} \times \Omega), \quad q \in [1, \infty) \quad \text{if} \quad n = 3,$$

$$u \in L_{q,\omega}(\mathbf{R} \times \Omega), \quad q = 2(n + 1)/(n - 3) \quad \text{if} \quad n > 3$$

and

$$u_t, u_{x_j} \in L_{q,\omega}(\mathbf{R} \times \Omega), \quad q = 2(n + 1)/(n - 1) \quad \text{if} \quad n > 1 \ (j = 1, \ldots, n).$$

Remark 1.1.7 The (t, s)-Fourier method is used in the papers PRODI [26], RABINOWITZ [32] and MAWHIN [22]. Only generalized solutions are found by means of this method.

1.2. *The weakly non-linear case*

Let us consider the problem $(\mathscr{P}^\varepsilon_\omega)$ given by

$$u_{tt}(t, x) - u_{xx}(t, x) + a\, u_t(t, x) + c\, u(t, x) = \varepsilon\, F(u, \varepsilon)\,(t, x)\,,$$

$$(t, x) \in Q\,, \quad \varepsilon \in [-\varepsilon_0, \varepsilon_0]\,, \quad (\varepsilon_0 > 0) \tag{1.2.1}$$

and by (1.1.2) and (1.1.3).

We keep the notation of the previous section. A function $u \in C([-\varepsilon^*, \varepsilon^*]; \mathscr{U})$ $(\varepsilon^* \in (0, \varepsilon_0])$ is called a solution of $(\mathscr{P}^\varepsilon_\omega)$ if $u(\varepsilon)$ for every $\varepsilon \in [-\varepsilon^*, \varepsilon^*]$ satisfies (1.2.1) almost everywhere in Q.

THEOREM 1.2.1 (*The non-critical case*). *Let $a \neq 0$ and $c \neq -k^2$ for all $k \in \mathbf{N}$. Suppose that the following assumptions hold:*

(*a*) *the operator $F(u, \varepsilon)$ maps $B(0, \varrho; \mathscr{U}) \times [-\varepsilon_0, \varepsilon_0]$ into $\mathscr{G}$ continuously for some $\varrho > 0$;*

(*b*) *there is a $\lambda > 0$ such that*

$$\|F(u_1, \varepsilon) - F(u_2, \varepsilon)\|_{\mathscr{G}} \leqq \lambda \|u_1 - u_2\|_{\mathscr{U}}$$

for all $u_1, u_2 \in B(0, \varrho; \mathscr{U})$ and $\varepsilon \in [-\varepsilon_0, \varepsilon_0]$.
Then the problem $(\mathscr{P}^\varepsilon_\omega)$ has a locally unique solution $u^*(\varepsilon) \in \mathscr{U}$ such that $u^*(0) = 0$.

Proof. According to Theorem 1.1.1 $(\mathscr{P}^\varepsilon_\omega)$ has a solution if and only if there exists a solution $u(\varepsilon) \in \mathscr{U}$, continuous in ε, of the equation

$$u = \varepsilon\, L^{-1}\, F(u, \varepsilon)\,,$$

where L^{-1} is given by (1.1.11).

Applying I: Corollary 3.4.1, we obtain immediately the required result.

THEOREM 1.2.2 (*The critical case*). *Let* $a \neq 0$ *and* $c = -k_0^2$ *for some* $k_0 \in \mathbf{N}$. *Suppose that*:

(*a*) *the equation*

$$\int_0^\omega \int_0^\pi F(d_0 \sin k_0 x, 0)(\tau, \xi) \sin k_0 \xi \, d\xi \, d\tau = 0$$

has a solution $d_0 = d_0^* \in \mathbf{R}$;

(*b*) *the operator* $F(u, \varepsilon)$ *and its G-derivative* $F'_u(u, \varepsilon)$ *for some* $\varrho > 0$ *map* $B(d_0^* \sin k_0 x, \varrho; \mathscr{U}) \times [-\varepsilon_0, \varepsilon_0]$ *continuously into* $\mathscr{G}$ *and* $\mathscr{L}(\mathscr{U}, \mathscr{G})$, *respectively*;

$$(c) \int_0^\omega \int_0^\pi F'_u(d_0^* \sin k_0 x, 0)(\sin k_0 x)(\tau, \xi) \sin k_0 \xi \, d\xi \, d\tau \neq 0 \, .$$

Then the problem $(\mathscr{P}_\omega^\varepsilon)$ *has a locally unique solution* $u^*(\varepsilon) \in \mathscr{U}$ *such that* $u^*(0) = d_0^* \sin k_0 x$.

Proof. We look for a solution in the form

$$u(t, x) = d \sin k_0 x + w(t, x) \, ,$$

where $d \in \mathbf{R}$ and $w \in \mathscr{U} \cap \mathscr{W}$. According to Theorem 1.1.2, $(\mathscr{P}_\omega^\varepsilon)$ is then equivalent to the system for $(w, d) \in (\mathscr{U} \cap \mathscr{W}) \times \mathbf{R}$, which depends continuously on ε,

$$G_1(w, d, \varepsilon) \equiv w(t, x) - \varepsilon L^{-1} F(d \sin k_0 x + w(t, x), \varepsilon)(t, x) = 0 \, ,$$

$$G_2(w, d, \varepsilon) \equiv \int_0^\omega \int_0^\pi F(d \sin k_0 x + w(t, x), \varepsilon)(\tau, \xi) \sin k_0 \xi \, d\xi \, d\tau = 0 \, ,$$

L^{-1} is given by (1.1.12).

Using the implicit function theorem I: Theorem 3.4.2 just as in the proof of III: Theorem 1.2.2 (p. 75), we readily obtain the assertion of the present theorem.

Remark 1.2.1 Let

$$F(u, \varepsilon)(t, x) = f(t, x, u(t, x), u_t(t, x), u_x(t, x), \varepsilon) \, .$$

To avoid conditions on the growth of $f = f(t, x, u_0, u_1, u_2, \varepsilon)$ with respect to u_1 and u_2 (see III: Remark 1.2.3 and III: Remark 1.2.4) we prefer to work with solutions belonging to $H_\omega^3(Q)$, which implies (by I: Theorem 2.7.2) that all the derivatives in f belong to $C_\omega(\bar{Q})$. Note that by Remark 1.1.2, $L^{-1} \in \mathscr{L}(\mathscr{G}_1, \mathscr{U}_1)$ (or $\mathscr{L}(\mathscr{G}_1 \cap \mathscr{W}, \mathscr{U}_1 \cap \mathscr{W})$), where $\mathscr{U}_1 = H_\omega^3(Q) \cap \dot{H}_\omega^1(Q)$, $\mathscr{G}_1 =$

$= H^{2,1}_{\omega}(Q)$, and consequently Theorems 1.2.1 and 1.2.2 remain valid with $\mathscr{U}_1$ and $\mathscr{G}_1$, instead of $\mathscr{U}$ and $\mathscr{G}$, respectively. The operator F satisfies the assumptions of Theorem 1.2.1 modified for $\mathscr{U}_1$ and $\mathscr{G}_1$ provided that f is ω-periodic in t, continuous on $\mathbf{R} \times [0, \pi] \times [-\varrho, \varrho]^3 \times [-\varepsilon_0, \varepsilon_0]$ for some $\varrho > 0$ together with its partial derivatives

$$D_t^{\alpha} D_x^{\beta} D_{u_0}^{\gamma_0} D_{u_1}^{\gamma_1} D_{u_2}^{\gamma_2} f\,, \quad \alpha + \beta + \sum_{j=0}^{2} \gamma_j \leqq 2\,, \quad \alpha + 2\beta \leqq 2\,,$$

and these derivatives are Lipschitz continuous in u_0, u_1 and u_2. The assumption (b) of Theorem 1.2.2 is satisfied provided that f is ω-periodic in t, continuous together with its partial derivatives

$$D_t^{\alpha} D_x^{\beta} D_{u_0}^{\gamma_0} D_{u_1}^{\gamma_1} D_{u_2}^{\gamma_2} f\,, \quad \alpha + \beta + \sum_{j=0}^{2} \gamma_j \leqq 3\,, \quad \alpha + 2\beta \leqq 2$$

on

$$\mathbf{R} \times [0, \pi] \times [-|d_0^*| - \varrho, |d_0^*| + \varrho] \times [-\varrho, \varrho] \times$$
$$\times [-k_0|d_0^*| - \varrho, k_0|d_0^*| + \varrho] \times [-\varepsilon_0, \varepsilon_0] \quad \text{for some} \quad \varrho > 0\,.$$

Remark 1.2.2 The corresponding non-linear problem in the non-critical case can also be treated by the Schauder fixed-point theorem (see, for example, MAWHIN [22]).

Problem 1.2.1 Investigate the weakly non-linear problem corresponding to the n-dimensional case.

1.3. *The t-Fourier method*

In the remaining two sections of this paragraph we proceed more briefly and leave proofs of most assertions to the reader as easy exercises.

Let us solve the problem given by (1.1.1)–(1.1.3) by means of the t-Fourier method. By a solution we now mean a function u from the space

$$\mathscr{U} = C(I; H^2_{\omega}(\mathbf{R})) \cap C^1(I; H^1_{\omega}(\mathbf{R})) \cap C^2(I; H^0_{\omega}(\mathbf{R}))\,, \quad I = [0, \pi]\,, \tag{1.3.1}$$

satisfying (1.1.1) for all $x \in (0, \pi)$ and (1.1.2) both in the sense of the space $H^0_{\omega}(\mathbf{R})$.

(*a*) We use the expansion of $u \in C(I; H^0_{\omega}(\mathbf{R}))$ in the form III: (2.1.3). We show that u belongs to $\mathscr{U}$ if and only if $u_j \in C^2(I; \mathbf{C})$ for all $j \in \mathbf{Z}$ and the series

$$\sum_{j=-\infty}^{\infty} j^4 |u_j(x)|^2\,, \quad \sum_{j=-\infty}^{\infty} j^2 |u_j'(x)|^2\,, \quad \text{and} \quad \sum_{j=-\infty}^{\infty} |u_j''(x)|^2 \tag{1.3.2}$$

converge uniformly on I. The norm on $\mathscr{U}$ is equivalent to

$$\|u\|_{\mathscr{U}} = \max \{\sup_{x\in I} (\sum_{j=-\infty}^{\infty} (1 + j^4) |u_j(x)|^2)^{1/2} ,$$

$$\sup_{x\in I} (\sum_{j=-\infty}^{\infty} (1 + j^2) |u_j'(x)|^2)^{1/2} , \quad \sup_{x\in I} (\sum_{j=-\infty}^{\infty} |u_j''(x)|^2)^{1/2}\} .$$

(b) Suppose that $g \in C(I; H^1_\omega(\mathbf{R}))$. When g is written in the form III: (2.1.6), the series

$$\sum_{j=-\infty}^{\infty} j^2 |g_j(x)|^2$$

converges uniformly and

$$\sup_{x\in I} (\sum_{j=-\infty}^{\infty} (1 + j^2) |g_j(x)|^2)^{1/2}$$

determines an equivalent norm of g in $C(I; H^1_\omega(\mathbf{R}))$.

(c) The coefficients u_j ($j \in \mathbf{Z}$) are defined as the solutions of the problems

$$-u_j''(x) + (-v^2 j^2 + c + avji)\, u_j(x) = g_j(x) , \quad x \in (0, \pi) , \tag{1.3.3}$$

$$u_j(0) = u_j(\pi) = 0 . \tag{1.3.4}$$

They are given by

$$u_j(x) = \int_0^\pi G_j(x, \xi)\, g_j(\xi)\, d\xi , \tag{1.3.5}$$

where the Green function is

$$G_j(x, \xi) = \varkappa_j^{-1} (\sinh \varkappa_j \pi)^{-1} \sinh \varkappa_j \xi \sinh \varkappa_j (\pi - x) , \quad 0 \leqq \xi \leqq x ,$$

$$G_j(x, \xi) = G_j(\xi, x) , \quad x < \xi \leqq \pi ,$$

$$\varkappa_j = (-v^2 j^2 + c + avji)^{1/2} = \mu_j + i v_j ,$$

$$\mu_j = 2^{-1/2} (-v^2 j^2 + c + ((-v^2 j^2 + c)^2 + a^2 v^2 j^2)^{1/2})^{1/2} ,$$

$$v_j = 2^{-1/2} (v^2 j^2 - c + ((-v^2 j^2 + c)^2 + a^2 v^2 j^2)^{1/2})^{1/2} \operatorname{sgn} j$$

if $j \in \mathbf{Z} \setminus \{0\}$ or $j = 0$ ($\operatorname{sgn} 0 = 1$) and $c \neq 0, -k^2, k \in \mathbf{N}$. For $c = 0, j = 0$ we have

$$G_0(x, \xi) = \pi^{-1} \xi (\pi - x) , \quad 0 \leqq \xi \leqq x .$$

If $c = -k_0^2$ for some $k_0 \in \mathbf{N}$ and $j = 0$ then (1.3.3), (1.3.4) has a solution if and only if

$$\int_0^\pi g_0(\xi) \sin k_0 \xi \, \mathrm{d}\xi = 0$$

When this condition is satisfied, then

$$u_0(x) = \delta \sin k_0 x - k_0^{-1} \int_0^x \sin k_0(x - \xi)\, g_0(\xi)\, \mathrm{d}\xi\,, \quad \delta \in \mathbf{R}\,.$$

(*d*) The next step is to show that $\lim_{|j| \to \infty} \mu_j = 2^{-1}|a| > 0$ (compare with the corresponding case studied in Chap. III).
One then proves that there exist positive constants c_1, c_2, c_3 and c_4 such that $c_1 \leqq \mu_j \leqq c_2$, $j \in \mathbf{Z} \setminus \{0\}$, $|\sinh \varkappa_j x| \leqq c_3$, $x \in I$, $j \in \mathbf{Z}$, $|\sinh \varkappa_j \pi| \gtrsim c_4$, $j \in \mathbf{Z} \setminus \{0\}$. Further, there exist positive constants c_5 and c_6 such that $c_5 i^2 \leqq \leqq |\varkappa_j|^2 \leqq c_6 j^2$, $j \in \mathbf{Z} \setminus \{0\}$.

(*e*) Now using (1.3.5) we derive

$$|u_j(x)|^2 \leqq c_7 j^{-2} \int_0^\pi |g_j(\xi)|^2 \, \mathrm{d}\xi\,, \quad j \in \mathbf{Z} \setminus \{0\}\,,$$

with $c_7 > 0$ independent of j and g. Hence, the series $\sum_{j=-\infty}^{\infty} j^4 |u_j(x)|^2$ is easily seen to be uniformly convergent on I. The uniform convergence of the other two series in (1.3.2) can be established similarly.

(*f*) Finally, one has to summarize the results and formulate existence theorems.

Remark 1.3.1 Using the fact that $H^1_\omega(\mathbf{R})$ is embedded in $C_\omega(\mathbf{R})$ we find that from $u \in \mathscr{U}$ it follows that $u \in C^1_\omega(\mathbf{R} \times I)$ (see III: Lemma 2.1.1). If $g \in C(I; H^2_\omega(\mathbf{R}))$, then the solution u obviously belongs to

$$C(I; H^3_\omega(\mathbf{R})) \cap C^1(I; H^2_\omega(\mathbf{R})) \cap C^2(I; H^1_\omega(\mathbf{R}))\,, \tag{1.3.6}$$

which implies that $u \in C^2_\omega(\mathbf{R} \times I)$.

Remark 1.3.2 Kopáčková [19] applies her more general result to the problem (1.1.1)–(1.1.3) and finds a solution in the space (1.3.6).

Problem 1.3.1 Investigate the above problem under the assumption $g \in \in C^1(I; H^0_\omega(\mathbf{R}))$.

Problem 1.3.2 Study the problem given by

$$u_{tt}(t, x) - u_{xx}(t, x) + a\, u_t(t, x) + c\, u(t, x) = g(t, x)\,,$$
$$(t, x) \in \mathbf{R}^2\,, \quad (a \neq 0)\,. \tag{1.3.7}$$

Suppose that $c > 0$ and $g \in C(\mathbf{R}; H^1_\omega(\mathbf{R}))$ and prove the existence of a solution in the space (1.3.1), where $I = \mathbf{R}$.

Hint: The only solution to the equation (1.3.3) that is of class $C^2(\mathbf{R}; \mathbf{C})$ is

$$u_j(x) = (2\varkappa_j)^{-1} \int_{-\infty}^{\infty} e^{-\varkappa_j|x-\xi|}\, g_j(\xi)\, \mathrm{d}\xi\;;$$

see III: Sec. 2.2.

Problem 1.3.3 Investigate the existence of a solution $u \in H^2_\omega(\mathbf{R}^2)$ to the problem (1.3.7) with $c > 0$ and $g \in H^{1,0}_\omega(\mathbf{R}^2)$ by means of the t-Fourier method and the Fourier transform.

Hint: See III: Sec. 2.2.

Problem 1.3.4 Study the corresponding weakly non-linear problems.

Problem 1.3.5 Prove Theorem 1.1.3 and Theorem 1.1.4 with $g \in H^{1,0}_\omega(\mathbf{R} \times \Omega)$ by means of the t-Fourier method.

Hint: see III: Remark 2.1.4.

1.4. *The s-Fourier method*

We use the s-Fourier method to study the n-dimensional problem $(\mathscr{P}_\omega)$ given by

$$u_{tt}(t, x) + 2a\, u_t(t, x) + E\, u(t, x) + c\, u(t, x) = g(t, x)\,,$$
$$(t, x) \in Q = \mathbf{R} \times \Omega\,, \tag{1.4.1}$$

(1.1.14) or (1.1.14′) and by (1.1.15) with E given by (1.1.16). The assumptions are the same as in Sec. 1.1 (p. 135). In addition we use the notation and results of II: Sec. 2.4.

By a solution to $(\mathscr{P}_\omega)$ we mean a function

$$u \in \mathscr{U} = C_\omega(\mathbf{R}; \mathscr{D}(E)) \cap C^1_\omega(\mathbf{R}; \mathscr{D}(E^{1/2})) \cap C^2_\omega(\mathbf{R}; H^0(\Omega))$$

satisfying the equation (1.4.1) for any $t \in \mathbf{R}$ in the sense of $H^0(\Omega)$.

(a) By II: Theorem 2.4.2 (p. 58) the corresponding initial boundary-value problem $(\mathscr{I}\mathscr{B})$ given by (1.4.1), (1.1.14) or (1.1.14′) and

$$u(0, x) = \varphi(x), \quad u_t(0, x) = \psi(x), \quad x \in \Omega$$

has a unique solution

$$u \in C([0, \omega]; \mathscr{D}(E)) \cap C^1([0, \omega]; \mathscr{D}(E^{1/2})) \cap C^2([0, \omega]; H^0(\Omega)),$$

provided that $g \in C([0, \omega]; \mathscr{D}(E^{1/2}))$, $\varphi \in \mathscr{D}(E)$, and $\psi \in \mathscr{D}(E^{1/2})$. Hence, assuming that $g \in C_\omega(\mathbf{R}; \mathscr{D}(E^{1/2}))$, we find that u is ω-periodic in t if and only if

$$u(\omega, x) = u(0, x) = \varphi(x),$$
$$u_t(\omega, x) = u_t(0, x) = \psi(x), \quad x \in \Omega.$$

This is equivalent to the infinite system for the Fourier coefficients

$$u_k(\omega) = \varphi_k,$$
$$\dot{u}_k(\omega) = \psi_k, \quad k \in \mathbf{N}.$$

(b) Using the formulae II: (2.4.29)–(2.4.31), respectively, we obtain for each $k \in \mathbf{N}$, a system for two unknowns φ_k and ψ_k:

$$\varphi_k(e^{-a\omega} \cos \beta\omega - 1 + a\beta^{-1} e^{-a\omega} \sin \beta\omega) + \psi_k\beta^{-1} e^{-a\omega} \sin \beta\omega =$$
$$= -\beta^{-1} \int_0^\omega e^{-a(\omega-\tau)} \sin \beta(\omega - \tau)\, g_k(\tau)\, d\tau,$$
$$\varphi_k(a\, e^{-a\omega} \cos \beta\omega - a - \beta\, e^{-a\omega} \sin \beta\omega) + \psi_k(e^{-a\omega} \cos \beta\omega - 1) =$$
$$= -\int_0^\omega e^{-a(\omega-\tau)} \cos \beta(\omega - \tau)\, g_k(\tau)\, d\tau \tag{1.4.2}$$

if $\lambda_k + c - a^2 > 0$, $(\beta = (\lambda_k + c - a^2)^{1/2})$;

$$\varphi_k((a\omega + 1)\, e^{-a\omega} - 1) + \psi_k\omega\, e^{-a\omega} = -\int_0^\omega e^{-a(\omega-\tau)} (\omega - \tau)\, g_k(\tau)\, d\tau,$$
$$\varphi_k\, a(e^{-a\omega} - 1) + \psi_k(e^{-a\omega} - 1) = -\int_0^\omega e^{-a(\omega-\tau)}\, g_k(\tau)\, d\tau \tag{1.4.3}$$

if $\lambda_k + c - a^2 = 0$;

$$\varphi_k(e^{-a\omega} \cosh \gamma\omega - 1 + a\gamma^{-1} e^{-a\omega} \sinh \gamma\omega) + \psi_k\gamma^{-1} e^{-a\omega} \sinh \gamma\omega =$$
$$= -\gamma^{-1} \int_0^\omega e^{-a(\omega-\tau)} \sinh \gamma(\omega - \tau)\, g_k(\tau)\, d\tau,$$

$$\varphi_k(a\, e^{-a\omega} \cosh \gamma\omega - a + \gamma\, e^{-a\omega} \sinh \gamma\omega) + \psi_k(e^{-a\omega} \cosh \gamma\omega - 1) =$$

$$= - \int_0^\omega e^{-a(\omega-\tau)} \cosh \gamma(\omega - \tau)\, g_k(\tau)\, \mathrm{d}\tau \tag{1.4.4}$$

if $a^2 - \lambda_k - c > 0$, $(\gamma = (a^2 - \lambda_k - c)^{1/2})$.

(*c*) The determinants of the systems (1.4.2)–(1.4.4) are

$$e^{-2a\omega} - 2e^{-a\omega} \cos\left[(\lambda_k + c - a^2)^{1/2}\, \omega\right] + 1\,, \tag{1.4.5}$$

$$(e^{-a\omega} - 1)^2\,, \tag{1.4.6}$$

and

$$e^{-2a\omega} - 2e^{-a\omega} \cosh\left[(a^2 - \lambda_k - c)^{1/2}\, \omega\right] + 1\,, \tag{1.4.7}$$

respectively.

The following assertions are now to be proved. The determinants (1.4.5), (1.4.6) are never zero. Moreover, these determinants are bounded from below with respect to k by $(e^{-a\omega} - 1)^2 > 0$.

The determinant (1.4.7) can vanish; this occurs exactly if $\lambda_k + c = 0$. If this is the case, then the system (1.4.4) has a solution if and only if the right-hand side is orthogonal to every solution of the corresponding homogeneous adjoint system (that is, to $(\delta, a^{-1}\delta)$, $\delta \in \mathbf{R}$). This is equivalent to the condition

$$\int_0^\omega g_k(\tau)\, \mathrm{d}\tau = 0\,.$$

If this condition is satisfied, then the system (1.4.4) has a one-parameter family of solutions $(\varphi_k, \psi_k) = \left(\delta, (1 - e^{-2a\omega})^{-1} \int_0^\omega e^{-2a(\omega-\tau)}\, g_k(\tau) \mathrm{d}\tau\right)$, where δ is an arbitrary real number.

(*d*) Let us distinguish two cases:

(i) $c \neq -\lambda_k$ for all $k \in \mathbf{N}$;

(ii) there exists a $k_0 \in \mathbf{N}$ such that (1.1.17) holds ($-c$ is an eigenvalue of multiplicity r).

In the first case the (φ_k, ψ_k) are uniquely determined for all $k \in \mathbf{N}$. Putting

$$\varphi(x) = \sum_{k=1}^\infty \varphi_k\, v_k(x)\,, \quad \psi(x) = \sum_{k=1}^\infty \psi_k\, v_k(x) \tag{1.4.8}$$

we obtain easily

$$\varphi \in \mathscr{D}(E)\,, \quad \psi \in \mathscr{D}(E^{1/2})$$

and

$$\|\varphi\|_{\mathscr{D}(E)} + \|\psi\|_{\mathscr{D}(E^{1/2})} \leqq \text{const.}\, \|g\|_{C_\omega(\mathbf{R};\mathscr{D}(E^{1/2}))}\,. \tag{1.4.9}$$

By II: Theorem 2.4.2 the functions φ and ψ determine a unique solution u of $(\mathscr{IB})$ belonging to $\mathscr{U}$. Hence u is a solution of $(\mathscr{P}_\omega)$ and $g \to u$ is a continuous linear mapping.

In the second case, using II: (2.4.31) we find that the homogeneous problem $(\mathscr{P}_\omega)$ has an r-parameter family of solutions

$$v(t, x) = \sum_{l=0}^{r-1} d_l \, v_{k_0+l}(x) \,, \quad d_l \in \mathbf{R} \,.$$

The problem $(\mathscr{P}_\omega)$ has a solution only if

$$\int_0^\omega \int_\Omega g(\tau, \xi) \, v_{k_0+l}(\xi) \, \mathrm{d}\xi \, \mathrm{d}\tau = 0 \,, \quad l = 0, \ldots, r - 1 \,.$$

We define the functions φ and ψ by (1.4.8) where φ_k for $k \in \mathbf{N} \setminus \{k_0, \ldots$ $\ldots, k_0 + r - 1\}$ and ψ_k for $k \in \mathbf{N}$ are uniquely determined and

$$\varphi_{k_0+l} = \frac{1}{2a\omega} \left\{ \int_0^\omega \left[t g_{k_0+l}(t) + \int_0^t e^{-2a(t-\tau)} \, g_{k_0+l}(\tau) \, \mathrm{d}\tau \right] \mathrm{d}t + \right.$$

$$\left. + \frac{1 - 2a\omega - e^{-2a\omega}}{2a(1 - e^{-2a\omega})} \int_0^\omega e^{-2a(\omega-\tau)} \, g_{k_0+l}(\tau) \, \mathrm{d}\tau \right\}, \, l = 0 \,, \ldots, r - 1 \,.$$

Again (1.4.9) holds, φ and ψ determine a unique solution w of $(\mathscr{P}_\omega)$ satisfying

$$\int_0^\omega \int_\Omega w(\tau, \xi) \, v_{k_0+l}(\xi) \, \mathrm{d}\xi \, \mathrm{d}\tau = 0 \,, \quad l = 0, \ldots, r - 1 \,,$$

and $g \to w$ is a continuous linear mapping.

(*e*) Finally, one has to summarize the results and formulate existence theorems.

Remark 1.4.1 The preceding results show (in contrast to the heat equation, see III: Remark 2.3.4) that in the spaces chosen above for none of the three Fourier methods the operator defined by the left-hand side of (1.1.1) (and by Dirichlet boundary conditions) has a closed range.

Remark 1.4.2 The Dirichlet problem for the equation (1.4.1), where a and c are ω-periodic functions of t, is mentioned in VAGHI [38].

§ 2. The Poincaré method

2.1. *The linear case*: $-\infty < x < \infty$

Let us consider the problem $(\mathscr{P}_\omega)$ given by

$$u_{tt}(t, x) - u_{xx}(t, x) + 2a\, u_t(t, x) + c\, u(t, x) = g(t, x), \quad (t, x) \in \mathbf{R}^2, \tag{2.1.1}$$

$$u(t + \omega, x) - u(t, x) = 0, \quad (t, x) \in \mathbf{R}^2. \tag{2.1.2}$$

We assume that $a > 0$, $c > 0$, and $g \in C^{0,1}_\omega(\mathbf{R}^2)$.

By a solution we mean a function

$$u \in \mathscr{U} = C^2_\omega(\mathbf{R}^2)$$

satisfying (2.1.1).

The corresponding initial value problem $(\mathscr{I})$ is given by (2.1.1) and by

$$u(0, x) = \varphi(x), \quad u_t(0, x) = \psi(x), \quad x \in \mathbf{R}$$

with

$$\varphi \in C^2(\mathbf{R}), \quad \psi \in C^1(\mathbf{R}).$$

According to II: Lemma 2.3.1 and II: Remark 2.3.1 (p. 50) the problem $(\mathscr{I})$ has a unique classical solution $u \in C^2_{\text{loc}}((0, \infty) \times \mathbf{R}) \cap C^{1,0}_{\text{loc}}(\mathbf{R}^+ \times \mathbf{R})$. (In what follows, we find that it belongs to $C^2(\mathbf{R}^+ \times \mathbf{R})$.) This solution is determined by

$$u(t, x) = \Theta(\varphi, \psi)(t, x) + G(g)(t, x), \tag{2.1.3}$$

where

$$\Theta(\varphi, \psi)(t, x) = \tfrac{1}{2} e^{-at} \Big\{ \varphi(x + t) + \varphi(x - t) + \\ + \int_{x-t}^{x+t} [J_0(d^{1/2}(t^2 - (x - \xi)^2)^{1/2})(\psi(\xi) + a\, \varphi(\xi)) + \\ + \frac{\partial J_0(d^{1/2}(t^2 - (x - \xi)^2)^{1/2})}{\partial t} \varphi(\xi)]\, \mathrm{d}\xi \Big\}, \tag{2.1.4}$$

$$G(g)(t, x) = \tfrac{1}{2} e^{-at} \int_0^t \int_{x-t+\tau}^{x+t-\tau} e^{a\tau} J_0(d^{1/2}((t - \tau)^2 - \\ - (x - \xi)^2)^{1/2})\, g(\tau, \xi)\, \mathrm{d}\xi\, \mathrm{d}\tau, \tag{2.1.5}$$

J_0 is the Bessel function of order zero, and

$$d = -a^2 + c .$$

Let us study the operators Θ and G in more detail. We need some estimates of the Bessel function and its derivatives. By II: [21] we have

$$J_0'(\zeta) = -J_1(\zeta), \quad J_1'(\zeta) = \frac{J_1(\zeta)}{\zeta} - J_2(\zeta),$$

$$J_2'(\zeta) = \frac{2J_2(\zeta)}{\zeta} - J_3(\zeta),$$

$$|J_n(\zeta)| \leqq \frac{|\zeta|^n}{2^n n!} e^{|\mathrm{Im}\zeta|}, \quad \zeta \in \mathbf{C}, \quad n \in \mathbf{N}$$

(where J_n is the Bessel function of order n). In the above integrals $\zeta = d^{1/2}((t-\tau)^2 - (x-\xi)^2)^{1/2}$ with $(t-\tau)^2 - (x-\xi)^2 > 0$, $t - \tau > 0$, and we have (omitting now the argument $\zeta = d^{1/2}((t-\tau)^2 - (x-\xi)^2)^{1/2}$):

$$|J_0| \leqq 1,$$

$$\left|\frac{\partial J_0}{\partial x}\right| = \left|J_0' d \frac{-(x-\xi)}{d^{1/2}((t-\tau)^2 - (x-\xi)^2)^{1/2}}\right| =$$

$$= \left|J_1 d \frac{x-\xi}{d^{1/2}((t-\tau)^2 - (x-\xi)^2)^{1/2}}\right| \leqq \frac{d}{2}|x-\xi|,$$

$$\left|\frac{\partial J_0}{\partial t}\right| = \left|J_0' d \frac{t-\tau}{d^{1/2}((t-\tau)^2 - (x-\xi)^2)^{1/2}}\right| =$$

$$= \left|-J_1 d \frac{t-\tau}{d^{1/2}((t-\tau)^2 - (x-\xi)^2)^{1/2}}\right| \leqq \frac{d}{2}(t-\tau),$$

$$\left|\frac{\partial^2 J_0}{\partial t\,\partial x}\right| = \left|J_1' d^2 \frac{(t-\tau)(x-\xi)}{d((t-\tau)^2 - (x-\xi)^2)} -\right.$$

$$\left. - J_1 d^2 \frac{(t-\tau)(x-\xi)}{d^{3/2}((t-\tau)^2 - (x-\xi)^2)^{3/2}}\right| =$$

$$= \left|-J_2 d^2 \frac{(t-\tau)(x-\xi)}{d((t-\tau)^2 - (x-\xi)^2)}\right| \leqq \frac{d^2}{8}(t-\tau)|x-\xi|,$$

$$\left|\frac{\partial^2 J_0}{\partial x^2}\right| = \left|-J_1' d^2 \frac{(x-\xi)^2}{d((t-\tau)^2-(x-\xi)^2)} + \right.$$

$$\left. + J_1 d^2 \frac{(t-\tau)^2}{d^{3/2}((t-\tau)^2-(x-\xi)^2)^{3/2}}\right| =$$

$$= \left|J_1 d \frac{1}{d^{1/2}((t-\tau)^2-(x-\xi)^2)^{1/2}} + \right.$$

$$\left. + J_2 d^2 \frac{(x-\xi)^2}{d((t-\tau)^2-(x-\xi)^2)}\right| \leqq \frac{d}{2}\left(1+\frac{d}{4}(x-\xi)^2\right),$$

$$\left|\frac{\partial^2 J_0}{\partial t^2}\right| = \left|-J_1' d^2 \frac{(t-\tau)^2}{d((t-\tau)^2-(x-\xi)^2)} + \right.$$

$$\left. + J_1 d^2 \frac{(x-\xi)^2}{d^{3/2}((t-\tau)^2-(x-\xi)^2)^{3/2}}\right| =$$

$$= \left|-J_1 d \frac{1}{d^{1/2}((t-\tau)^2-(x-\xi)^2)^{1/2}} + \right.$$

$$\left. + J_2 d^2 \frac{(t-\tau)^2}{d((t-\tau)^2-(x-\xi)^2)}\right| \leqq \frac{d}{2}\left(1+\frac{d}{4}(t-\tau)^2\right),$$

$$\left|\frac{\partial^3 J_0}{\partial t^3}\right| = \left|-J_1' d^2 \frac{t-\tau}{d((t-\tau)^2-(x-\xi)^2)} + \right.$$

$$+ J_1 d^2 \frac{t-\tau}{d^{3/2}((t-\tau)^2-(x-\xi)^2)^{3/2}} +$$

$$+ J_2' d^3 \frac{(t-\tau)^3}{d^{3/2}((t-\tau)^2-(x-\xi)^2)^{3/2}} -$$

$$\left. - J_2 d^3 \frac{2(x-\xi)^2(t-\tau)}{d^2((t-\tau)^2-(x-\xi)^2)^2}\right| =$$

$$= \left|3J_2 d^2 \frac{t-\tau}{d((t-\tau)^2-(x-\xi)^2)} - \right.$$

$$\left. - J_3 d^3 \frac{(t-\tau)^3}{d^{3/2}((t-\tau)^2-(x-\xi)^2)^{3/2}}\right| \leqq$$

$$\leqq \frac{3}{8} d^2(t-\tau)\left(1+\frac{1}{18} d(t-\tau)^2\right) \tag{2.1.6}$$

for $d \geqq 0$. Writing $|d|$ instead of d and multiplying the right-hand sides by $e^{|d|^{1/2}((t-\tau)^2-(x-\xi)^2)^{1/2}}$, we obtain the corresponding estimates for $d < 0$.

Substituting $x + \chi$ for ξ in (2.1.4) and (2.1.5) and using the estimates (2.1.6) we obtain for $d \geqq 0$

$$\left|\Theta(\varphi, \psi)(t, x)\right| \leqq e^{-at} \sup_{x\in\mathbf{R}} |\varphi(x)| +$$

$$+ \tfrac{1}{2} e^{-at} \left\{ \int_{-t}^{t} \left| J_0(d^{1/2}(t^2 - \chi^2)^{1/2}) (\psi(x + \chi) + a\, \varphi(x + \chi)) + \right.\right.$$

$$\left.\left. + \frac{\partial J_0(d^{1/2}(t^2 - \chi^2)^{1/2})}{\partial t} \varphi(x + \chi) \right| \mathrm{d}\chi \right\} \leqq$$

$$\leqq e^{-at} \left(1 + at + \frac{dt^2}{2}\right) \sup_{x\in\mathbf{R}} |\varphi(x)| + e^{-at} t \sup_{x\in\mathbf{R}} |\psi(x)|$$

and

$$\left|G(g)(t, x)\right| \leqq \tfrac{1}{2} e^{-at} \int_0^t \int_{-t+\tau}^{t-\tau} e^{a\tau} \left| J_0(d^{1/2}((t - \tau)^2 - \right.$$

$$\left. - \chi^2)^{1/2})\, g(\tau, x + \chi) \right| \mathrm{d}\chi\, \mathrm{d}\tau \leqq \sup_{(t,x)\in\mathbf{R}^+\times\mathbf{R}} |g(t, x)| \int_0^t e^{-a(t-\tau)} (t - \tau)\, \mathrm{d}\tau\,.$$

For $d < 0$ we have

$$\left|\Theta(\varphi, \psi)(t, x)\right| \leqq e^{-(a-|d|^{1/2})t} \left(1 + at + \frac{|d|\, t^2}{2}\right) \sup_{x\in\mathbf{R}} |\varphi(x)| +$$

$$+ e^{-(a-|d|^{1/2})t}\, t \sup_{x\in\mathbf{R}} |\psi(x)|$$

and

$$\left|G(g)(t, x)\right| \leqq \sup_{(t,x)\in\mathbf{R}^+\times\mathbf{R}} |g(t, x)| \int_0^t e^{-(a-|d|^{1/2})(t-\tau)} (t - \tau)\, \mathrm{d}\tau\,.$$

Similar calculations lead to similar estimates for the derivatives of $\Theta(\varphi, \psi)$ and $G(g)$, so that we can conclude that

$$\max_{\alpha+\beta\leqq 2} \{\sup_{x\in\mathbf{R}} \left|D_t^\alpha D_x^\beta\, \Theta(\varphi, \psi)(t, x)\right|\} \leqq$$

$$\leqq E(t) \left(P_1(t) \|\varphi\|_{C^2(\mathbf{R})} + P_2(t) \|\psi\|_{C^1(\mathbf{R})}\right), \quad t \in \mathbf{R}^+ \tag{2.1.7}$$

and

$$\max_{\alpha+\beta\leqq 2} \{\sup_{x\in\mathbf{R}} \left|D_t^\alpha D_x^\beta\, G(g)(t, x)\right|\} \leqq$$

$$\leqq (1 + E(t))\, P_3(t) \|g\|_{C^{0,1}(\mathbf{R}^+\times\mathbf{R})}, \quad t \in \mathbf{R}^+, \tag{2.1.8}$$

where $E(t) = e^{-at}$ for $d \geqq 0$, $E(t) = e^{-(a-|d|^{1/2})t}$ for $d < 0$, and the $P_j(t)$ $(j = 1, 2, 3)$ are certain polynomials with positive coefficients. Taking into account that

$$\lim_{t \to \infty} e^{-\varkappa t} P(t) = 0 \tag{2.1.9}$$

for $\varkappa > 0$ and any polynomial P, we obtain, as a consequence of (2.1.7) and (2.1.8), $\Theta \in \mathscr{L}(C^2(\mathbf{R}) \times C^1(\mathbf{R}), C^2(\mathbf{R}^+ \times \mathbf{R}))$ and $G \in \mathscr{L}(C^{0,1}(\mathbf{R}^+ \times \mathbf{R}), C^2(\mathbf{R}^+ \times \mathbf{R}))$.

Since the solution to $(\mathscr{I})$ is unique, the existence of a solution of the problem $(\mathscr{P}_\omega)$ is equivalent to the existence of a fixed point $(\varphi, \psi) \in C^2(\mathbf{R}) \times C^1(\mathbf{R})$ of the translation operator T which is defined by

$$T(\varphi, \psi) = \Gamma(\omega)(\varphi, \psi) + \gamma ,$$

where

$$\Gamma(t)(\varphi, \psi)(\cdot) = \left(\Theta(\varphi, \psi)(t, \cdot), \ \frac{\partial}{\partial t} \Theta(\varphi, \psi)(t, \cdot)\right), \tag{2.1.10}$$

$$\gamma(\cdot) = \left(G(g)(\omega, \cdot), \ \frac{\partial}{\partial t} G(g)(\omega. \cdot)\right). \tag{2.1.11}$$

The norm of the linear operator $\Gamma(\omega)$ acting on $C^2(\mathbf{R}) \times C^1(\mathbf{R})$ need not be less than 1, so that we cannot use the fixed-point theorem, I: Theorem 3.1.1 for contracting operators. However, we can apply I: Theorem 3.1.2.; for we know that the operator $\Gamma(t)$ has the semigroup properties, that is

$$\Gamma(t_1 + t_2) = \Gamma(t_2)\Gamma(t_1), \quad \Gamma(0) = I, \quad t_1, t_2 \in \mathbf{R}^+ ,$$

hence $\Gamma^n(\omega) = \Gamma(n\omega)$, $n \in \mathbf{N}$.

It follows from (2.1.7) and (2.1.9) that there is an $m \in \mathbf{N}$ such that $\|\Gamma^m(\omega)\| < 1$, where the norm is taken in the space $\mathscr{L}(C^2(\mathbf{R}) \times C^1(\mathbf{R}), C^2(\mathbf{R}) \times C^1(\mathbf{R}))$. Since $T^m(\varphi, \psi) = \Gamma^m(\omega)(\varphi, \psi) + \sum_{k=0}^{m-1} \Gamma^k(\omega)\gamma$, we obtain

$$\|T^m(\varphi_1, \psi_1) - T^m(\varphi_2, \psi_2)\|_{C^2(\mathbf{R}) \times C^1(\mathbf{R})} \leqq$$

$$\leqq \|\Gamma^m(\omega)\| \, \|(\varphi_1, \psi_1) - (\varphi_2, \psi_2)\|_{C^2(\mathbf{R}) \times C^1(\mathbf{R})}$$

for any $(\varphi_1, \psi_1), (\varphi_2, \psi_2) \in C^2(\mathbf{R}) \times C^1(\mathbf{R})$ which shows that T^m is contracting. By I: Theorem 3.1.2, T has a unique fixed point, consequently, the problem

$(\mathscr{P}_\omega)$ has a unique solution $u \in \mathscr{U}$. Using (2.1.3) and (2.1.7) we can find it explicitly:

$$u(t, x) = \lim_{n\to\infty, n\in\mathbf{N}} u(t + n\omega, x) =$$

$$= \lim_{n\to\infty} \tfrac{1}{2} e^{-a(t+n\omega)} \int_0^{t+n\omega} \int_{x-t-n\omega+\tau}^{x+t+n\omega-\tau} e^{a\tau} J_0(d^{1/2}((t + n\omega - \tau)^2 -$$

$$- (x - \xi)^2)^{1/2})\, g(\tau, \xi)\, \mathrm{d}\xi\, \mathrm{d}\tau =$$

$$= \lim_{n\to\infty} \tfrac{1}{2} e^{-a(t+n\omega)} \int_0^{t+n\omega} \int_{x-\sigma}^{x+\sigma} e^{a(t+n\omega-\sigma)} J_0(d^{1/2}(\sigma^2 -$$

$$- (x - \xi)^2)^{1/2})\, g(t - \sigma, \xi)\, \mathrm{d}\xi\, \mathrm{d}\sigma =$$

$$= \frac{1}{2} \int_0^{\infty} \int_{x-\tau}^{x+\tau} e^{-a\tau} J_0(d^{1/2}(\tau^2 - (x - \xi)^2)^{1/2})\, g(t - \tau, \xi)\, \mathrm{d}\xi\, \mathrm{d}\tau\, .$$

The estimates (2.1.6) yield $\|u\|_{\mathscr{U}} \leqq C\|g\|_{C_\omega{}^{0,1}(\mathbf{R}^2)}$ with $C > 0$ independent of g. Denoting by L the operator $Lu = u_{tt} - u_{xx} + 2au_t + cu$ on $\mathscr{U}$, we have the following theorem.

THEOREM 2.1.1 (*The non-critical case*). *Let* $a > 0$, $c > 0$ *and* $g \in C_\omega^{0,1}(\mathbf{R}^2)$. *Then the problem* $(\mathscr{P}_\omega)$ *has a unique solution* $u = L^{-1}g \in \mathscr{U}$ *and* $L^{-1} \in \mathscr{L}(C_\omega^{0,1}(\mathbf{R}^2), \mathscr{U})$.

Remark 2.1.1 We may suppose that $g \in C_\omega^{1,0}(\mathbf{R}^2)$ instead of $g \in C_\omega^{0,1}(\mathbf{R}^2)$. It is only necessary to change the order of integration when estimating (2.1.5).

Remark 2.1.2 If $a < 0$ in (2.1.1), then by setting $t = -s$ we come back to the case above.

Remark 2.1.3 If the problem is given by

$$v_{tt}(t, x) - v_{xx}(t, x) + 2a\, v_t(t, x) + 2b\, v_x(t, x) + c_1\, v(t, x) =$$

$$= g_1(t, x)\,, \quad (t, x) \in \mathbf{R}^2\,, \tag{2.1.12}$$

$$v(t + \omega, x) - v(t, x) = 0\,, \quad (t, x) \in \mathbf{R}^2 \tag{2.1.13}$$

$(a, b \in \mathbf{R}, a \neq 0)$, then by putting $v(t, x) = e^{bx} u(t, x)$ we arrive at the problem $(\mathscr{P}_\omega)$ for u with $c = c_1 + b^2$ and $g(t, x) = e^{-bx} g_1(t, x)$. If the assumptions on a, c and g of Theorem 2.1.1 are satisfied, we can establish an existence theorem for (2.1.12), (2.1.13). Now v belongs to the Banach space of functions for which

$$e^{-bx} v(t, x) \in C_\omega^2(\mathbf{R}^2)\,.$$

2.2. *The linear case*: $0 \leqq x \leqq \pi$

We consider the problem $(\mathscr{P}_\omega)$ given by

$$u_{tt}(t, x) - u_{xx}(t, x) + 2a\, u_t(t, x) + c\, u(t, x) = g(t, x)\,,$$

$$(t, x) \in Q = \mathbf{R} \times (0, \pi)\,, \tag{2.2.1}$$

$$u(t, 0) = u(t, \pi) = 0\,, \quad t \in \mathbf{R}\,, \tag{2.2.2}$$

$$u(t + \omega, x) - u(t, x) = 0\,, \quad (t, x) \in Q\,, \tag{2.2.3}$$

where $a > 0$, $c > 0$ and $g \in C_\omega^{0,1}(\overline{Q})$ (or $g \in C_\omega^{1,0}(\overline{Q})$).

By a solution we mean a function

$$u \in \mathscr{U} = C_\omega^2(\overline{Q})$$

satisfying (2.2.1) and (2.2.2).

The corresponding initial boundary-value problem $(\mathscr{I}\mathscr{B})$ is given by (2.2.1), (2.2.2) and by

$$u(0, x) = \varphi(x)\,, \quad u_t(0, x) = \psi(x)\,, \quad x \in [0, \pi]$$

with $\varphi \in C^2([0, \pi])$, $\psi \in C^1([0, \pi])$.

By II: Theorem 2.3.1 the problem $(\mathscr{I}\mathscr{B})$ has a unique classical solution which is determined by (2.1.3) provided that

$$\varphi(0) = \varphi(\pi) = \varphi''(0) = \varphi''(\pi) = \psi(0) = \psi(\pi) = 0\,,$$

$$g(t, 0) = g(t, \pi) = 0\,, \quad t \in \mathbf{R}$$

and φ, ψ, and g are extended in x to the whole x-axis as odd and 2π-periodic functions (the extended functions are denoted by the same symbols as the original ones).

Let $\Phi = C^2_{2\pi(o)}(\mathbf{R}) \times C^1_{2\pi(o)}(\mathbf{R})$. It can be verified that the operator $\Gamma(t)$ given by (2.1.10) maps Φ into itself and that the pair γ given by (2.1.11) belongs to Φ. The same arguments as in the previous section yield the following theorem.

THEOREM 2.2.1 (*The non-critical case*). *Let $a > 0$, $c > 0$, and $g \in \dot{C}_\omega^{(\alpha,1-\alpha)}(\overline{Q})$ with $\alpha = 0$ or 1.*
Then the problem $(\mathscr{P}_\omega)$ has a unique solution $u = L^{-1}g \in \mathscr{U}$ and $L^{-1} \in \mathscr{L}(\dot{C}_\omega^{(\alpha,1-\alpha)}(\overline{Q}), \mathscr{U})$.

Remark 2.2.1 Since it is our intention to acquaint the reader with another approach to the solution of the problem where $x \in (0, \pi)$, we have proceeded in a different way from that in III: Sec. 3.1; namely, we have used the known solution of the problem for $x \in \mathbf{R}$. An unpleasant consequence is that we are compelled to retain the assumption $c > 0$. We could get rid of this restriction by proceeding as in III: Sec. 3.1 and using the Fredholm theory of systems of integral equations.

Remark 2.2.2 A similar approach can be used to study the problem given by

$$u_{tt}(t, x) - u_{xx}(t, x) + 2a\, u_t(t, x) + c\, u(t, x) = g(t, x)\,,$$

$$(t, x) \in \mathbf{R} \times (0, \infty)\,,$$

$$u(t, 0) = 0\,, \quad t \in \mathbf{R}$$

and by (2.2.3).

2.3. *The weakly non-linear case*

Let us deal with the problem $(\mathscr{P}^{\varepsilon}_{\omega})$ given by

$$u_{tt}(t, x) - u_{xx}(t, x) + 2a\, u_t(t, x) + c\, u(t, x) =$$
$$= \varepsilon\, F(u, \varepsilon)\,(t, x)\,, \quad (t, x) \in \mathbf{R}^2\,, \quad \varepsilon \in [-\varepsilon_0, \varepsilon_0]\,, \quad (\varepsilon_0 > 0) \tag{2.3.1}$$

and by (2.1.2).

A function $u \in C([-\varepsilon^*, \varepsilon^*]; C^2_{\omega}(\mathbf{R}^2))$ $(\varepsilon^* \in (0, \varepsilon_0])$ is called a solution of $(\mathscr{P}^{\varepsilon}_{\omega})$ if $u(\varepsilon)$ satisfies (2.3.1) for every $\varepsilon \in [-\varepsilon^*, \varepsilon^*]$.

The usual procedure yields the following theorem.

THEOREM 2.3.1 (*The non-critical case*). *Let* $a \neq 0$, $c > 0$. *Suppose that*

(*a*) *the operator* $F(u, \varepsilon)$ *maps* $B(0, \varrho; C^2_{\omega}(\mathbf{R}^2)) \times [-\varepsilon_0, \varepsilon_0]$ *into* $C^{0,1}_{\omega}(\mathbf{R}^2)$ *continuously for some* $\varrho > 0$;

(*b*) *there is a* $\lambda > 0$ *such that*

$$\|F(u_1, \varepsilon) - F(u_2, \varepsilon)\|_{C_{\omega}^{0,1}(\mathbf{R}^2)} \leqq \lambda \|u_1 - u_2\|_{C_{\omega}^{2}(\mathbf{R}^2)}$$

for all $u_1, u_2 \in B(0, \varrho; C^2_{\omega}(\mathbf{R}^2))$ *and* $\varepsilon \in [-\varepsilon_0, \varepsilon_0]$.
Then the problem $(\mathscr{P}^{\varepsilon}_{\omega})$ *has a locally unique solution* $u^*(\varepsilon) \in C^2_{\omega}(\mathbf{R}^2)$ *such that* $u^*(0) = 0$.

Remark 2.3.1 Let

$$F(u, \varepsilon)\,(t, x) = f(t, x, u(t, x), u_t(t, x), u_x(t, x), \varepsilon)\,.$$

The assumptions (a) and (b) are satisfied if $f = f(t, x, u_0, u_1, u_2, \varepsilon)$ is ω-periodic in t, and $f, f_x, f_{u_j}, j = 0, 1, 2$, are continuous and bounded on $\mathbf{R}^2 \times [-\varrho, \varrho]^3 \times [-\varepsilon_0, \varepsilon_0]$ for some $\varrho > 0$, Lipschitz continuous in u_0, u_1, and u_2 uniformly with respect to $(t, x, \varepsilon) \in \mathbf{R}^2 \times [-\varepsilon_0, \varepsilon_0]$, and continuous in ε uniformly with respect to $(t, x, u_0, u_1, u_2) \in \mathbf{R}^2 \times [-\varrho, \varrho]^3$.

Problem 2.3.1 Investigate the weakly non-linear problem corresponding to the problem studied in Sec. 2.2.

§ 3. Singularly perturbed problems

3.1. *General considerations*

In this paragraph we deal with periodic solutions of the so-called hyperbolic heat equation, which describes the transfer of heat more precisely than the heat equation in Chap. III. Let us investigate the problem $(\mathscr{P}_\omega^{\mu,\varepsilon})$ given by

$$\mu\, u_{tt}(t, x) + u_t(t, x) + E\, u(t, x) + c\, u(t, x) =$$

$$= \varepsilon\, F(u, \varepsilon)(t, x)\,, \quad (t, x) \in Q = \mathbf{R} \times \Omega\,, \tag{3.1.1}$$

$$u(t, x) = 0\,, \quad (t, x) \in \mathbf{R} \times \partial\Omega\,, \tag{3.1.2}$$

$$u(t + \omega, x) - u(t, x) = 0\,, \quad (t, x) \in Q\,, \tag{3.1.3}$$

where $\mu \in \mathbf{R}^+$, E given by (1.1.16) is uniformly elliptic and positive definite with a_{jk}, $a_0 \in C^\infty(\Omega)$, $a_{jk} = a_{kj}$, $\varepsilon \in [-\varepsilon_0, \varepsilon_0]$, $\varepsilon_0 > 0$, $\Omega \subset \mathbf{R}^n$ is a bounded domain of class C^∞ and c is a constant (the assumptions on smoothness of a_{jk}, a_0 and Ω can be weakened substantially; see II: Remark 1.1.1).

We are interested, above all, in the problem of convergence of the solutions $u^\mu(\varepsilon)$ of $(\mathscr{P}_\omega^{\mu,\varepsilon})$ to the solution $u^0(\varepsilon)$ of $(\mathscr{P}_\omega^{0,\varepsilon})$ as $\mu \to 0+$. We use the (t, s)-Fourier method. Note that results similar to those established below can be proved by applying the same procedure to the problem with the Neumann or the Newton boundary conditions. More generally, we can derive without substantial changes similar results for the abstract problem

$$\mu\, u''(t) + u'(t) + A\, u(t) = \varepsilon\, F(u, \varepsilon)(t)\,, \quad t \in \mathbf{R}\,,$$

$$u(t + \omega) - u(t) = 0\,, \quad t \in \mathbf{R}\,,$$

where $u : \mathbf{R} \to H$, H is a Hilbert space, and $A : \mathscr{D}(A) \subset H \to H$ is a linear selfadjoint positive definite operator (see [36]). This abstract problem is also examined by another method in VII: [64].

By II: Theorem 1.3.2 the operator E is self-adjoint in $H^0(\Omega)$ with $\mathscr{D}(E) = H^2(\Omega) \cap \dot{H}^1(\Omega)$ and has an orthonormal base $\{v_k\}_{k \in \mathbf{N}}$ of eigenfunctions for the corresponding positive eigenvalues λ_k, $(k \in \mathbf{N})$. A function $u^\mu : [-\varepsilon^*, \varepsilon^*] \to \mathscr{U}_\mu = H^2_\omega(Q) \cap \dot{H}^1_\omega(Q)$, $(\mu > 0, \varepsilon^* > 0)$ is called a solution of $(\mathscr{P}_\omega^{\mu,\varepsilon})$ if $u^\mu(\varepsilon)$

satisfies (3.1.1) for every $\varepsilon \in [-\varepsilon^*, \varepsilon^*]$ almost everywhere in Q. For $\mu = 0$ we set $\mathscr{U}_0 = H^{1,2}_\omega(Q) \cap \dot{H}^1_\omega(Q)$ and define a solution to $(\mathscr{P}^{0,\varepsilon}_\omega)$ quite similarly. Now II: Theorem 1.3.4 implies that $u \in \mathscr{U}_\mu$ for $\mu > 0$ if and only if $\sum_{j,k}(j^4 + \lambda_k^2)\,|u_{jk}|^2 < \infty$, where

$$u_{jk} = \omega^{-1/2} \int_0^\omega \int_\Omega u(t, x)\, e^{-ij\nu t}\, v_k(x)\, \mathrm{d}x\, \mathrm{d}t\,, \quad j \in \mathbf{Z}\,, \quad k \in \mathbf{N}\,.$$

(For $u \in \mathscr{U}_0$ we have $\sum_{j,k}(j^2 + \lambda_k^2)\,|u_{jk}|^2 < \infty$.) In what follows we distinguish the cases

(i) $\lambda_k \neq -c$ for all $k \in \mathbf{N}$ (the non-critical case);
(ii) $\lambda_{k_0} = -c$ for some $k_0 \in \mathbf{N}$ (the critical case).

(For simplicity we assume λ_{k_0} to be simple.)

We denote by $\mathscr{V}$ the subspace of $H^0_\omega(Q)$ of functions of the form $u(t, x) = d\, v_{k_0}(x)$, $d \in \mathbf{R}$, and by $\mathscr{W}$ its orthogonal complement in $H^0_\omega(Q)$, that is, the space of functions in $H^0_\omega(Q)$ of the form

$$u(t, x) = \omega^{-1/2} \sum_{\substack{j,k \\ (j,k) \neq (0,k_0)}} u_{jk} e^{ij\nu t}\, v_k(x)\,.$$

We define an operator L_μ by $\mathscr{D}(L_\mu) = \mathscr{U}_\mu$ and $L_\mu u = \mu u_{tt} + u_t + Eu + cu$. By III: Sec. 1.1 (Theorems 1.1.3 and 1.1.4) and Sec. 1.1 of the present chapter (Theorems 1.1.3 and 1.1.4), we have

$$L_0^{-1} \in \mathscr{L}(H^0_\omega(Q), \mathscr{U}_0)\,, \quad L_\mu^{-1} \in \mathscr{L}(H^\alpha_\omega(\mathbf{R}; \mathscr{D}(E^{(1-\alpha)/2})), \mathscr{U}_\mu)$$

or

$$L_0^{-1} \in \mathscr{L}(\mathscr{W}, \mathscr{U}_0 \cap \mathscr{W})\,, \quad L_\mu^{-1} \in \mathscr{L}(H^\alpha_\omega(\mathbf{R}; \mathscr{D}(E^{(1-\alpha)/2})) \cap \mathscr{W}, \mathscr{U}_\mu \cap \mathscr{W})\,,$$

if (i) or (ii) respectively, holds, where $\mu > 0$, $\alpha \in [0, 1]$.

In what follows we derive some properties of the family of operators L_μ, $\mu \geqq 0$, which describe adequately the behaviour with respect to μ of the solution to the linear problem $L_\mu u = g$ and are useful in the treatment of the non-linear problem (3.1.1)–(3.1.3).

THEOREM 3.1.1 *Suppose that an operator E and a domain Ω satisfy the above assumptions and that (i) or (ii) holds.*
Then for any $\mu_0 > 0$ there exists a constant K such that

$$\|L_\mu^{-1}\|_{\mathscr{L}(H_\omega{}^{1,0}(Q), \mathscr{U}_\mu)} \leqq K \tag{3.1.4}$$

or

$$\|L_\mu^{-1}\|_{\mathscr{L}(H_\omega{}^{1,0}(Q) \cap \mathscr{W}, \mathscr{U}_\mu \cap \mathscr{W})} \leqq K \quad \textit{for all} \quad \mu \in [0, \mu_0]\,.$$

Proof. Suppose, for example, that (i) is satisfied (the case (ii) is treated similarly).

Let $g \in H^{1,0}_{\omega}(Q)$, $L_\mu u^\mu = g$. We also suppose that $\nu = (2\pi/\omega) = 1$, $c = 0$ without loss of generality. Then

$$\|u^\mu\|^2_{\mathscr{U}_\mu} = \sum_{j,k} \frac{(j^4 + \lambda_k^2)\,|g_{jk}|^2}{j^2 + (\lambda_k - \mu j^2)^2},$$

where

$$g_{jk} = (2\pi)^{-1/2} \int_0^{2\pi} \int_\Omega g(t, x)\, e^{-ijt}\, v_k(x)\, \mathrm{d}x\, \mathrm{d}t\,, \quad j \in \mathbf{Z}\,, \quad k \in \mathbf{N}\,.$$

To prove (3.1.4) it suffices to show that

$$\frac{j^4 + \lambda_k^2}{j^2 + (\lambda_k - \mu j^2)^2} \leqq \text{const}\,.\, j^2$$

for $j \in \mathbf{Z}, j \neq 0, k \in \mathbf{N}$.

Let $\alpha \geqq 0$, $\beta \geqq 0$, $\alpha + \beta = 2$ and $\lambda_k = j^2 \Theta_{jk}$ (put $\Theta = \Theta_{jk}$). Then we have

$$\frac{j^{2\alpha}\lambda_k^\beta}{j^2 + (\lambda_k - \mu j^2)^2} = \frac{j^2 \Theta^\beta}{1 + j^2(\Theta - \mu)^2} \leqq j^2\, \varphi(\Theta, \mu)\,,$$

where

$$\varphi(\Theta, \mu) = \frac{\Theta^\beta}{1 + (\Theta - \mu)^2}\,.$$

As $\lim_{\Theta \to \infty} \varphi(\Theta, \mu) = 0$ or 1, uniformly with respect to $\mu \in [0, \mu_0]$, there exists a Θ_0 such that

$$\sup \{\varphi(\Theta, \mu);\, \Theta \in \mathbf{R}^+,\, \mu \in [0, \mu_0]\} =$$
$$= \max \{\varphi(\Theta, \mu),\, \varphi(\infty, \mu);\, \Theta \in [0, \Theta_0],\, \mu \in [0, \mu_0]\} < \infty\,.$$

We complete the proof by putting $\alpha = 2$, $\beta = 0$ or $\alpha = 0$, $\beta = 2$.

THEOREM 3.1.2 *Suppose that E and Ω satisfy the above assumptions and that (i) or (ii) holds.*
Then for any $\mu_0 > 0$ there exists a constant K such that

$$\|L_\mu^{-1}\|_{\mathscr{L}(\dot{H}_\omega{}^{0,1}(Q), \mathscr{U}_\mu)} \leqq \frac{K}{\mu}$$

or

$$\|L_\mu^{-1}\|_{\mathscr{L}(\dot{H}_\omega{}^{0,1}(Q) \cap \mathscr{W}, \mathscr{U}_\mu \cap \mathscr{W})} \leqq \frac{K}{\mu} \quad \textit{for} \quad \mu \in (0, \mu_0]\,. \tag{3.1.5}$$

Proof. Proceeding as in the proof of Theorem 3.1.1, we find it sufficient to show that (again we put $v = 1$, $c = 0$)

$$\frac{j^4 + \lambda_k^2}{j^2 + (\lambda_k - \mu j^2)^2} \leqq \frac{\text{const. } \lambda_k}{\mu^2}, \quad k \in \mathbf{N} .$$

Setting $j^2 = \Theta\lambda_k$ we find for $\alpha \geqq 0$, $\beta \geqq 0$, $\alpha + \beta = 2$

$$\frac{|j|^{2\alpha}\lambda_k^\beta}{j^2 + (\lambda_k - \mu j^2)^2} = \frac{\Theta^\alpha \lambda_k}{\Theta + \lambda_k(1 - \mu\Theta)^2} \leqq$$

$$\leqq \lambda_k(\min_{k\in\mathbf{N}} (1, \lambda_k))^{-1} \varphi(\Theta, \mu) ,$$

where

$$\varphi(\Theta, \mu) = \frac{\Theta^\alpha}{\Theta + (1 - \mu\Theta)^2} . \tag{3.1.6}$$

Let $\mu \in (0, \mu_0]$ be arbitrary but fixed. Then $\varphi(0, \mu) = 0$ or 1 and

$$\lim_{\Theta\to\infty} \varphi(\Theta, \mu) = \begin{cases} 0 & \text{if} \quad \alpha < 2 \\ \mu^{-2} & \text{if} \quad \alpha = 2 . \end{cases}$$

Further, it is clear that $\Theta + (1 - \mu\Theta)^2 \geqq 1$, $\Theta \in \mathbf{R}^+$ if $\mu \leqq \frac{1}{2}$ and

$$\min_{\Theta\in\mathbf{R}^+} (\Theta + (1 - \mu\Theta)^2) = \min [1, (4\mu - 1)/4\mu^2] \geqq (2\mu_0)^{-2}$$

if $\mu > \frac{1}{2}$. Moreover, $\varphi(\Theta, \mu) \leqq \Theta^{\alpha-1}$ for $\Theta > 0$. Thus, for $\alpha \in [0, 1]$ we have $\sup_{\Theta\in\mathbf{R}^+} \varphi(\Theta, \mu) \leqq$ const. $< \infty$, independently of μ. Now let $\alpha \in [1, 2]$. Then $\varphi(\Theta, \mu) \leqq \Theta^{\alpha-1} \leqq \mu^{2(1-\alpha)}$ provided that $0 \leqq \Theta \leqq \mu^{-2}$. On the other hand if $\Theta > \mu^{-2}$ then it can be verified easily that $\inf \{\mu^{-2}\Theta^{-2}[\Theta + (1 - \mu\Theta)^{-2}]; \Theta > 0, \mu \in (0, \mu_0]\} = \varkappa > 0$. Hence $\varphi(\Theta, \mu) \leqq \varkappa^{-1}\mu^{-2}\Theta^{\alpha-2} \leqq \varkappa^{-1}\mu^{-2}(\mu^{-2})^{\alpha-2} = \varkappa^{-1}\mu^{2(1-\alpha)}$. We complete the proof by putting $\alpha = 2$, $\beta = 0$ or $\alpha = 0$, $\beta = 2$.

THEOREM 3.1.3. *Suppose that E and Ω satisfy the above assumptions and that (i) or (ii) holds.*

Then there exists a constant $M > 0$ such that

$$\|L_\mu^{-1} - L_0^{-1}\| \leqq M\mu , \quad \mu \in \mathbf{R}^+ , \tag{3.1.7}$$

where the norm in (3.1.7) *is taken in* $\mathscr{L}(H_\omega^{1,0}(Q), \mathscr{U}_0)$ *or in* $\mathscr{L}(H_\omega^{1,0}(Q) \cap \mathscr{W}, \mathscr{U}_0 \cap \mathscr{W})$.

Proof. Suppose, for example, that (i) is satisfied. Let $\mu > 0$ be arbitrary and let u^μ and u^0 satisfy, respectively, $L_\mu u^\mu = g$ and $L_0 u^0 = g$ with some $g \in H^{1,0}_\omega(Q)$. Subtracting the last two equalities we get

$$u^\mu - u^0 = -\mu L_0^{-1} u^\mu_{tt} \,. \tag{3.1.8}$$

This equality rewritten in terms of the Fourier coefficients means that

$$u^\mu_{jk} - u^0_{jk} = \mu \frac{v^2 j^2 g_{jk}}{(ijv + \lambda_k + c)(ijv + \lambda_k + c - \mu j^2 v^2)}\,, \tag{3.1.9}$$

where

$$u^\mu_{jk} = \omega^{-1/2} \int_0^\omega \int_\Omega u^\mu(t, x)\, e^{-ijvt}\, v_k(x)\, dx\, dt\,,$$

$$g_{jk} = \omega^{-1/2} \int_0^\omega \int_\Omega g(t, x)\, e^{-ijvt}\, v_k(x)\, dx\, dt\,, \quad j \in \mathbf{Z}\,, \quad k \in \mathbf{N}\,.$$

It suffices to prove that

$$\sum_{j,k} (j^2 + \lambda_k^2) \left|u^\mu_{jk} - u^0_{jk}\right|^2 \leqq \text{const.}\, \mu^2 \sum_{j,k} j^2 \left|g_{jk}\right|^2 .$$

Since $u^\mu_{0k} = u^0_{0k}$, this inequality is a consequence of the estimates

$$j^2 \left|u^\mu_{jk} - u^0_{jk}\right|^2 = \frac{\mu^2 v^4 j^6 \left|g_{jk}\right|^2}{[v^2 j^2 + (\lambda_k + c)^2]\,[v^2 j^2 + (\lambda_k + c - \mu v^2 j^2)^2]} \leqq$$

$$\leqq \mu^2 v^4 \frac{j^6 \left|g_{jk}\right|^2}{(j^2 v^2)(j^2 v^2)} = \mu^2 j^2 \left|g_{jk}\right|^2\,, \quad j \neq 0$$

and

$$\lambda_k^2 \left|u^\mu_{jk} - u^0_{jk}\right|^2 = \mu^2 v^4 \frac{j^4 \lambda_k^2 \left|g_{jk}\right|^2}{[v^2 j^2 + (\lambda_k + c)^2]\,[v^2 j^2 + (\lambda_k + c - \mu v^2 j^2)^2]} \leqq$$

$$\leqq \mu^2 v^4 \frac{j^4 \lambda_k^2 \left|g_{jk}\right|^2}{(\lambda_k + c)^2 (j^2 v^2)} \leqq \mu^2 v^2 \sup_{k \in \mathbf{N}} \frac{\lambda_k^2}{(\lambda_k + c)^2} j^2 \left|g_{jk}\right|^2\,, \quad j \neq 0\,.$$

THEOREM 3.1.4 *Suppose that E, and Ω satisfy the above assumptions and that (i) or (ii) holds.*
Then there exists a constant $M > 0$ such that

$$\left\|(L_\mu^{-1} - L_0^{-1})\, g\right\|_{H_\omega^{\alpha_1}(\mathbf{R}; H^0(\Omega))} \leqq M \mu^{1-\alpha_1} \left\|g\right\|_{H_\omega^0(\mathbf{R}; \mathscr{D}(E^{\alpha_1/2}))} \tag{3.1.10}$$

for $0 \leqq \alpha_1 < 1$ *and* $g \in H^0_\omega(\mathbf{R}; \mathscr{D}(E^{\alpha_1/2}))$ *or* $g \in H^0_\omega(\mathbf{R}; \mathscr{D}(E^{\alpha_1/2})) \cap \mathscr{W}$, *or*

$$\left\|(L_\mu^{-1} - L_0^{-1})\, g\right\|_{H_\omega^0(\mathbf{R}; \mathscr{D}(E^{\alpha_2/2}))} \leqq M \mu^\delta \left\|g\right\|_{H_\omega^0(\mathbf{R}; \mathscr{D}(E^{(\alpha_2 - 1 + \delta)/2}))} \tag{3.1.11}$$

for $0 \leqq \alpha_2 < 2$, $\delta > 0$, $1 - \alpha_2 \leqq \delta \leqq 1$ *and* $g \in H^0_\omega(\mathbf{R}; \mathcal{D}(E^{(\alpha_2 - 1 + \delta)/2}))$ *or* $g \in H^0_\omega(\mathbf{R}; \mathcal{D}(E^{(\alpha_2 - 1 + \delta)/2})) \cap \mathcal{W}$.

Proof. Suppose that (i) is satisfied (the other case is quite similar). It is clear from (3.1.9) that

$$\left\|\left(L_\mu^{-1} - L_0^{-1}\right) g\right\|^2_{H_\omega^{\alpha_1}(\mathbf{R}; H^0(\Omega))} =$$

$$= \mu^2 v^4 \sum_{j,k} \frac{|j|^{2\alpha_1 + 4} |g_{jk}|^2}{[v^2 j^2 + (\lambda_k + c)^2][v^2 j^2 + (\lambda_k + c - \mu v^2 j^2)^2]}$$

and

$$\left\|\left(L_\mu^{-1} - L_0^{-1}\right) g\right\|^2_{H_\omega^0(\mathbf{R}; \mathcal{D}(E^{\alpha_2/2}))} =$$

$$= \mu^2 v^4 \sum_{j,k} \frac{j^4 \lambda_k^{\alpha_2} |g_{jk}|^2}{[v^2 j^2 + (\lambda_k + c)^2][v^2 j^2 + (\lambda_k + c - \mu v^2 j^2)^2]} .$$

Taking the terms with the same indices j, k in the two series and putting $j^2 = = v^{-2} \Theta \lambda_k$, we obtain

$$\mu^2 v^{-2\alpha_1} \frac{\Theta^{\alpha_1 + 2} \lambda_k^{\alpha_1 + 2} |g_{jk}|^2}{\lambda_k^2 \left[\Theta + \frac{(\lambda_k + c)^2}{\lambda_k}\right]\left[\Theta + \lambda_k \left(\frac{\lambda_k + c}{\lambda_k} - \mu\Theta\right)^2\right]},$$

$$\mu^2 \frac{\Theta^2 \lambda_k^{\alpha_2 + 2} |g_{jk}|^2}{\lambda_k^2 \left[\Theta + \frac{(\lambda_k + c)^2}{\lambda_k}\right]\left[\Theta + \lambda_k \left(\frac{\lambda_k + c}{\lambda_k} - \mu\Theta\right)^2\right]} . \tag{3.1.12}$$

In what follows we restrict ourselves to the case $c = 0$, because the general case involves only formal complications. Since $a^p b^{1-p} \leqq pa + (1 - p) b$ for any positive real numbers a, b, $0 \leqq p \leqq 1$, we have $\Theta + \lambda_k \geqq \Theta^p \lambda_k^{1-p}$. Thus, the expressions (3.1.12) have the upper estimate

$$\mu^2 v^{-2\alpha_1} \left(\min_{k \in \mathbf{N}} (1, \lambda_k)\right)^{-1} \frac{\Theta^{\alpha_1 + 2 - p} \lambda_k^{\alpha_1 - 1 + p}}{\Theta + (1 - \mu\Theta)^2} |g_{jk}|^2 \tag{3.1.13}$$

and

$$\mu^2 \left(\min_{k \in \mathbf{N}} (1, \lambda_k)\right)^{-1} \frac{\Theta^{2 - p} \lambda_k^{\alpha_2 - 1 + p}}{\Theta + (1 - \mu\Theta)^2} |g_{jk}|^2 , \tag{3.1.14}$$

respectively. We can obtain appropriate estimates of these expressions if we only know the value of $\sup_{\Theta \in \mathbf{R}^+} \varphi(\Theta, \mu)$ given by (3.1.6). It was shown in the proof of Theorem 3.1.2 that $\sup_{\Theta \in \mathbf{R}^+} \varphi(\Theta, \mu) \leqq \text{const.}/\mu^{2(\alpha - 1)}$ if $1 < \alpha < 2$.

Setting $p = 1$, $\alpha = 1 + \alpha_1$ and $p = \delta$, $\alpha = 2 - \delta$ in (3.1.13) and (3.1.14) respectively, we obtain (3.1.10) and (3.1.11) after some rearrangements.

3.2. *The main theorems*

Let α_1, α_2 and δ be the numbers in Theorem 3.1.4. We write $\mathcal{G} = H_\omega^{1,0}(Q)$ or $\mathcal{G} = H_\omega^0(\mathbf{R}; \mathcal{D}(E^{\alpha_1/2}))$ or $\mathcal{G} = H_\omega^0(\mathbf{R}; \mathcal{D}(E^{(\alpha_2 - 1 + \delta)/2}))$ and still suppose that Ω, a_{jk}, and a_0 satisfy the assumptions at the beginning of Sec. 3.1.

THEOREM 3.2.1 (*The non-critical case*). *Let $\lambda_k \neq -c$ for all $k \in \mathbf{N}$. Suppose that:*

(*a*) *for some $\varrho > 0$, the operator $F(u, \varepsilon)$ maps $B(0, \varrho; \mathcal{U}_\mu) \times [-\varepsilon_0, \varepsilon_0]$ continuously into $\mathcal{G}$ if $\mu > 0$ and into $H_\omega^0(Q)$ if $\mu = 0$;*

(*b*) *there is a $\lambda > 0$ such that*

$$\|F(u_1, \varepsilon) - F(u_2, \varepsilon)\|_{\mathcal{G}} \leqq \lambda \|u_1 - u_2\|_{\mathcal{U}_\mu}$$

for all $u_1, u_2 \in B(0, \varrho; \mathcal{U}_\mu)$ and $\varepsilon \in [-\varepsilon_0, \varepsilon_0]$, ($\varepsilon_0 > 0$),

$$\|F(u_1, \varepsilon) - F(u_2, \varepsilon)\|_{H_\omega{}^0(Q)} \leqq \lambda \|u_1 - u_2\|_{H_\omega{}^\alpha(\mathbf{R}; \mathcal{D}(E^{\beta/2}))}$$

for all $u_1, u_2 \in B(0, \varrho; \mathcal{U}_0)$ and $\varepsilon \in [-\varepsilon_0, \varepsilon_0]$, where $(\alpha, \beta) = (1, 2)$ or $(\alpha, \beta) = (\alpha_1, 0)$ or $(\alpha, \beta) = (0, \alpha_2)$ provided that $\mathcal{G} = H_\omega^{1,0}(Q)$ or $\mathcal{G} = H_\omega^0(\mathbf{R}; \mathcal{D}(E^{\alpha_1/2}))$ or $\mathcal{G} = H_\omega^0(\mathbf{R}; \mathcal{D}(E^{(\alpha_2 - 1 + \delta)/2}))$, respectively;

(*c*) $\sup \{\|F(u, \varepsilon)\|_{H_\omega{}^\alpha(\mathbf{R}; \mathcal{D}(E^{\beta/2}))}; \|u\|_{\mathcal{U}_\mu} \leqq \varrho, \ |\varepsilon| \leqq \varepsilon_0\} = K_0 < \infty$, *$\alpha$ and β as above.*

Then for any $\mu \in [0, \mu_0]$ there is an $\varepsilon(\mu) > 0$ such that the problem $(\mathcal{P}_\omega^{\mu,\varepsilon})$ has a unique solution $u^\mu(\varepsilon) \in \mathcal{U}_\mu$ for any $\varepsilon \in [-\varepsilon(\mu), \varepsilon(\mu)]$, and

$$\|u^\mu(\varepsilon) - u^0(\varepsilon)\|_{H_\omega{}^{1,2}(Q)} \leqq c_1 |\varepsilon| \mu, \quad \mu \in [0, \mu_0],$$

$$|\varepsilon| \leqq \min(\varepsilon(\mu), (2\lambda K)^{-1}) \equiv m(\mu) \quad \text{if} \quad \mathcal{G} = H_\omega^{1,0}(Q); \tag{3.2.1}$$

$$\|u^\mu(\varepsilon) - u^0(\varepsilon)\|_{H_\omega{}^{\alpha_1}(\mathbf{R}; H^0(\Omega))} \leqq c_2 |\varepsilon| \mu^{1-\alpha_1},$$

$$\mu \in [0, \mu_0], \quad |\varepsilon| \leqq m(\mu) \quad \text{if} \quad \mathcal{G} = H_\omega^0(\mathbf{R}; \mathcal{D}(E^{\alpha_1/2})); \tag{3.2.2}$$

and

$$\|u^\mu(\varepsilon) - u^0(\varepsilon)\|_{H_\omega{}^0(\mathbf{R}; \mathcal{D}(E^{\alpha_2/2}))} \leqq c_3 |\varepsilon| \mu^\delta, \quad \mu \in [0, \mu_0],$$

$$|\varepsilon| \leqq m(\mu) \quad \text{if} \quad \mathcal{G} = H_\omega^0(\mathbf{R}; \mathcal{D}(E^{(\alpha_2 - 1 + \delta)/2})), \tag{3.2.3}$$

where c_2 does not depend on α_1 and c_3 does not depend on α_2 or δ and $K = \|L_0^{-1}\|_{\mathcal{L}(H_\omega{}^0(Q), H_\omega{}^{1,2}(Q))}$. *Moreover, if $\mathcal{G} = H_\omega^{1,0}(Q)$, then $\varepsilon(\mu)$ can be chosen to be independent of μ; otherwise $\varepsilon(\mu)$ can be chosen such that* $\lim_{\mu \to 0+} [\varepsilon(\mu)/\mu] > 0$.

Proof. To prove the existence of a solution we apply I: Theorem 3.4.1 to the equation $u = \varepsilon L_\mu^{-1} F(u, \varepsilon)$. Proceeding as in the proof of III: Theorem 1.2.1 ($\mu = 0$) we obtain the assertion concerning the existence and the property of $\varepsilon(\mu)$. It remains to prove (3.2.1)–(3.2.3). For $\alpha \in [0, 1]$, $\beta \in [0, 2]$ we can write

$$\|u^\mu - u^0\|_{H_\omega{}^\alpha(\mathbf{R};\mathscr{D}(E^{\beta/2}))} = |\varepsilon| \, \|L_\mu^{-1} F(u^\mu, \varepsilon) - L_0^{-1} F(u^0, \varepsilon)\|_{H_\omega{}^\alpha(\mathbf{R};\mathscr{D}(E^{\beta/2}))} \leqq$$
$$\leqq |\varepsilon| \, [\|(L_\mu^{-1} - L_0^{-1}) F(u^\mu, \varepsilon)\|_{H_\omega{}^\alpha(\mathbf{R};\mathscr{D}(E^{\beta/2}))} +$$
$$+ \|L_0^{-1}(F(u^\mu, \varepsilon) - F(u^0, \varepsilon))\|_{H_\omega{}^\alpha(\mathbf{R};\mathscr{D}(E^{\beta/2}))}] .$$

Let us suppose, for example, that $(\alpha, \beta) = (\alpha_1, 0)$, $0 \leqq \alpha_1 < 1$. Using (3.1.10), and the assumptions (b) and (c), we find that

$$\|u^\mu(\varepsilon) - u^0(\varepsilon)\|_{H_\omega{}^{\alpha_1}(\mathbf{R};H^0(\Omega))} \leqq |\varepsilon| \, [M \, \mu^{1-\alpha_1} \|F(u^\mu(\varepsilon), \varepsilon)\|_{H_\omega{}^0(\mathbf{R};\mathscr{D}(E^{\alpha_1/2}))} +$$
$$+ \lambda K \|u^\mu(\varepsilon) - u^0(\varepsilon)\|_{H_\omega{}^{\alpha_1}(\mathbf{R};H^0(\Omega))}] \leqq$$
$$\leqq |\varepsilon| \, \lambda K \|u^\mu(\varepsilon) - u^0(\varepsilon)\|_{H_\omega{}^{\alpha_1}(\mathbf{R};H^0(\Omega))} + M K_0 |\varepsilon| \, \mu^{1-\alpha_1} ,$$

which yields

$$\|u^\mu(\varepsilon) - u^0(\varepsilon)\|_{H_\omega{}^{\alpha_1}(\mathbf{R};H^0(\Omega))} \leqq M K_0 |\varepsilon| \, \mu^{1-\alpha_1} (1 - |\varepsilon| \, \lambda K)^{-1} \leqq$$
$$\leqq c_2 |\varepsilon| \, \mu^{1-\alpha_1} \quad \text{for} \quad \mu \in [0, \mu_0] , \quad |\varepsilon| \leqq m(\mu) .$$

Similarly in the other two cases.

THEOREM 3.2.2 (*The critical case*). *Let* $c = -\lambda_{k_0}$ *for some* $k_0 \in \mathbf{N}$. *Suppose that*

(*a*) *the equation*

$$\int_0^\omega \int_\Omega F(d_0 \, v_{k_0}, 0)(\tau, \xi) \, v_{k_0}(\xi) \, d\xi \, d\tau = 0$$

has a solution $d_0 = d_0^* \in \mathbf{R}$;

(*b*) *for some* $\varrho > 0$, $\varepsilon_0 > 0$ *the operator* $F(u, \varepsilon)$ *maps* $B(d_0^* v_{k_0}, \varrho; \mathscr{U}_\mu) \times \times [-\varepsilon_0, \varepsilon_0]$ *continuously into* $\mathscr{G}$ *if* $\mu > 0$ *and into* $H_\omega^0(Q)$ *if* $\mu = 0$, *where* $\mathscr{G}$ *is one of the spaces in Theorem* 3.2.1;

(*c*) *the Gâteaux derivative* $F_u'(u, \varepsilon)$ *exists and maps* $B(d_0^* v_k, \varrho; \mathscr{U}_\mu) \times \times [-\varepsilon_0, \varepsilon_0]$ *continuously into* $\mathscr{L}(\mathscr{U}_\mu, \mathscr{G})$ *for* $\mu > 0$;

(*d*) $F_u'(u, \varepsilon)$ *maps* $B(d_0^* v_{k_0}, \varrho; H_\omega^\alpha(\mathbf{R}; \mathscr{D}(E^{\beta/2})) \times [-\varepsilon_0, \varepsilon_0]$ *continuously into* $\mathscr{L}(H_\omega^\alpha(\mathbf{R}; \mathscr{D}(E^{\beta/2})), H_\omega^0(Q))$;

$$(e) \quad a \equiv \int_0^\omega \int_\Omega [F_u'(d_0^* v_{k_0}, 0) \, v_{k_0}](\tau, \xi) \, v_{k_0}(\xi) \, d\xi \, d\tau \neq 0 .$$

Then for any $\mu \in [0, \mu_0]$ there is an $\varepsilon(\mu) > 0$ such that the problem $(\mathscr{P}_\omega^{\mu,\varepsilon})$ has a unique solution $u^\mu(\varepsilon) \in \mathscr{U}_\mu$ for any $\varepsilon \in [-\varepsilon(\mu), \varepsilon(\mu)]$ such that $u^\mu(0)(t, x) = d_0^* v_{k_0}(x)$ $(t \in \mathbf{R}, x \in \Omega)$. The solution $u^\mu(\varepsilon)$ satisfies (3.2.1) *or* (3.2.2) *or* (3.2.3) *in accordance with the alternatives for $\mathscr{G}$ in Theorem* 3.2.1. *The function $\varepsilon(\mu)$ has the same property as in Theorem* 3.2.1.

Proof. In the proof of the existence of a solution we use I: Theorem 3.4.2 as in the proof of III: Theorem 1.2.2. The property of $\varepsilon(\mu)$ can be established by giving a more detailed proof of the existence of a solution, repeating the proof of I: Theorem 3.4.2 to the present case.

The solution $u^\mu(\varepsilon)$ is of the form

$$u^\mu(\varepsilon) = d^\mu(\varepsilon)\, v_{k_0} + w^\mu(\varepsilon)\,, \tag{3.2.4}$$

where the pair $(d^\mu(\varepsilon), w^\mu(\varepsilon))$ is a solution of the system

$$\int_0^\omega \int_\Omega F(d\, v_{k_0} + w, \varepsilon)(\tau, \xi)\, v_{k_0}(\xi)\, \mathrm{d}\xi\, \mathrm{d}\tau = 0\,, \tag{3.2.5}$$

$$w = \varepsilon\, L_{\mu_*}^{-1}\, F(d\, v_{k_0} + w, \varepsilon)\,. \tag{3.2.6}$$

Clearly, $d^\mu(0) = d_0^*$ for $\mu \in [0, \mu_0]$. By (3.2.5)

$$d^\mu(\varepsilon) = d^\mu(\varepsilon) - \frac{1}{a} \int_0^\omega \int_\Omega F(d^\mu(\varepsilon)\, v_{k_0} + w^\mu(\varepsilon), \varepsilon)(\tau, \xi)\, v_{k_0}(\xi)\, \mathrm{d}\xi\, \mathrm{d}\tau\,.$$

Hence,

$$\begin{aligned} |d^\mu(\varepsilon) - d^0(\varepsilon)| &= \frac{1}{|a|} \left| a(d^\mu(\varepsilon) - d^0(\varepsilon)) - \int_0^\omega \int_\Omega [F(d^\mu(\varepsilon)\, v_{k_0} + \right. \\ &\left. + w^\mu(\varepsilon), \varepsilon)(\tau, \xi) - F(d^0(\varepsilon)\, v_{k_0} + w^0(\varepsilon), \varepsilon)(\tau, \xi)]\, v_{k_0}(\xi)\, \mathrm{d}\xi\, \mathrm{d}\tau \right| \leqq \\ &\leqq \frac{1}{|a|} \left\{ \left| \int_0^\omega \int_\Omega [F'_u(d_0^*\, v_{k_0}, 0)\, ((d^\mu(\varepsilon) - d^0(\varepsilon))\, v_{k_0} + w^\mu(\varepsilon) - \right. \right. \\ &- w^0(\varepsilon))(\tau, \xi) - (F(d^\mu(\varepsilon)\, v_{k_0} + w^\mu(\varepsilon), \varepsilon)(\tau, \xi) - F(d^0(\varepsilon)\, v_{k_0} + \\ &\left. + w^0(\varepsilon), \varepsilon)(\tau, \xi))]\, v_{k_0}(\xi)\, \mathrm{d}\xi\, \mathrm{d}\tau \right| + \left| \int_0^\omega \int_\Omega F'_u(d_0^*\, v_{k_0}, 0)\, (w^\mu(\varepsilon) - \right. \\ &\left.\left. - w^0(\varepsilon))(\tau, \xi)\, v_{k_0}(\xi)\, \mathrm{d}\xi\, \mathrm{d}\tau \right| \right\}. \end{aligned}$$

Using the Schwarz inequality in $H^0_\omega(Q)$, we conclude that

$$\left|d^\mu(\varepsilon) - d^0(\varepsilon)\right| \leqq \frac{1}{|a|} \|v_{k_0}\|_{H_\omega{}^0(Q)} \|F'_u(d_0^* v_{k_0}, 0)\left(\left(d^\mu(\varepsilon) - d^0(\varepsilon)\right) v_{k_0} + \right.$$

$$+ w^\mu(\varepsilon) - w^0(\varepsilon)) - \left(F(d^\mu(\varepsilon) v_{k_0} + w^\mu(\varepsilon), \varepsilon) - F(d^0(\varepsilon) v_{k_0} + \right.$$

$$+ w^0(\varepsilon), \varepsilon))\|_{H_\omega{}^0(Q)} + \frac{1}{|a|} \|v_{k_0}\|_{H_\omega{}^0(Q)} \|F'_u(d_0^* v_{k_0}, 0)\left(w^\mu(\varepsilon) - \right.$$

$$- w^0(\varepsilon))\|_{H_\omega{}^0(Q)} .$$

The continuity in $\mathscr{L}(H^\alpha_\omega(\mathbf{R}; \mathscr{D}(E^{\beta/2})), H^0_\omega(Q))$ of the derivative $F'_u(u, \varepsilon)$ in the neighbourhood of $(d_0^* v_{k_0}, 0)$ implies that

$$\|F'_u(d_0^* v_{k_0}, 0)(u_1 - u_2) - (F(u_1, \varepsilon) - F(u_2, \varepsilon))\|_{H_\omega{}^0(Q)} \leqq$$

$$\leqq c_1(r) \|u_1 - u_2\|_{H_\omega{}^\alpha(\mathbf{R}; \mathscr{D}(E^{\beta/2}))}$$

for $u_1, u_2 \in B(d_0^* v_{k_0}, r; H^\alpha_\omega(\mathbf{R}; \mathscr{D}(E^{\beta/2})))$, where $r > 0$, $\lim_{r \to 0+} c_1(r) = 0$. For any $r > 0$ we can achieve that $u^\mu(\varepsilon) \in B(d_0^* v_{k_0}, r; H^\alpha_\omega(\mathbf{R}; \mathscr{D}(E^{\beta/2}))$ by choosing $\varepsilon_1 \in (0, \varepsilon(\mu))$ sufficiently small. Thus, we can conclude that

$$\left|d^\mu(\varepsilon) - d^0(\varepsilon)\right| \leqq |a|^{-1} c_1(r) \left[\left|d^\mu(\varepsilon) - d^0(\varepsilon)\right| \|v_{k_0}\|_{H_\omega{}^\alpha(\mathbf{R}; \mathscr{D}(E^{\beta/2}))} + \right.$$

$$\left. + \|w^\mu(\varepsilon) - w^0(\varepsilon)\|_{H_\omega{}^\alpha(\mathbf{R}; \mathscr{D}(E^{\beta/2}))}\right] +$$

$$+ |a|^{-1} \|F'_u(d_0^* v_{k_0}, 0)\|_{\mathscr{L}(H_\omega{}^\alpha(\mathbf{R}; \mathscr{D}(E^{\beta/2})), H^0(Q))} \|w^\mu(\varepsilon) - w^0(\varepsilon)\|_{H_\omega{}^\alpha(\mathbf{R}; \mathscr{D}(E^{\beta/2}))} \leqq$$

$$\leqq c_2(r) \left|d^\mu(\varepsilon) - d^0(\varepsilon)\right| + c_3 \|w^\mu(\varepsilon) - w^0(\varepsilon)\|_{H_\omega{}^\alpha(\mathbf{R}; \mathscr{D}(E^{\beta/2}))} .$$

Choosing r so small that $c_2(r) < 1$, we obtain

$$\left|d^\mu(\varepsilon) - d^0(\varepsilon)\right| \leqq c_4 \|w^\mu(\varepsilon) - w^0(\varepsilon)\|_{H_\omega{}^\alpha(\mathbf{R}; \mathscr{D}(E^{\beta/2}))} . \tag{3.2.7}$$

Let us suppose, for example, that $(\alpha, \beta) = (0, \alpha_2)$.

Then we find from (3.2.6) and (3.1.11) that

$$\|w^\mu(\varepsilon) - w^0(\varepsilon)\|_{H_\omega{}^0(\mathbf{R}; \mathscr{D}(E^{\alpha_2/2}))} = |\varepsilon| \|L_\mu^{-1} F(d^\mu(\varepsilon) v_{k_0} + w^\mu(\varepsilon), \varepsilon) -$$

$$- L_0^{-1} F(d^0(\varepsilon) v_{k_0} + w^0(\varepsilon), \varepsilon)\|_{H_\omega{}^0(\mathbf{R}; \mathscr{D}(E^{\alpha_2/2}))} \leqq$$

$$\leqq |\varepsilon| \left[\|(L_\mu^{-1} - L_0^{-1}) F(d^\mu(\varepsilon) v_{k_0} + w^\mu(\varepsilon), \varepsilon)\|_{H_\omega{}^0(\mathbf{R}; \mathscr{D}(E^{\alpha_2/2}))} + \right.$$

$$+ \|L_0^{-1}(F(d^\mu(\varepsilon) v_{k_0} + w^\mu(\varepsilon), \varepsilon) - F(d^0(\varepsilon) v_{k_0} +$$

$$\left. + w^0(\varepsilon), \varepsilon))\|_{H_\omega{}^0(\mathbf{R}; \mathscr{D}(E^{\alpha_2/2}))}\right] \leqq |\varepsilon| \left[M K_0 \mu^\delta + \right.$$

$$+ \|L_0^{-1}\|_{\mathscr{L}(H_\omega{}^0(Q), \mathscr{U}_0)} \|F(d^\mu(\varepsilon) v_{k_0} + w^\mu(\varepsilon), \varepsilon) -$$

$$\left. - F(d^0(\varepsilon) v_{k_0} + w^0(\varepsilon), \varepsilon)\|_{H_\omega{}^0(\mathbf{R}; H^0(\Omega))}\right] \leqq |\varepsilon| \left[M K_0 \mu^\delta + \right.$$

$$+ \|L_0^{-1}\|_{\mathscr{L}(H_\omega{}^0(Q), \mathscr{U}_0)} c_5\left(\left|d^\mu(\varepsilon) - d^0(\varepsilon)\right| \|v_{k_0}\|_{H_\omega{}^0(\mathbf{R}; \mathscr{D}(E^{\alpha_2/2}))} + \right.$$

$$\left.\left. + \|w^\mu(\varepsilon) - w^0(\varepsilon)\|_{H_\omega{}^0(\mathbf{R}; \mathscr{D}(E^{\alpha_2/2}))}\right)\right] .$$

Here K_0 is the same as in the condition (c) of Theorem 3.2.1 which is implied by the assumption (c). According to (3.2.7) by taking $|\varepsilon|$ sufficiently small (evidently without loss of the property of $\varepsilon(\mu)$) we find that

$$\|w^\mu(\varepsilon) - w^0(\varepsilon)\|_{H_\omega{}^0(\mathbf{R};\mathscr{D}(E^{\alpha_2/2}))} \leqq c_6|\varepsilon|\,\mu^\delta\,,$$

which together with (3.2.7) yields

$$|d^\mu(\varepsilon) - d^0(\varepsilon)| \leqq c^7|\varepsilon|\,\mu^\delta\,.$$

This completes the proof by virtue of (3.2.4).

Remark 3.2.1 Using the Poincaré method KOPÁČKOVÁ [18] examines the problem $(\mathscr{P}^{\mu,\varepsilon}_\omega)$ for $\varepsilon = \mu$, $E = -D_x^2$, $\Omega = (0, \pi)$ and $F(u)(t, x) = f(t, x, u, u_t, u_x)$, where $f \in C^{0,1,1,1,1}(\mathbf{R} \times \Omega \times [-\varrho, \varrho]^3)$ and the first derivatives of f with respect to u, u_t, and u_x are Lipschitz continuous in $\mathbf{R} \times \Omega \times [-\varrho, \varrho]^3$, ($\varrho > 0$ is arbitrary). She then shows that

$$\lim_{\mu \to 0+} \|u^\mu(\mu) - u^0(0)\|_{C_\omega{}^2(\mathbf{R}\times\Omega)} = 0\,.$$

In [23], MUSUKAEVA and NAMAZOV investigate the asymptotic behaviour as $\varepsilon \to 0+$ of a periodic solution u_ε to the equation

$$\varepsilon u_{tt} + \beta(t)\,u_t = \sum_{j,k=1}^{n} (a_{jk}(x)\,u_{x_j})_{x_k} + c(x)\,u + f(t, x)\,, \quad t \in [0, T]\,, \quad x \in \Omega$$

with the Dirichlet boundary condition.

They assume that $\beta(t) \geqq \beta_0 > 0$, $\sum_{j,k=1}^{n} a_{jk}(x)\,\xi_j\xi_k \geqq 0$, β, c, and f are sufficiently regular, $c(x) < 0$, $-c(x) - 2^{-1}\beta'(t) > 0$, and find an approximation of u_ε of order $0(\varepsilon)$ in $L_2((0, T) \times \Omega)$.

§ 4. Supplements and comments on linear and weakly non-linear problems

4.1. *The adjoint problem method*

Let us derive necessary conditions for the solubility of the problems $(\mathscr{P}^j_\omega)$, $j = 1, 2, 3, 4$, given by

$$u_{tt}(t, x) - u_{xx}(t, x) + a\,u_t(t, x) + c\,u(t, x) = g(t, x)\,,$$
$$(t, x) \in \mathbf{R} \times (0, \pi)\,, \quad (a, c \in \mathbf{R}, a \neq 0) \tag{4.1.1}$$

$$u(0, x) - u(\omega, x) = 0,$$

$$u_t(0, x) - u_t(\omega, x) = 0, \quad x \in [0, \pi] \tag{4.1.2}$$

and (4.1.12), (4.1.13), (4.1.14) or (4.1.15) from III: Sec. 4.1.

Proceeding as in that section we find that the formally adjoint problems $(\mathscr{P}^{j*}_{\omega})$ are defined by

$$v_{tt}(t, x) - v_{xx}(t, x) - a\, v_t(t, x) + c\, v(t, x) = 0, \quad (t, x) \in \mathbf{R} \times (0, \pi), \tag{4.1.3}$$

$$v(0, x) - v(\omega, x) = 0,$$

$$v_t(0, x) - v_t(\omega, x) = 0, \quad x \in [0, \pi] \tag{4.1.4}$$

and by III: (4.1.18)–(4.1.21), respectively, while the complementary adjoint boundary forms are determined, respectively, by III: (4.1.22)–(4.1.25).

It is easily seen that the adjoint problems $(\mathscr{P}^{j*}_{\omega})$ have only the trivial solution in the non-critical cases and a solution depending only on x in the critical cases.

Hence, in the latter case their solutions are the same as those found for the corresponding problems in III: Sec. 4.1, and III: Lemmas 4.1.1–4.1.4 are valid for the problems $(\mathscr{P}^{j}_{\omega})$ of the present section without any change.

4.2. *The Ficken-Fleishman method*

The first authors who dealt with the problem (2.3.1), (2.1.2) were FLEISHMAN [10]–[12] and FICKEN [8]. Their best result is included in a common paper [9]. They treated the problem for the special case $F(u) = u^3$. Using the same method (sometimes called the *limit solution method*), HAVLOVÁ [14] generalizes their result to a general Nemyckiĭ operator $F(u, \varepsilon) = f(.,\, ., u, u_t, u_x, \varepsilon)$.

First, one studies the initial problem $(\mathscr{I})$ given by

$$u_{tt}(t, x) - u_{xx}(t, x) + 2a\, u_t(t, x) + c\, u(t, x) =$$

$$= g(t, x) + \varepsilon f(t, x, u(t, x), u_t(t, x), u_x(t, x), \varepsilon),$$

$$(t, x) \in Q = \mathbf{R}^+ \times \mathbf{R}, \quad \varepsilon \in [-\varepsilon_0, \varepsilon_0], \quad (\varepsilon_0 > 0), \tag{4.2.1}$$

$$u(0, x) = \varphi(x), \quad u_t(0, x) = \psi(x), \quad x \in \mathbf{R} \tag{4.2.2}$$

under the following assumptions:

$$a > 0, \quad c > 0, \tag{4.2.3}$$

$$g \in C^{0,1}(Q), \tag{4.2.4}$$

$f = f(t, x, u_0, u_1, u_2, \varepsilon)$ and its partial derivatives f_x, and f_{u_j}, $j = 0, 1, 2$

are continuous in t, x, u_0, u_1 and u_2 on $\mathbf{R}^+ \times \mathbf{R}^4 \times [-\varepsilon_0, \varepsilon_0]$. Further, for any $\varrho \geqq 0$ there are constants $\varkappa(\varrho)$, $\lambda(\varrho)$ such that the functions f, f_x, and f_{u_j} are bounded by $\varkappa(\varrho)$, and are Lipschitz continuous in u_0, u_1, and u_2 with the Lipschitz constant $\lambda(\varrho)$ on $Q \times [-\varrho, \varrho]^3 \times [-\varepsilon_0, \varepsilon_0]$, (4.2.5)

$$\varphi \in C^2(\mathbf{R}), \quad \psi \in C^1(\mathbf{R}). \tag{4.2.6}$$

A solution is a function $u \in \mathscr{U} = C^2(Q)$ satisfying (4.2.1), (4.2.2) for any sufficiently small $|\varepsilon|$. The problem $(\mathscr{I})$ is equivalent to that of finding a fixed point $u \in \mathscr{U}$ of the operator

$$U(u)(t, x) = \tfrac{1}{2}e^{-at}\Big\{\varphi(x + t) + \varphi(x - t) +$$

$$+ \int_{x-t}^{x+t} [J_0(d^{1/2}(t^2 - (x - \xi)^2)^{1/2})(\psi(\xi) + a\,\varphi(\xi)) +$$

$$+ \frac{\partial J_0(d^{1/2}(t^2 - (x - \xi)^2)^{1/2})}{\partial t}\,\varphi(\xi)]\,\mathrm{d}\xi +$$

$$+ \int_0^t \int_{x-t+\tau}^{x+t-\tau} J_0(d^{1/2}((t - \tau)^2 - (x - \xi)^2)^{1/2})\,e^{a\tau}[g(\tau, \xi) +$$

$$+ \varepsilon f(\tau, \xi, u(\tau, \xi), u_t(\tau, \xi), u_x(\tau, \xi), \varepsilon)]\,\mathrm{d}\xi\,\mathrm{d}\tau\Big\}$$

(see (2.1.3)). Estimates like those in Sec. 2.1 enable us to use the Banach fixed point theorem, I: Theorem 3.1.1, for contracting operators and to obtain the following result: for any sufficiently large ϱ there is an $\varepsilon_1 \in (0, \varepsilon_0]$ such that for all $\varepsilon \in [-\varepsilon_1, \varepsilon_1]$ the problem $(\mathscr{I})$ has a unique solution $u \in \mathscr{U}$ with $\|u\|_{\mathscr{U}} \leqq \varrho$.

In particular, there is a unique solution with $\|u\|_{\mathscr{U}} \leqq \varrho$ that satisfies the initial conditions

$$u(0, x) = 0\,, \quad u_t(0, x) = 0\,.$$

For this u let us write

$$\varphi_n(x) = u(n\omega, x)\,, \quad \psi_n(x) = u_t(n\omega, x)\,,$$

$$u_n(t, x) = u(t + n\omega, x)\,, \quad n \in \mathbf{N}\,.$$

Assuming now that g and f are ω-periodic in t, we can prove that $\{u_n\}_{n=1}^{\infty}$ forms a Cauchy sequence in $\mathscr{U}$, and since $\mathscr{U}$ is complete, there is a $\bar{u} \in \mathscr{U}$ such that $\|u_n - \bar{u}\|_{\mathscr{U}} \to 0$ as $n \to \infty$ and $\|\bar{u}\|_{\mathscr{U}} \leqq \varrho$. The sequences $\{\varphi_n\}_{n=1}^{\infty}$ and $\{\psi_n\}_{n=1}^{\infty}$ also have limits φ and ψ in $C^2(\mathbf{R})$ and $C^1(\mathbf{R})$ respectively. Finally, it can be shown that the function $\bar{u}$ satisfies equation (4.2.1) for the initial data φ, ψ. The function $\bar{u}$ is clearly ω-periodic:

$$\bar{u}(t + \omega, x) = \lim_{n\to\infty} u_n(t + \omega, x) = \lim_{n\to\infty} u_{n+1}(t, x) = \bar{u}(t, x) .$$

Thus, the following theorem is valid.

THEOREM 4.2.1 *Suppose that the assumptions* (4.2.3)–(4.2.5) *hold, that* g *and* f *are* ω*-periodic in* t, *and that* ϱ *is sufficiently large. Then for a suitable* $\varepsilon^* \in (0, \varepsilon_0]$ *and all* $\varepsilon \in [-\varepsilon^*, \varepsilon^*]$ *there is a unique* $u \in \mathscr{U}$, $\|u\|_{\mathscr{U}} \leqq \varrho$, *satisfying* (4.2.1) *and* ω*-periodic in* t.

VÍTEK [39] obtains similar results for the equation

$$u_{tt}(t, x) - \Delta_n u(t, x) + 2a\, u_t(t, x) + 2\sum_{j=1}^{n} b_j\, u_{x_j}(t, x) + c\, u(t, x) =$$

$$= g(t, x) + \varepsilon f(t, x, u(t, x), \varepsilon) , \quad (t, x) \in \mathbf{R}^+ \times \mathbf{R}^n ,$$

where $n = 2$ or 3, $a, b_j, c \in \mathbf{R}$, $a > 0$, $\sum_{j=1}^{n} b_j^2 + c > 0$, g and f are sufficiently smooth, periodic in t and $|\varepsilon|$ is sufficiently small.

4.3. *Results of Rabinowitz*

In the second part of [32], RABINOWITZ investigates the existence of an ω-periodic solution to the problem

$$u_{tt}(t, x) - u_{xx}(t, x) + a\, u_t(t, x) = g(t, x) , \quad (t, x) \in \mathbf{R} \times (0, \pi) , \quad (a \neq 0),$$
$$u(t, 0) = u(t, \pi) = 0 , \quad t \in \mathbf{R} , \tag{4.3.1}$$

and to the corresponding weakly non-linear problem given by

$$u_{tt}(t, x) - u_{xx}(t, x) + a\, u_t(t, x) = \varepsilon f(t, x, u(t, x), u_t(t, x), u_x(t, x))$$

and (4.3.1). He uses two methods. The first is based on the (t, s)-Fourier expansion (see Remark 1.1.7). The second, using negative norms (see I: Sec. 1.7, p. 15) and the theory of elliptic boundary-value problems, is applied to the more general n-dimensional case

$$u_{tt}(t, x) + E\, u(t, x) + a\, u_t(t, x) =$$

$$= g(t, x) + \varepsilon f(t, x, u(t, x), u_t(t, x), \operatorname{grad} u(t, x)) ,$$

$$(t, x) \in Q = \mathbf{R} \times \Omega , \quad (a \neq 0) ,$$

$$u(t, x) = 0 , \quad (t, x) \in \mathbf{R} \times \partial\Omega ,$$

where $\Omega \subset \mathbf{R}^n$ is a bounded domain,

$$E\, v(x) = - \sum_{j,k=1}^{n} (a_{jk}(x)\, v_{x_k}(x))_{x_j} + c(x)\, v(x)\,,$$

Ω, a_{jk}, and c are of class C^∞, and

$$\int_\Omega \varphi(x)\, E\, \varphi(x)\, \mathrm{d}x \geqq \gamma \int_\Omega \sum_{j=1}^{n} \varphi_{x_j}^2(x)\, \mathrm{d}x\,, \quad \varphi \in C_0^\infty(\Omega)\,, \quad (\gamma > 0)\,.$$

Provided that $g \in H_\omega^r(Q)$, that f is of class C^r and ω-periodic in t, and that $|\varepsilon|$ is sufficiently small, a solution $u \in H_\omega^{r+1}(Q) \cap \dot{H}_\omega^1(Q)$ is found. For $a > 0$ the exponential stability of the solution is proved.

In [33], the same author deals with the existence of an ω-periodic solution to the problem defined by

$$u_{tt}(t, x) - u_{xx}(t, x) + a\, u_t(t, x) =$$

$$= \varepsilon\, f(t, x, u(t, x), u_t(t, x), u_x(t, x), u_{tt}(t, x), u_{tx}(t, x), u_{xx}(t, x))\,,$$

$$(t, x) \in \mathbf{R} \times (0, \pi)$$

and (4.3.1), provided that $a \neq 0$ and f is ω-periodic in t. By a clever use of a deep theorem on accelerated convergence due to MOSER (see I: [25]) he proves that the above problem has a regular solution for sufficiently small $|\varepsilon|$ and sufficiently smooth f.

These results are briefly summarized in [31].

4.4. *The telegraph equation with $a = 0$*

We mention at least briefly the very complicated problem of the existence of periodic solutions to the problem

$$Lu \equiv u_{tt}(t, x) - u_{xx}(t, x) + c\, u(t, x) = g(t, x)\,, \quad (t, x) \in Q = \mathbf{R} \times (0, \pi),$$

$$u(t, 0) = u(t, \pi) = 0\,, \quad t \in \mathbf{R}\,,$$

$$u(t + \omega, x) = u(t, x)\,, \quad (t, x) \in \mathbf{R} \times [0, \pi]\,,$$

with $c \neq 0$. Using the (t, s)-Fourier method we find immediately that the problem reduces to the investigation of the following system of equations

$$(-\nu^2 j^2 + k^2 + c)\, u_{jk} = g_{jk}\,, \quad j \in \mathbf{Z}\,, \quad k \in \mathbf{N}\,,$$

where $\nu = 2\pi/\omega$ and u_{jk}, g_{jk} are the Fourier coefficients of the functions u and g. The problem shows different behaviour for different values of ω and c.

For example, when $\omega = 2\pi$, i.e. $v = 1$, we have $\dim \mathcal{N}(L) < \infty$ (in fact, if $k \neq |j|$ and $k + |j| > |c|$, then $|-j^2 + k^2| = |k - |j|| (k + |j|) > |c|$). In this sense the problem is similar to that investigated in §§ 1 and 2. But a difference is that L^{-1} is an element of $\mathscr{L}(H^2_{2\pi}(Q) \cap \dot{H}^1_{2\pi}(Q), H^2_{2\pi}(Q) \cap \cap \dot{H}^1_{2\pi}(Q))$ in the non-critical case.

On the other hand, if $v^2 = p$, p is not a perfect square and $c = -1$, then it is known that the Pell equation

$$-pj^2 + k^2 = 1$$

has infinitely many solutions, that is $\dim \mathcal{N}(L) = \infty$. Thus the problem has many points in common with problems for the wave equation which is investigated in Chap. V. For a special case of a strongly non-linear problem, see RABINOWITZ V: [60].

§ 5. Comments on strongly non-linear problems

In this paragraph let $\Omega \subset \mathbf{R}^n$ be a sufficiently regular bounded domain. All functions considered are supposed to be sufficiently smooth and ω-periodic in t.

5.1. *Results of Prodi, Prouse, Krylová, Buzzetti, Lions, Toušek, Biroli and Nakao. (The wave equation with strongly non-linear damping)*

Much attention has been paid to the problem given by an equation of the type

$$u_{tt}(t, x) - \Delta_n u(t, x) + \alpha(u_t(t, x)) = g(t, x), \quad (t, x) \in \mathbf{R} \times \Omega \tag{5.1.1}$$

and by

$$u(t, x) = 0, \quad (t, x) \in \mathbf{R} \times \partial\Omega, \tag{5.1.2}$$

$$u(t + \omega, x) - u(t, x) = 0, \quad (t, x) \in \mathbf{R} \times \Omega, \tag{5.1.3}$$

where α is a function on $\mathbf{R}$. This function, which from the physical point of view represents the damping of the system, is here non-linear in contrast to the preceding paragraphs.

PRODI [26] (a preliminary communication is in [25]) proves the existence of a solution $u \in \dot{H}^1_\omega(\mathbf{R} \times \Omega)$ to the Dirichlet problem for the equation

$$u_{tt}(t, x) - \Delta_n u(t, x) + \alpha(t, x, u_t(t, x)) = F(u)(t, x), \quad (t, x) \in \mathbf{R} \times \Omega,$$

where

$$m \leqq \frac{\alpha(t, x, p_1) - \alpha(t, x, p_2)}{p_1 - p_2} \leqq M, \quad p_1, p_2 \in \mathbf{R}, \quad p_1 \neq p_2,$$

($M \geqq m > 0$) and either $F(u)(t, x) = f(t, x, u)$ with f growing in u at most linearly, or $F(u)(t, x) = f(t, x, u, \operatorname{grad} u)$ with f Lipschitz continuous in u and grad u with a sufficiently small Lipschitz constant. The Banach contraction principle and the Leray-Schauder theorem are used.

PROUSE [29] proves the existence of a solution $u \in W^1_{\infty,\omega}(\mathbf{R}; \dot{H}^1(\Omega)) \cap \cap W^2_{\infty,\omega}(\mathbf{R}; H^0(\Omega))$ to (5.1.1)–(5.1.3) provided that $n \leqq 2$ and

$$\alpha(\xi) = \xi + |\xi|\,\xi\,. \tag{5.1.4}$$

PRODI [28] (a preliminary communication is in [27]) treats the problem (5.1.1)–(5.1.3) with n arbitrary, and $\alpha(\xi)$ continuous, increasing and behaving asymptotically like

$$|\xi|^{\varrho-1}\xi\,, \quad (\varrho \geqq 1)\,.$$

His method is based on the decomposition of the required solution in the form

$$u = \bar{u} + v\,, \quad \bar{u} \text{ independent of } t\,, \quad \int_0^\omega v\,dt = 0\,. \tag{5.1.5}$$

He finds

$$v \in W^0_{\infty,\omega}(\mathbf{R}; \dot{H}^1(\Omega)) \cap W^1_{\infty,\omega}(\mathbf{R}; H^0(\Omega)) \cap W^1_{\varrho+1,\omega}(\mathbf{R}; L_{\varrho+1}(\Omega))\,,$$

$$\bar{u} \in W^1_{(\varrho+1)/\varrho}(\Omega)\,.$$

The results of these two authors are presented in [30].

Using Prouse's technique KRYLOVÁ [20] investigates the existence of a solution $u \in W^1_{\infty,\omega}(\mathbf{R}; V) \cap W^2_{\infty,\omega}(\mathbf{R}; H^0(\Omega))$ ($\dot{H}^m(\Omega) \subset V \subset H^m(\Omega)$, V closed in $H^m(\Omega)$) to the problem given by

$$u_{tt}(t, x) + E\,u(t, x) + \alpha(u_t(t, x)) = g(t, x)\,, \quad (t, x) \in \mathbf{R} \times \Omega \tag{5.1.6}$$

with rather general boundary conditions. Here

$$E\,v(x) = \sum_{|\alpha|,|\beta| \leqq m} (-1)^{|\alpha|} D^\alpha(a_{\alpha\beta}(x)\,D^\beta v(x))\,, \tag{5.1.7}$$

and α is a function of the form (5.1.4).

In particular, for the problem (5.1.1)–(5.1.3), she obtains even for $n = 3$ the same results as PROUSE [29].

BUZZETTI studies the problem (5.1.1)–(5.1.3) for $n = 1$. In [5] and [6] he obtains a solution almost everywhere, supposing that $\alpha(\xi) = \xi + \beta(\xi)$ with $\beta \in C^1_{\text{loc}}(\mathbf{R})$ increasing and $\beta(0) = 0$, or with β non-decreasing and $\beta(0-) \leqq 0$, $\beta(0+) \geqq 0$.

In all papers (except for [26]) quoted so far the Faedo-Galerkin method combined with a compactness argument is used. On the other hand, LIONS [21] investigates the problem (5.1.1)–(5.1.3) (with n arbitrary) for

$$\alpha(\xi) = |\xi|^{\varrho-1}\,\xi\,, \quad (\varrho > 1)$$

by means of the method of elliptic regularization, and finds a solution in the form (5.1.5), where $\bar{u} \in \dot{H}^1(\Omega) + W^2_{(\varrho+1)/\varrho}(\Omega) \cap \dot{W}^1_{(\varrho+1)/\varrho}(\Omega)$, $v \in H^0_\omega(\mathbf{R}; \dot{H}^1(\Omega)) \cap$ $\cap\, W^1_{\varrho+1,\omega}(\mathbf{R}; L_{\varrho+1}(\Omega))$.

TOUŠEK [37] treats the Dirichlet problem for the equation (5.1.6) with

$$\alpha(\xi) = \xi + |\xi|^{\varrho-1}\xi \quad \text{or} \quad \alpha(\xi) = |\xi|^{\varrho-1}\xi\,, \quad (\varrho \geqq 1)\,.$$

He uses the theory of maximal monotone operators and the Ficken-Fleishman method.

As a consequence of the results on the existence of bounded solutions, BIROLI [2], [3] obtains solutions to (5.1.1)–(5.1.3) with α increasing, $\alpha(0) = 0$.

NAKAO [24] deals with the Dirichlet problem for the equation

$$u_{tt}(t, x) + E\,u(t, x) + \alpha(x, u_t(t, x)) + \beta(x, u(t, x)) =$$

$$= \varepsilon\, g(t, x)\,, \quad (t, x) \in \mathbf{R} \times \Omega$$

without any monotonicity assumptions on β and with α monotone of a certain growth in u_t. Here E is given by (1.1.16) and $|\varepsilon|$ is sufficiently small.

5.2. *Results of Mawhin, Fučík, Brézis, Nirenberg, Biroli, Horáček and Zecca. (Problems with non-linear forcing term)*

MAWHIN [22] proves the existence of 2π-periodic solutions in both variables $u \in H^0((0, 2\pi) \times (0, 2\pi))$ to the equation

$$u_{tt}(t, x) - u_{xx}(t, x) + a\,u_t(t, x) + \beta(u(t, x)) = g(t, x)\,, \quad (t, x) \in \mathbf{R}^2\,,$$

where $a \neq 0$ and β is a continuous function on $\mathbf{R}$ of at most linear growth, satisfying one of the following conditions

(a) $\gamma \leqq u^{-1}\,\beta(u) \leqq \delta$ for all sufficiently large $|u|$ with $\gamma, \delta \in (\mu_1, \mu_2)$;

(b) $\lim\limits_{u\to-\infty} u^{-1}\,\beta(u) = \gamma$, $\lim\limits_{u\to\infty} u^{-1}\,\beta(u) = \delta$ and $\gamma, \delta \in [\mu_1, \mu_2]$, $\gamma \neq \delta$;

(c) $\lim\limits_{|u|\to\infty} u^{-1}\,\beta(u) = \mu_1$, μ_1 is not a perfect square;

(d) $\lim\limits_{|u|\to\infty} u^{-1}\,\beta(u) = q^2$ for some $q \in \mathbf{Z}$ provided that β and g satisfy certain additional requirements.

Here μ_1 and μ_2 are consecutive elements of the set $\Sigma = \{k^2 - j^2; k, j \in \mathbf{Z}\} = (2\mathbf{Z} + 1) \cup 4\mathbf{Z}$. Conditions (a)–(c) correspond to the non-critical, and (d) to the critical case. The author utilizes the compactness properties and the degree theory.

MAWHIN [22], and FUČÍK and MAWHIN [13] deal with 2π-periodic solutions (in both variables) to the equation

$$u_{tt}(t, x) - u_{xx}(t, x) + a\, u_t(t, x) - \mu\, u^+(t, x) + \nu\, u^-(t, x) + \\ + \beta(u(t, x)) = g(t, x)\,, \quad (t, x) \in \mathbf{R}^2\,,$$

where $a, \mu, \nu \in \mathbf{R}$, $a \neq 0$, and β is a continuous bounded function. The authors decompose the set $\mathbf{R}^2$ of all pairs (μ, ν) (see III: [83]) and obtain existence results of different types on different components.

BRÉZIS and NIRENBERG [4] apply their general results on ranges of non-linear operators, for example, to the Dirichlet problem for the equation

$$u_{tt}(t, x) + E\, u(t, x) + a\, u_t(t, x) + \beta(t, x, u(t, x)) = \\ = g(t, x)\,, \quad (t, x) \in \mathbf{R} \times \Omega\,,$$

where E is given by (5.1.7) and $a \neq 0$.

The function β is supposed to be continuous in u and to satisfy III: (5.6.1), (5.6.2) (p. 121), and for some $c_0 > 0$

$$\int_0^\omega \int_\Omega \beta_+ v^+ \,\mathrm{d}x\,\mathrm{d}t - \int_0^\omega \int_\Omega \beta_- v^- \,\mathrm{d}x\,\mathrm{d}t \geqq \\ \geqq \int_0^\omega \int_\Omega g v \,\mathrm{d}x\,\mathrm{d}t + c_0 \left(\int_0^\omega \int_\Omega |v|^2 \,\mathrm{d}x\,\mathrm{d}t \right)^{1/2}$$

for all $v \in \mathcal{N}(E)$. The authors prove the existence of a generalized 2π-periodic solution, provided that one of the conditions (a) and (b) in III: Sec. 5.6 is satisfied (for (b) we now have

$$\alpha = \inf \left\{ \left| j^2 - \lambda_k + \frac{a^2 j^2}{j^2 - \lambda_k} \right| ; j, k \in \mathbf{Z}, \lambda_k < j^2 \right\}$$

and the λ_k are the eigenvalues of E).

The regularity properties of solutions are also investigated.

BIROLI [1] treats the Dirichlet problem for the equation

$$u_{tt}(t, x) - \Delta_n u(t, x) + a\, u_t(t, x) + \beta(u(t, x)) = g(t, x)\,, \quad (t, x) \in \mathbf{R} \times \Omega\,,$$

where $a > 0$ and β is a monotone function on $\mathbf{R}$, $\beta(0) = 0$. He proves the existence of a solution $u \in W^0_{\infty,\omega}(\mathbf{R}; H^1(\Omega)) \cap W^1_{\infty,\omega}(\mathbf{R}; H^0(\Omega))$.

HORÁČEK [15], [16] asserts (without proof) that the Dirichlet problem for the equation

$$u_{tt}(t, x) - \Delta_n u(t, x) + u_t(t, x) + u^3(t, x) = g(t, x), \quad (t, x) \in \mathbf{R} \times \Omega,$$

has a generalized or a classical ω-periodic solution provided that $n = 2$ or $n = 1$ respectively.

ZECCA [41] studies generalized 2π-periodic solutions (in both variables) of a "many-valued" equation

$$u_{tt}(t, x) - u_{xx}(t, x) + a\, u_t(t, x) + c\, u(t, x) \in f(t, x, u), \quad (a \neq 0)$$

in the non-critical case.

5.3. *Results of v. Wahl, Clements, Kakita and Şowunmi. (Problems with non-linear elliptic part and other problems)*

v. WAHL [40] deals with the equation

$$u_{tt}(t, x) + \sum_{|\alpha|, |\beta| \leqq m} D_x^\alpha (A_{\alpha\beta}(t, x, u(t, x))\, D_x^\beta u(t, x)) + a\, u_t(t, x) =$$

$$= \varepsilon\, g(t, x), \quad (t, x) \in \mathbf{R}^{n+1},$$

provided that $a \neq 0$, $|\varepsilon|$ is sufficiently small, and $A_{\alpha\beta}$ are sufficiently regular and ensure positivity in a certain sense. Using the Poincaré method, and some results due to him for the initial value problem, and applying the Tihonov fixed-point theorem (see I: Theorem 3.2.3), he proves the existence of a classical ω-periodic solution.

CLEMENTS [7] proves by the Faedo-Galerkin and compactness methods the existence of a solution $u \in W^0_{\infty,\omega}(\mathbf{R}; \dot{H}^m(\Omega)) \cap W^1_{\infty,\omega}(\mathbf{R}; H^0(\Omega))$ to the problem given by the equation of the form

$$u_{tt}(t, x) + \sum_{|\alpha| \leqq m} D_x^\alpha \Big(\sum_{|\beta| \leqq m} a_{\alpha\beta}(t, x)\, D_x^\beta u(t, x) +$$

$$+ B_\alpha(t, x, u(t, x), \ldots, D_x^{m-1} u(t, x)) \Big) + a\, u_t(t, x) =$$

$$= g(t, x), \quad (t, x) \in \mathbf{R} \times \Omega, \quad (a \neq 0)$$

and by the Dirichlet boundary conditions

$$D_x^\alpha u(t, x) = 0, \quad (t, x) \in \mathbf{R} \times \partial\Omega, \quad |\alpha| \leqq m - 1.$$

Similar methods combined with monotonicity argument are used by KAKITA [17] to study the existence of a solution $u \in W^0_{\infty,\omega}(\mathbf{R}; \dot{W}^m_p(\Omega)) \cap H^1_\omega(\mathbf{R}; H^0(\Omega))$ to the Dirichlet problem for the equation

$$u_{tt}(t, x) + \sum_{|\alpha| \leq m} (-1)^{|\alpha|} D^\alpha_x A_\alpha(x, u(t, x), \ldots, D^m_x u(t, x)) +$$

$$+ b_0'(|u(t, x)|^2) u_t(t, x) - \Delta u_t(t, x) = g(t, x), \quad (t, x) \in \mathbf{R} \times \Omega,$$

where $b_0'(s^2)$ is a non-negative function on $\mathbf{R}$ of a polynomial growth.

Making use of the t-Fourier method, ŞOWUNMI [34] investigates the existence of ω-periodic solutions to the Dirichlet problem for the equation

$$u_{tt}(t, x) - u_{xx}(t, x) - \int_{-\infty}^t K(t - \tau) u_{xx}(\tau, x) \, d\tau - \lambda u_{txx}(t, x) =$$

$$= g(t, x), \quad (t, x) \in Q = \mathbf{R} \times (0, \pi),$$

where $K \in H^0(0, \infty)$, $\lambda \in \mathbf{R}$ and $g \in H^0_\omega(Q)$. He shows that for $\lambda \neq 0$ the problem has a unique solution $u \in H^1_\omega(\mathbf{R}; \dot{H}^1(0, \pi))$ if and only if

$$-v^2 j^2 + k^2 \left(1 + vj\lambda i + \int_0^\infty K(\tau) e^{-ijv\tau} \, d\tau\right) \neq 0$$

for all $j \in \mathbf{Z}$, $k \in \mathbf{N}$ $(v = 2\pi/\omega)$.

In [35] the same author deals with the existence of ω-periodic solutions to the Dirichlet problem for the equation

$$\varrho u_{tt}(t, x) - (\sigma(u_x(t, x)))_x - \lambda u_{txx}(t, x) = g(t, x), \quad (t, x) \in \mathbf{R} \times (0, 1),$$

where $\varrho > 0$, $\lambda > 0$ and σ is an increasing function on $\mathbf{R}$.

Chapter V

The wave equation

§ 1. The Dirichlet boundary conditions; the Poincaré method

1.1. *The linear case; general considerations*

Let $(\mathscr{P}^0_\omega)$ be the problem given by the equations

$$u_{tt}(t, x) - u_{xx}(t, x) = g(t, x), \quad (t, x) \in Q, \tag{1.1.1}$$

$$u(t, 0) = 0, \quad u(t, \pi) = 0, \quad t \in \mathbf{R}, \tag{1.1.2}$$

$$u(t + \omega, x) = u(t, x), \quad (t, x) \in Q, \tag{1.1.3}$$

where

$$Q = \mathbf{R} \times [0, \pi]$$

and $g \in \dot{C}^{(\alpha, 1-\alpha)}_\omega(Q)$ with fixed $\alpha = 1$ or 0 (this means that g and $D^\alpha_t D^{1-\alpha}_x g$ belong to $C_\omega(Q)$ and that

$$g(t, 0) = g(t, \pi) = 0, \quad t \in \mathbf{R}; \tag{1.1.4}$$

see II: Sec. 2.2.).

By a (classical) solution we mean a function

$$u \in \mathscr{U} \equiv \dot{C}^2_\omega(Q)$$

satisfying (1.1.1). Here we use the following notation:

$$\dot{C}^2(I \times [0, \pi]) = \{u \in C^2(I \times [0, \pi]);\, u(t, 0) = u(t, \pi) = 0,\, t \in I\},$$

I being a closed interval.

Remark 1.1.1 Instead of $g \in \dot{C}_\omega^{(\alpha,1-\alpha)}(Q)$, $\alpha = 0$ or 1, we could take $g \in \dot{C}_\omega^{(1,0)}(Q) + \dot{C}_\omega^{(0,1)}(Q)$. The function

$$\|g\| = \inf\{\|g_1\|_{C_\omega^{(1,0)}(Q)} + \|g_2\|_{C_\omega^{(0,1)}(Q)} ;\ g_1 \in \dot{C}_\omega^{(1,0)}(Q),\ g_2 \in \dot{C}_\omega^{(0,1)}(Q),\ g = g_1 + g_2\}$$

is a norm in this space.

Remark 1.1.2 Clearly, for a classical solution or for a generalized solution having the first derivatives u_t and u_x, the condition (1.1.3) is equivalent to

$$u(\omega, x) - u(0, x) = 0\,, \quad u_t(\omega, x) - u_t(0, x) = 0\,, \quad x \in [0, \pi]\,. \tag{1.1.3*}$$

For since g is ω-periodic in t, by (1.1.3*), the function $v(t, x) = u(t + \omega, x)$ is a solution of (1.1.1) and (1.1.2) with the same initial conditions, and by the uniqueness of this initial boundary-value problem, $v(t, x) = u(t, x)$ (see II: Theorem 2.2.1). Furthermore, (1.1.3*) is equivalent to

$$u_x(\omega, x) - u_x(0, x) = 0\,, \quad u_t(\omega, x) - u_t(0, x) = 0\,, \quad x \in [0, \pi] \tag{1.1.3**}$$

because $u(\omega, 0) = u(0, 0) = 0$ (cf. also Remark 1.1.4).

By II: Theorem 2.2.1 (p. 48) any classical solution of (1.1.1), (1.1.2) can be written in the form

$$u(t, x) = \sigma(x + t) - \sigma(-x + t) + \frac{1}{2}\int_0^t \int_{x-t+\tau}^{x+t-\tau} g_e(\tau, \xi)\, d\xi\, d\tau\,, \tag{1.1.5}$$

where $\sigma \in C_{2\pi}^2(\mathbf{R})$ and g_e is the 2π-periodic odd extension in x of the original function g.

Clearly, $g_e \in C_{\omega,2\pi(o)}^{(\alpha,1-\alpha)}(\mathbf{R}^2)$. By (1.1.5) u is defined for $x \in \mathbf{R}$ and

$$u(t, x) = u(t, x + 2\pi) = -u(t, -x)\,, \quad (t, x) \in \mathbf{R}^2\,.$$

Hence a solution u to the problem $(\mathscr{P}_\omega^0)$ given by (1.1.5) is also a solution to the "extended" problem $(\mathscr{P}_{\omega,e}^0)$ given by

$$(u_e)_{tt}(t, x) - (u_e)_{xx}(t, x) = g_e(t, x)\,, \quad (t, x) \in \mathbf{R}^2\,, \tag{1.1.1'}$$

$$u_e(t, x) = u_e(t, x + 2\pi) = -u_e(t, -x)\,, \quad (t, x) \in \mathbf{R}^2\,, \tag{1.1.2'}$$

$$u_e(t + \omega, x) = u_e(t, x)\,, \quad (t, x) \in \mathbf{R}^2\,. \tag{1.1.3'}$$

On the other hand, the restriction of u_e to $\mathbf{R} \times [0, \pi]$ is a solution u of $(\mathscr{P}_\omega^0)$. Thus the two problems are equivalent, and in what follows we usually denote u_e simply by u and g_e by g.

Remark 1.1.3 The problem

$$u_{tt}(t, x) - u_{xx}(t, x) = g(t, x), \quad (t, x) \in \mathbf{R}^2,$$
$$u(t, x) = u(t, x + 2\pi), \quad (t, x) \in \mathbf{R}^2,$$
$$u(t + \omega, x) = u(t, x), \quad (t, x) \in \mathbf{R}^2,$$

with $g(t + \omega, x) = g(t, x + 2\pi) = g(t, x)$ can be also studied.

A solution u given by (1.1.5) satisfies (1.1.3) if and only if

$$\sigma(x + t + \omega) - \sigma(x + t) - \sigma(-x + t + \omega) + \sigma(-x + t) +$$
$$+ \frac{1}{2}\int_0^{t+\omega}\int_{x-t-\omega+\tau}^{x+t+\omega-\tau} g(\tau, \xi)\, d\xi\, d\tau - \frac{1}{2}\int_0^{t}\int_{x-t+\tau}^{x+t-\tau} g(\tau, \xi)\, d\xi\, d\tau = 0, \tag{1.1.6}$$

that is,

$$\sigma(x + t + \omega) - \sigma(x + t) - \sigma(-x + t + \omega) + \sigma(-x + t) +$$
$$+ \frac{1}{2}\int_{-\omega}^{0}\int_{x-t+\tau}^{x+t-\tau} g(\tau, \xi)\, d\xi\, d\tau = 0, \quad (t, x) \in \mathbf{R}^2. \tag{1.1.6'}$$

However,

$$\int_{x-t+\tau}^{-x+t-\tau} g(\tau, \xi)\, d\xi = 0$$

since g is odd in ξ. Hence the relation (1.1.6′) is now

$$\sigma(x + t + \omega) - \sigma(x + t) - \sigma(-x + t + \omega) + \sigma(-x + t) +$$
$$+ \frac{1}{2}\int_{-\omega}^{0}\int_{-x+t}^{0} g(\tau, \xi - \tau)\, d\xi\, d\tau +$$
$$+ \frac{1}{2}\int_{-\omega}^{0}\int_{0}^{x+t} g(\tau, \xi - \tau)\, d\xi\, d\tau = 0. \tag{1.1.6''}$$

Lemma 1.1.1 (1.1.6) *holds if and only if there is a constant c such that*

$$\sigma(x + \omega) - \sigma(x) + \frac{1}{2}\int_{-\omega}^{0}\int_{0}^{x} g(\tau, \xi - \tau)\, d\xi\, d\tau = c \tag{1.1.7}$$

for $x \in \mathbf{R}$.

Proof. Sufficiency: Substituting the corresponding integrals for $\sigma(x + t + \omega) - \sigma(x + t)$ and $\sigma(-x + t + \omega) - \sigma(-x + t)$ in (1.1.7) in the left-hand side of (1.1.6′) we obtain

$$-\frac{1}{2}\int_{-\omega}^{0}\int_{0}^{x+t} g(\tau, \xi - \tau)\, \mathrm{d}\xi\, \mathrm{d}\tau + \frac{1}{2}\int_{-\omega}^{0}\int_{0}^{-x+t} g(\tau, \xi - \tau)\, \mathrm{d}\xi\, \mathrm{d}\tau +$$

$$+\frac{1}{2}\int_{-\omega}^{0}\int_{x-t+\tau}^{x+t-\tau} g(\tau, \xi)\, \mathrm{d}\xi\, \mathrm{d}\tau + c - c =$$

$$= \frac{1}{2}\int_{-\omega}^{0}\int_{x+t}^{-x+t} g(\tau, \xi - \tau)\, \mathrm{d}\xi\, \mathrm{d}\tau + \frac{1}{2}\int_{-\omega}^{0}\int_{x-t+\tau}^{x+t-\tau} g(\tau, \xi)\, \mathrm{d}\xi\, \mathrm{d}\tau =$$

$$= -\frac{1}{2}\int_{-\omega}^{0}\int_{-x+t-\tau}^{x+t-\tau} g(\tau, \xi)\, \mathrm{d}\xi\, \mathrm{d}\tau + \frac{1}{2}\int_{-\omega}^{0}\int_{x-t+\tau}^{x+t-\tau} g(\tau, \xi)\, \mathrm{d}\xi\, \mathrm{d}\tau = 0\,.$$

Necessity: Putting $x + t = \alpha$, $-x + t = \beta$, we can rewrite the relation (1.1.6″) as

$$\sigma(\alpha + \omega) - \sigma(\alpha) + \frac{1}{2}\int_{-\omega}^{0}\int_{0}^{\alpha} g(\tau, \xi - \tau)\, \mathrm{d}\xi\, \mathrm{d}\tau =$$

$$= \sigma(\beta + \omega) - \sigma(\beta) + \frac{1}{2}\int_{-\omega}^{0}\int_{0}^{\beta} g(\tau, \xi - \tau)\, \mathrm{d}\xi\, \mathrm{d}\tau\,,$$

from which (1.1.7) follows readily when the independence of α and β is taken into account.

Remark 1.1.4 If (1.1.3) can be replaced by (1.1.3**), then (1.1.7) readily follows from

$$(\sigma'(x + \omega) - \sigma'(x)) + (\sigma'(-x + \omega) - \sigma'(-x)) +$$

$$+\frac{1}{2}\int_{0}^{\omega} [g(\tau, x + \omega - \tau) - g(\tau, x - \omega + \tau)]\, \mathrm{d}\tau = 0\,,$$

$$(\sigma'(x + \omega) - \sigma'(x)) - (\sigma'(-x + \omega) - \sigma'(-x)) +$$

$$+\frac{1}{2}\int_{0}^{\omega} [g(\tau, x + \omega - \tau) + g(\tau, x - \omega + \tau)]\, \mathrm{d}\tau = 0\,, \quad x \in [0, \pi]\,.$$

Now it is useful to distinguish three cases:

(a) $\omega = 2\pi n$, (b) $\omega = 2\pi p/q$, (c) $\omega = 2\pi\alpha$, n, p, $q \in \mathbf{N}$, p, q relatively prime, α irrational. We are not able to solve the case (c) by the present method. For this case see Sec. 6.3.

1.2. *The case $\omega = 2\pi n$*

Setting $\omega = 2\pi n$ in (1.1.7), we obtain

$$\int_{-2\pi n}^{0} \int_{0}^{x} g(\tau, \xi - \tau)\, d\xi\, d\tau = c\,, \quad x \in \mathbf{R}\,. \tag{1.2.1}$$

Differentiating with respect to x, we have

$$\int_{-2\pi n}^{0} g(\tau, x - \tau)\, d\tau = 0\,, \quad x \in \mathbf{R}\,, \tag{1.2.2}$$

which is equivalent to

$$\int_{0}^{2\pi n} g(\tau, x - \tau)\, d\tau = 0\,, \quad x \in \mathbf{R}\,, \tag{1.2.2'}$$

because g is 2π-periodic in x and $2\pi n$-periodic in t. Thus, we have proved the following result.

THEOREM 1.2.1 *Let $g \in \dot{C}_{2\pi n}^{(\alpha,1-\alpha)}(Q)$ with fixed $\alpha = 0$ or 1.*
Then the problems $(\mathscr{P}_{2\pi n}^{0})$ and $(\mathscr{P}_{2\pi n,e}^{0})$ have solutions if and only if (1.2.2′) holds. If this condition is satisfied, then the solutions $u^*(t, x)$ and $u_e^*(t, x)$ are given by $\sigma(x + t) - \sigma(-x + t) + z(t, x)$, where

$$z(t, x) = \frac{1}{2} \int_{0}^{t} \int_{x-t+\tau}^{x+t-\tau} g(\tau, \xi)\, d\xi\, d\tau$$

and σ is an arbitrary element of $C_{2\pi}^2(\mathbf{R})$. Writing $z = \Lambda g$ we see that there is a constant c such that

$$\|\Lambda g\|_{\mathscr{U}} \leqq c\|g\|_{C_{2\pi n}^{(\alpha,1-\alpha)}(Q)}\,.$$

Problem 1.2.1 Show that (1.2.2) under the assumptions of Theorem 1.2.1 is equivalent to

$$\int_{0}^{2\pi n} \int_{0}^{\pi} g(\tau, \xi)\, e^{ik\tau} \sin k\xi\, d\xi\, d\tau = 0\,, \quad k \in \mathbf{N} \tag{1.2.2''}$$

(see Sec. 6.2).

1.3. *The case $\omega = 2\pi p/q$*

Let $p, q \in \mathbf{N}$ be relatively prime. By Lemma 1.1.1 our problem is reduced to the investigation of the existence of a solution $\sigma(x) \in C_{2\pi}^2(\mathbf{R})$ to the equation

$$\sigma(x + 2\pi p/q) - \sigma(x) + \frac{1}{2} \int_{-2\pi p/q}^{0} \int_{0}^{x} g(\tau, \xi - \tau)\, d\xi\, d\tau = c\,, \quad x \in \mathbf{R}\,. \tag{1.3.1}$$

This is evidently equivalent to the system of equations

$$\sigma(2\pi p/q) - \sigma(0) = c\,, \tag{1.3.2}$$

$$\sigma'(x + 2\pi p/q) - \sigma'(x) + \frac{1}{2}\int_{-2\pi p/q}^{0} g(\tau, x - \tau)\,\mathrm{d}\tau = 0\,, \quad x \in \mathbf{R}\,. \tag{1.3.3}$$

Since c is arbitrary, (1.3.2) can always be satisfied. ((1.3.3) has already been studied, see II: [17].) Putting successively $x + 2\pi pj/q$, $j = 0, 1, \ldots, q - 1$ instead of x in (1.3.3), summing all these equations, and making use of $\sigma(x + 2\pi p) - \sigma(x) = 0$, we obtain as a necessary condition for the solubility of (1.3.3)

$$\sum_{j=0}^{q-1}\int_{-2\pi p/q}^{0} g(\tau, x + 2\pi pj/q - \tau)\,\mathrm{d}\tau = 0\,, \quad x \in \mathbf{R}\,, \tag{1.3.4}$$

which is equivalent to

$$\int_0^{2\pi p} g(\tau, x - \tau)\,\mathrm{d}\tau = 0\,, \quad x \in \mathbf{R}\,. \tag{1.3.4'}$$

It is easy to see that the series

$$\frac{1}{2}\sum_{j=0}^{\infty}\int_{-2\pi p/q}^{0} g(\tau, x + 2\pi pj/q - \tau)\,\mathrm{d}\tau \tag{1.3.5}$$

is a formal solution of (1.3.3). Denoting its partial sums by $S_k(x)$ we have $S_q(x) \equiv 0$ by (1.3.4) and, since g is 2π-periodic in x, $S_{k+q}(x) = S_k(x)$, $k = 1, 2, \ldots$. This implies that the series (1.3.5) is summable in the sense of Cesaro's arithmetic means, namely

$$\sigma_2'(x) = \lim_{n\to\infty}\frac{1}{n}\left(S_1(x) + S_2(x) + \ldots + S_n(x)\right) = \frac{1}{q}\sum_{k=1}^{q} S_k(x) =$$

$$= \frac{1}{2q}\int_{-\omega}^{0}\sum_{j=1}^{q-1}(q - j)\,g(\tau, x + (j - 1)\,\omega - \tau)\,\mathrm{d}\tau\,, \quad x \in \mathbf{R}\,. \tag{1.3.6}$$

Because

$$\int_0^{2\pi}\int_{-\omega}^{0}\frac{1}{2q}\sum_{j=1}^{q-1}(q - j)\,g(\tau, x + (j - 1)\,\omega - \tau)\,\mathrm{d}\tau\,\mathrm{d}x = 0\,,$$

$$\sigma_2(x) = \int_0^{x}\int_{-\omega}^{0}\frac{1}{2q}\sum_{j=1}^{q-1}(q - j)\,g(\tau, \xi + (j - 1)\,\omega - \tau)\,\mathrm{d}\tau\,\mathrm{d}\xi =$$

$$= \int_0^{x}\int_0^{\omega}\frac{1}{2q}\sum_{j=1}^{q-1}(q - j)\,g(\tau, \xi + j\omega - \tau)\,\mathrm{d}\tau\,\mathrm{d}\xi \tag{1.3.7}$$

is again an element of $C^2_{2\pi}(\mathbf{R})$ satisfying (1.3.1). To obtain the general solution of (1.3.1) it suffices to put $\sigma(x) = \sigma_1(x) + \sigma_2(x)$, where σ_1 is an arbitrary function in $C^2(\mathbf{R})$ having both the periods 2π and $2\pi p/q$, so that σ_1 must be of the period $2\pi/q$.

THEOREM 1.3.1 *Let $g \in \dot{C}^{(\alpha,1-\alpha)}_{2\pi p/q}(Q)$ with fixed $\alpha = 0$ or 1. Then $(\mathscr{P}^0_{2\pi p/q})$ and $(\mathscr{P}^0_{2\pi p/q,e})$ have solutions if and only if (1.3.4') holds. If this condition is satisfied, then the solutions $u^*(t, x)$ and $u^*_e(t, x)$ are given by $\sigma_1(x + t) - \sigma_1(-x + t) + z(t, x)$, where*

$$z(t, x) = \sigma_2(x + t) - \sigma_2(-x + t) + \frac{1}{2}\int_0^t \int_{x-t+\tau}^{x+t-\tau} g(\tau, \xi)\, d\xi\, d\tau\,, \qquad (1.3.8)$$

while σ_2 is defined by (1.3.7) and σ_1 is an arbitrary element of $C^2_{2\pi/q}(\mathbf{R})$. Writing $z = Ag$ we see that there is a constant c such that

$$\|Ag\|_{\mathscr{U}} \leqq c\|g\|_{C^{(\alpha,1-\alpha)}{}_{2\pi p/q}(Q)}\,.$$

COROLLARY 1.3.1 *Suppose that $g \in \dot{C}^{(\alpha,1-\alpha)}_{2\pi p/q}(Q)$ with fixed $\alpha = 0$ or 1, $p = 2k - 1$, $q = 2l$, $k, l \in \mathbf{N}$ and that $g(t, x) = g(t, \pi - x)$. Then $(\mathscr{P}^0_{2\pi p/q})$ and $(\mathscr{P}^0_{2\pi p/q,e})$ always have solutions.*

Proof.

$$\int_0^{2\pi(2k-1)} g(\tau, x - \tau)\, d\tau =$$

$$= \int_0^{\pi(2k-1)} g(\tau, x - \tau)\, d\tau + \int_{\pi(2k-1)}^{2\pi(2k-1)} g(\tau, x - \tau)\, d\tau =$$

$$= \int_0^{\pi(2k-1)} [g(\tau, x - \tau) + g(\tau, x - \tau - \pi)]\, d\tau =$$

$$= \int_0^{\pi(2k-1)} [g(\tau, x - \tau) + g(\tau, -x + \tau)]\, d\tau = 0\,.$$

THEOREM 1.3.2 *Suppose that g satisfies the assumptions of Corollary 1.3.1. Then the problems $(\mathscr{P}^0_{2\pi p/q})$ and $(\mathscr{P}^0_{2\pi p/q,e})$ have a unique solution u^* with the property*

$$u^*(t, x) = u^*(t, \pi - x)\,. \qquad (1.3.9)$$

This solution is given by

$$u^*(t, x) = (\Lambda g)(t, x) =$$

$$= \frac{1}{2q} \int_{-x+t}^{x+t} \int_0^\omega \sum_{j=1}^{q-1} (q - j)\, g(\tau, \xi + j\omega - \tau)\, d\tau\, d\xi +$$

$$+ \frac{1}{2} \int_0^t \int_{x-t+\tau}^{x+t-\tau} g(\tau, \xi)\, d\xi\, d\tau \tag{1.3.10}$$

and there is a constant c such that

$$\|\Lambda g\|_{\mathscr{U}} \leqq c \|g\|_{C_{2\pi p/q}^{(\alpha, 1-\alpha)}(Q)} .$$

Proof.

$$\sigma_2(\pi - x + t) - \sigma_2(-\pi + x + t) =$$

$$= \int_{-\pi+x+t}^{\pi-x+t} \int_0^\omega \frac{1}{2q} \sum_{j=1}^{q-1} (q - j)\, g(\tau, \xi + j\omega - \tau)\, d\tau\, d\xi =$$

$$= \int_{-\pi+x+t}^{\pi+x+t} \int_0^\omega \frac{1}{2q} \sum_{j=1}^{q-1} (q - j)\, g(\tau, \xi + j\omega - \tau)\, d\tau\, d\xi +$$

$$+ \int_{\pi+x+t}^{\pi-x+t} \int_0^\omega \frac{1}{2q} \sum_{j=1}^{q-1} (q - j)\, g(\tau, \xi + j\omega - \tau)\, d\tau\, d\xi =$$

$$= \int_{x+t}^{-x+t} \int_0^\omega \frac{1}{2q} \sum_{j=1}^{q-1} (q - j)\, g(\tau, \eta + \pi + j\omega - \tau)\, d\tau\, d\eta =$$

$$= \sigma_2(x + t) - \sigma_2(-x + t) ,$$

$$\int_0^t \int_{\pi-x-t+\tau}^{\pi-x+t-\tau} g(\tau, \xi)\, d\xi\, d\tau = \int_0^t \int_{x-t+\tau}^{x+t-\tau} g(\tau, \xi)\, d\xi\, d\tau$$

so that σ_1 must satisfy $\sigma_1(\pi - x + t) - \sigma_1(-\pi + x + t) = \sigma_1(x + t) - \sigma_1(-x + t)$. Since $\sigma_1 \in C^2_{\pi/l}(\mathbf{R})$, this relation gives $\sigma_1(x + t) - \sigma_1(-x + t) = 0$ identically. This completes the proof.

COROLLARY 1.3.2 *Let* $g \in \dot{C}^{(\alpha, 1-\alpha)}_{2\pi p/q}(Q)$ *with* $\alpha = 0$ *or* 1, $p = 2k$, $q = 2l - 1$, $k, l \in \mathbf{N}$ *and* $g(t + \omega/2, x) = -g(t, x)$.

Then $(\mathscr{P}^0_{2\pi p/q})$ *and* $(\mathscr{P}^0_{2\pi p/q, e})$ *always have solutions.*

Proof.

$$\int_0^{4k\pi} g(\tau, x - \tau)\, d\tau = \int_0^{2k\pi} g(\tau, x - \tau)\, d\tau + \int_{2k\pi}^{4k\pi} g(\tau, x - \tau)\, d\tau =$$

$$= \int_0^{2k\pi} [g(\tau, x - \tau) + g(\tau + (2l - 1)\, \omega/2, x - \tau - 2k\pi)]\, d\tau = 0 .$$

THEOREM 1.3.3 *Suppose that g satisfies the assumptions of Corollary* 1.3.2. *Then $(\mathscr{P}^0_{2\pi p/q})$ and $(\mathscr{P}^0_{2\pi p/q,e})$ have a unique solution u^* with the property*

$$u^*(t + \omega/2, x) = -u^*(t, x). \tag{1.3.11}$$

This solution is given by the formula

$$u^*(t, x) = (\Lambda g)(t, x) = \frac{1}{4}\int_{-x+t}^{x+t}\int_{-2k\pi}^{0} g(\tau, \xi - \tau)\,\mathrm{d}\tau\,\mathrm{d}\xi + $$

$$+ \frac{1}{2}\int_0^t\int_{x-t+\tau}^{x+t-\tau} g(\tau, \xi)\,\mathrm{d}\xi\,\mathrm{d}\tau \tag{1.3.12}$$

and there is a constant c such that

$$\|\Lambda g\|_{\mathscr{U}} \leqq c\|g\|_{C_{2\pi p/q}(\alpha, 1-\alpha)(Q)}.$$

Proof. The condition of periodicity is equivalent to the conditions

$$u(0, x) + u(\omega/2, x) = 0,$$
$$u_t(0, x) + u_t(\omega/2, x) = 0,$$

which in turn are equivalent to

$$u_x(0, x) + u_x(\omega/2, x) = 0,$$
$$u_t(0, x) + u_t(\omega/2, x) = 0. \tag{1.3.13}$$

Substituting (1.1.5) in (1.3.13) we obtain

$$\sigma'(x + \omega/2) + \sigma'(x) + \frac{1}{2}\int_0^{\omega/2} g(\tau, x + \omega/2 - \tau)\,\mathrm{d}\tau = 0, \quad x \in \mathbf{R}. \tag{1.3.14}$$

Replacing x in succession by $x + j\omega/2$, $j = 0, 1, 2, \ldots, 2l - 2$, multiplying by $(-1)^j$, and summing all these relations, we obtain

$$2\sigma'(x) = -\sum_{j=0}^{2l-2}\frac{(-1)^j}{2}\int_0^{\omega/2} g(\tau, x + (j + 1)\,\omega/2 - \tau)\,\mathrm{d}\tau =$$

$$= \frac{1}{2}\int_{-2k\pi}^{0} g(\tau, x - \tau)\,\mathrm{d}\tau.$$

It can be easily verified that this function satisfies (1.3.14). Also, it can be shown that this solution is unique in the class of 2π-periodic functions.
Since

$$\int_0^{2\pi}\int_{-2k\pi}^{0} g(\tau, x - \tau)\,\mathrm{d}\tau\,\mathrm{d}x = 0,$$

the function

$$\sigma(x) = \frac{1}{4}\int_0^x \int_{-2k\pi}^0 g(\tau, \xi - \tau)\,\mathrm{d}\tau\,\mathrm{d}\xi$$

is again 2π-periodic, and the assertion of the theorem follows readily.

Remark 1.3.1 In [17] HALE studies the problem $(\mathscr{P}^0_{2\pi})$ from the geometrical point of view. We briefly explain his approach for the extended problem $(\mathscr{P}^0_{2\pi p/q,e})$ given by (1.1.1′)–(1.1.3′) with $\omega = 2\pi p/q$, $p, q \in \mathbf{N}$ relatively prime. Let R be the operator mapping $C_{\omega,2\pi(o)}(\mathbf{R}^2)$ into $C_{2\pi/q}(\mathbf{R})$ given by

$$(Rg)(x) = \sum_{j=0}^{q-1} \int_{-\omega}^0 g(\tau, x + j\omega - \tau)\,\mathrm{d}\tau\,, \quad x \in \mathbf{R}$$

and let P be the operator mapping $C_{\omega,2\pi(o)}(\mathbf{R}^2)$ into $C_{2\pi/q,2\pi/q}(\mathbf{R}^2)$ given by

$$(Pg)(t, x) = (2\pi p)^{-1}\,[(Rg)(x + t) - (Rg)(-x + t)]\,, \quad (t, x) \in \mathbf{R}^2\,.$$

Suppose that $\mathscr{V}$ and $\mathscr{W}$ are two subspaces of $C_{\omega,2\pi(o)}(\mathbf{R}^2)$:

$$\begin{aligned}\mathscr{V} &= \{u \in C_{\omega,2\pi(o)}(\mathbf{R}^2);\ \text{there is an } s \in C_{2\pi/q}(\mathbf{R}) \text{ such that} \\ &\qquad u(t, x) = s(x + t) - s(-x + t),\ (t, x) \in \mathbf{R}^2\}\,, \\ \mathscr{W} &= \{u \in C_{\omega,2\pi(o)}(\mathbf{R}^2);\ Ru = 0\}\,.\end{aligned}$$

We can verify the following properties:

(1) $Pg = 0$ if and only if $Rg = 0$.

(2) $Rg = 0$ if and only if

$$\int_0^\omega \int_0^\pi g(t, x)\, v(t, x)\,\mathrm{d}x\,\mathrm{d}t = 0$$

for all $v \in \mathscr{V}$ (see Sec. 8.1),

(3) $Pg = g$ for $g \in \mathscr{V}$,

(4) $Pg = 0$ for $g \in \mathscr{W}$,

(5) P maps $C_{\omega,2\pi(o)}(\mathbf{R}^2)$ into $\mathscr{V}$.

Using these properties, we can deduce immediately that every $g \in C_{\omega,2\pi(o)}(\mathbf{R}^2)$ can be written in the form $g = g_{\mathscr{V}} + g_{\mathscr{W}}$, with $g_{\mathscr{V}} \in \mathscr{V}$, $g_{\mathscr{W}} \in \mathscr{W}$ uniquely determined. Of course, $g_{\mathscr{V}} = Pg$, $g_{\mathscr{W}} = g - Pg$. Note that the spaces $\mathscr{V}$ and $\mathscr{W}$ and the mappings P and R depend on ω. It turns out that Theorem 1.3.1 can be put in the following form.

THEOREM 1.3.1′ *Let $g \in C^{(\alpha,1-\alpha)}_{\omega,2\pi(o)}(\mathbf{R}^2)$ with fixed $\alpha = 0$ or 1, $\omega = 2\pi p/q$, $p, q \in \mathbf{N}$ relatively prime.*

Then the problem $(\mathscr{P}^0_{\omega,e})$ has a solution if and only if $g \in \mathscr{W}$. If

$g \in C^{(\alpha,1-\alpha)}_{\omega,2\pi(o)}(\mathbf{R}^2) \cap \mathscr{W}$, then the solution can be written in the form $u = v + w$, where $v \in C^2_{\omega,2\pi(o)}(\mathbf{R}^2) \cap \mathscr{V}$ is arbitrary and the element $w \in C^2_{\omega,2\pi(o)}(\mathbf{R}^2) \cap \mathscr{W}$ is uniquely determined by $w = (I - P)\, z = \Lambda g$, z being given by (1.3.8). The operator Λ is a member of $\mathscr{L}(C^{(\alpha,1-\alpha)}_{\omega,2\pi(o)}(\mathbf{R}^2) \cap \mathscr{W},\ C^2_{\omega,2\pi(o)}(\mathbf{R}^2) \cap \mathscr{W})$.

The application of this method to the non-linear case is contained in Remark 1.4.9.

Remark 1.3.2 From Theorem 1.3.1 it is clear that the operator L, $Lu = u_{tt} - u_{xx}$, $\mathscr{D}(L) = \dot{C}^2_{2\pi p/q}(Q)$ has an infinitely dimensional null-space $\mathscr{N}(L)$ and thus the corresponding problem $(\mathscr{P}^0_{2\pi p/q})$ is critical. On the other hand, in Theorems 1.3.2 and 1.3.3, $\mathscr{D}(L)$ are chosen in such a way that the null-spaces are trivial and the corresponding problems $(\mathscr{P}^0_{2\pi p/q})$ are non-critical.

Problem 1.3.1 Show that (1.3.4′) under the assumptions of Theorem 1.3.1 is equivalent to

$$\int_0^{2\pi p/q} \int_0^{\pi} g(\tau, \xi)\, e^{ilq\tau} \sin lq\xi \,\mathrm{d}\xi \,\mathrm{d}\tau = 0\,, \quad l \in \mathbf{N}\,.$$

1.4. *The non-linear case*

Let us investigate the problem $(\mathscr{P}_\omega)$ given by

$$u_{tt}(t, x) - u_{xx}(t, x) = F(u)(t, x)\,, \quad (t, x) \in Q\,, \tag{1.4.1}$$

$$u(t, 0) = u(t, \pi) = 0\,, \quad t \in \mathbf{R}\,, \tag{1.4.2}$$

$$u(t + \omega, x) = u(t, x)\,, \quad (t, x) \in Q\,, \tag{1.4.3}$$

where

$$F \text{ maps } \dot{C}^1_\omega(Q) \text{ into } C_\omega(Q)\,, \tag{1.4.4$_1$}$$

$$F \text{ maps } \dot{C}^2([0, \omega] \times [0, \pi]) \text{ into}$$

$$\dot{C}^{(\alpha,1-\alpha)}([0, \omega] \times [0, \pi])\,, \quad \alpha = 0 \text{ or } 1\,, \tag{1.4.4$_2$}$$

$$F(u) \mid [0, \omega] \times [0, \pi] = F(u \mid [0, \omega] \times [0, \pi]) \tag{1.4.5}$$

for every $u \in \dot{C}^1_\omega(Q)$ such that $u \mid [0, \omega] \times [0, \pi] \in \dot{C}^2([0, \omega] \times [0, \pi])$.

The function $F(u)$ defined on Q or $[0, \omega] \times [0, \pi]$ can be extended to a function $F_e(u)$ defined on $\mathbf{R} \times \mathbf{R}$ or $[0, \omega] \times \mathbf{R}$, respectively, in such a way that

$$F_e(u)(t, x) = F_e(u)(t, x + 2\pi) = -F_e(u)(t, -x)\,, \quad x \in \mathbf{R}$$

(see II: Sec. 2.2). Then by Sec. 1.1 it is natural to replace $(\mathscr{P}_\omega)$ by the problem $(\mathscr{P}_{\omega,e})$:

$$(u_e)_{tt}(t, x) - (u_e)_{xx}(t, x) = F_e(u_e \mid Q)(t, x), \quad (t, x) \in \mathbf{R}^2, \tag{1.4.1'}$$

$$u_e(t, x) = u_e(t, x + 2\pi) = -u_e(t, -x), \quad (t, x) \in \mathbf{R}^2, \tag{1.4.2'}$$

$$u_e(t + \omega, x) = u_e(t, x), \quad (t, x) \in \mathbf{R}^2. \tag{1.4.3'}$$

If u_e is a solution of $(\mathscr{P}_{\omega,e})$, then its restriction to $x \in [0, \pi]$ is a solution of $(\mathscr{P}_\omega)$. To simplify the later expressions we write $F_e(u)$ instead of $F_e(u \mid Q)$. Putting

$$\mathscr{U}_{1,\omega} = C^2_{\omega,2\pi(o)}(\mathbf{R}^2),$$

$$\mathscr{U}_1([0, \omega]) = C^2_{0,2\pi(o)}([0, \omega] \times \mathbf{R})$$

we obtain the following theorem.

THEOREM 1.4.1 *Let $\omega = 2\pi n$, $n \in \mathbf{N}$. Suppose that F satisfies (1.4.4) and (1.4.5) with $\omega = 2\pi n$.*
Then $(\mathscr{P}_{2\pi n,e})$ has a solution $u = u^* \in \mathscr{U}_{1,2\pi n}$ if and only if the system of equations

$$-u(t, x) + \sigma(x + t) - \sigma(-x + t) + $$
$$+ \frac{1}{2}\int_0^t \int_{x-t+\tau}^{x+t-\tau} F_e(u)(\tau, \xi)\, d\xi\, d\tau = 0, \quad t \in [0, 2\pi n], \quad x \in \mathbf{R}, \tag{1.4.6}$$

$$\int_0^{2\pi n} F_e(u)(\tau, x - \tau)\, d\tau = 0, \quad x \in \mathbf{R} \tag{1.4.7}$$

has a solution (u^{**}, σ^{**}), where $u^{**} \in \mathscr{U}_1([0, 2\pi n])$ and $\sigma^{**} \in C^2_{2\pi}(\mathbf{R})$.

Proof. Suppose that $(\mathscr{P}_{2\pi n,e})$ has a solution $u^* \in \mathscr{U}_{1,2\pi n}$. Then (1.4.6) evidently holds with $u^{**} = u^* \mid [0, 2\pi n] \times \mathbf{R}$ and

$$\sigma^{**}(x) = \frac{1}{2}\int_0^x [u_x^*(0, \xi) + u_t^*(0, \xi)]\, d\xi.$$

By Theorem 1.2.1, we see that (1.4.7) is satisfied. On the other hand, if the system (1.4.6)–(1.4.7) has a solution (u^{**}, σ^{**}), then we can extend u^{**} to $\mathbf{R}^2$ as an $2\pi n$-periodic function in t. This extended function u^* belongs to $C^1(\mathbf{R}^2)$, and $u^*_{tx}, u^*_{xx} \in C(\mathbf{R}^2)$. According to (1.4.4) and (1.4.5), a further calculation shows that also $u^*_{tt} \in C(\mathbf{R}^2)$. Hence u^* is a solution of $(\mathscr{P}_{2\pi n,e})$.

The following theorem can be proved similarly.

THEOREM 1.4.2 *Let $\omega = 2\pi p/q$, $p, q \in \mathbf{N}$ relatively prime. Suppose that F satisfies (1.4.4) and (1.4.5) with $\omega = 2\pi p/q$.*
Then $(\mathscr{P}_{2\pi p/q,e})$ has a solution $u = u^* \in \mathscr{U}_{1,2\pi p/q}$ if and only if the system of equations

$$-u(t,x) + \sigma_1(x+t) + \sigma_2(x+t) - \sigma_1(-x+t) - \sigma_2(-x+t) + \\ + \frac{1}{2}\int_0^t \int_{x-t+\tau}^{x+t-\tau} F_e(u)(\tau,\xi)\,\mathrm{d}\xi\,\mathrm{d}\tau = 0\,, \quad t \in [0, 2\pi p/q]\,, \quad x \in \mathbf{R}\,, \tag{1.4.8}$$

$$-\sigma_2(x) + \frac{1}{2q}\int_0^x \int_0^\omega \sum_{j=1}^{q-1} (q-j)\,F_e(u)(\tau, \xi + j\omega - \tau)\,\mathrm{d}\tau\,\mathrm{d}\xi = 0\,, \quad x \in \mathbf{R}\,, \tag{1.4.9}$$

$$\sum_{j=1}^{q} \int_0^\omega F_e(u)(\tau, x + j\omega - \tau)\,\mathrm{d}\tau = 0\,, \quad x \in \mathbf{R} \tag{1.4.10}$$

has a solution $(u^{**}, \sigma_1^{**}, \sigma_2^{**})$, where $u^{**} \in \mathscr{U}_1([0, 2\pi p/q])$, $\sigma_1^{**} \in C^2_{2\pi/q}(\mathbf{R})$, $\sigma_2^{**} \in C^2_{2\pi}(\mathbf{R})$.

From the practical point of view we come to some more instructive theorems by supposing that the perturbation F is weakly non-linear. Let $\varepsilon_0 > 0$. We investigate the problem $(\mathscr{P}^\varepsilon_\omega)$ given by

$$u_{tt}(t,x) - u_{xx}(t,x) = \varepsilon\, F(u,\varepsilon)(t,x)\,, \quad t \in \mathbf{R}\,, \quad x \in [0,\pi]\,, \tag{1.4.11}$$

(1.4.2) and (1.4.3) supposing that

$F(\cdot, \varepsilon)$ satisfies the assumptions (1.4.4) and (1.4.5) for every

$$\varepsilon \in [-\varepsilon_0, \varepsilon_0]\,. \tag{1.4.12}$$

Moreover, we assume that

F maps $C^2([0,\omega] \times [0,\pi]) \times [-\varepsilon_0, \varepsilon_0]$ into $\dot{C}^{(\alpha,1-\alpha)}([0,\omega] \times [0,\pi])$ continuously, $\alpha = 0$ or 1. (1.4.13)

As in the linear case we replace our problem by the "extended" problem $(\mathscr{P}^\varepsilon_{\omega,e})$ given by

$$u_{tt}(t,x) - u_{xx}(t,x) = \varepsilon\, F_e(u,\varepsilon)(t,x)\,, \quad (t,x) \in \mathbf{R}^2\,, \quad \varepsilon \in [-\varepsilon_0, \varepsilon_0]\,,$$

(1.4.2′), and (1.4.3′). (1.4.11′)

A function $u \in C([-\varepsilon^*, \varepsilon^*]; \mathscr{U}_{1,\omega})$, $\varepsilon_0 \geqq \varepsilon^* > 0$ is called a solution of $(\mathscr{P}^\varepsilon_{\omega,e})$ if $u(\varepsilon)$ satisfies (1.4.11′) for every $\varepsilon \in [-\varepsilon^*, \varepsilon^*]$. Evidently, the restriction of a solution of $(\mathscr{P}^\varepsilon_{\omega,e})$ to $x \in [0,\pi]$ is a solution of $(\mathscr{P}^\varepsilon_\omega)$. In what follows, we write simply $F(u, \varepsilon)$ instead of $F_e(u, \varepsilon)$. Closed subspaces of a Banach space B are denoted by B^* and B_*. (Here and in the next two paragraphs the star does not denote a dual space). Finally, for a given space $\mathcal{U}_1^*([0, \omega])$ we define

$$\mathcal{U}_{1,\omega}^* = \{u \in \mathcal{U}_{1,\omega};\ u \mid [0, \omega] \times \mathbf{R} \in \mathcal{U}_1^*([0, \omega])\} .$$

The following two lemmas are immediate consequences of Theorems 1.4.1 and 1.4.2, respectively.

LEMMA 1.4.1 *Suppose that* (1.4.12) *and* (1.4.13) *hold* with $\omega = 2\pi n$. *Then* $(\mathcal{P}_{2\pi n,e}^{\varepsilon})$ *has a solution only if*

$$\Gamma(\sigma)(\xi) \equiv \int_0^{2\pi n} F(\sigma(x + t) - \sigma(-x + t), 0)(\tau, \xi - \tau)\,\mathrm{d}\tau = 0\,, \quad \xi \in \mathbf{R} \tag{1.4.14}$$

has a solution $\sigma = \sigma_0^* \in C_{2\pi}^2(\mathbf{R})$.

LEMMA 1.4.2 *Suppose that* (1.4.12) *and* (1.4.13) *hold* with $\omega = 2\pi p/q$. *Then* $(\mathcal{P}_{2\pi p/q,e}^{\varepsilon})$ *has a solution only if*

$${}^1\Gamma(\sigma)(\xi) \equiv \sum_{j=1}^{q} \int_0^{2\pi p/q} F(\sigma(x + t) - \sigma(-x + t), 0)(\tau, \xi + j\omega - \tau)\,\mathrm{d}\tau = 0, \quad \xi \in \mathbf{R} \tag{1.4.15}$$

has a solution $\sigma = \sigma_1^* \in C_{2\pi/q}^2(\mathbf{R})$.

The next few theorems give sufficient conditions for the solubility of the problem $(\mathcal{P}_{\omega,e}^{\varepsilon})$.

THEOREM 1.4.3 *Suppose that*:

(*i*) (1.4.14) *has a solution* $\sigma_0^* \in C_{2\pi}^{2*}(\mathbf{R})$;

(*ii*) *the operator* $F(u, \varepsilon)$ *satisfies* (1.4.12) *with* $\omega = 2\pi n$.

The mapping $F(u, \varepsilon)$ *and its* G-*derivative* $F_u'(u, \varepsilon)$ *are continuous from* $B(u_0^*, \varrho; \mathcal{U}_1^*([0, 2\pi n])) \times [-\varepsilon_0, \varepsilon_0]$ *into* $C_{0,2\pi(o)}^{(\alpha,r-\alpha)}([0, \omega] \times \mathbf{R})$ *and* $\mathcal{L}(\mathcal{U}_1^*([0, 2\pi n]), C_{0,2\pi(o)}^{(\alpha,r-\alpha)}([0, \omega] \times \mathbf{R}))$, *respectively*, *where* $u_0^*(t, x) = \sigma_0^*(x + t) - \sigma_0^*(-x + t)$, $r = 1$ *or* 2, $\alpha \in \{0, 1, r\}$ *and* $u_0^* \in \mathcal{U}_{1,2\pi n}^*$;

(*iii*) *the operator*

$$G_1(u, \sigma, \varepsilon)(t, x) = -u(t, x) + \sigma(x + t) - \sigma(-x + t) + \frac{\varepsilon}{2}\int_0^t \int_{x-t+\tau}^{x+t-\tau} F(u, \varepsilon)(\tau, \xi)\,\mathrm{d}\xi\,\mathrm{d}\tau$$

maps $B(u_0^*, \varrho; \mathcal{U}_1^*([0, 2\pi n])) \times B(\sigma_0^*, \varrho; C_{2\pi}^{2*}(\mathbf{R})) \times [-\varepsilon_0, \varepsilon_0]$

into $\mathscr{U}_1^*([0, 2\pi n])$ *and the operator*

$$G_2(u, \varepsilon)(x) = \int_0^{2\pi n} F(u, \varepsilon)(\tau, x - \tau)\, d\tau$$

maps $B(u_0^*, \varrho; \mathscr{U}_1^*([0, 2\pi n])) \times [-\varepsilon_0, \varepsilon_0]$ *into* $C_{2\pi*}^r(\mathbf{R})$;

(iv) $[\Gamma'_\sigma(\sigma_0^*)]^{-1} \in \mathscr{L}(C_{2\pi*}^r(\mathbf{R}), C_{2\pi}^{2*}(\mathbf{R}))$.

Then the problem $(\mathscr{P}_{2\pi n, e}^{\varepsilon})$ *has a locally unique solution*

$u^*(\varepsilon) \in \mathscr{U}_{1,2\pi n}^*$ *such that* $u^*(0) = u_0^*$.

Proof. We put $B_1 = \mathscr{U}_1^*([0, 2\pi n]) \times C_{2\pi}^{2*}(\mathbf{R})$, $B_2 = \mathbf{R}$, $B = \mathscr{U}_1^*([0, 2\pi n]) \times \times C_{2\pi*}^r(\mathbf{R})$, $x = (u, \sigma)$, $p = \varepsilon$, $x_0 = (u_0^*, \sigma_0^*)$, $p_0 = 0$ and $G = (G_1, G_2)$. We claim that all the assumptions of I: Theorem 3.4.2 are satisfied.

(1) is an immediate consequence of (ii) and (iii).

(2) is obvious.

We recall that $D_x G = (G'_{1(u,\sigma)}, G'_{2(u,\sigma)})$, where

$$G'_{1(u,\sigma)}(u, \sigma, \varepsilon)(\bar{u}, \bar{\sigma})(t, x) = -\bar{u}(t, x) + \bar{\sigma}(x + t) - \bar{\sigma}(-x + t) +$$

$$+ \frac{\varepsilon}{2} \int_0^t \int_{x-t+\tau}^{x+t-\tau} F'_u(u, \varepsilon)\, \bar{u}(\tau, \xi)\, d\xi\, d\tau$$

and

$$G'_{2(u,\sigma)}(u, \varepsilon)(\bar{u}, \bar{\sigma})(x) = \int_0^{2\pi n} F'_u(u, \varepsilon)\, \bar{u}(\tau, x - \tau)\, d\tau\,.$$

Hence, we can deduce from (ii) that (3) holds.

Let $(v, \eta) \in B$. The second equation of the system

$$-\bar{u}(t, x) + \bar{\sigma}(x + t) - \bar{\sigma}(-x + t) = v(t, x)\,,$$

$$\int_0^{2\pi n} F'_u(u_0^*, 0)\, \bar{u}(\tau, x - \tau)\, d\tau = \eta(x)$$

is equivalent to

$$\int_0^{2\pi n} F'_u(u_0^*, 0)(Z\bar{\sigma})(\tau, \xi - \tau)\, d\tau =$$

$$= \eta(\xi) + \int_0^{2\pi n} F'_u(u_0^*, 0)\, v(\tau, \xi - \tau)\, d\tau\,,$$

where

$$(Z\bar{\sigma})(t, x) = \bar{\sigma}(x + t) - \bar{\sigma}(-x + t)\,.$$

Thus, by (*iv*) the system has a unique solution $(\bar{u}, \bar{\sigma})$ for which

$$\|(\bar{u}, \bar{\sigma})\|_{B_1} \leqq c\|(v, \eta)\|_B ,$$

with c independent of (v, η). Hence (4) is satisfied, and this completes the proof.

THEOREM 1.4.4 *Suppose that:*

(*i*) (1.4.15) *has a solution* $\sigma_1^* \in C^{2*}_{2\pi/q}(\mathbf{R})$;

(*ii*) *the operator* $F(u, \varepsilon)$ *satisfies* (1.4.12) *with* $\omega = 2\pi p/q$. *The mapping* $F(u, \varepsilon)$ *and its G-derivative* $F'_u(u, \varepsilon)$ *are continuous from* $B(u_0^*, \varrho; \mathscr{U}_1^*([0, \omega])) \times \times [-\varepsilon_0, \varepsilon_0]$ *into* $C^{(\alpha, r-\alpha)}_{0,2\pi(o)}([0, \omega] \times \mathbf{R})$ *and* $\mathscr{L}(\mathscr{U}_1^*([0, \omega]), C^{(\alpha, r-\alpha)}_{0,2\pi(o)}([0, \omega] \times \times \mathbf{R}))$, *respectively, where* $u_0^*(t, x) = \sigma_1^*(x + t) - \sigma_1^*(-x + t)$, $r = 1$ *or* 2, $\alpha \in \{0, 1, r\}$ *and* $u_0^* \in \mathscr{U}^*_{1,\omega}$;

(*iii*) *the operator*

$$G_3(u, \sigma_1, \sigma_2, \varepsilon)(t, x) = -u(t, x) + \sigma_1(x + t) - \sigma_1(-x + t) +$$

$$+ \sigma_2(x + t) - \sigma_2(-x + t) + \frac{\varepsilon}{2}\int_0^t \int_{x-t+\tau}^{x+t-\tau} F(u, \varepsilon)(\tau, \xi)\,d\xi\,d\tau$$

maps $B(u_0^*, \varrho; \mathscr{U}_1^*([0, \omega])) \times B(\sigma_1^*, \varrho; C^{2*}_{2\pi/q}(\mathbf{R})) \times B(0, \varrho; C^{2*}_{2\pi}(\mathbf{R})) \times [-\varepsilon_0, \varepsilon_0]$ *into* $\mathscr{U}_1^*([0, \omega])$, *the operator*

$$G_4(u, \sigma_2, \varepsilon)(x) = -\sigma_2(x) +$$

$$+ \frac{\varepsilon}{2q}\int_0^x \int_0^\omega \sum_{j=1}^{q-1} (q - j)\, F(u, \varepsilon)(\tau, \xi + j\omega - \tau)\,d\tau\,d\xi$$

maps $B(u_0^*, \varrho; \mathscr{U}_1^*([0, \omega])) \times B(0, \varrho; C^{2*}_{2\pi}(\mathbf{R})) \times [-\varepsilon_0, \varepsilon_0]$ *into* $C^{2*}_{2\pi}(\mathbf{R})$, *and the operator*

$$G_5(u, \varepsilon)(x) = \sum_{j=1}^{q} \int_0^{2\pi p/q} F(u, \varepsilon)(\tau, x + j\omega - \tau)\,d\tau$$

maps $\mathscr{U}_1^*([0, \omega]) \times [-\varepsilon_0, \varepsilon_0]$ *into* $C^r_{2\pi/q^*}(\mathbf{R})$;

(*iv*) $[{}^1\Gamma'_\sigma(\sigma_1^*)]^{-1} \in \mathscr{L}(C^r_{2\pi/q^*}(\mathbf{R}); C^{2*}_{2\pi/q}(\mathbf{R}))$.

Then the problem $(\mathscr{P}^\varepsilon_{2\pi p/q,e})$ *has a locally unique solution* $u^*(\varepsilon) \in \mathscr{U}^*_{1,2\pi p/q}$ *such that* $u^*(0) = u_0^*$.

The *proof* is similar to that of the preceding theorem.

We now define two spaces

$$\mathscr{U}_{2,\omega} = \{u \in \mathscr{U}_{1,\omega};\, u(t, x) = u(t, \pi - x), (t, x) \in \mathbf{R}^2\} ,$$

$$\mathscr{U}_{3,\omega} = \{u \in \mathscr{U}_{1,\omega};\, u(t, x) = -u(t + \omega/2, x), (t, x) \in \mathbf{R}^2\}$$

and prove the following result.

THEOREM 1.4.5 *Let* $\omega = 2\pi p/q$, $p = 2k - 1$, $q = 2l$, $k, l \in \mathbf{N}$ *and suppose that*:

(*i*) $F(u, \varepsilon)$ *maps* $B(0, \varrho; \mathscr{U}_{2,\omega}) \times [-\varepsilon_0, \varepsilon_0]$ *continuously into the space*

$$\{u \in C^{(\alpha,1-\alpha)}_{\omega,2\pi(o)}(\mathbf{R}^2);\ u(t, \pi - x) = u(t, x), (t, x) \in \mathbf{R}^2\}$$

where $\alpha = 0$ *or* 1;

(*ii*) *there is a constant* λ *such that*

$$\|F(u_1, \varepsilon) - F(u_2, \varepsilon)\|_{C^{(\alpha,1-\alpha)}(Q)} \leqq \lambda \|u_1 - u_2\|_{\mathscr{U}_{2,\omega}}$$

for all $u_1, u_2 \in B(0, \varrho; \mathscr{U}_{2,\omega})$ *and* $\varepsilon \in [-\varepsilon_0, \varepsilon_0]$.
Then the problem $(\mathscr{P}^{\varepsilon}_{\omega,e})$ *has a locally unique solution* $u^*(\varepsilon) \in \mathscr{U}_{2,\omega}$ *such that* $u^*(0) = 0$.

Proof. Let Λ be the operator defined in Theorem 1.3.2. If we find a solution $u(\varepsilon)$ of the equation

$$u = \varepsilon \Lambda F(u, \varepsilon), \tag{1.4.16}$$

then $u(\varepsilon)$ is a solution of $(\mathscr{P}^{\varepsilon}_{\omega,e})$. Applying I: Corollary 3.4.1 (p. 29) to (1.4.16), we obtain the required assertion.

The following theorem can be proved similarly.

THEOREM 1.4.6 *Let* $\omega = 2\pi p/q$, $p = 2k$, $q = 2l - 1$, $k, l \in \mathbf{N}$ *and suppose that*:

(*i*) $F(u, \varepsilon)$ *maps* $B(0, \varrho; \mathscr{U}_{3,\omega}) \times [-\varepsilon_0, \varepsilon_0]$ *continuously into*

$$\{u \in C^{(\alpha,1-\alpha)}_{\omega,2\pi(o)}(\mathbf{R}^2);\ u(t + \omega/2, x) = -u(t, x), (t, x) \in \mathbf{R}^2\};$$

(*ii*) *there is a constant* λ *such that*

$$\|F(u_1, \varepsilon) - F(u_2, \varepsilon)\|_{C^{(\alpha,1-\alpha)}(Q)} \leqq \lambda \|u_1 - u_2\|_{\mathscr{U}_{3,\omega}}$$

for all $u_1, u_2 \in B(0, \varrho; \mathscr{U}_{3,\omega})$ *and* $\varepsilon \in [-\varepsilon_0, \varepsilon_0]$.
Then $(\mathscr{P}^{\varepsilon}_{\omega,e})$ *has a locally unique solution* $u^*(\varepsilon) \in \mathscr{U}_{3,\omega}$ *such that* $u^*(0) = 0$.

Remark 1.4.1 Let us treat a special case of the operator $F(u, \varepsilon)$ in the form of a substitution operator $f(t, x, u, u_t, u_x, \varepsilon)$. We make the following assumptions:

Let $k \in \mathbf{N}$ and suppose that $f(t, x, u_0, u_1, u_2, \varepsilon)$ is ω-periodic in t and continuous, together with its derivatives
$D_t^{\alpha} D_x^{\beta} D_{u_0}^{\gamma_0} D_{u_1}^{\gamma_1} D_{u_2}^{\gamma_2} f$, $\quad \alpha + \beta + \gamma_0 + \gamma_1 + \gamma_2 \leqq k$, $\quad \alpha + \beta \leqq k - 1$
on the set
$\{(t, x, u_0, u_1, u_2, \varepsilon);\ t \in \mathbf{R}, x \in [0, \pi], u_0, u_1, u_2 \in \mathbf{R}, \varepsilon \in [-\varepsilon_0, \varepsilon_0]\}$.
Suppose also that f and all its derivatives of even order vanish on the hyperplanes $x = u_0 = u_1 = 0$ and $x = \pi$, $u_0 = u_1 = 0$. (1.4.17)

If $F(u, \varepsilon) = f(t, x, u, \varepsilon)$, then $F(u, \varepsilon)$ satisfies the assumption (*ii*) of Theorem 1.4.3 and Theorem 1.4.4 with $r = 2$, provided that $f(t, x, u_0, \varepsilon)$ satisfies (1.4.17) with $k = 3$.

If $F(u, \varepsilon) = f(t, x, u, u_t, u_x, \varepsilon)$, then $F(u, \varepsilon)$ satisfies the assumption (*ii*) of Theorem 1.4.3 and Theorem 1.4.4 with $r = 1$, provided that $f(t, x, u_0, u_1, u_2, \varepsilon)$ satisfies (1.4.17) with $k = 2$ and that

$$\left|\frac{\partial f}{\partial u_1}\right| + \left|\frac{\partial f}{\partial u_2}\right| \not\equiv 0 .$$

Remark 1.4.2 The more general problem given by

$$u_{tt}(t, x) - u_{xx}(t, x) = g(t, x) + \varepsilon F(u, \varepsilon)(t, x)$$

with (1.4.2) and (1.4.3) can be investigated in the following way. If the corresponding linear problem $(\mathscr{P}^0_\omega)$ has no solution, then the given problem has no solution either. If such solutions exist, take an arbitrary but fixed one and denote it by u_0. Setting $u = u_0 + v$, we obtain for v the problem $(\mathscr{P}^\varepsilon_\omega)$ with $F_1(v, \varepsilon) = F(u_0 + v, \varepsilon)$.

Remark 1.4.3 If $F(\cdot, \varepsilon)$ maps $C^1(\mathbf{R}^2)$ into $C(\mathbf{R}^2)$, or $C(\mathbf{R}^2)$ into $C(\mathbf{R}^2)$ respectively, then, roughly speaking, Theorems 1.4.3, 1.4.4, 1.4.5 and 1.4.6 can be rewritten for the generalized solutions $u \in C^1(\mathbf{R}^2)$ and $u \in C^0(\mathbf{R}^2)$, respectively.

Remark 1.4.4 The problem $(\mathscr{P}^\varepsilon_{2\pi(2k-1)/2l})$ was studied by different methods and under stronger assumptions than here by several Soviet mathematicians (see ARTEM'EV [2], KARP [26]–[30], MITRJAKOV [39], [40], SOLOV'EV [67]). MUSTAFAZADE deals with the same problem in $\mathbf{R}^{n+1}$ in [41]. The problem $(\mathscr{P}_{2\pi 2k/(2l-1)})$ was first investigated by SOKOLOV in [64].

Remark 1.4.5 $(\mathscr{P}^\varepsilon_{2\pi n})$ and $(\mathscr{P}^\varepsilon_{2\pi p/q})$ were studied earlier by VEJVODA by the Poincaré method. In [78] he finds first the solutions $u = U(\sigma, \varepsilon)$ of (1.4.6) and $u = U(\sigma_1, \sigma_2, \varepsilon)$ of (1.4.8), and substitutes them in (1.4.7) and (1.4.9), (1.4.10), respectively. (Preliminary communications on these results are contained in [76] and [77]. Unfortunately, in the English version of [77] the assertion concerning the case $\omega = 2\pi\alpha$ is false.)

In [80] the operator $F(u, \varepsilon)$ is extended in x from $[0, \pi]$ to $\mathbf{R}$ in a fairly arbitrary way and the function σ is extended depending on the extension of $F(u, \varepsilon)$. This approach makes it possible to treat inhomogeneous boundary conditions. On the other hand, without imposing additional assumptions on $F(u, \varepsilon)$ one can ensure in this way the continuity of u but not, in general, of its first and second derivatives on the characteristics passing through the points $(0, k\pi)$, $k \in \mathbf{Z}$. Here, in fact, these derivatives may have jumps. Such solutions are called quasiclassical. Let us add here that in [80] also the other boundary-value problems described in § 4 are investigated by the same method.

Remark 1.4.6 STRAŠKRABA [73] investigates the problem $(\mathscr{P}^{\varepsilon}_{2\pi})$ with $F(u, \varepsilon) = \alpha u + \beta u^{2n+1} + h(t, x)$ and $F(u, \varepsilon) = (\alpha + \beta u^2) u_t + h(t, x)$, provided that $\alpha\beta \neq 0$ and the sufficiently smooth function h satisfies certain estimates.

Remark 1.4.7 KLIMPEROVÁ [31] deals with the problem $(\mathscr{P}^{\varepsilon}_{2\pi})$ with $F(u, \varepsilon) = \sin u + h(t, x)$, where the function h is supposed to be smooth and sufficiently small.

Remark 1.4.8 The question arises as to whether there are reasonable sufficient conditions for the existence of an ω-periodic solution to the problem

$$u_{tt} - \Delta_n u = f \quad \text{for} \quad (t, x) \in \mathbf{R} \times \Omega, \quad \Omega \subset \mathbf{R}^n,$$
$$u(t, x) = 0 \quad \text{for} \quad x \in \partial\Omega$$

with $n > 1$. It turns out that, in general, we come across insurmountable difficulties. The special case of a spherically symmetric problem in $\mathbf{R}^3$ was treated by VEJVODA in [79].

Remark 1.4.9 The method of HALE [17] that was described in Remark 1.3.1 for the linear case can also be used in the non-linear case. Keeping the notation, we define two operators

$$G_1(v, w, \varepsilon) = w - \Lambda(I - P) F(v + w, \varepsilon),$$
$$G_2(v, w, \varepsilon) = P F(v + w, \varepsilon) \tag{1.4.18}$$

for $v \in \mathscr{U}_{1,\omega} \cap \mathscr{V}$, $w \in \mathscr{U}_{1,\omega} \cap \mathscr{W}$. Then $G_1(v, w, \varepsilon) \in \mathscr{U}_{1,\omega} \cap \mathscr{W}$ and $G_2(v, w, \varepsilon) \in C^r_{\omega,2\pi(o)}(\mathbf{R}^2) \cap \mathscr{V}$, where $r = 1$ or 2 in accordance with the assumptions on $F(u, \varepsilon)$. Applying I: Theorem 3.4.2 to the equations $G_1(v, w, \varepsilon) = 0$ and $G_2(v, w, \varepsilon) = 0$, we come to a similar conclusion on the existence of a $2\pi p/q$-periodic solution as in Theorem 1.4.4. Note that we work here in the spaces of functions that are periodic in t. This is possible because the operator G_1 defined by $(1.4.18_1)$ maps $2\pi p/q$-periodic functions in t into themselves, in contrast to the operator G_3 defined in (iii) of Theorem 1.4.4.

§ 2. The Dirichlet boundary conditions; the Günzler method

2.1. *The linear case; general considerations*

Whereas in the Poincaré method one starts from the corresponding initial boundary value problem, in the method to be explained in this paragraph the point of departure is the unilateral problem with boundary data given at $x = 0$. One advantage of this approach, which was apparently used for the first time by GÜNZLER (see [18] in the "Bibliography of papers on related topics") in a study of almost periodic solutions of the wave equation, is the fact that one need not impose unnecessary conditions on g for $x = 0$ and $x = \pi$. Another advantage is that this procedure makes it possible to investigate problems with inhomogeneous boundary conditions easily. In this paragraph we again use the notation

$$Q = \mathbf{R} \times [0, \pi].$$

Let $(\mathscr{P}_\omega^0)$ be the problem given by the equations

$$u_{tt}(t, x) - u_{xx}(t, x) = g(t, x), \quad (t, x) \in Q, \tag{2.1.1}$$

$$u(t, 0) = h_0(t), \quad u(t, \pi) = h_1(t), \quad t \in \mathbf{R}, \tag{2.1.2}$$

$$u(t + \omega, x) = u(t, x), \quad (t, x) \in Q, \tag{2.1.3}$$

where $g \in C_\omega^{(\alpha, 1-\alpha)}(Q)$, $\alpha = 0$ or 1, and $h_0, h_1 \in C_\omega^2(\mathbf{R})$. We again look for a classical solution $u \in C_\omega^2(Q)$.

By II: Lemma 2.1.2, the solution of the unilateral problem given by (2.1.1) and

$$u(t, 0) = a(t), \quad u_x(t, 0) = b(t) \tag{2.1.4}$$

with $a \in C_\omega^2(\mathbf{R})$, $b \in C_\omega^1(\mathbf{R})$ is determined by the formula

$$u(t, x) = \frac{1}{2}(a(x + t) + a(-x + t)) + $$
$$+ \frac{1}{2}\int_{-x+t}^{x+t} b(\tau)\,\mathrm{d}\tau - \frac{1}{2}\int_0^x \int_{t-x+\xi}^{t+x-\xi} g(\tau, \xi)\,\mathrm{d}\tau\,\mathrm{d}\xi. \tag{2.1.5}$$

It is readily verified that the function u given by (2.1.5) is ω-periodic in t. Thus, this u is a solution of $(\mathscr{P}_\omega^0)$ if and only if

$$a(t) = h_0(t), \tag{2.1.6}$$

$$\frac{1}{2}(a(t+\pi) + a(t-\pi)) + \frac{1}{2}\int_{t-\pi}^{t+\pi} b(\tau)\,\mathrm{d}\tau -$$

$$-\frac{1}{2}\int_0^{\pi}\int_{t-\pi+\xi}^{t+\pi-\xi} g(\tau, \xi)\,\mathrm{d}\tau\,\mathrm{d}\xi = h_1(t) \tag{2.1.7}$$

for $t \in \mathbf{R}$.

Now let us examine separately the cases $\omega = 2\pi$ and $\omega = 2\pi p/q$. (The case $\omega = 2\pi\alpha$, α irrational, can hardly be handled by the method of this paragraph. Here the Fourier methods are more adequate, see § 6.)

2.2. *The case* $\omega = 2\pi$

If we substitute $h_0(t)$ for $a(t)$ in (2.1.7) and $t + \pi$ for t, take into account that $h_0(t)$ is 2π-periodic, and use the fact that

$$\int_t^{t+2\pi} b(\tau)\,\mathrm{d}\tau = \int_0^{2\pi} b(\tau)\,\mathrm{d}\tau,$$

we obtain

$$\int_0^{2\pi} b(\tau)\,\mathrm{d}\tau = 2h_1(t+\pi) - 2h_0(t) + \int_0^{\pi}\int_{t+\xi}^{t+2\pi-\xi} g(\tau, \xi)\,\mathrm{d}\tau\,\mathrm{d}\xi.$$

This equation is equivalent to the pair of equations

$$\int_0^{2\pi} b(\tau)\,\mathrm{d}\tau = 2h_1(\pi) - 2h_0(0) + \int_0^{\pi}\int_{\xi}^{2\pi-\xi} g(\tau, \xi)\,\mathrm{d}\tau\,\mathrm{d}\xi, \tag{2.2.1}$$

$$0 = h_1'(t+\pi) - h_0'(t) + \frac{1}{2}\int_0^{\pi} [g(t-\xi, \xi) - g(t+\xi, \xi)]\,\mathrm{d}\xi, \tag{2.2.2}$$

of which the former imposes a limitation only on the average of $b(t)$. Hence the following theorem holds.

THEOREM 2.2.1 *Let* $g \in C_{2\pi}^{(\alpha,1-\alpha)}(Q)$, $\alpha = 0$ *or* 1, *and* $h_0, h_1 \in C_{2\pi}^2(\mathbf{R})$. *Then* $(\mathscr{P}_{2\pi}^0)$ *has a solution if and only if* (2.2.2) *holds. If this condition is satisfied, then the solution* $u^*(t, x)$ *is given by* (2.1.5), *where* $a(t) = h_0(t)$ *and* $b(t)$ *is an arbitrary element of* $C_{2\pi}^1(\mathbf{R})$ *satisfying* (2.2.1).

Remark 2.2.1 The condition (2.2.2) for $h_0 = 0$, $h_1 = 0$, and $g \in C_{2\pi,2\pi(o)}^{(\alpha,1-\alpha)}(\mathbf{R}^2)$ is equivalent to (1.2.2′) for $n = 1$.

2.3. *The case* $\omega = 2\pi p/q$

Let $p, q \in \mathbf{N}$ be relatively prime. If we again substitute $h_0(t)$ for $a(t)$ in (2.1.7) and $t + \pi$ for t, we obtain

$$\int_t^{t+2\pi} b(\tau)\,\mathrm{d}\tau = 2h_1(t+\pi) - h_0(t+2\pi) - h_0(t) + \\ + \int_0^\pi \int_{t+\xi}^{t+2\pi-\xi} g(\tau, \xi)\,\mathrm{d}\tau\,\mathrm{d}\xi\,.$$

This equation is equivalent to the pair of equations

$$\int_0^{2\pi} b(\tau)\,\mathrm{d}\tau = 2h_1(\pi) - h_0(2\pi) - h_0(0) + \\ + \int_0^\pi \int_\xi^{2\pi-\xi} g(\tau, \xi)\,\mathrm{d}\tau\,\mathrm{d}\xi\,, \tag{2.3.1}$$

$$b(t+2\pi) - b(t) = 2h_1'(t+\pi) - h_0'(t+2\pi) - h_0'(t) + \\ + \int_0^\pi [g(t+2\pi-\xi, \xi) - g(t+\xi, \xi)]\,\mathrm{d}\xi\,, \tag{2.3.2}$$

the former of which is easily satisfied. Treating (2.3.2) like (1.3.3), we obtain the following theorem.

THEOREM 2.3.1 *Let* $g \in C^{(\alpha,1-\alpha)}_{2\pi p/q}(Q)$, $\alpha = 0$ *or* 1, *and* $h_0, h_1 \in C^2_{2\pi p/q}(\mathbf{R})$. *Then* $(\mathscr{P}^0_{2\pi p/q})$ *has a solution if and only if*

$$2\sum_{j=0}^{p-1} h_1'(t+\pi+2\pi j) - 2\sum_{j=0}^{p-1} h_0'(t+2\pi j) + \\ + \sum_{j=0}^{p-1} \int_0^\pi [g(t+2\pi(j+1)-\xi, \xi) - g(t+2\pi j+\xi, \xi)]\,\mathrm{d}\xi = 0\,, \quad t \in \mathbf{R}\,. \tag{2.3.3}$$

If this condition is satisfied, then the solution $u^*(t, x)$ *of* $(\mathscr{P}^0_{2\pi p/q})$ *is given by* (2.1.5), *where* $a(t) = h_0(t)$, $b(t) = b_1(t) + b_2(t)$,

$$b_2(t) = -\frac{1}{p}\sum_{j=0}^{p-1}(p-j)\,\alpha(t+2\pi j)\,, \tag{2.3.4}$$

$$\alpha(t) = 2h_1'(t+\pi) - h_0'(t+2\pi) - h_0'(t) + \\ + \int_0^\pi [g(t+2\pi-\xi, \xi) - g(t+\xi, \xi)]\,\mathrm{d}\xi$$

and $b_1 \in C^1_{2\pi/q}(\mathbf{R})$ *is such that* $b(t)$ *satisfies* (2.3.1).

Remark 2.3.1 The condition (2.3.3) for $h_0 = h_1 = 0$, and $g \in C^{(\alpha,1-\alpha)}_{\omega,2\pi(o)}(R^2)$ is equivalent to (1.3.4′).

Problem 2.3.1 Prove the assertions of Theorems 1.3.2 and 1.3.3 by means of the Günzler method.

2.4. *The non-linear case*

We deal with the problem $(\mathscr{P}_\omega)$ given by

$$u_{tt}(t, x) - u_{xx}(t, x) = F(u)(t, x), \quad (t, x) \in Q, \tag{2.4.1}$$

$$u(t, 0) = X_0(u)(t),$$

$$u(t, \pi) = X_1(u)(t), \quad t \in \mathbf{R} \tag{2.4.2}$$

and (2.1.3).

Let us assume that

$$F \text{ maps } C^2_\omega(Q) \text{ into } C^{(\alpha,1-\alpha)}_\omega(Q) \text{ with } \alpha = 0 \text{ or } 1, \tag{2.4.3$_1$}$$

$$X_j \text{ maps } C^2_\omega(Q) \text{ into } C^s_\omega(\mathbf{R}) \text{ for } j = 0, 1 \text{ and some } s \geqq 2. \tag{2.4.3$_2$}$$

The following two theorems are easy to prove using Theorems 2.2.1 and 2.3.1, respectively.

THEOREM 2.4.1 *Suppose that the operators F, X_0, and X_1 satisfy (2.4.3) with $\omega = 2\pi$.*
Then $(\mathscr{P}_{2\pi})$ has a solution $u^* \in C^2_{2\pi}(Q)$ if and only if the system

$$-u(t, x) + \tfrac{1}{2}(X_0(u)(x + t) + X_0(u)(-x + t)) +$$

$$+ \frac{1}{2}\int_{-x+t}^{x+t} b(\tau)\,d\tau - \frac{1}{2}\int_0^x \int_{t-x+\xi}^{t+x-\xi} F(u)(\tau, \xi)\,d\tau\,d\xi = 0, \quad (t, x) \in Q,$$

$$-\int_0^{2\pi} b(\tau)\,d\tau + 2X_1(u)(\pi) - 2X_0(u)(0) +$$

$$+ \int_0^\pi \int_\xi^{2\pi-\xi} F(u)(\tau, \xi)\,d\tau\,d\xi = 0,$$

$$2\frac{d}{dt}[X_1(u)(t + \pi) - X_0(u)(t)] +$$

$$+ \int_0^\pi [F(u)(t - \xi, \xi) - F(u)(t + \xi, \xi)]\,d\xi = 0, \quad t \in \mathbf{R} \tag{2.4.4}$$

has a solution (u^*, b^*), $u^* \in C^2_{2\pi}(Q)$, $b^* \in C^1_{2\pi}(\mathbf{R})$.

THEOREM 2.4.2 *Suppose that the operators* F, X_0, *and* X_1 *satisfy* (2.4.3) *with* $\omega = 2\pi p/q$, $p, q \in \mathbf{N}$ *relatively prime.*
Then $(\mathscr{P}_{2\pi p/q})$ *has a solution* $u^* \in C^2_{2\pi p/q}(Q)$ *if and only if the system*

$$
\begin{aligned}
&-u(t, x) + \tfrac{1}{2}(X_0(u)(x + t) + X_0(u)(-x + t)) + \\
&+ \frac{1}{2}\int_{-x+t}^{x+t} [b_1(\tau) + b_2(\tau)]\, d\tau - \\
&- \frac{1}{2}\int_0^x \int_{t-x+\xi}^{t+x-\xi} F(u)(\tau, \xi)\, d\tau\, d\xi = 0, \quad (t, x) \in Q, \\
&b_2(t) + \frac{1}{p}\sum_{j=0}^{p-1} (p - j)\left\{\frac{d}{dt}[2X_1(u)(t + \pi + 2\pi j) - \right. \\
&- X_0(u)(t + 2\pi(j + 1)) - X_0(u)(t + 2\pi j)] + \\
&+ \int_0^\pi [F(u)(t + 2\pi(j + 1) - \xi, \xi) - \\
&\left. - F(u)(t + 2\pi j + \xi, \xi)]\, d\xi\right\} = 0, \quad t \in \mathbf{R}, \\
&- \int_0^{2\pi} (b_1(\tau) + b_2(\tau))\, d\tau + 2X_1(u)(\pi) - X_0(u)(2\pi) - \\
&- X_0(u)(0) + \int_0^\pi \int_\xi^{2\pi-\xi} F(u)(\tau, \xi)\, d\tau\, d\xi = 0, \\
&2\frac{d}{dt}\sum_{j=0}^{p-1} [X_1(u)(t + \pi + 2\pi j) - X_0(u)(t + 2\pi j)] + \\
&+ \sum_{j=0}^{p-1} \int_0^\pi [F(u)(t + 2\pi(j + 1) - \xi, \xi) - \\
&- F(u)(t + 2\pi j + \xi, \xi)]\, d\xi = 0, \quad t \in \mathbf{R},
\end{aligned}
\tag{2.4.5}
$$

has a solution

$$(u^*, b_1^*, b_2^*), \quad u^* \in C^2_{2\pi p/q}(Q), \quad b_1^* \in C^1_{2\pi/q}(\mathbf{R}), \quad b_2^* \in C^1_{2\pi p/q}(\mathbf{R}).$$

Now let us deal with the weakly non-linear problem $(\mathscr{P}^\varepsilon_\omega)$ given by the equations

$$
\begin{aligned}
&u_{tt}(t, x) - u_{xx}(t, x) = \varepsilon F(u, \varepsilon)(t, x), \quad (t, x) \in Q, \\
&u(t, 0) = \varepsilon X_0(u, \varepsilon)(t), \quad t \in \mathbf{R}, \\
&u(t, \pi) = \varepsilon X_1(u, \varepsilon)(t), \quad t \in \mathbf{R}, \\
&u(t + \omega, x) = u(t, x), \quad (t, x) \in Q.
\end{aligned}
\tag{2.4.6}
$$

where

$$F \in C(C^2_\omega(Q) \times [-\varepsilon_0, \varepsilon_0]; C^{(\alpha,1-\alpha)}_\omega(Q)),$$
$$X_j \in C(C^2_\omega(Q) \times [-\varepsilon_0, \varepsilon_0]; C^2_\omega(\mathbf{R}))$$

with $\varepsilon_0 > 0$ and $\alpha = 0$ or 1.

Equations $(2.4.4_2)$ and $(2.4.5_{2,3})$ suggest that in problems $(\mathscr{P}^\varepsilon_{2\pi})$ and $(\mathscr{P}^\varepsilon_{2\pi p/q})$ the average of $b(t)$ and of $b_1(t)$ is of order ε. Hence we write

$$b(t) = c(t) + \varepsilon d\,, \quad \text{where} \int_0^{2\pi} c(\tau)\,\mathrm{d}\tau = 0\,,$$

and

$$b_1(t) = c_1(t) + \varepsilon d_1\,, \quad b_2(t) = \varepsilon\, c_2(t)\,, \quad \text{where} \int_0^{2\pi/q} c_1(\tau)\,\mathrm{d}\tau = 0\,.$$

Then equations (2.4.4) and (2.4.5) take the form (after dividing by ε whenever possible)

$$G_1(u, c, d, \varepsilon)(t, x) \equiv -u(t, x) + \frac{1}{2}\int_{-x+t}^{x+t} c(\tau)\,\mathrm{d}\tau +$$
$$+ \frac{\varepsilon}{2}\left\{2dx + X_0(u, \varepsilon)(x + t) + X_0(u, \varepsilon)(-x + t) - \right.$$
$$\left. - \int_0^x \int_{t-x+\xi}^{t+x-\xi} F(u, \varepsilon)(\tau, \xi)\,\mathrm{d}\tau\,\mathrm{d}\xi\right\} = 0\,, \quad (t, x) \in Q\,,$$
$$G_2(u, \varepsilon)(t) \equiv 2\frac{\mathrm{d}}{\mathrm{d}t}\left[X_1(u, \varepsilon)(t + \pi) - X_0(u, \varepsilon)(t)\right] +$$
$$+ \int_0^\pi \left[F(u, \varepsilon)(t - \xi, \xi) - F(u, \varepsilon)(t + \xi, \xi)\right]\mathrm{d}\xi = 0\,, \quad t \in \mathbf{R}\,,$$
$$G_3(u, d, \varepsilon) \equiv -2\pi d + 2X_1(u, \varepsilon)(\pi) - 2X_0(u, \varepsilon)(0) +$$
$$+ \int_0^\pi \int_\xi^{2\pi-\xi} F(u, \varepsilon)(\tau, \xi)\,\mathrm{d}\tau\,\mathrm{d}\xi = 0 \tag{2.4.7}$$

and

$$G_4(u, c_1, c_2, d_1, \varepsilon)(t, x) \equiv -u(t, x) + \frac{1}{2}\int_{-x+t}^{x+t} c_1(\tau)\,\mathrm{d}\tau +$$
$$+ \frac{\varepsilon}{2}\left\{\int_{-x+t}^{x+t} c_2(\tau)\,\mathrm{d}\tau + 2d_1 x + X_0(u, \varepsilon)(x + t) + X_0(u, \varepsilon)(-x + t) - \right.$$
$$\left. - \int_0^x \int_{t-x+\xi}^{t+x-\xi} F(u, \varepsilon)(\tau, \xi)\,\mathrm{d}\tau\,\mathrm{d}\xi\right\} = 0\,, \quad (t, x) \in Q\,,$$

$$G_5(u, c_2, \varepsilon)(t) \equiv c_2(t) +$$

$$+ \frac{1}{p} \sum_{j=0}^{p-1} (p - j) \left\{ \frac{\mathrm{d}}{\mathrm{d}t} [2X_1(u, \varepsilon)(t + \pi + 2\pi j) - \right.$$

$$- X_0(u, \varepsilon)(t + 2\pi(j + 1)) - X_0(u, \varepsilon)(t + 2\pi j)] +$$

$$+ \int_0^{\pi} [F(u, \varepsilon)(t + 2\pi(j + 1) - \xi, \xi) -$$

$$\left. - F(u, \varepsilon)(t + 2\pi j + \xi, \xi)] \,\mathrm{d}\xi \right\} = 0 , \quad t \in \mathbf{R} ,$$

$$G_6(u, \varepsilon)(t) \equiv 2 \frac{\mathrm{d}}{\mathrm{d}t} \sum_{j=0}^{p-1} [X_1(u, \varepsilon)(t + \pi + 2\pi j) -$$

$$- X_0(u, \varepsilon)(t + 2\pi j)] + \sum_{j=0}^{p-1} \int_0^{\pi} [F(u, \varepsilon)(t + 2\pi(j + 1) - \xi, \xi) -$$

$$- F(u, \varepsilon)(t + 2\pi j + \xi, \xi)] \,\mathrm{d}\xi = 0 , \quad t \in \mathbf{R} ,$$

$$G_7(u, c_1, c_2, d_1, \varepsilon) \equiv -2\pi d_1 - \int_0^{2\pi} c_2(\tau) \,\mathrm{d}\tau +$$

$$+ 2X_1(u, \varepsilon)(\pi) - X_0(u, \varepsilon)(2\pi) - X_0(u, \varepsilon)(0) +$$

$$+ \int_0^{\pi} \int_{\xi}^{2\pi - \xi} F(u, \varepsilon)(\tau, \xi) \,\mathrm{d}\tau \,\mathrm{d}\xi = 0 . \tag{2.4.8}$$

As in Sec. 1.4, a function $u \in C([-\varepsilon^*, \varepsilon^*]; C^2_{\omega}(Q))$, $\varepsilon_0 \geqq \varepsilon^* > 0$ is called a solution of $(\mathscr{P}^{\varepsilon}_{\omega})$ if $u(\varepsilon)$ satisfies (2.4.6) for every $\varepsilon \in [-\varepsilon^*, \varepsilon^*]$.

Noting that $(2.4.7_3)$ and $(2.4.8_4)$ are solvable with respect to d and d_1 respectively, the following lemmas and theorems can easily be proved with the help of I: Theorem 3.4.2 (p. 30), as in Sec. 1.4.

LEMMA 2.4.1 *Suppose the following assumptions hold.*

(*i*) $F(u, \varepsilon)$ *maps* $C^2_{2\pi}(Q) \times [-\varepsilon_0, \varepsilon_0]$ *continuously into* $C^{(\alpha, 1-\alpha)}_{2\pi}(Q)$, *where* $\alpha = 0$ *or* 1.

(*ii*) $X_j(u, \varepsilon)$ *maps* $C^2_{2\pi}(Q) \times [-\varepsilon_0, \varepsilon_0]$ *continuously into* $C^2_{2\pi}(\mathbf{R})$, $j = 0, 1$. *Then the problem* $(\mathscr{P}^{\varepsilon}_{2\pi})$ *has a solution only if the equation*

$$\Gamma(c)(t) \equiv 2\frac{\mathrm{d}}{\mathrm{d}t}\left[X_1\left(\frac{1}{2}\int_{-x+t}^{x+t} c(\sigma)\,\mathrm{d}\sigma, 0\right)(t+\pi) - \right.$$

$$\left. - X_0\left(\frac{1}{2}\int_{-x+t}^{x+t} c(\sigma)\,\mathrm{d}\sigma, 0\right)(t)\right] + \int_0^{\pi}\left[F\left(\frac{1}{2}\int_{-x+t}^{x+t} c(\sigma)\,\mathrm{d}\sigma, 0\right)(t-\xi, \xi) - \right.$$

$$\left. - F\left(\frac{1}{2}\int_{-x+t}^{x+t} c(\sigma)\,\mathrm{d}\sigma, 0\right)(t+\xi, \xi)\right]\mathrm{d}\xi = 0\,, \quad t \in \mathbf{R} \tag{2.4.9}$$

has a solution $c = c^* \in C^1_{2\pi}(\mathbf{R})$ *for which*

$$\int_0^{2\pi} c^*(\sigma)\,\mathrm{d}\sigma = 0\,.$$

THEOREM 2.4.3 *Suppose that the following assumptions hold.*

(*i*) *Equation* (2.4.9) *has a solution*

$$c = c^* \in C^{1*}_{2\pi}(\mathbf{R}) = \left\{c \in C^1_{2\pi}(\mathbf{R}); \int_0^{2\pi} c(\sigma)\,\mathrm{d}\sigma = 0\right\}.$$

(*ii*) $F(u, \varepsilon)$ *is a continuous mapping of* $B(u_0^*, \varrho_1; C^2_{2\pi}(Q)) \times [-\varepsilon_0, \varepsilon_0]$ *into* $C^{(\alpha, r-\alpha)}_{2\pi}(Q)$, *and* $X_j(u, \varepsilon)$ *for* $j = 0, 1$ *maps* $B(u_0^*, \varrho_1; C^2_{2\pi}(Q)) \times [-\varepsilon_0, \varepsilon_0]$ *continuously into* $C^s_{2\pi}(\mathbf{R})$, *where* $r = 1$ or 2, $s \geqq 2$, $\alpha \in \{0, 1, r\}$ *and*

$$u_0^*(t, x) = \frac{1}{2}\int_{-x+t}^{x+t} c^*(\tau)\,\mathrm{d}\tau\,.$$

(*iii*) *The G-derivative* $F'_u(u, \varepsilon)$ *maps* $B(u_0^*, \varrho_1; C^2_{2\pi}(Q)) \times [-\varepsilon_0, \varepsilon_0]$ *continuously into* $\mathscr{L}(C^2_{2\pi}(Q), C^{(\alpha, r-\alpha)}_{2\pi}(Q))$ *and for* $j = 0, 1$ *the G-derivative* $X'_{j,u}(u, \varepsilon)$ *maps* $B(u_0^*, \varrho_1; C^2_{2\pi}(Q)) \times [-\varepsilon_0, \varepsilon_0]$ *continuously into* $\mathscr{L}(C^2_{2\pi}(Q), C^s_{2\pi}(\mathbf{R}))$.

(*iv*) *The operator* $G_2(u, \varepsilon)$ (*defined by* (2.4.7$_2$)) *maps* $B(u_0^*, \varrho_1; C^2_{2\pi}(Q)) \times \times [-\varepsilon_0, \varepsilon_0]$ *into* $C^{\min(r, s-1)}_{2\pi *}(\mathbf{R})$.

(*v*) $[\Gamma'_c(c^*)]^{-1} \in \mathscr{L}(C^{\min(r, s-1)}_{2\pi *}(\mathbf{R}); C^{1*}_{2\pi}(\mathbf{R}))$.

Then the problem $(\mathscr{P}^{\varepsilon}_{2\pi})$ *has a locally unique solution* $u^*(\varepsilon) \in C^2_{2\pi}(Q)$ *such that* $u^*(0) = u_0^*$.

LEMMA 2.4.2 *Suppose that the following assumptions hold.*

(*i*) $F(u, \varepsilon)$ *maps* $C^2_{2\pi p/q}(Q) \times [-\varepsilon_0, \varepsilon_0]$ *continuously into* $C^{(\alpha, 1-\alpha)}_{2\pi p/q}(Q)$ *with* $\alpha = 0$ *or* 1.

(*ii*) $X_j(u, \varepsilon)$ *maps* $C^2_{2\pi p/q}(Q) \times [-\varepsilon_0, \varepsilon_0]$ *continuously into* $C^2_{2\pi p/q}(\mathbf{R})$, $j = 0, 1$.

Then the problem $(\mathscr{P}^{\varepsilon}_{2\pi p/q})$ *has a solution only if the equation*

$$^{1}\Gamma(c_1)(t) \equiv 2\frac{\mathrm{d}}{\mathrm{d}t}\sum_{j=0}^{p-1}\left[X_1\left(\frac{1}{2}\int_{-x+t}^{x+t} c_1(\sigma)\,\mathrm{d}\sigma, 0\right)(t + \pi + 2\pi j) - \right.$$

$$\left. - X_0\left(\frac{1}{2}\int_{-x+t}^{x+t} c_1(\sigma)\,\mathrm{d}\sigma, 0\right)(t + 2\pi j)\right] +$$

$$+ \sum_{j=0}^{p-1}\int_0^{\pi}\left[F\left(\frac{1}{2}\int_{-x+t}^{x+t} c_1(\sigma)\,\mathrm{d}\sigma, 0\right)(t + 2\pi(j+1) - \xi, \xi) - \right.$$

$$\left. - F\left(\frac{1}{2}\int_{-x+t}^{x+t} c_1(\sigma)\,\mathrm{d}\sigma, 0\right)(t + 2\pi j + \xi, \xi)\right]\mathrm{d}\xi = 0 \tag{2.4.10}$$

has a solution $c_1 = c_1^* \in C^1_{2\pi/q}(\mathbf{R})$ *for which*

$$\int_0^{2\pi/q} c_1^*(\sigma)\,\mathrm{d}\sigma = 0\,.$$

THEOREM 2.4.4 *Suppose that the following assumptions hold.*

(*i*) *Equation* (2.4.10) *has a solution*

$$c_1 = c_1^* \in C^{1*}_{2\pi/q}(\mathbf{R}) = \left\{c_1 \in C^1_{2\pi/q};\ \int_0^{2\pi/q} c_1(\sigma)\,\mathrm{d}\sigma = 0\right\}.$$

(*ii*) $F(u, \varepsilon)$ *is a continuous mapping of* $B(u_0^*, \varrho_1; C^2_{2\pi p/q}(Q)) \times [-\varepsilon_0, \varepsilon_0]$ *into* $C^{(\alpha, r-\alpha)}_{2\pi p/q}(Q)$, *and* $X_j(u, \varepsilon)$ *for* $j = 0, 1$ *maps* $B(u_0^*, \varrho_1; C^2_{2\pi p/q}(Q)) \times [-\varepsilon_0, \varepsilon_0]$ *continuously into* $C^s_{2\pi p/q}(\mathbf{R})$, *where* $r = 1$ *or* 2, $s \geqq 2$, $\alpha \in \{0, 1, r\}$ *and*

$$u_0^*(t, x) = \frac{1}{2}\int_{-x+t}^{x+t} c_1^*(\sigma)\,\mathrm{d}\sigma\,.$$

(*iii*) *The G-derivative* $F'_u(u, \varepsilon)$ *maps* $B(u_0^*, \varrho_1; C^2_{2\pi p/q}(Q)) \times [-\varepsilon_0, \varepsilon_0]$ *continuously into* $\mathscr{L}(C^2_{2\pi p/q}(Q), C^{(\alpha, r-\alpha)}_{2\pi p/q}(Q))$ *and the G-derivative* $X'_{j,u}(u, \varepsilon)$ *maps* $B(u_0^*, \varrho_1; C^2_{2\pi p/q}(Q)) \times [-\varepsilon_0, \varepsilon_0]$ *continuously into* $\mathscr{L}(C^2_{2\pi p/q}(Q), C^s_{2\pi p/q}(\mathbf{R}))$.

(*iv*) *The operator* $G_6(u, \varepsilon)$ (*defined by* (2.4.8$_3$)) *maps* $B(u_0^*, \varrho_1; C^2_{2\pi p/q}(Q)) \times$ $\times [-\varepsilon_0, \varepsilon_0]$ *into* $C^{\min(r, s-1)}_{2\pi/q^*}(\mathbf{R})$.

(*v*) $[{}^1\Gamma'_{c_1}(c_1^*)]^{-1} \in \mathscr{L}(C^{\min(r, s-1)}_{2\pi/q^*}(\mathbf{R}), C^{1*}_{2\pi/q}(\mathbf{R}))$.

Then the problem $(\mathscr{P}^{\varepsilon}_{2\pi p/q})$ *has a locally unique solution* $u^*(\varepsilon) \in C^2_{2\pi p/q}(Q)$ *such that* $u^*(0) = u_0^*$.

2.5. *The case of monotone perturbations*

In Theorems 1.4.5 and 1.4.6 of the preceding paragraph we have introduced two classes of right-hand sides for which the corresponding problem $(\mathscr{P}^{\varepsilon}_{\omega})$, with special values of ω, always has a solution. Here we present two other classes (namely those whose right-hand sides are, roughly speaking, monotone in u or in u_t) that have the same property and are probably much more important from a practical point of view.

Let us define operators $P : C^j_{2\pi}(Q) \to C^j_{2\pi}(Q)$ and $\tilde{Q} : C^j_{2\pi}(Q) \to C^j_{2\pi}(\mathbf{R})$, $j \in \mathbf{N}$, by

$$(Pg)(t, x) = (2\pi)^{-1} \int_0^{\pi} [g(t + x - \xi, \xi) - g(t - x - \xi, \xi) -$$
$$- g(t + x + \xi, \xi) + g(t - x + \xi, \xi)] \, \mathrm{d}\xi ,$$
$$(\tilde{Q}g)(t) = \int_0^{\pi} [g(t - \sigma, \sigma) - g(t + \sigma, \sigma)] \, \mathrm{d}\sigma .$$

These operators P and $\tilde{Q}$ have some simple properties that are described in the following lemmas.

LEMMA 2.5.1 *The operators P and $\tilde{Q}$ defined above satisfy $\tilde{Q}P = \tilde{Q}$.*

Proof. For $g \in C^j_{2\pi}(Q)$ we see that

$$(\tilde{Q}Pg)(t) = (2\pi)^{-1}$$
$$\int_0^{\pi}\int_0^{\pi} [g(t - \xi, \xi) - g(t - 2\sigma - \xi, \xi) - g(t + \xi, \xi) + g(t - 2\sigma + \xi, \xi) -$$
$$- g(t + 2\sigma - \xi, \xi) + g(t - \xi, \xi) + g(t + 2\sigma + \xi, \xi) -$$
$$- g(t + \xi, \xi)] \, \mathrm{d}\xi \, \mathrm{d}\sigma =$$
$$= \int_0^{\pi} [g(t - \xi, \xi) - g(t + \xi, \xi)] \, \mathrm{d}\xi = (\tilde{Q}g)(t) .$$

LEMMA 2.5.2 *Let $g \in C^j_{2\pi}(Q), j \in \mathbf{N}$ fixed, be given. Then $(Pg)(t, x) = 0$ for all $(t, x) \in Q$ if and only if $(\tilde{Q}g)(t) = 0$ for all $t \in \mathbf{R}$.*

Proof. We have $2\pi(Pg)(t, x) = (\tilde{Q}g)(t + x) - (\tilde{Q}g)(t - x)$ from the definition of P and $\tilde{Q}$. By Lemma 2.5.1, $(\tilde{Q}g)(t) = (\tilde{Q}Pg)(t)$. These two relations prove our lemma.

LEMMA 2.5.3 *Let $F(u)$ be a mapping satisfying* $(2.4.3_1)$ with $\omega = 2\pi$, *let $u \in C^2_{2\pi}(Q)$, and let $b \in C^2_{2\pi}(\mathbf{R})$. Suppose that*

$$u(t, x) = b(t + x) - b(t - x) -$$
$$- \frac{1}{2}\int_0^x \int_{t-x+\xi}^{t+x-\xi} (I - P) \, F(u) \, (\vartheta, \xi) \, \mathrm{d}\vartheta \, \mathrm{d}\xi +$$
$$+ \frac{x}{2\pi}\int_0^{\pi} \int_{t-\pi+\xi}^{t+\pi-\xi} (I - P) \, F(u) \, (\vartheta, \xi) \, \mathrm{d}\vartheta \, \mathrm{d}\xi \qquad (2.5.1)$$

holds for $(t, x) \in Q$.

Then u solves the problem

$$u_{tt}(t, x) - u_{xx}(t, x) = (I - P) F(u)(t, x), \quad (t, x) \in Q, \tag{2.5.2}$$

$$u(t, 0) = u(t, \pi) = 0, \quad t \in \mathbf{R},$$

$$u(t + 2\pi, x) = u(t, x), \quad (t, x) \in Q. \tag{2.5.3}$$

Proof. We put

$$u_1(t, x) = \frac{x}{2\pi} \int_0^\pi \int_{t-\pi+\xi}^{t+\pi-\xi} (I - P) F(u)(\vartheta, \xi) \, d\vartheta \, d\xi.$$

By Lemma 2.5.1, $\tilde{Q}(I - P) F(u) = 0$. This implies that u_1 has the form $u_1(t, x) = \alpha x$, where α is a constant. We can now easily verify (see II: Sec. 2.1) that (2.5.2) and (2.5.3) hold.

In what follows we investigate the problem $(\mathscr{P}_{2\pi}^{\varepsilon})$ given by

$$u_{tt}(t, x) - u_{xx}(t, x) = \varepsilon f(t, x, u(t, x), \varepsilon), \quad (t, x) \in Q.$$

$$u(t, 0) = u(t, \pi) = 0, \quad t \in \boldsymbol{R},$$

$$u(t + 2\pi, x) = u(t, x), \quad (t, x) \in Q. \tag{2.5.4}$$

THEOREM 2.5.1 (see LOVICAROVÁ [35]). *Let $f(t, x, u, \varepsilon)$ defined on $Q \times [-2r_0 - 1, 2r_0 + 1] \times [-\varepsilon_0, \varepsilon_0]$ be 2π-periodic in t and suppose that:*

(i) $D_t^j D_u^k f$, $j + k \leqq 3$ are continuous on $Q \times [-2r_0 - 1, 2r_0 + 1] \times [-\varepsilon_0, \varepsilon_0]$;

(ii) there is a $\gamma > 0$ such that $D_u^1 f \leqq -\gamma$ on $Q \times [-2r_0, 2r_0] \times [-\varepsilon_0, \varepsilon_0]$, where $\gamma r_0 > \sup \{|f(t, x, 0, 0)|; (t, x) \in Q\}$.

Then there is a $b_0^* \in C_{2\pi}^2(\mathbf{R})$ such that the problem $(\mathscr{P}_{2\pi}^{\varepsilon})$ has a locally unique solution $u^* \in C([-\varepsilon^*, \varepsilon^*]; C_{2\pi}^2(Q))$ $(0 < \varepsilon^* \leqq \varepsilon_0$, ε^* sufficiently small) with $u^*(0)(t, x) = b_0^*(t + x) - b_0^*(t - x)$.

Proof. We set

$$F(u, \varepsilon)(t, x) = f(t, x, u(t, x), \varepsilon),$$

$$(Zb)(t, x) = b(t + x) - b(t - x), \quad (t, x) \in Q,$$

$$G_1(u, b, \varepsilon)(t, x) \equiv -u(t, x) + Zb(t, x) -$$

$$- \frac{\varepsilon}{2} \int_0^x \int_{t-x+\xi}^{t+x-\xi} (I - P) F(u, \varepsilon)(\vartheta, \xi) \, d\vartheta \, d\xi +$$

$$+ \frac{\varepsilon}{2\pi} x \int_0^\pi \int_{t-\pi+\xi}^{t+\pi-\xi} (I - P) F(u, \varepsilon)(\vartheta, \xi) \, d\vartheta \, d\xi, \quad (t, x) \in Q,$$

and

$$G_2(u, \varepsilon)(t) \equiv (\tilde{Q} F(u, \varepsilon))(t), \quad t \in \mathbf{R}.$$

Now, putting $G(u, b, \varepsilon) \equiv (G_1(u, b, \varepsilon), G_2(u, \varepsilon))$ and

$$C^{2*}_{2\pi}(\mathbf{R}) = \left\{ b \in C^2_{2\pi}(\mathbf{R}); \int_0^{2\pi} b(\sigma)\, d\sigma = 0 \right\},$$

we find an element $(u_0^*, b_0^*) \in C^2_{2\pi}(Q) \times C^{2*}_{2\pi}(\mathbf{R})$ such that

$$G(u_0^*, b_0^*, 0) = 0. \tag{2.5.5}$$

From the definition of G_1 and G_2 we conclude that this is equivalent to the existence of a function b_0^* for which

$$\tilde{Q} F(Zb_0^*, 0) = 0. \tag{2.5.6}$$

Now instead of looking for a $b \in C^{2*}_{2\pi}(\mathbf{R})$ satisfying (2.5.6), we define

$$T_\delta b \equiv b + \delta \tilde{Q} F(Zb, 0),$$

and search for a fixed point of the mapping T_δ for an appropriate $\delta > 0$. We write

$$A(p, \varrho) = \max \{ |D_t^j D_u^k f(t, x, u, 0)|; j + k \leqq p, (t, x) \in Q, |u| \leqq \varrho \}.$$

For positive numbers r_1, r_2, and r_3 (r_0 is in (*ii*)) we define ($l = 2, 3$)

$$M_l = \{ b \in C^{2*}_{2\pi}(\mathbf{R}) \cap C^l_{2\pi}(\mathbf{R}); \max \{ |D^i b(x)|; x \in \mathbf{R} \} \leqq r_i, i = 0, 1, 2, l \}.$$

The set M_3 is convex and relatively compact in $C^{2*}_{2\pi}(\mathbf{R})$. We can show easily that T_δ is a continuous mapping from M_2 into $C^{2*}_{2\pi}(\mathbf{R})$ for every pair r_1, r_2. In what follows we prove that numbers r_1, r_2, r_3, and δ can be determined so that T_δ maps M_3 into M_3. Let us take $b \in M_3$. We must show that $T_\delta b \in M_3$ for an appropriate $\delta > 0$ independent of b. We fix $\eta \in (0, r_0)$ and take any $y \in \mathbf{R}$. If $|b(y)| \leqq r_0 - \eta$, then $|(T_\delta b)(y)| \leqq r_0 - \eta + 2\pi\delta A(0, 2r_0)$. If $r_0 - \eta < b(y) \leqq r_0$, then we can write

$$(T_\delta b)(y) = b(y) +$$

$$+ \delta \int_0^\pi [g_1(y, s)(b(y) - b(y - 2s)) + g_2(y, s)(b(y) - b(y + 2s))]\, ds +$$

$$+ \delta \int_0^\pi [F(0, 0)(y - s, s) - F(0, 0)(y + s, s)]\, ds,$$

where

$$g_j(y, s) = \int_0^1 D_u^1 f(y + (-1)^j s, s, \varrho(Zb)(y + (-1)^j s, s), 0)\, d\varrho$$

are defined on Q and $g_j(y, s) \leqq -\gamma < 0$ $(j = 1, 2)$. With these facts in mind we derive the estimate

$$(T_\delta b)(y) \leqq b(y) - 2\pi\delta\gamma\, b(y) +$$

$$+ \delta \sum_{j=1}^{2} \int_{\Pi_j^-} (g_j(y, s) + \gamma)(b(y) - b(y + (-1)^j 2s))\, ds + 2\pi\delta\, A(0, 0) \leqq$$

$$\leqq b(y) - 2\pi\delta(\gamma\, b(y) - \eta\, A(1, 2r_0) - A(0, 0)),$$

where $\Pi_j^- = \{s \in [0, \pi];\ b(y) - b(y + 2(-1)^j s) < 0\}$. Obviously, $T_\delta\, b(y) \geqq \geqq r_0 - \eta - 2\pi\delta\, A(0, 2r_0)$ holds for $r_0 - \eta < b(y) \leqq r_0$. If $-r_0 \leqq b(y) < < -r_0 + \eta$, we can find similar estimates. Choosing $\eta > 0$ sufficiently small, we find a $\delta_0 > 0$ such that $\|T_\delta b\|_{C_{2\pi}(\mathbf{R})} \leqq r_0$ for every $\delta \in (0, \delta_0]$, since $\gamma r_0 > > A(0, 0)$ (by (*ii*)). To prove that $\|D_y^j(T_\delta b)\|_{C_{2\pi}(\mathbf{R})} \leqq r_j$ $(j = 1, 2, 3)$ we take into account the relation

$$D_y^j(T_\delta b)(y) = D_y^j\, b(y) +$$

$$+ \sum_{\varkappa=1}^{2} \delta \int_0^\pi D_u^1 f(y - (-1)^\varkappa s, s,\ (Zb)(y - (-1)^\varkappa s, s), 0)(D_y^j\, b(y) -$$

$$- D_y^j\, b(y - 2(-1)^\varkappa s))\, ds + \delta\, X_j(y),$$

where each $\|X_j\|_{C_{2\pi}(\mathbf{R})}$ can be estimated by a constant that depends only on $r_0, \ldots, r_{j-1}$, $A(j, 2r_0)$. Thus repeating the same procedure as above we can choose r_j $(j = 1, 2, 3)$ so that from $\|D_y^j b\|_{C_{2\pi}(\mathbf{R})} \leqq r_j$ it follows that $\|D_y^j(T_\delta b)\|_{C_{2\pi}(\mathbf{R})} \leqq r_j$ for every sufficiently small $\delta > 0$. Applying I: Corollary 3.2.1 to the sets M_2, M_3, and to the mapping T_δ (for sufficiently small δ), we obtain a function b_0^* for which $\tilde{Q}\, F(Zb_0^*, 0) = 0$. Since f is monotonic we can prove that this b_0^* is unique in $C({}^{2*}_{2\pi}\mathbf{R})$. Hence, setting $u_0^* = Zb_0^*$, we see that (u_0^*, b_0^*) satisfies (2.5.5). Putting $B = B_1 = C_{2\pi}^2(Q) \times C_{2\pi}^{2*}(\mathbf{R})$, $B_2 = \mathbf{R}$, $p = \varepsilon$, $x = (u, b)$, $p_0 = 0$, $x_0 = (u_0^*, b_0^*)$, we easily find that the assumptions (1)–(3) of I: Theorem 3.4.2 are satisfied. (4) is an immediate consequence of the following argument.

Writing for a moment $V(b) = \tilde{Q}\, F(Zb, 0)$, we prove that $W \equiv V_b'(b_0^*) \in \mathscr{L}(C_{2\pi}^{2*}(\mathbf{R}), C_{2\pi}^{2*}(\mathbf{R}))$ has a bounded inverse. First of all, we prove that W is one-to-one. Suppose that $Wb = 0$ for $b \in C_{2\pi}^{2*}(\mathbf{R})$. Let $y_0 \in \mathbf{R}$ be a point for which $b(y_0) = \max\{b(y); y \in \mathbf{R}\}$. If for some $s \in [0, \pi]$ we have $b(y_0 - 2s) < < b(y_0)$, then

$$0 = (Wb)(y_0) =$$

$$= \sum_{j=1}^{2} (-1)^{j-1} \int_0^\pi D_u^1 f(y_0 + (-1)^j s, s, (Zb_0^*)(y_0 + (-1)^j s, s), 0) \,.$$

$$. (Zb)(y_0 + (-1)^j s, s) \, ds < 0 \,.$$

This is a contradiction. Hence $b = \text{const} = 0$, because

$$\int_0^{2\pi} b(y) \, dy = 0 \,.$$

Secondly, we prove that the operator W can be written in the form $W = = W_1 + W_2$, where W_1 is an isomorphism of $C_{2\pi}^{2*}(\mathbf{R})$ onto itself and W_2 is a compact operator. By I: [43], Chap. X, § 5 Theorem 2 (p. 284) we immediately find that $W^{-1} \in \mathscr{L}(C_{2\pi}^{2*}(\mathbf{R}), C_{2\pi}^{2*}(\mathbf{R}))$ exists. Therefore, we define a function

$$g_0(y) = \sum_{j=1}^{2} \int_0^\pi D_u^1 f(y + (-1)^j s, s, (Zb_0^*)(y + (-1)^j s, s), 0) \, ds$$

and then an operator W_1 mapping $C_{2\pi}^{2*}(\mathbf{R})$ into itself by

$$W_1 \, b(y) = g_0(y) \, b(y) - (2\pi)^{-1} \int_0^{2\pi} g_0(s) \, b(s) \, ds \,.$$

The operator W_1^{-1} can be written as

$$W_1^{-1} b \, (y) = \frac{1}{g_0(y)} \left(b(y) - \left(\int_0^{2\pi} \frac{1}{g_0(s)} \, ds \right)^{-1} \int_0^{2\pi} \frac{b(s)}{g_0(s)} \, ds \right).$$

These two formulae show that W_1 is a linear homeomorphism of $C_{2\pi}^{2*}(\mathbf{R})$ onto itself. Finally, the operator W_2 from $C_{2\pi}^{2*}(\mathbf{R})$ into itself is defined by

$$W_2 \, b(y) = (2\pi)^{-1} \int_0^{2\pi} g_0(s) \, b(s) \, ds -$$

$$- \sum_{j=1}^{2} \int_0^\pi D_u^1 f(y + (-1)^j s, s, (Zb_0^*)(y + (-1)^j s, s), 0) \,.$$

$$. \, b(y + 2(-1)^j s) \, ds \,,$$

and we can prove that for every bounded subset $U \subset C^{2*}_{2\pi}(\mathbf{R})$ the set $W_2 U$ is a relatively compact subset of $C^{2*}_{2\pi}(\mathbf{R})$. This follows from the Arzelà-Ascoli theorem if we take into account that for $k = 1, 2$

$$D^k_y W_2\, b(y) =$$

$$= -\sum_{j=1}^{2} \int_0^{\pi} D^1_u f(y + (-1)^j s, s, (Zb^*_0)(y + (-1)^j s, s), 0) \,.$$

$$. \, D^k_y\, b(y + (-1)^j 2s)\, ds + Y_k(y) =$$

$$= -\tfrac{1}{2} \int_{y-2\pi}^{y} D^1_u f\left(\frac{y+s}{2}, \frac{y-s}{2}, (Zb^*_0)\left(\frac{y+s}{2}, \frac{y-s}{2}\right), 0\right) .$$

$$. \, D^k\, b(s)\, ds -$$

$$-\tfrac{1}{2} \int_{y}^{y+2\pi} D^1_u f\left(\frac{y+s}{2}, \frac{s-y}{2}, (Zb^*_0)\left(\frac{y+s}{2}, \frac{s-y}{2}\right), 0\right) .$$

$$. \, D^k\, b(s)\, ds + Y_k(y) \,,$$

where Y_k contains only the derivatives of b up to order $k - 1$ and is therefore equicontinuous. It is easy to verify that the first two integrals are also equicontinuous. Hence, W_2 is compact. This completes the proof.

Using ideas similar to those in the proof of the preceding theorem we can prove the following result.

THEOREM 2.5.2 (*see* LOVICAROVÁ [35]). *Suppose that* $f(t, x, u_0, u_1, u_2)$ *defined on* $\mathbf{R} \times [0, \pi] \times \mathbf{R}^3$ *satisfies the assumptions:*

(*i*) f *is* 2π*-periodic in* t, *and the derivatives* $D^j_t D^{k_0}_{u_0} D^{k_1}_{u_1} D^{k_2}_{u_2} f$ *for* $j + k_0 + k_1 + k_2 \leqq 3$ *are continuous;*

(*ii*) *there exist an* $r_0 > 0$ *and a* $\gamma > 0$ *such that*

(*a*) $D^1_{u_1} f \leqq -\gamma$ *on* $G_2 \equiv Q \times [-\pi r_0, \pi r_0] \times [-2r_0, 2r_0]^2$,

(*b*) $\sup_{G_2} D^1_{u_2} f - \inf_{G_2} D^1_{u_2} f - \gamma < 0$;

(*iii*) $\gamma r_0 - \sup\{|f(t, x, u_0, 0, u_2)|;\ (t, x) \in Q, |u_0| \leqq \pi r_0, |u_2| \leqq 2r_0\} > 0$.
Then for every sufficiently small ε *there exists a solution in* $C^2(Q)$ *to the problem*

$$u_{tt}(t, x) - u_{xx}(t, x) = \varepsilon f(t, x, u, u_t, u_x) \,,$$

$$u(t, 0) = u(t, \pi) = 0 \,, \quad u(t + 2\pi, x) = u(t, x) \,.$$

Remark 2.5.1 RABINOWITZ was the first to realize that the problem $(\mathscr{P}^{\varepsilon}_{2\pi})$ with $F(u, \varepsilon)(t, x) = f(t, x, u)$ has a solution, provided that $f_u \geqq \beta > 0$. In his first paper [55] (in a preliminary form in [56] and in an extended form in [57]) he shows that instead of looking for a function u solving the bifurcation equation (see Theorem 8.1.2)

$$\int_0^{2\pi} \int_0^{\pi} f(t, x, u(t, x)) (\sigma(x + t) - \sigma(-x + t)) \, dx \, dt = 0 ,$$

$\sigma \in H^0_{2\pi}(\mathbf{R})$ arbitrary, we can look for a $v \in \mathscr{V}$ ($\mathscr{V} = \{v; v(t, x) = s(x + t) - s(-x + t)\}$) that minimizes the functional

$$\int_0^{2\pi} \int_0^{\pi} \Phi(t, x, v(t, x) + w(t, x)) \, dx \, dt$$

($w \in \mathscr{V}^{\perp}$ fixed, $\Phi_u = f$). Making use of the L_2-technique he constructs in a sophisticated way a bounded sequence of successive approximations whose convergence is proved by the Banach-Sachs theorem, see II: [2] (p. 181). He also obtains a similar result for the problem $(\mathscr{P}^{\varepsilon}_{2\pi})$ with $F(u, \varepsilon) = \alpha u_t + f(t, x, u)$, under the hypothesis that f is "small" compared with the dissipation coefficient α.

In [58] RABINOWITZ generalizes the former result to the problem $(\mathscr{P}^{\varepsilon}_{2\pi})$ with $F(u)(t, x) = u^{2k+1} + g(t, x, u)$, $g(t, x, u_1) \geqq g(t, x, u_2)$ for $u_1 \geqq u_2$ and obtains a more precise result concerning the behaviour in ε, by showing that the problem has an unbounded set of classical solutions (ε, u) intersecting $(0, \mathscr{V})$. The last assertion for the strictly monotone case is included in [59] as an application of a general theory of non-linear eigenvalue problems.

In [63] DE SIMON and TORELLI first prove the existence of a solution in L_2 of $(\mathscr{P}^{\varepsilon}_{2\pi})$ with $F(u, \varepsilon)(t, x) = f(t, x, u)$ assuming that

$$0 < h \leqq (f(t, x, u'') - f(t, x, u'))(u'' - u')^{-1} \leqq H$$

and making use of the Browder-Minty theorem (I: Theorem 3.3.1). When f has a continuous derivative $f_u(t, x, u) \geqq h > 0$, they then show, by applying the implicit function theorem, that there is a weak solution in $C_{2\pi}(Q)$. Finally, they carry over the first result to the equation

$$u_{tt} - u_{xx} - u_{yy} = \varepsilon f(t, x, y, u)$$

which is to be investigated on the rectangle $[0, a] \times [0, b]$. TORELLI in [74] discusses the existence of a solution to the problem $(\mathscr{P}^{\varepsilon}_{2\pi})$ with $F(u, \varepsilon)(t, x) = f(t, x, u)$, where f satisfies the conditions

$$f(t, x, u_2) - f(t, x, u_1) \geqq k(u_2 - u_1)^{\varrho} ,$$

$$|f(t, x, u)| \leqq H|u|^{\varrho} + M .$$

HALL [18]–[19] deals with the existence of a 2π-periodic function u satisfying

$$Lu \equiv D_{tt}\, u(t, x) + (-1)^p\, D_x^{2p}\, u(t, x) = \varepsilon\, F(u)\,(t, x)\,,$$

and being odd and 2π-periodic in x. The assumptions imposed on F are shown to be satisfied for some special types of F, for example, $F(u(t, x)) = |u|^{r-1}$ $\operatorname{sgn} u - g(t, x)$. The solution u of each of the problems described is sought in an appropriate Hilbert space H in the form $u = v + w$, where $v \in \mathscr{V} \equiv \mathscr{N}(L)$ and $w \in \mathscr{V}^{\perp}$, $\mathscr{V}^{\perp}$ being the complement of $\mathscr{N}(L)$ in H. Denoting by P_1 and P_2 the projections on $\mathscr{V}$ and $\mathscr{V}^{\perp}$, we verify easily that the problem in question is equivalent to the system

$$\begin{aligned} &P_1\, F(v + w) = 0\,, \\ &Lw = \varepsilon P_2\, F(v + w)\,. \end{aligned} \tag{2.5.7}$$

The bifurcation equation $(2.5.7_1)$ is solved in [18] with the help of the implicit function theorem whereas in [19] the solution is found as the limit of the solutions of equation $(2.5.7_1)$ in the finite-dimensional subspaces of H.

SOVA in [68] develops a technique for solving a general equation $Lu = \varepsilon f(u)$ in some appropriate quotient space of a Banach space, and applies it to an investigation of the existence of solutions to the problem $(\mathscr{P}^{\varepsilon}_{2\pi})$ under rather general assumptions on the monotone behaviour of $F(u, \varepsilon)$ in u. Similar problems are treated in [69]. This approach has the advantage that the projections P_1 and P_2 mentioned above need not exist. In this context, see also HALL [20].

NAKAO in [42] proves the existence of a weak solution to the problem $(\mathscr{P}^{\varepsilon}_{2\pi})$ with $F(u, \varepsilon)\,(t, x) = g(t, x)\,|u|^{\alpha}\, u + f(t, x)$, where g is positive almost everywhere. Regularity, especially Hölder continuity, of this solution is treated in [44]. Regularity of solutions to more general equations is dealt with in [43], see VI: Sec. 2.3. The existence and regularity of 2π-periodic solutions to the wave equation with a non-linearity satisfying only mild assumptions on smoothness and monotonicity is investigated in [45].

The dependence of the "monotone" non-linearity on a small parameter in dealing with the existence of 2π-periodic solutions has been dropped by BRÉZIS and NIRENBERG in the paper [7], which is mainly concerned with characterizations of the ranges of non-linear operators. The authors, among many other applications, study the existence of a weak 2π-periodic solution to the problem

$$\begin{aligned} &u_{tt} - u_{xx} + f(t, x, u) = 0\,, \quad x \in (0, \pi)\,, \\ &u(t, 0) = u(t, \pi) = 0\,. \end{aligned} \tag{2.5.8}$$

See also [6]. In [8] the existence of 2π-periodic solutions to (2.5.8) is proved, provided that $f(t, x, \cdot)$ is non-decreasing, $f(t, x, u_0(t, x)) \equiv 0$ for a function $u_0 \in L^2((0, 2\pi) \times (0, \pi))$, and f satisfies certain growth conditions.

RABINOWITZ [60] deals with the existence of a weak solution to the problem

$$u_{tt} - u_{xx} = cu + g(t, x, u), \quad (t, x) \in \mathbf{R} \times (0, \pi),$$

$$u(t, 0) = u(t, \pi) = 0, \quad t \in \mathbf{R},$$

$$u(t + 2\pi, x) = u(t, x), \quad (t, x) \in \mathbf{R} \times (0, \pi)$$

provided that $c \neq 0$ and the function g satisfies certain regularity and growth conditions (e.g. $|g(t, x, u)| \leqq M$ and $|c^{-1} g_u(t, x, u)| < 1$ for $(t, x, u) \in \mathbf{R} \times \times [0, \pi] \times \mathbf{R}$ and

$$\int_0^u g(t, x, v)\,\mathrm{d}v \to \infty \text{ as } |u| \to \infty).$$

§ 3. Examples

3.1. *The problem $(\mathscr{P}_{2\pi}^{\varepsilon})$ with $F(u, \varepsilon)(t, x) = h(t, x) + \alpha\, u(t, x) + \beta\, u^3(t, x)$ (the Poincaré method)*

Let $(\mathscr{P}_{2\pi}^{\varepsilon})$ be the problem

$$u_{tt}(t, x) - u_{xx}(t, x) = \varepsilon[h(t, x) + \alpha\, u(t, x) + \beta\, u^3(t, x)],$$

$$(t, x) \in Q = \mathbf{R} \times [0, \pi], \tag{3.1.1}$$

$$u(t, 0) = 0, \quad u(t, \pi) = 0, \quad t \in \mathbf{R}, \tag{3.1.2}$$

$$u(t + 2\pi, x) = u(t, x), \quad t \in \mathbf{R}, \quad x \in [0, \pi]. \tag{3.1.3}$$

We assume that

(a) α and β are constants, $\beta \neq 0$,

(b) $h \in C_{2\pi}^{0,2}(Q)$,

$$h(t, 0) = h(t, \pi) = h_{xx}(t, 0) = h_{xx}(t, \pi) = 0, \quad t \in \mathbf{R}, \tag{3.1.4}$$

$$h(t, \pi - x) = h(t, x), \quad t \in \mathbf{R}, \quad x \in [0, \pi], \tag{3.1.5}$$

$$\int_0^{2\pi} h_e(\tau, x - \tau)\,\mathrm{d}\tau \not\equiv 0, \tag{3.1.6}$$

where h_e is the odd 2π-periodic extension in the variable x of h.

This function h_e satisfies the relation

$$h_e(t, x + \pi) = -h_e(t, x). \tag{3.1.7}$$

Let us verify that the "extended" problem $(\mathscr{P}^{\varepsilon}_{2\pi,e})$ defined by

$$u_{tt}(t, x) - u_{xx}(t, x) = \varepsilon[h_e(t, x) + \alpha\, u(t, x) + \beta\, u^3(t, x)], \quad (t, x) \in \mathbf{R}^2, \tag{3.1.1'}$$

$$u(t, x) = u(t, x + 2\pi) = -u(t, -x), \quad (t, x) \in \mathbf{R}^2, \tag{3.1.2'}$$

$$u(t + 2\pi, x) = u(t, x), \quad (t, x) \in \mathbf{R}^2 \tag{3.1.3'}$$

satisfies all the assumptions of Theorem 1.4.3 when we set $n = 1$, $r = 2$, $\alpha = 0$,

$$\mathscr{U}_1^*([0, 2\pi]) = \{u \in \mathscr{U}_1([0, 2\pi]);\, u(t, \pi - x) = u(t, x),$$
$$t \in [0, 2\pi],\, x \in \mathbf{R}\},$$

$$C^{2*}_{2\pi}(\mathbf{R}) = \{\sigma \in C^2_{2\pi}(\mathbf{R});\, \sigma(x + \pi) = -\sigma(x),\, x \in \mathbf{R}\},$$

$$C^2_{2\pi*}(\mathbf{R}) = C^{2*}_{2\pi}(\mathbf{R}).$$

Evidently,

$$u(t, x + \pi) = -u(t, x) \quad \text{for} \quad u \in \mathscr{U}_1^*([0, 2\pi]), \quad t \in [0, 2\pi], \quad x \in \mathbf{R}. \tag{3.1.8}$$

(1) We verify (i).
Note that for any $\sigma \in C^{2*}_{2\pi}(\mathbf{R})$

$$\int_0^{2\pi} \sigma^k(-x + 2\tau)\, d\tau = \int_0^{2\pi} \sigma^k(\xi)\, d\xi, \quad x \in \mathbf{R},$$

$$\int_0^{2\pi} \sigma(\xi)\, d\xi = 0, \int_0^{2\pi} \sigma^3(\xi)\, d\xi = 0.$$

Hence after putting

$$\frac{\alpha}{3\beta} = \gamma \quad \text{and} \quad \frac{1}{4\pi\beta}\int_0^{2\pi} h_e(\tau, x - \tau)\, d\tau \equiv H(x)$$

(1.4.14) can be written in the form

$$\frac{1}{2\pi\beta}\,\Gamma(\sigma)(x) = \sigma^3(x) +$$
$$+ 3\left(\gamma + \frac{1}{2\pi}\int_0^{2\pi} \sigma^2(\xi)\, d\xi\right)\sigma(x) + 2H(x) = 0, \quad x \in \mathbf{R}. \tag{3.1.9}$$

Let

$$I = \frac{1}{2\pi}\int_0^{2\pi} \sigma^2(\xi)\,d\xi$$

and $p = I + \gamma$. We study the modified equation

$$y^3 + 3py + 2H(x) = 0\,. \tag{3.1.10}$$

Since the form of the solution of (3.1.10) depends on the sign of its discriminant $D(x) = p^3 + H^2(x)$, we have to distinguish several cases.

First let us suppose that

$$D(x) = p^3 + H^2(x) \geqq 0\,, \quad x \in \mathbf{R}\,. \tag{3.1.11}$$

Then the solution of (3.1.10) is given by Cardano's formula

$$\begin{aligned} y(x, I) &= (-H(x) + (H^2(x) + p^3)^{1/2})^{1/3} + \\ &+ (-H(x) - (H^2(x) + p^3)^{1/2})^{1/3}\,. \end{aligned} \tag{3.1.12}$$

Now (3.1.11) is equivalent to

$$p \geqq 0\,. \tag{3.1.11'}$$

For since h is 2π-periodic in x, we have by (3.1.7)

$$\begin{aligned} 4\pi\beta \int_0^{2\pi} H(\xi)\,d\xi &= \int_0^{2\pi}\int_0^{2\pi} h_e(\tau, \xi - \tau)\,d\tau\,d\xi = \\ &= \int_0^{2\pi}\int_0^{2\pi} h_e(\tau, \xi - \tau)\,d\xi\,d\tau = \int_0^{2\pi}\int_0^{2\pi} h_e(\tau, \xi)\,d\xi\,d\tau = 0\,, \end{aligned}$$

so that there is an x_0 such that $H(x_0) = 0$.

The solution (3.1.12) is evidently 2π-periodic, since $H(x)$ is. It is of class C^2 if we strengthen the assumption (3.1.11') to

$$p > 0\,. \tag{3.1.13}$$

Moreover, y satisfies

$$\begin{aligned} y(x + \pi, I) &= (H(x) + (H^2(x) + p^3)^{1/2})^{1/3} + \\ &+ (H(x) - (H^2(x) + p^3)^{1/2})^{1/3} = -y(x, I) \end{aligned}$$

since by (3.1.7) $H(x)$ satisfies the same relation. Consequently, (3.1.12) is a solution of (3.1.9) and lies in $C^{2*}_{2\pi}(\mathbf{R})$ if and only if there is an $I^* \in \mathbf{R}^+$ such that

$$2\pi I^* = \int_0^{2\pi} y^2(\xi; I^*)\,d\xi \tag{3.1.14}$$

and $p^* = \gamma + I^* > 0$.

Let

$$\zeta(I) = \int_0^{2\pi} y^2(\xi, I)\, d\xi \,,$$

$$K_\pm(x) = (-H(x) \pm (H^2(x) + p^3)^{1/2})^{1/3} \,.$$

Then we have

$$\frac{d\zeta}{dI} = p^2 \int_0^{2\pi} \frac{(K_+ + K_-)^2 (K_- - K_+)}{(K_+K_-)^2 (H^2(\xi) + p^3)^{1/2}} \, d\xi < 0$$

for $I > \max(0, -\gamma)$. If $\gamma < 0$, we assume that

$$\gamma > -\frac{1}{2\pi} \int_0^{2\pi} [2H(\xi)]^{2/3}\, d\xi \,, \tag{3.1.15}$$

that is,

$$\zeta(-\gamma) = \int_0^{2\pi} [2H(\xi)]^{2/3}\, d\xi > -2\pi\gamma \,.$$

If $\gamma \geqq 0$, we have

$$\zeta(0) > 0 \,.$$

The continuity of ζ implies that there is a unique $I^* > \max(0, -\gamma)$ such that (3.1.14) is satisfied. Hence (3.1.9) has a solution $\sigma^* \in C^{2*}_{2\pi}(\mathbf{R})$.

Secondly, let us suppose that

$$D(x) < 0 \quad \text{for all} \quad x \in \mathbf{R} \,, \tag{3.1.17}$$

so that necessarily $\gamma < 0$. Obviously, (3.1.17) is equivalent to

$$p < -m = -\max\{H^{2/3}(x);\, x \in [0, \pi]\} \,. \tag{3.1.17'}$$

(3.1.10) has three distinct solutions

$$y_j(x, p) = 2(-p)^{1/2} \cos\frac{1}{3}\left(\arccos\left(\frac{-H(x)}{(-p)^{3/2}}\right) + 2\pi j\right), \quad j = 0, 1, 2 \,. \tag{3.1.18}$$

We easily find that the functions y_0 and y_1 have constant signs so that they cannot be π-antiperiodic. On the other hand, the function y_2, which we denote simply by y, is an element of $C^2_{2\pi}(\mathbf{R})$ and is π-antiperiodic.

To make the notation simpler, let

$$\lambda(x, p) = \frac{1}{3}\left(\arccos\left(\frac{-H(x)}{(-p)^{3/2}}\right) + 4\pi\right).$$

Let us find now sufficient conditions for the existence of a constant I^* satisfying (3.1.14), which is equivalent to

$$p = \gamma + \frac{1}{2\pi}\int_0^{2\pi} 4(-p)\cos^2 \lambda(x, p)\,\mathrm{d}x\,. \tag{3.1.19}$$

We denote by $\zeta(p)$ the function on the right-hand side of the equation (3.1.19). Provided that

$$4\cos^2 \lambda(x, p) \leqq \delta\,, \quad \delta \in (0, 1)\,, \tag{3.1.20}$$

we then obtain $\gamma < \zeta(\gamma) \leqq \gamma(1 - \delta)$. (The first inequality follows easily from (3.1.6).) According to (3.1.10) we have

$$\frac{\partial y}{\partial p}(x, p) = -\frac{y(x, p)}{y^2(x, p) + p}\,.$$

Then

$$\frac{\mathrm{d}\zeta}{\mathrm{d}p}(p) = \frac{4}{\pi}\int_0^{2\pi} \frac{\cos^2 \lambda(x, p)}{1 - 4\cos^2 \lambda(x, p)}\,\mathrm{d}x$$

and when (3.1.20) is taken into account

$$0 < \frac{\mathrm{d}\zeta}{\mathrm{d}p}(p) \leqq \frac{2\delta}{1 - \delta}\,.$$

Hence the curve $z = \zeta(p)$ lies under the straight line

$$\zeta_1(p) = \gamma(1 - \delta) + \frac{2\delta}{1 - \delta}(p - \gamma)\,. \tag{3.1.21}$$

Let us suppose that $2\delta/(1 - \delta) < 1$, that is, $\delta < 1/3$. When we denote by $P^* = (p^*, \zeta^*)$ the point of intersection of the straight line $\zeta_2(p) = p$ and the curve $\zeta(p)$, and by $\tilde{P} = (\tilde{p}, \tilde{\zeta})$ the point of intersection of ζ_2 and ζ_1, we have

$$p^* \leqq \tilde{p} = \gamma\,\frac{1 + \delta^2 - 4\delta}{1 - 3\delta}\,.$$

The number p^* satisfies (3.1.17′) if $\gamma(1 + \delta^2 - 4\delta)/(1 - 3\delta) < -m$. Let $\mu = -m/\gamma$ and suppose that $0 < \mu < 1$, that is, $\gamma < -m$. Then (3.1.17′) holds for $0 < \delta < 1/3$, if

$$\delta^2 + \delta(3\mu - 4) + 1 - \mu > 0\,. \tag{3.1.22}$$

Let $\delta_1 = \delta_1(\gamma)$ and $\delta_2 = \delta_2(\gamma)$ be the roots of $\delta^2 + \delta(3\mu - 4) + 1 - \mu = 0$. We easily see that $0 < \delta_1 < 2 - \sqrt{3} < 1/3 < \delta_2$ and $\lim_{\gamma \to -\infty} \delta_1(\gamma) = 2 - \sqrt{3}$. Now (3.1.22) holds on $(-\infty, \delta_1) \cup (\delta_2, +\infty)$. The intersection of this set with $(0, 1/3)$ is the interval $(0, \delta_1)$. Hence we can assert that if

$$\cos^2 \lambda(x, p^*(\gamma, H(x))) \leqq \delta/4 \text{ for some } \quad \delta \in (0, \delta_1), \tag{3.1.23}$$

then (3.1.9) has a unique solution $\sigma^*(x) = y(x, p^*) \in C_{2\pi}^{2*}(\mathbf{R})$. Here (3.1.23) is satisfied if

$$|H(x)| \leqq -\cos(3 \arccos \delta^{1/2}/2)(-p)^{3/2} .$$

For $\mu \in (0, 1)$ this condition holds if

$$\gamma < \min(-m, -mK), \tag{3.1.24}$$

where

$$K = \min_{\delta \in [0, \delta_1(\gamma)]} \left(\frac{1 + \delta^2 - 4\delta}{1 - 3\delta} \right)^{-1} \cos^{-2/3} 3 \left(\arccos \frac{\delta^{1/2}}{2} \right).$$

(We omit other possibilities corresponding to the case where $D(x)$ changes its sign on $\mathbf{R}$. Qualitative arguments show that in this case we would find, in general, only solutions whose derivatives have jumps.)

(2) (*ii*) is satisfied according to (b).

(3) It is easy to verify that $G_1(u, \sigma, \varepsilon)$ and $G_2(u, \varepsilon)$ for $u \in \mathscr{U}_1^*([0, 2\pi])$ and $\sigma \in C_{2\pi}^{2*}(\mathbf{R})$ lie in $C^2([0, 2\pi] \times \mathbf{R})$ and $C_{2\pi}^2(\mathbf{R})$ respectively. The relations

$$G_1(u, \sigma, \varepsilon)(t, \pi - x) = -u(t, \pi - x) + \sigma(\pi - x + t) - \sigma(x - \pi + t) +$$
$$+ \frac{\varepsilon}{2} \int_0^t \int_{\pi - x - t + \tau}^{\pi - x + t - \tau} [h_e(\tau, \xi) + \alpha\, u(\tau, \xi) + \beta\, u^3(\tau, \xi)] \, d\xi \, d\tau =$$
$$= -u(t, x) - \sigma(-x + t) + \sigma(x + t) + \frac{\varepsilon}{2} \int_0^t \int_{x - t + \tau}^{x + t - \tau} [h_e(\tau, \pi - \eta) +$$
$$+ \alpha\, u(\tau, \pi - \eta) + \beta\, u^3(\tau, \pi - \eta)] \, d\eta \, d\tau = G_1(u, \sigma, \varepsilon)(t, x),$$
$$G_1(u, \sigma, \varepsilon)(t, x) = -G_1(u, \sigma, \varepsilon)(t, -x),$$
$$G_1(u, \sigma, \varepsilon)(t, x + 2\pi) = G_1(u, \sigma, \varepsilon)(t, x)$$

and

$$G_2(u, \varepsilon)(x + \pi) = \int_0^{2\pi} [h_e(\tau, x + \pi - \tau) + \alpha\, u(\tau, x + \pi - \tau) +$$
$$+ \beta\, u^3(\tau, x + \pi - \tau)] \, d\tau = -G_2(u, \varepsilon)(x)$$

prove (*iii*).

(4) We have to show that the equation

$$\frac{1}{6\pi\beta}\,\Gamma'_\sigma(\sigma^*)\,\bar\sigma(x) \equiv \left(\gamma + \sigma^{*2}(x) + \frac{1}{2\pi}\int_0^{2\pi}\sigma^{*2}(\xi)\,\mathrm{d}\xi\right)\bar\sigma(x) +$$

$$+\frac{1}{\pi}\,\sigma^*(x)\int_0^{2\pi}\sigma^*(\xi)\,\bar\sigma(\xi)\,\mathrm{d}\xi = \varrho(x) \tag{3.1.25}$$

has for any $\varrho \in C^2_{2\pi*}(\mathbf{R})$ a unique solution $\bar\sigma^* \in C^{2*}_{2\pi}(\mathbf{R})$ such that

$$\|\bar\sigma^*\|_{C_{2\pi}{}^2(\mathbf{R})} \leqq k\|\varrho\|_{C_{2\pi}{}^2(\mathbf{R})} \tag{3.1.26}$$

for a certain constant k. First let us suppose that (3.1.13) holds. Putting

$$\bar I = \frac{1}{\pi}\int_0^{2\pi}\sigma^*(\xi)\,\bar\sigma(\xi)\,\mathrm{d}\xi\,,$$

we have

$$\bar\sigma(x) = \frac{\varrho(x) - \sigma^*(x)\,\bar I}{\gamma + \sigma^{*2}(x) + I^*(\gamma)}\,.$$

Note that by virtue of (3.1.13)

$$\gamma + \sigma^{*2}(x) + I^*(\gamma) \geqq \gamma + I^*(\gamma) > 0\,.$$

Let $\bar I^*$ be the solution of the equation

$$\pi\bar I = \int_0^{2\pi}\sigma^*(\xi)\,\bar\sigma(\xi)\,\mathrm{d}\xi = \int_0^{2\pi}\sigma^*(\xi)\,\frac{\varrho(\xi) - \sigma^*(\xi)\,\bar I}{\gamma + \sigma^{*2}(\xi) + I^*(\gamma)}\,\mathrm{d}\xi\,,$$

that is,

$$\bar I^* = \left[\pi + \int_0^{2\pi}\frac{\sigma^{*2}(\xi)}{\gamma + \sigma^{*2}(\xi) + I^*(\gamma)}\,\mathrm{d}\xi\right]^{-1}\int_0^{2\pi}\frac{\sigma^*(\xi)\,\varrho(\xi)}{\gamma + \sigma^{*2}(\xi) + I^*(\gamma)}\,\mathrm{d}\xi\,.$$

Then

$$\bar\sigma^*(x) = (\gamma + \sigma^{*2}(x) + I^*(\gamma))^{-1}\left\{\varrho(x) - \sigma^*(x)\left[\pi +\right.\right.$$

$$\left.\left.+\int_0^{2\pi}\frac{\sigma^{*2}(\xi)}{\gamma + \sigma^{*2}(\xi) + I^*(\gamma)}\,\mathrm{d}\xi\right]^{-1}\int_0^{2\pi}\frac{\sigma^*(\xi)\,\varrho(\xi)}{\gamma + \sigma^{*2}(\xi) + I^*(\gamma)}\,\mathrm{d}\xi\right\}.$$

Now, writing

$$[\gamma + \sigma^{*2}(x) + I^*(\gamma)]^{-1} = a(x) > 0$$

we easily find that (3.1.26) holds with

$$k = \max_{x \in [0,\pi]} \left(a(x) + 2|a'(x)| + |a''(x)|\right) .$$

$$. \left\{1 + \|\sigma^*\|_{C_{2\pi}^2(\mathbf{R})} \left[\pi + \int_0^{2\pi} a(\xi)\, \sigma^{*2}(\xi)\, d\xi\right]^{-1} \int_0^{2\pi} |a(\xi)|\, |\sigma^*(\xi)|\, d\xi\right\} .$$

Secondly, let (3.1.17′) hold. At this point of the proof we observe only that

$$\sigma^{*2}(x) + \gamma + I^*(\gamma) \neq 0$$

(because $\sigma^{*2}(x) + \gamma + I^*(\gamma) \leqq p^*(1 - \delta) < 0$), and

$$\pi + \int_0^{2\pi} \frac{\sigma^{*2}(\xi)}{\sigma^{*2}(\xi) + \gamma + I^*(\gamma)}\, d\xi \neq 0$$

(because

$$\pi + \int_0^{2\pi} \frac{\sigma^{*2}(\xi)}{\sigma^{*2}(\xi) + \gamma + I^*(\gamma)}\, d\xi \geqq \pi \frac{1 - 3\delta}{1 - \delta} > 0$$

for $\delta \in (0, \delta_1)$).

Since the restriction of u^* to x in $[0, \pi]$ is a solution to the problem $(\mathscr{P}_{2\pi}^{\varepsilon})$, we have proved the following theorem.

THEOREM 3.1.1 *Suppose that* (*a*), (*b*) *and* (3.1.15) *or* (3.1.24) *are satisfied. Then for sufficiently small* ε *the problem* (3.1.1)–(3.1.3) *has a locally unique solution* $u^*(\varepsilon)$ *that is continuous in* ε *and is such that* $u^*(0)(t, x) = \sigma^*(x + t) - \sigma^*(-x + t)$.

Problem 3.1.1 Using the Poincaré method investigate the problem $(\mathscr{P}_{2\pi/3}^{\varepsilon})$ with

$$F(u, \varepsilon) = h(t, x) + (\alpha + \beta u^2)\, u_t .$$

3.2. *The problem* $(\mathscr{P}_{2\pi}^{\varepsilon})$ *with* $F(u, \varepsilon)(t, x) = h(t, x) + \alpha\, u(t, x) + \beta\, u^3(t, x)$ (*the Günzler method*)

Let $(\mathscr{P}_{2\pi}^{\varepsilon})$ be the problem (3.1.1)–(3.1.3) and assume that

(a) α and β are constants, $\beta \neq 0$;

(b) $h \in C_{2\pi}^{2,0}(Q)$ satisfies

$$\int_0^{\pi} [h(t - \xi, \xi) - h(t + \xi, \xi)]\, d\xi \not\equiv 0 .$$

Let us verify that our problem satisfies all the assumptions of Theorem 2.4.3 when $r = 2$, $s = 3$, $\alpha = 2$, $X_1 = X_2 = 0$, and

$$C_{2\pi*}^{\min(r,s-1)}(\mathbf{R}) = \left\{\varrho \in C_{2\pi}^2(\mathbf{R});\ \int_0^{2\pi} \varrho(t)\,\mathrm{d}t = 0\right\}.$$

We use the notation

$$\gamma = \frac{4\alpha}{3\beta} \quad \text{and} \quad H(t) = \frac{4}{\beta\pi}\int_0^{\pi} [h(t-\xi,\xi) - h(t+\xi,\xi)]\,\mathrm{d}\xi\,.$$

(1) We now verify (i). We put

$$C(t) = \int^t c(\tau)\,\mathrm{d}\tau\,.$$

Then (2.4.9) is equivalent to the system

$$C^3(t) + 3(\gamma + I_2)\,C(t) - I_3 + H(t) = 0\,, \tag{3.2.1}$$

$$I_1 = 0\,, \tag{3.2.2}$$

where

$$I_k = \frac{1}{2\pi}\int_0^{2\pi} C^k(\tau)\,\mathrm{d}\tau\,, \quad k = 1, 2, 3\,.$$

Let us suppose that

$$\gamma + I_2 > 0\,. \tag{3.2.3}$$

If I_2 and I_3 are constants, then there is a unique real solution $C^*(t) = C^*(t)(I_2, I_3) \in C_{2\pi}^2(\mathbf{R})$ of equation (3.2.1). If $I_3 > \max_{t\in[0,2\pi]} H(t)$, or $I_3 < \min_{t\in[0,2\pi]} H(t)$, then $C^*(t)(I_2, I_3) > 0$, or $C^*(t)(I_2, I_3) < 0$, respectively and correspondingly $I_1 > 0$, or $I_1 < 0$. Because $C^*(t)(I_2, I_3)$ is continuous in I_3, we can find an $I_3^* \in (\min_{t\in[0,2\pi]} H(t), \max_{t\in[0,2\pi]} H(t))$ such that

$$I_1^* = \int_0^{2\pi} C^*(\tau)(I_2, I_3^*)\,\mathrm{d}\tau = 0\,.$$

The uniqueness of I_3^* follows readily from (3.2.1). Integrating (3.2.1), we obtain easily

$$I_3^* = \frac{1}{2\pi}\int_0^{2\pi} C^{*3}(\tau)(I_2, I_3^*)\,\mathrm{d}\tau\,.$$

(Note that $I_3^* = I_3^*(I_2)$.) Thus it remains to satisfy the condition

$$2\pi I_2 = \int_0^{2\pi} C^{*2}(\tau)(I_2, I_3^*)\,d\tau \equiv \zeta(I_2)\,. \tag{3.2.4}$$

Suppose that

$$\gamma > \frac{-1}{2\pi} \int_0^{2\pi} [I_3^*(-\gamma) - H(\tau)]^{2/3}\,d\tau\,. \tag{3.2.5}$$

If $\gamma \geqq 0$ or $\gamma < 0$, then $\zeta(0) > 0$ (by virtue of (b)) or $\zeta(-\gamma) > -2\pi\gamma$ (by virtue of (3.2.5)). Hence, from (3.2.1),

$$\frac{\partial}{\partial I_2} C^*(t)(I_2, I_3) = -\frac{C^*(t)(I_2, I_3)}{C^{*2}(t)(I_2, I_3) + \gamma + I_2}\,,$$

$$\frac{\partial}{\partial I_3} C^*(t)(I_2, I_3) = \frac{1}{3[C^{*2}(t)(I_2, I_3) + \gamma + I_2]}\,.$$

Further, it follows from the relation

$$I_3^*(I_2) = \frac{1}{2\pi} \int_0^{2\pi} C^{*3}(\tau)(I_2, I_3^*(I_2))\,d\tau$$

that

$$\frac{dI_3^*}{dI_2}(I_2) = \left[\frac{-3}{2\pi} \int_0^{2\pi} \frac{C^{*3}(\tau)(I_2, I_3^*(I_2))}{C^{*2}(\tau)(I_2, I_3^*(I_2)) + \gamma + I_2}\,d\tau\right] \cdot$$

$$\cdot \left[1 - \frac{1}{2\pi} \int_0^{2\pi} \frac{C^{*2}(\tau)(I_2, I_3^*(I_2))}{C^{*2}(\tau)(I_2, I_3^*(I_2)) + \gamma + I_2}\,d\tau\right]^{-1}\,.$$

Hence, using the relations

$$\int_0^{2\pi} \frac{C^{*3}(\tau)(I_2, I_3^*(I_2))}{C^{*2}(\tau)(I_2, I_3^*(I_2)) + \gamma + I_2}\,d\tau =$$

$$= -(\gamma + I_2) \int_0^{2\pi} \frac{C^*(\tau)(I_2, I_3^*(I_2))}{C^{*2}(\tau)(I_2, I_3^*(I_2)) + \gamma + I_2}\,d\tau\,,$$

$$\int_0^{2\pi} \frac{C^{*2}(\tau)(I_2, I_3^*(I_2))}{C^{*2}(\tau)(I_2, I_3^*(I_2)) + \gamma + I_2}\,d\tau =$$

$$= 2\pi - (\gamma + I_2) \int_0^{2\pi} \frac{1}{C^{*2}(\tau)(I_2, I_3^*(I_2)) + \gamma + I_2}\,d\tau$$

and the Hölder inequality, we obtain

$$\frac{d\zeta}{dI_2}(I_2) = 2\int_0^{2\pi} C^*(\tau)(I_2, I_3^*(I_2))\left[\frac{\partial}{\partial I_2}C^*(\tau)(I_2, I_3^*(I_2)) + \right.$$

$$\left. + \frac{\partial}{\partial I_3}C^*(\tau)(I_2, I_3^*(I_2))\frac{\partial I_3^*}{\partial I_2}(I_2)\right] d\tau \leqq 0 .$$

Thus, there exists a unique point I_2^* satisfying (3.2.3) and (3.2.4). The function

$$c^*(t) = \frac{d}{dt}C^*(t)(I_2^*, I_3^*)$$

is a solution of (2.4.9), and $c^* \in C_{2\pi}^{1*}(\mathbf{R})$.

(2) (*ii*) and (*iii*) are satisfied according to (b).

(3) For the proof of (*iv*) we observe only that

$$\int_0^{2\pi}\int_0^{\pi}(F(u)(\tau - \xi, \xi) - F(u)(\tau + \xi, \xi))\, d\xi\, d\tau = 0 .$$

(4) To prove (*v*) it suffices to show that the equation

$$\Gamma_c'(c^*)(\bar{c})(t) = \varrho(t) \tag{3.2.6}$$

has for any $\varrho \in C_{2\pi*}^2(\mathbf{R})$ a unique solution $\bar{c}^* \in C_{2\pi}^{1*}(\mathbf{R})$ such that

$$\|\bar{c}^*\|_{C_{2\pi}{}^1(\mathbf{R})} \leqq k\|\varrho\|_{C_{2\pi}{}^2(\mathbf{R})} \tag{3.2.7}$$

for a certain constant k. We put

$$\bar{C}(t) = \int^t \bar{c}(\tau)\, d\tau ,$$

so that

$$\int_0^{2\pi}\bar{C}(\tau)\, d\tau = 0 ,$$

and let

$$R(t) = \frac{4}{3\pi\beta}\varrho(t) , \quad \bar{I}_2 = \frac{1}{\pi}\int_0^{2\pi} C^*(\tau)\,\bar{C}(\tau)\, d\tau ,$$

$$\bar{I}_3 = \frac{1}{2\pi}\int_0^{2\pi} C^{*2}(\tau)\,\bar{C}(\tau)\, d\tau , \text{ and } \gamma + I_2^* = p^* .$$

Then (3.2.6) takes the form

$$\bar{C}(t)(C^{*2}(t) + p^*) + C^*(t)\,\bar{I}_2 - \bar{I}_3 = R(t) . \tag{3.2.8}$$

Again, if $\bar{I}_2$ and $\bar{I}_3$ are constants, then

$$\bar{C}(t) = \bar{C}(t)\,(\bar{I}_2, \bar{I}_3) = \frac{R(t) + \bar{I}_3 - C^*(t)\,\bar{I}_2}{C^{*2}(t) + p^*} .$$

Obviously, $\bar{C}(t)$ is a 2π-periodic function.

We look for constants $\bar{I}_2$ and $\bar{I}_3$ such that

$$\int_0^{2\pi} \bar{C}(\tau)\,(\bar{I}_2, \bar{I}_3)\,\mathrm{d}\tau = 0 , \tag{3.2.9}$$

$$\bar{I}_2 = \frac{1}{\pi}\int_0^{2\pi} C^*(\tau)\,\bar{C}(\tau)\,(\bar{I}_2, \bar{I}_3)\,\mathrm{d}\tau . \tag{3.2.10}$$

With the notation $g(t) = [C^{*2}(t) + p^*]^{-1}$, these conditions give the following system for $\bar{I}_2^*$ and $\bar{I}_3^*$:

$$\bar{I}_2 \int_0^{2\pi} C^*(\tau)\,g(\tau)\,\mathrm{d}\tau + \bar{I}_3\left(-\int_0^{2\pi} g(\tau)\,\mathrm{d}\tau\right) = \int_0^{2\pi} R(\tau)\,g(\tau)\,\mathrm{d}\tau ,$$

$$\bar{I}_2\left(\pi + \int_0^{2\pi} C^{*2}(\tau)\,g(\tau)\,\mathrm{d}\tau\right) + \bar{I}_3\left(-\int_0^{2\pi} C^*(\tau)\,g(\tau)\,\mathrm{d}\tau\right) =$$

$$= \int_0^{2\pi} R(\tau)\,C^*(\tau)\,g(\tau)\,\mathrm{d}\tau . \tag{3.2.11}$$

By means of the Hölder inequality we can prove that the determinant D of (3.2.11) does not vanish. Hence, there are (uniquely determined) constants $\bar{I}_2^*$ and $\bar{I}_3^*$ satisfying (3.2.9) and (3.2.10). From (3.2.8) we also see that

$$\bar{I}_3^* = \frac{1}{2\pi}\int_0^{2\pi} C^{*2}(\tau)\,\bar{C}(\tau)\,(\bar{I}_2^*, \bar{I}_3^*)\,\mathrm{d}\tau .$$

The function $\bar{c}^*(t) = (\mathrm{d}/\mathrm{d}t)\,\bar{C}(t)\,(\bar{I}_2^*, \bar{I}_3^*)$ satisfies (3.2.6), and $\bar{c}^*(t) \in C_{2\pi}^{1*}(\mathbf{R})$. The estimate (3.2.7) is a consequence of the explicit form of $\bar{c}^*$. Thus, we have proved the following result.

THEOREM 3.2.1 *Suppose that the conditions (a), (b), and (3.2.5) hold. Then for sufficiently small ε the problem (3.1.1)–(3.1.3) has a locally unique solution $u^*(\varepsilon) \in C_{2\pi}^2(Q)$ that is continuous in ε and is such that*

$$u^*(0)\,(t, x) = \frac{1}{2}\int_{-x+t}^{x+t} c^*(\tau)\,\mathrm{d}\tau.$$

§ 4. The Newton and combined boundary conditions; the Günzler method

4.1. *General considerations*

Let us consider the problem $(\mathscr{P}_\omega^0)$ given by

$$u_{tt}(t, x) - u_{xx}(t, x) = g(t, x), \quad (t, x) \in Q = \mathbf{R} \times [0, \pi], \tag{4.1.1}$$

$$\begin{aligned} &u_x(t, 0) + \sigma_0(t)\, u(t, 0) = h_0(t), \\ &\beta\, u_x(t, \pi) + \sigma_1(t)\, u(t, \pi) = h_1(t), \quad t \in \mathbf{R}, \end{aligned} \tag{4.1.2}$$

$$u(t + \omega, x) = u(t, x), \quad (t, x) \in Q. \tag{4.1.3}$$

(Here β is a constant; it suffices to distinguish the cases $\beta = 0$ or 1. If $\beta = 0$, let us put $\sigma_1(t) \equiv 1$.) We suppose that $g \in C_\omega^{(\alpha, 1-\alpha)}(Q)$ with fixed $\alpha = 0$ or 1, and that $h_0, h_1, \sigma_0, \sigma_1 \in C_\omega^1(\mathbf{R})$.

By a solution of the problem $(\mathscr{P}_\omega^0)$ we mean a function $u \in C_\omega^2(Q)$ satisfying (4.1.1)–(4.1.3).

When we put $u(t, 0) = a(t)$ and $u_x(t, 0) = b(t)$, by virtue of II: Lemma 2.1.2 we can write a solution of (4.1.1) and (4.1.3) in the form

$$u(t, x) = 2^{-1} \Big\{ a(x + t) + a(-x + t) + \\ + \int_{-x+t}^{x+t} b(\tau)\, d\tau - \int_0^x \int_{t-x+\xi}^{t+x-\xi} g(\tau, \xi)\, d\tau\, d\xi \Big\}, \quad (t, x) \in Q. \tag{4.1.4}$$

The conditions (4.1.2) are satisfied if and only if

$$b(t) + \sigma_0(t)\, a(t) = h_0(t), \tag{4.1.5}$$

$$\beta \Big\{ a'(\pi + t) - a'(-\pi + t) + b(\pi + t) + b(-\pi + t) - \\ - \int_0^\pi (g(t + \pi - \xi, \xi) + g(t - \pi + \xi, \xi))\, d\xi \Big\} + \\ + \sigma_1(t) \Big\{ a(\pi + t) + a(-\pi + t) + \int_{-\pi+t}^{\pi+t} b(\tau)\, d\tau - \\ - \int_0^\pi \int_{t-\pi+\xi}^{t+\pi-\xi} g(\tau, \xi)\, d\tau\, d\xi \Big\} = 2h_1(t), \quad t \in \mathbf{R}. \tag{4.1.6}$$

Let us investigate separately the case $\beta = 1$ (the Newton boundary conditions) and the case $\beta = 0$ (the combined boundary conditions).

4.2. *The case* $\beta = 1$, $\omega = 2\pi$

In this case the conditions (4.1.5) and (4.1.6) reduce to

$$2a(t)\,(\sigma_1(t+\pi) - \sigma_0(t)) - \sigma_1(t+\pi)\int_0^{2\pi} \sigma_0(\tau)\,a(\tau)\,d\tau = H_1(t)\,, \qquad (4.2.1)$$

where

$$H_1(t) = 2h_1(t+\pi) - 2h_0(t) - \sigma_1(t+\pi)\int_0^{2\pi} h_0(\tau)\,d\tau +$$

$$+ \int_0^{\pi} (g(t-\xi,\xi) + g(t+\xi,\xi))\,d\xi +$$

$$+ \sigma_1(t+\pi)\int_0^{\pi}\int_{t+\xi}^{t+2\pi-\xi} g(\tau,\xi)\,d\tau\,d\xi\,.$$

First, let

$$\sigma_1(t+\pi) - \sigma_0(t) = 0 \quad \text{for all} \quad t \in \mathbf{R}\,. \qquad (4.2.2)$$

Then the following theorem is evident.

THEOREM 4.2.1 *Let* $\beta = 1$. *Let* $g \in C_{2\pi}^{(\alpha,1-\alpha)}(Q)$ *with fixed* $\alpha = 0$ *or* 1, σ_0, $\sigma_1 \in C_{2\pi}^2(\mathbf{R})$, h_0, $h_1 \in C_{2\pi}^1(\mathbf{R})$. *Let* (4.2.2) *be satisfied.*

(*i*) *Then the problem* $(\mathscr{P}_{2\pi}^0)$ *has a solution if and only if there exists a constant K such that*

$$K\,\sigma_0(t) + 2h_1(t+\pi) - 2h_0(t) + \int_0^{\pi} (g(t-\xi,\xi) + g(t+\xi,\xi))\,d\xi +$$

$$+ \sigma_0(t)\int_0^{\pi}\int_{t+\xi}^{t+2\pi-\xi} g(\tau,\xi)\,d\tau\,d\xi = 0 \qquad (4.2.3)$$

for all $t \in \mathbf{R}$.

(*ii*) *Let* (4.2.3) *be satisfied. Put*

$$k = \begin{cases} 0 & \text{if} \quad \sigma_0(t) \equiv 0\,, \\ \left(K + \int_0^{2\pi} h_0(\tau)\,d\tau\right)\left(\int_0^{2\pi} \sigma_0^2(\tau)\,d\tau\right)^{-1} & \text{otherwise}\,, \end{cases}$$

$$w(t,x) = 2^{-1}\Big\{k\,\sigma_0(x+t) + k\,\sigma_0(-x+t) - k\int_{-x+t}^{x+t} \sigma_0^2(\tau)\,d\tau +$$

$$+ \int_{-x+t}^{x+t} h_0(\tau)\,d\tau - \int_0^{x}\int_{t-x+\xi}^{t+x-\xi} g(\tau,\xi)\,d\tau\,d\xi\Big\}\,,$$

$$v(t, x) = 2^{-1}\left\{a(x + t) + a(-x + t) - \int_{-x+t}^{x+t} \sigma_0(\tau)\, a(\tau)\, d\tau\right\},$$

where the function a is an arbitrary element of $C^2_{2\pi}(\mathbf{R})$ such that

$$\int_0^{2\pi} a(\tau)\, \sigma_0(\tau)\, d\tau = 0\,.$$

Then the solution u of the problem $(\mathscr{P}^0_{2\pi})$ is given by $u = v + w$.

(iii) Put $w = \Lambda(g, h_0, h_1)$. Then

$$\|\Lambda(g, h_0, h_1)\|_{C^2(Q)} \leqq c(\|g\|_{C^{(\alpha, 1-\alpha)}(Q)} + \|h_0\|_{C^1(\mathbf{R})} + \|h_1\|_{C^1(\mathbf{R})})$$

holds with a suitable c independent of g, h_0, and h_1.

Suppose further, that

$$\sigma_1(t + \pi) - \sigma_0(t) \neq 0 \quad \text{for all} \quad t \in \mathbf{R}\,. \tag{4.2.4}$$

Let

$$J = \int_{-\pi}^{\pi} \sigma_0(\tau)\, a(\tau)\, d\tau\,.$$

Then the function

$$a(t) = 2^{-1}(\sigma_1(t + \pi) - \sigma_0(t))^{-1}\,(H_1(t) + J\,\sigma_1(t + \pi)) \tag{4.2.5}$$

is a solution of (4.2.1) if and only if

$$J\left(1 - 2^{-1}\int_{-\pi}^{\pi} \sigma_1(\tau + \pi)\, \sigma_0(\tau)\, (\sigma_1(\tau + \pi) - \sigma_0(\tau))^{-1}\, d\tau\right) =$$

$$= 2^{-1}\int_{-\pi}^{\pi} H_1(\tau)\, \sigma_0(\tau)\, (\sigma_1(\tau + \pi) - \sigma_0(\tau))^{-1}\, d\tau\,. \tag{4.2.6}$$

Our result can be summarized in the following two theorems.

THEOREM 4.2.2 *Let $\beta = 1$. Let $g \in C^{(\alpha, 2-\alpha)}_{2\pi}(Q)$ with fixed $\alpha = 0$, 1 or 2, $\sigma_0, \sigma_1, h_0, h_1 \in C^2_{2\pi}(\mathbf{R})$, let (4.2.4) be satisfied and*

$$\int_{-\pi}^{\pi} \sigma_1(\tau + \pi)\, \sigma_0(\tau)\, (\sigma_1(\tau + \pi) - \sigma_0(\tau))^{-1}\, d\tau \neq 2\,.$$

Then the problem $(\mathscr{P}^0_{2\pi})$ has a unique solution u given by (4.1.4), (4.1.5), (4.2.5), and (4.2.6). Put $u = \Lambda(g, h_0, h_1)$. Then

$$\|\Lambda(g, h_0, h_1)\|_{C^2(Q)} \leqq c(\|g\|_{C^{(\alpha, 2-\alpha)}(Q)} + \|h_0\|_{C^2(\mathbf{R})} + \|h_1\|_{C^2(\mathbf{R})})$$

holds with a suitable c independent of g, h_0, and h_1.

THEOREM 4.2.3 *Let* $\beta = 1$. *Let* $g \in C_{2\pi}^{(\alpha, 2-\alpha)}(Q)$ *with fixed* $\alpha = 0$, 1 *or* 2, let $\sigma_0, \sigma_1, h_0, h_1 \in C_{2\pi}^2(\mathbf{R})$, *let* (4.2.4) *be satisfied and*

$$\int_{-\pi}^{\pi} \sigma_1(\tau + \pi)\, \sigma_0(\tau)\, (\sigma_1(\tau + \pi) - \sigma_0(\tau))^{-1}\, d\tau = 2\,.$$

(*i*) *Then the problem* $(\mathscr{P}_{2\pi}^0)$ *has a solution if and only if*

$$\int_{-\pi}^{\pi} H_1(\tau)\, \sigma_0(\tau)\, (\sigma_1(\tau + \pi) - \sigma_0(\tau))^{-1}\, d\tau = 0\,. \tag{4.2.7}$$

(*ii*) *Let* (4.2.7) *be satisfied. Put*

$$v(t, x) = 2^{-1} \left\{ a_1(x + t) + a_1(-x + t) + \int_{-x+t}^{x+t} b_1(\tau)\, d\tau \right\} J\,,$$

$$w(t, x) = 2^{-1} \left\{ a_2(x + t) + a_2(-x + t) + \int_{-x+t}^{x+t} b_2(\tau)\, d\tau - \right.$$

$$\left. - \int_0^x \int_{t-x+\xi}^{t+x-\xi} g(\tau, \xi)\, d\tau\, d\xi \right\},$$

where

$$a_1(t) = 2^{-1}\, \sigma_1(t + \pi)\, (\sigma_1(t + \pi) - \sigma_0(t))^{-1}\,,$$
$$b_1(t) = -\sigma_0(t)\, a_1(t)\,,$$
$$a_2(t) = 2^{-1}\, H_1(t)\, (\sigma_1(t + \pi) - \sigma_0(t))^{-1}\,,$$
$$b_2(t) = h_0(t) - \sigma_0(t)\, a_2(t)\,,$$

and $J \in \mathbf{R}$ *is arbitrary.*
Then the solution of the problem $(\mathscr{P}_{2\pi}^0)$ *can be written in the form* $u = v + w$.

(*iii*) *Put* $w = A(g, h_0, h_1)$. *Then*

$$\|A(g, h_0, h_1)\|_{C^2(Q)} \leqq c(\|g\|_{C^{(\alpha, 2-\alpha)}(Q)} + \|h_0\|_{C^2(\mathbf{R})} + \|h_1\|_{C^2(\mathbf{R})})$$

holds with a suitable c independent of g, h_0, *and* h_1.

Remark 4.2.1 The case when the set $\{t \in \mathbf{R};\ \sigma_1(t + \pi) - \sigma_0(t) \neq 0\}$ is neither empty nor $\mathbf{R}$ offers many possibilities and requires raising the smoothness of g, h_j, and σ_j. In general, the second circumstance prevents us from applying our methods to study these possibilities in the non-linear case, therefore we leave them aside.

Problem 4.2.1 Formulate corresponding theorems for the existence of solutions to the weakly non-linear problem

$$\begin{aligned} &u_{tt}(t, x) - u_{xx}(t, x) = \varepsilon F(u, \varepsilon)(t, x), \quad (t, x) \in Q, \\ &u_x(t, 0) + \sigma_0(t) u(t, 0) = \varepsilon X_0(u, \varepsilon)(t), \\ &u_x(t, \pi) + \sigma_1(t) u(t, \pi) = \varepsilon X_1(u, \varepsilon)(t), \quad t \in \mathbf{R}, \\ &u(t, x) = u(t + 2\pi, x), \quad (t, x) \in Q. \end{aligned}$$

4.3. *The case* $\beta = 0, \sigma_1(t) = 1, \omega = 2\pi$

In this case conditions (4.1.5) and (4.1.6) are equivalent to

$$2a(t) - \int_0^{2\pi} \sigma_0(\tau) a(\tau) \, d\tau = H_2(t), \quad t \in \mathbf{R}, \tag{4.3.1}$$

where

$$H_2(t) = 2h_1(t + \pi) - \int_0^{2\pi} h_0(\tau) \, d\tau + \int_0^{\pi} \int_{t+\xi}^{t+2\pi-\xi} g(\tau, \xi) \, d\tau \, d\xi .$$

Put

$$J = \int_0^{2\pi} \sigma_0(\tau) a(\tau) \, d\tau .$$

Then

$$a(t) = 2^{-1}(J + H_2(t)) \tag{4.3.2}$$

is a solution of (4.3.1) if and only if

$$J\left(1 - 2^{-1} \int_{-\pi}^{\pi} \sigma_0(\tau) \, d\tau\right) = 2^{-1} \int_{-\pi}^{\pi} \sigma_0(\tau) H_2(\tau) \, d\tau . \tag{4.3.3}$$

This leads immediately to the next two theorems.

THEOREM 4.3.1 *Let* $\beta = 0$, $\sigma_1(t) \equiv 1$. *Let* $g \in C_{2\pi}^{(\alpha, 1-\alpha)}(Q)$ *with fixed* $\alpha = 0$ *or* 1, $\sigma_0, h_0 \in C_{2\pi}^1(\mathbf{R})$, $h_1 \in C_{2\pi}^2(\mathbf{R})$ *and*

$$\int_{-\pi}^{\pi} \sigma_0(\tau) \, d\tau \neq 2 .$$

Then the problem $(\mathscr{P}_{2\pi}^0)$ *has a unique solution* u *given by* (4.1.4), (4.1.5), (4.3.2), *and* (4.3.3). *Put* $u = \Lambda(g, h_0, h_1)$. *Then*

$$\|\Lambda(g, h_0, h_1)\|_{C^2(Q)} \leqq c(\|g\|_{C^{(\alpha, 1-\alpha)}(Q)} + \|h_0\|_{C^1(\mathbf{R})} + \|h_1\|_{C^2(\mathbf{R})})$$

holds with a suitable c *independent of* g, h_0, *and* h_1.

THEOREM 4.3.2 *Let* $\beta = 0$, $\sigma_1(t) \equiv 1$. *Let* $g \in C_{2\pi}^{(\alpha, 1-\alpha)}(Q)$ *with fixed* $\alpha = 0$ *or* 1, $\sigma_0, h_0 \in C^1_{2\pi}(\mathbf{R})$, $h_1 \in C^2_{2\pi}(\mathbf{R})$ *and*

$$\int_{-\pi}^{\pi} \sigma_0(\tau)\,\mathrm{d}\tau = 2\,.$$

(*i*) *Then the problem* $(\mathscr{P}^0_{2\pi})$ *has a solution u if and only if*

$$\int_{-\pi}^{\pi} \sigma_0(\tau)\,H_2(\tau)\,\mathrm{d}\tau = 0\,. \tag{4.3.4}$$

(*ii*) *Let* (4.3.4) *be satisfied. Put*

$$v(t, x) = \left\{1 - 2^{-1}\int_{-x+t}^{x+t} \sigma_0(\tau)\,\mathrm{d}\tau\right\} J\,,$$

$$w(t, x) = 2^{-1}\left\{a_1(x + t) + a_1(-x + t) + \int_{-x+t}^{x+t} b_1(\tau)\,\mathrm{d}\tau - \right.$$

$$\left. - \int_0^x \int_{t-x+\xi}^{t+x-\xi} g(\tau, \xi)\,\mathrm{d}\tau\,\mathrm{d}\xi\right\},$$

where

$$a_1(t) = 2^{-1}\,H_2(t)\,,$$

$$b_1(t) = h_0(t) - \sigma_0(t)\,a_1(t)\,,$$

and $J \in \mathbf{R}$ *is arbitrary. Then the solution u of the problem* $(\mathscr{P}^0_{2\pi})$ *is given by* $u = v + w$.

(*iii*) *Put* $w = \Lambda(g, h_0, h_1)$. *Then*

$$\|\Lambda(g, h_0, h_1)\|_{C^2(Q)} \leqq c(\|g\|_{C^{(\alpha, 1-\alpha)}(Q)} + \|h_0\|_{C^1(\mathbf{R})} + \|h_1\|_{C^2(\mathbf{R})})$$

holds with a suitable c independent of g, h_0, *and* h_1.

Problem 4.3.1 Formulate corresponding theorems for the existence of solutions to the weakly non-linear problem

$$u_{tt}(t, x) - u_{xx}(t, x) = \varepsilon\,F(u, \varepsilon)\,(t, x)\,, \quad (t, x) \in Q\,,$$

$$u_x(t, 0) + \sigma_0(t)\,u(t, 0) = \varepsilon\,X_0(u, \varepsilon)\,(t)\,,$$

$$u(t, \pi) = \varepsilon\,X_1(u, \varepsilon)\,(t)\,, \quad t \in \mathbf{R}\,,$$

$$u(t, x) = u(t + 2\pi, x)\,, \quad (t, x) \in Q\,.$$

Remark 4.3.1 For $\omega = 2\pi p/q$ $(p, q \in \mathbf{N})$ or $\omega = 2\pi\alpha$ (α irrational) the use of this method is not simple and the corresponding problems are studied by the (t, s)-Fourier method in Sec. 6.5 under less general assumptions.

Remark 4.3.2 The problem $(\mathscr{P}^0_\omega)$ has also been studied by VEJVODA [80] (see Remark 1.4.5) and PEŠL [47] (see Remark 6.4.5) under somewhat different hypotheses.

§ 5. Entrainment of frequency

5.1. *Introduction*

We assume that $\lambda > 0$ is fixed throughout this paragraph. We investigate the existence of solutions that are $(2\pi + \varepsilon\lambda)$-periodic in t of the problem

$$u_{tt}(t, x) - u_{xx}(t, x) = \varepsilon F(u, \varepsilon)(t, x), \quad (t, x) \in \mathbf{R}^2, \tag{5.1.1}$$

$$u(t, x) = u(t, x + 2\pi) = -u(t, -x), \quad (t, x) \in \mathbf{R}^2. \tag{5.1.2}$$

For the sake of simplicity we restrict ourselves to the case

$$F(u, \varepsilon)(t, x) = f(t, x, u(t, x), u_t(t, x), u_x(t, x), \varepsilon) \tag{5.1.3}$$

where the function $f = f(t, x, u_0, u_1, u_2, \varepsilon)$ is defined on $\mathbf{R}^5 \times [-\varepsilon_0, \varepsilon_0]$ and is $(2\pi + \varepsilon\lambda)$-periodic in t for every $x, u_0, u_1, u_2 \in \mathbf{R}$ and $\varepsilon \in [-\varepsilon_0, \varepsilon_0]$, $\varepsilon_0 > 0$. We shall show that under some additional general assumptions on f the problem (5.1.1)–(5.1.2) has for every small ε a solution that is $(2\pi + \varepsilon\lambda)$-periodic in t. Every result of this kind is of great importance from the practical point of view. When we deal with any realization of the process described by (5.1.1) and appropriate boundary conditions (which are replaced here by (5.1.2)), we cannot determine the frequency of the forcing term with absolute accuracy. Nonetheless, we are able to approximate the required frequency (2π, in our case) with an arbitrarily small error. Further, as we shall see (see § 6), the existence of ω-periodic solutions depends heavily on every change of the period ω. In this context we regard it very interesting and valuable that an existence theorem for the above-mentioned problem can be proved.

The first to attack this problem were FINK and HALL in [15], where they introduced also the term "entrainment of frequency". Their general approach is illustrated in Sec. 5.2, where the existence of $(2\pi + \varepsilon\lambda)$-periodic solutions of a second-order equation of the type (5.1.1) is proved under general assumptions on F (see also VII: 5.5). In Sec. 5.3 these assumptions are shown to be satisfied by a certain family of f.

5.2. *The existence of $(2\pi + \varepsilon\lambda)$-periodic solutions*

Let us introduce the notation that is kept throughout this paragraph. We put

$$\mathscr{H}^k = \left\{ s \in H^k_{2\pi}(\mathbf{R});\ \int_0^{2\pi} s(x)\,\mathrm{d}x = 0 \right\},$$

$$H^k_{2\pi(o)}(\mathbf{R}) = \{ s \in \mathscr{H}^k;\ s(x) = -s(-x)\,,\quad x \in \mathbf{R} \}$$

for $k \in \mathbf{Z}^+$. We equip the space $\mathscr{H}^k$ with the norm

$$\|s\|_k = \left(\int_0^{2\pi} |s^{(k)}(x)|^2\,\mathrm{d}x \right)^{1/2}.$$

It is easy to verify that $\mathscr{H}^k$ with the norm $\|\cdot\|_k$ is a Hilbert space whose scalar product is given by

$$\langle s_1, s_2 \rangle_k = \int_0^{2\pi} s_1^{(k)}(x)\, s_2^{(k)}(x)\,\mathrm{d}x\,.$$

For the sake of brevity, we put $T = 2\pi + 1$ and

$$\mathscr{U}^T = C^2([0, T];\ H^0_{2\pi(o)}(\mathbf{R})) \cap C^1([0, T];\ H^1_{2\pi(o)}(\mathbf{R})) \cap$$
$$\cap\ C^0([0, T];\ H^2_{2\pi(o)}(\mathbf{R}))\,,$$
$$\mathscr{U}^\infty_\omega = C^2_\omega(\mathbf{R};\ H^0_{2\pi(o)}(\mathbf{R})) \cap C^1_\omega(\mathbf{R};\ H^1_{2\pi(o)}(\mathbf{R})) \cap C^0_\omega(\mathbf{R};\ H^2_{2\pi(o)}(\mathbf{R}))\,,$$

where $\omega = \omega(\varepsilon) = 2\pi + \varepsilon\lambda$. The space $\mathscr{U}^T$ equipped with the norm

$$\|u\|_{\mathscr{U}^T} = \sum_{j=0}^{2} \|u\|_{C^j([0,T];H_{2\pi(o)}{}^{2-j}(\mathbf{R}))}$$

is a Banach space. Next, let F be an operator defined by (5.1.3) and finally, let $Z : \mathscr{H}^2 \to \mathscr{U}^T$ be an operator defined by

$$Z\,s(t, x) = s(x + t) - s(-x + t)\,,\quad t \in [0, T]\,,\quad x \in \mathbf{R}\,.$$

Let $\varepsilon_0 > 0$ be fixed. We assume that a function $f(t, x, u_0, u_1, u_2, \varepsilon)$ defined on $\mathbf{R}^5 \times [-\varepsilon_0, \varepsilon_0]$ satisfies the following conditions:

$$f(t, x, u_0, u_1, u_2, \varepsilon) = f(t, x + 2\pi, u_0, u_1, u_2, \varepsilon) =$$
$$= -f(t, -x, -u_0, -u_1, u_2, \varepsilon) \quad \text{on} \quad \mathbf{R}^5 \times [-\varepsilon_0, \varepsilon_0]\,, \tag{5.2.1}$$

$$f(t + 2\pi + \varepsilon\lambda, x, u_0, u_1, u_2, \varepsilon) =$$
$$= f(t, x, u_0, u_1, u_2, \varepsilon) \quad \text{on} \quad \mathbf{R}^5 \times [-\varepsilon_0, \varepsilon_0]\,, \tag{5.2.2}$$

$$D f(t, x, u_0, u_1, u_2, \varepsilon) \text{ is continuous on } \mathbf{R}^5 \text{ for every } \varepsilon \in [-\varepsilon_0, \varepsilon_0], \tag{5.2.3}$$

$$\limsup_{\varepsilon \to 0} \{|D f(t, x, u_0, u_1, u_2, \varepsilon) - D f(t, x, u_0, u_1, u_2, 0)|;$$

$$t \in [0, T],\ x \in \mathbf{R},\ |u_j| \leqq \varrho,\ j = 0, 1, 2\} = 0 \quad \text{for every} \quad \varrho > 0, \tag{5.2.4}$$

$$\limsup_{\varrho \to 0} \{|D f(t, x, u_0, u_1, u_2, \varepsilon) - D f(t, x, \bar{u}_0, \bar{u}_1, \bar{u}_2, \varepsilon)|;$$

$$t \in [0, T],\ x \in \mathbf{R},\ |u_j - \bar{u}_j| \leqq \varrho,\ 0 < |\varepsilon| < \varepsilon_0\} = 0$$

$$\text{for every } \bar{u}_0, \bar{u}_1, \bar{u}_2 \in \mathbf{R}, \tag{5.2.5}$$

where the differential operator D has the form

$$D = D_x^\alpha D_{u_0}^{\beta_0} D_{u_1}^{\beta_1} D_{u_2}^{\beta_2} \quad \text{and} \quad \alpha + \beta_0 + \beta_1 + \beta_2 \leqq 2, \quad \alpha \leqq 1.$$

Now let $(u, s) \in \mathscr{U}^T \times \mathscr{H}^2$ be a pair of functions satisfying

$$G_1(u, s, \varepsilon) = 0, \quad G_2(u, s, \varepsilon) = 0$$

for some ε, $0 < |\varepsilon| < \min(\varepsilon_0, \lambda^{-1})$, where

$$G_1(u, s, \varepsilon)(t, x) = -u(t, x) + Z s(t, x) +$$

$$+ \frac{\varepsilon}{2} \int_0^t \int_{x-t+\tau}^{x+t-\tau} F(u, \varepsilon)(\tau, \xi)\, d\xi\, d\tau, \quad t \in [0, T], \quad x \in \mathbf{R},$$

$$G_2(u, s, \varepsilon)(x) = \varepsilon^{-1}(s'(x) - s'(x - \varepsilon\lambda)) +$$

$$+ \frac{1}{2} \int_0^{2\pi + \varepsilon\lambda} F(u, \varepsilon)(\tau, x - \tau)\, d\tau, \quad x \in \mathbf{R}. \tag{5.2.6}$$

Lemma 1.1.1 and Remark 1.1.4 yield immediately that the u can be extended to $\mathbf{R} \times \mathbf{R}$ so that the resulting function is $(2\pi + \varepsilon\lambda)$-periodic in t, belongs to $\mathscr{U}^\infty_{\omega(\varepsilon)}$, and satisfies (5.1.1)–(5.1.2). However, the extended function is not a classical solution, but a solution almost everywhere.

THEOREM 5.2.1 *Let $\lambda > 0$. Suppose that a function f satisfies (5.2.1)–(5.2.5) and that the following two assumptions are satisfied:*

(i) There is an $s_0 \in \mathscr{H}^3$ such that $M s_0 = 0$, where

$$M s(x) = \lambda s''(x) + \frac{1}{2} \int_0^{2\pi} F(Zs, 0)(\tau, x - \tau)\, d\tau, \quad x \in \mathbf{R}. \tag{5.2.7}$$

(ii) The operator $V^\varepsilon \in \mathscr{L}(\mathscr{H}^2, \mathscr{H}^1)$ *given by*

$$V^\varepsilon \sigma(x) = |\varepsilon|^{-1} (\sigma'(x) - \sigma'(x - |\varepsilon| \lambda)) +$$

$$+ \frac{1}{2} \int_0^{2\pi} F'_u(Zs_0, 0)(Z\sigma)(\tau, x - \tau) \, d\tau, \quad x \in \mathbf{R}, \tag{5.2.8}$$

has an inverse $(V^\varepsilon)^{-1} \in \mathscr{L}(\mathscr{H}^1, \mathscr{H}^2)$ *such that*

$$\|(V^\varepsilon)^{-1}\|_{\mathscr{L}(\mathscr{H}^1, \mathscr{H}^2)} \leqq m \text{ for every } \quad 0 < |\varepsilon| \leqq \varepsilon_0, \tag{5.2.9}$$

with m *independent of* ε.
Then there is an $\varepsilon_1 \in (0, \varepsilon_0]$ *and an* $r > 0$ *with the following property: For any* ε, $0 < |\varepsilon| \leqq \varepsilon_1$, *there is a unique* $u(\varepsilon) \in \mathscr{U}^\infty_{\omega(\varepsilon)}$ *satisfying* (5.1.1)–(5.1.2) *and* $\|u(\varepsilon) - Zs_0\|_{\mathscr{U}^T} \leqq r$. *Moreover,*

$$\lim_{\varepsilon \to 0} \|u(\varepsilon) - Zs_0\|_{\mathscr{U}^T} = 0.$$

The following two simple lemmas are used in the proof of the theorem.

LEMMA 5.2.1 *Let* X *and* Y *be Banach spaces and let* $A \in \mathscr{L}(X, Y)$. *Suppose that* $A^{-1} \in \mathscr{L}(Y, X)$ *exists.*
Then $(A + B)^{-1} \in \mathscr{L}(Y, X)$ *exists for every* $B \in \mathscr{L}(X, Y)$ *satisfying*

$$\|B\|_{\mathscr{L}(X,Y)} \leqq (2\|A^{-1}\|_{\mathscr{L}(Y,X)})^{-1}.$$

Also,

$$\|(A + B)^{-1}\|_{\mathscr{L}(Y,X)} \leqq 2\|A^{-1}\|_{\mathscr{L}(Y,X)}.$$

The *proof* is obvious.

LEMMA 5.2.2 *Let* X *and* Y *be Banach spaces,* $\bar{m}$ *and* $\bar{\varepsilon}$ *positive numbers, and* $x_0 \in X$. *Let* $G_\varepsilon : X \to Y$, $\varepsilon \in (0, \bar{\varepsilon}]$, *be a mapping satisfying the following assumptions:*

(i) For every $\varepsilon \in (0, \bar{\varepsilon}]$ *the mapping* $G_\varepsilon : X \to Y$ *is continuous, and its F-derivative* $G'_\varepsilon : X \to \mathscr{L}(X, Y)$ *exists.*

(ii) $\lim_{\varrho \to 0} \sup\{\|G'_\varepsilon(x) - G'_\varepsilon(x_0)\|_{\mathscr{L}(X,Y)}; \varepsilon \in (0, \bar{\varepsilon}], x \in B(x_0, \varrho; X)\} < (\bar{m})^{-1}$,

$$\lim_{\varepsilon \to 0+} \|G_\varepsilon(x_0)\|_Y = 0.$$

(iii) $[G'_\varepsilon(x_0)]^{-1} \in \mathscr{L}(Y, X)$ *and* $\|[G'_\varepsilon(x_0)]^{-1}\|_{\mathscr{L}(Y,X)} \leqq \bar{m}$ *for every* $\varepsilon \in (0, \bar{\varepsilon}]$.
Then there are two numbers $\varepsilon_1 \in (0, \bar{\varepsilon}]$ *and* $r > 0$ *with the following property: For every* $\varepsilon \in (0, \varepsilon_1]$ *there is a unique* $x_\varepsilon \in X$ *satisfying* $G_\varepsilon(x_\varepsilon) = 0$ *and* $\|x_\varepsilon - x_0\|_X \leqq r$. *Moreover,*

$$\lim_{\varepsilon \to 0+} \|x_\varepsilon - x_0\|_X = 0.$$

This lemma is a slightly modified version of the implicit function theorem. Its *proof* is easy.

Proof of Theorem 5.2.1. We put $X = \mathscr{U}^T \times \mathscr{H}^2$, $Y = \mathscr{U}^T \times \mathscr{H}^1$ and $G(u, s, \varepsilon) = (G_1(u, s, \varepsilon), G_2(u, s, \varepsilon))$, where G_1 and G_2 are defined by (5.2.6). Assuming that $\varepsilon \in (0, \varepsilon_0]$, we shall prove that the mapping $G(\cdot, \cdot, \varepsilon)$ satisfies the assumptions of Lemma 5.2.2. Straightforward but lengthy calculations show that the mapping $G(\cdot, \cdot, \varepsilon) : X \to Y$ is continuous, its F-derivative $G'(\cdot, \cdot, \varepsilon) : X \to \mathscr{L}(X, Y)$ exists for every fixed $\varepsilon \in (0, \varepsilon_0]$, and

$$\lim_{\varrho \to 0} \sup \{\|G'(u, s, \varepsilon) - G'(u_0, s_0, \varepsilon)\|_{\mathscr{L}(X,Y)};$$

$$\varepsilon \in (0, \varepsilon_0],\ \|(u, s) - (u_0, s_0)\|_X \leqq \varrho\} = 0\,, \tag{5.2.10}$$

with

$$u_0 = Z s_0\,.$$

Assumption (i) yields

$$\lim_{\varepsilon \to 0} \|G(u_0, s_0, \varepsilon)\|_Y = 0\,. \tag{5.2.11}$$

The F-derivative G' of G with respect to (u, s) is

$$G'(u, s, \varepsilon) = (G_1'(u, s, \varepsilon)\,,\ G_2'(u, s, \varepsilon))\,,$$

where

$$G_1'(u, s, \varepsilon)\,(v, \sigma)\,(t, x) = -v(t, x) + (Z\sigma)\,(t, x) +$$

$$+ \frac{\varepsilon}{2}\int_0^t \int_{x-t+\tau}^{x+t-\tau} F_u'(u, \varepsilon)\, v(\tau, \xi)\, \mathrm{d}\xi\, \mathrm{d}\tau\,, \quad t \in [0, T]\,, \quad x \in \mathbf{R}\,,$$

$$G_2'(u, s, \varepsilon)\,(v, \sigma)\,(x) = \varepsilon^{-1}(\sigma'(x) - \sigma'(x - \varepsilon\lambda)) +$$

$$+ \frac{1}{2}\int_0^{2\pi+\varepsilon\lambda} F_u'(u, \varepsilon)\, v(\vartheta, x - \vartheta)\, \mathrm{d}\vartheta\,, \quad x \in \mathbf{R}$$

for $(v, \sigma) \in X$. It is easy to verify that $G'(u, s, \varepsilon) \in \mathscr{L}(X, Y)$ for every $(u, s) \in X$ and $\varepsilon \in (0, \varepsilon_0]$. Further, introducing two auxiliary operators by

$$H_1(v, \sigma)\,(t, x) = -v(t, x) + Z\,\sigma(t, x)\,, \quad t \in [0, T]\,, \quad x \in \mathbf{R}\,,$$

$$H_2(\varepsilon)\,(v, \sigma)\,(x) = \varepsilon^{-1}(\sigma'(x) - \sigma'(x - \varepsilon\lambda)) +$$

$$+ \frac{1}{2}\int_0^{2\pi} F_u'(u_0, 0)\, v(\tau, x - \tau)\, \mathrm{d}\tau\,, \quad x \in \mathbf{R}\,,$$

and putting $H(\varepsilon)(v, \sigma) = (H_1(v, \sigma), H_2(\varepsilon)(v, \sigma))$, we have $H(\varepsilon) \in \mathscr{L}(X, Y)$ for $\varepsilon \in (0, \varepsilon_0]$ and

$$\lim_{\varepsilon \to 0} \| G'(u_0, s_0, \varepsilon) - H(\varepsilon) \|_{\mathscr{L}(X,Y)} = 0 . \tag{5.2.12}$$

Let $(w, r) \in Y$. Setting

$$\sigma(x) = (V^\varepsilon)^{-1} \left(r(x) + \frac{1}{2} \int_0^{2\pi} F'(u_0, 0)\, w(\tau, x - \tau)\, d\tau \right), \quad x \in \mathbf{R},$$

$$v = Z\sigma - w ,$$

we obtain $H(\varepsilon)(v, \sigma) = (w, r)$. Hence $[H(\varepsilon)]^{-1}$ exists and assumption (ii) implies that there is an m_1 such that

$$\| [H(\varepsilon)]^{-1} \|_{\mathscr{L}(Y,X)} \leqq m_1 \quad \text{for every} \quad \varepsilon \in (0, \varepsilon_0] . \tag{5.2.13}$$

By Lemma 5.2.1 there are numbers $\overline{m}$ and $\bar{\varepsilon} \in (0, \varepsilon_0]$ such that for every $\varepsilon \in (0, \bar{\varepsilon}]$

$$\| [G'(u_0, s_0, \varepsilon)]^{-1} \|_{\mathscr{L}(Y,X)} \leqq \overline{m} . \tag{5.2.14}$$

Applying Lemma 5.2.2 to the mapping $G(\cdot, \cdot, \varepsilon) : X \to Y$, we obtain the assertion of Theorem 5.2.1 for $\varepsilon > 0$. For $\varepsilon < 0$ we replace $G_2(u, s, \varepsilon) = 0$ by $G_3(u, s, \varepsilon) = 0$, where

$$G_3(u, s, \varepsilon)(x) = G_2(u, s, \varepsilon)(x + \varepsilon\lambda) .$$

Applying the same procedure as above to the pair of operators $(G_1(u, s, \varepsilon), G_3(u, s, \varepsilon))$, we obtain the assertion of Theorem 5.2.1 for $\varepsilon < 0$. This completes the proof.

5.3. *Equations with monotone right-hand sides*

Let $\varepsilon_0 > 0$. We formulate the following assumptions on functions g and h:

$$g = g(u_0, u_1, u_2) \in C^2_{\mathrm{loc}}(\mathbf{R}^3) , \tag{5.3.1}$$

$$g(u_0, u_1, u_2) = -g(-u_0, -u_1, u_2) ,$$

$$g_{u_1}(u_0, u_1, u_2) \geqq \gamma_1 , \quad |g_{u_0}(u_0, u_1, u_2)| \leqq \gamma_0 ,$$

$$|g_{u_2}(u_0, u_1, u_2)| \leqq \gamma_2 \quad \text{for} \quad (u_0, u_1, u_2) \in \mathbf{R}^3 , \tag{5.3.2}$$

$$\gamma_1 - \gamma_2 - 2\gamma_0 > 0 , \tag{5.3.3}$$

$$h(t, x, \varepsilon) = h(t, x + 2\pi, \varepsilon) = -h(t, -x, \varepsilon) ,$$

$$h(t + 2\pi + \varepsilon\lambda, x, \varepsilon) = h(t, x, \varepsilon)$$

$$\text{for} \quad (t, x) \in \mathbf{R}^2 \quad \text{and} \quad \varepsilon \in [-\varepsilon_0, \varepsilon_0], \tag{5.3.4}$$

$$h(\cdot, \cdot, \varepsilon) \in C^{0,1}(\mathbf{R}^2) \quad \text{for every} \quad \varepsilon \in [-\varepsilon_0, \varepsilon_0], \tag{5.3.5}$$

$$\lim_{\varepsilon \to 0} \sup \{|h_x(t, x, \varepsilon) - h_x(t, x, 0)|;\ t \in [0, T], x \in \mathbf{R}\} = 0. \tag{5.3.6}$$

For example, if $g(u_0, u_1, u_2) = u_1 + P(u_1) + \Phi(u_0)\,\Psi(u_2)$, then g fulfils (5.3.1)–(5.3.3), provided that P is an odd polynomial with non-negative coefficients, Φ, $\Psi \in C^2_{\text{loc}}(\mathbf{R})$, Φ is odd, and $\|\Phi\|_{C^1(\mathbf{R})} \leqq 1/\sqrt{c}$, $\|\Psi\|_{C^1(\mathbf{R})} \leqq 1/\sqrt{c}$ with $c > 3$.

In this section we investigate the existence of a $(2\pi + \varepsilon\lambda)$-periodic solution of the problem

$$u_{tt}(t, x) - u_{xx}(t, x) = \varepsilon(g(u, u_t, u_x) + h(t, x, \varepsilon)), \quad (t, x) \in \mathbf{R}^2, \tag{5.3.7}$$

$$u(t, x) = u(t, x + 2\pi) = -u(t, -x), \quad (t, x) \in \mathbf{R}^2. \tag{5.3.8}$$

THEOREM 5.3.1 (*see* ŠTĚDRÝ [70]). *Let g and h be two functions satisfying* (5.3.1)–(5.3.6).
Then there is an $\varepsilon_1 > 0$, a $\varrho > 0$ and an $s_0 \in \mathscr{H}^3$ such that for every ε, $0 < |\varepsilon| \leqq \varepsilon_1$, there is a unique $u(\varepsilon) \in \mathscr{U}^\infty_{2\pi+\varepsilon\lambda}$ satisfying (5.3.7)–(5.3.8) *and $\|u(\varepsilon) - Zs_0\|_{\mathscr{U}^T} \leqq \varrho$. Moreover,*

$$\lim_{\varepsilon \to 0} \|u(\varepsilon) - Zs_0\|_{\mathscr{U}^T} = 0.$$

Proof. This theorem follows readily from Theorem 5.2.1 if $f = g + h$ and

(i) there is a function $s_0 \in \mathscr{H}^3$ satisfying $Ms_0 = 0$ (M defined by (5.2.7));

(ii) there is an operator $(V^\varepsilon)^{-1} \in \mathscr{L}(\mathscr{H}^1, \mathscr{H}^2)$ satisfying (5.2.9) (V^ε defined by (5.2.8)).

To simplify the subsequent formulae, we define an operator J by

$$J\,s(x) = \int_0^x s(\xi)\,d\xi + (2\pi)^{-1} \int_0^{2\pi} \xi\, s(\xi)\,d\xi. \tag{5.3.9}$$

Obviously, $J \in \mathscr{L}(\mathscr{H}^l, \mathscr{H}^{l+1})$ for every $l \in \mathbf{Z}^+$. We now show that (i) holds. Suppose that

$$\sigma(x) + (2\lambda)^{-1} J\left(\int_0^{2\pi} F(ZJ\sigma, 0)(\tau, x - \tau)\,d\tau\right) = 0, \quad x \in \mathbf{R}, \tag{5.3.10}$$

for some $\sigma = \sigma_0 \in \mathscr{H}^1$. Then we verify readily that the function $s_0 = J\sigma_0 \in \mathscr{H}^2$ satisfies $Ms_0 = 0$. The assumptions (5.3.1) and (5.3.5) used in (5.3.10) show that $s_0 \in \mathscr{H}^3$. Hence it suffices to look for a function $\sigma \in \mathscr{H}^1$ for which (5.3.10) holds.

Let us write

$$\Phi(\sigma)(x) = \int_0^{2\pi} F(ZJ\sigma, 0)(\tau, x - \tau)\,\mathrm{d}\tau\,, \quad x \in \mathbf{R}\,,$$

$$J_1 = (2\lambda)^{-1} J\,.$$

Then (5.3.10) takes the form

$$\sigma + J_1\,\Phi(\sigma) = 0\,. \tag{5.3.11}$$

We investigate (5.3.11) in the space $\mathscr{H}^1$. The mapping $J_1 \in \mathscr{L}(\mathscr{H}^1, \mathscr{H}^1)$ is compact and

$$\langle J_1\sigma, \sigma\rangle_1 = (2\lambda)^{-1}\,\langle (J\sigma)', \sigma'\rangle_0 = (2\lambda)^{-1}\,\langle \sigma, \sigma'\rangle_0 = 0\,. \tag{5.3.12}$$

Under our assumptions on F, the mapping Φ has the form

$$\Phi(\sigma)(x) = \int_0^{2\pi} g\left(\int_{-x+2\tau}^{x} \sigma(\eta)\,\mathrm{d}\eta, \sigma(x) - \sigma(-x + 2\tau),\right.$$

$$\left.\sigma(x) + \sigma(-x + 2\tau)\right)\mathrm{d}\tau + \int_0^{2\pi} h(\tau, x - \tau, 0)\,\mathrm{d}\tau\,. \tag{5.3.13}$$

Setting

$$g_j(x, \xi) = g_{u_j}\left(\int_\xi^x \sigma(\eta)\,\mathrm{d}\eta, \sigma(x) - \sigma(\xi), \sigma(x) + \sigma(\xi)\right),$$

$$H(x) = \int_0^{2\pi} h(\tau, x - \tau, 0)\,\mathrm{d}\tau\,,$$

and taking into account that the integrand in (5.3.13) is 2π-periodic we can write

$$(\Phi(\sigma))'(x) = \int_0^{2\pi} (g_0(x, \xi)\,\sigma(x) + (g_1(x, \xi) + g_2(x, \xi))\,\sigma'(x))\,\mathrm{d}\xi +$$

$$+ H'(x)\,, \quad x \in \mathbf{R}\,.$$

Thus,

$$\langle \Phi(\sigma), \sigma\rangle_1 = \langle (\Phi(\sigma))', \sigma'\rangle_0 =$$

$$= \int_0^{2\pi}\int_0^{2\pi} ((g_1(x, \xi) + g_2(x, \xi))\,(\sigma'(x))^2 +$$

$$+ g_0(x, \xi)\,\sigma(x)\,\sigma'(x))\,\mathrm{d}\xi\,\mathrm{d}x + \int_0^{2\pi} H'(x)\,\sigma'(x)\,\mathrm{d}x \geqq$$

$$\geqq 2\pi(\gamma_1 - \gamma_2)\,\|\sigma'\|_0^2 - 2\pi\,\gamma_0\|\sigma\|_0\,\|\sigma'\|_0 - \|H'\|_0\,\|\sigma'\|_0\,.$$

As $\|\sigma\|_0 \leqq \|\sigma'\|_0$, we obtain $\langle \Phi\sigma, \sigma\rangle_1 > 0$ for all σ, such that $\|\sigma\|_1 > (2\pi(\gamma_1 - \gamma_2 - \gamma_0))^{-1} \|H'\|_0 \equiv R_1$. Using (5.3.12), we can show that the equation $\sigma + \mu J_1 \Phi(\sigma) = 0$ has no solution for $\mu \in [0, 1]$ and $\|\sigma\|_1 > R_1$. By I: Theorem 3.5.2, there is a solution σ_0, $\|\sigma_0\|_1 \leqq R_1$ of the equation (5.3.11). Together with what was said above, this proves that there is an $s_0 \in \mathscr{H}^3$ for which $Ms_0 = 0$.

Next we show that (*ii*) holds. Putting

$$\bar{g}_j(x, \xi) = g_{u_j}(s_0(x) - s_0(\xi), s_0'(x) - s_0'(\xi), s_0'(x) + s_0'(\xi)),$$

we can write

$$V^\varepsilon \sigma(x) = |\varepsilon|^{-1}(\sigma'(x) - \sigma'(x - |\varepsilon|\lambda)) +$$
$$+ \frac{1}{2}\int_0^{2\pi} (\bar{g}_1(x, \xi)(\sigma'(x) - \sigma'(\xi)) + \bar{g}_2(x, \xi)(\sigma'(x) + \sigma'(\xi)) +$$
$$+ \bar{g}_0(x, \xi)(\sigma(x) - \sigma(\xi)))\, d\xi .$$

Let $\theta \in C_{2\pi}^\infty(\mathbf{R}) \cap \mathscr{H}^0$. Then

$$\langle V^\varepsilon\theta, -\theta'''\rangle_0 = |\varepsilon|^{-1}\left(\|\theta''\|_0^2 - \int_0^{2\pi} \theta''(x)\,\theta''(x - |\varepsilon|\lambda)\, dx\right) +$$
$$+ \frac{1}{2}\int_0^{2\pi}\int_0^{2\pi}\Big((\bar{g}_1(x, \xi) + \bar{g}_2(x, \xi))(\theta''(x))^2 +$$
$$+ \bar{g}_0(x, \xi)\,\theta'(x)\,\theta''(x)\Big)\, d\xi\, dx +$$
$$+ \frac{1}{2}\int_0^{2\pi}\left(\int_0^{2\pi}(\bar{g}_{1x}(x, \xi)(\theta'(x) - \theta'(\xi)) + \bar{g}_{2x}(x, \xi)(\theta'(x) + \theta'(\xi)) +\right.$$
$$\left. + \bar{g}_{0x}(x, \xi)(\theta(x) - \theta(\xi)))\, d\xi\right)\theta''(x)\, dx .$$

As

$$\|\theta\|_0 \leqq \|\theta'\|_0 \leqq \|\theta''\|_0 \quad \text{and} \quad \int_0^{2\pi} \theta''(x)\,\theta''(x - |\varepsilon|\lambda)\, dx \leqq \|\theta''\|_0^2$$

we have

$$\langle V^\varepsilon\theta, -\theta'''\rangle_0 \geqq \pi(\gamma_1 - \gamma_2 - \gamma_0)\|\theta''\|_0^2 -$$
$$- c_1\|\theta'\|_0\|\theta''\|_0 \geqq 2^{-1}\pi(\gamma_1 - \gamma_2 - \gamma_0)\|\theta''\|_0^2 -$$
$$- c_1^2(2\pi(\gamma_1 - \gamma_2 - \gamma_0))^{-1}\|\theta'\|_0^2 . \tag{5.3.13}$$

On the other hand,

$$\langle V^\varepsilon\theta, \theta'\rangle_0 \geqq \frac{1}{2}\int_0^{2\pi}\int_0^{2\pi} \gamma_1(\theta'(x) - \theta'(\xi))\,\theta'(x)\,\mathrm{d}x\,\mathrm{d}\xi +$$

$$+ \frac{1}{2}\int_0^{2\pi}\int_0^{2\pi} (\bar{g}_1(x, \xi) - \gamma_1)\,(\theta'(x) - \theta'(\xi))\,\theta'(x)\,\mathrm{d}x\,\mathrm{d}\xi +$$

$$+ \frac{1}{2}\int_0^{2\pi}\int_0^{2\pi} \bar{g}_2(x, \xi)\,(\theta'(x) + \theta'(\xi))\,\theta'(x)\,\mathrm{d}x\,\mathrm{d}\xi +$$

$$+ \frac{1}{2}\int_0^{2\pi}\int_0^{2\pi} \bar{g}_0(x, \xi)\,(\theta(x) - \theta(\xi))\,\theta'(x)\,\mathrm{d}x\,\mathrm{d}\xi =$$

$$= I_1 + I_2 + I_3 + I_4\,.$$

Interchanging the variables x and ξ in I_2 and using the relations $\bar{g}_1(\xi, x) = \bar{g}_1(x, \xi)$, $\bar{g}_1(x, \xi) - \gamma_1 \geqq 0$, we can write

$$2I_2 = 2^{-1}\int_0^{2\pi}\int_0^{2\pi} (\bar{g}_1(x, \xi) - \gamma_1)\,(\theta'(x) - \theta'(\xi))^2\,\mathrm{d}x\,\mathrm{d}\xi \geqq 0\,,$$

that is $I_2 \geqq 0$. A standard estimation of I_3 and I_4 yields

$$\langle V^\varepsilon\theta, \theta'\rangle_0 \geqq \pi\gamma_3\|\theta'\|_0^2\,,$$

where $\gamma_3 = \gamma_1 - \gamma_2 - 2\gamma_0 > 0$. Hence the operator Λ defined by

$$\Lambda\theta = \theta'' - c_2\theta\,,$$

$$c_2 = c_1^2(2\pi^2\,\gamma_3(\gamma_1 - \gamma_0 - \gamma_2))^{-1}\,,$$

satisfies

$$\langle (V^\varepsilon\theta)', \Lambda\theta\rangle_0 \geqq \gamma_4\|\theta''\|_0^2 \quad \text{for} \quad \theta \in \mathscr{H}^2\,, \tag{5.3.14}$$

where $\gamma_4 = 2^{-1}\pi(\gamma_1 - \gamma_2 - \gamma_0)$.

This inequality implies

$$\|V^\varepsilon\theta\|_1 \geqq \gamma_4(1 + c_2)^{-1}\,\|\theta\|_2\,, \quad \theta \in \mathscr{H}^2\,.$$

Hence $\mathscr{R}(V^\varepsilon)$ is a closed subspace of $\mathscr{H}^1$. If $\mathscr{R}(V^\varepsilon)$ should not be equal to $\mathscr{H}^1$, then we would find a $h_0 \in \mathscr{H}^1$, $h_0 \neq 0$ such that

$$\langle (V^\varepsilon\theta)', h_0'\rangle_0 = 0 \quad \text{for all} \quad \theta \in \mathscr{H}^2\,. \tag{5.3.15}$$

Using the Fourier series, we would find a $\theta_0 \in \mathscr{H}^2$, $\theta_0 \neq 0$, for which $\Lambda\theta_0 = h_0'$. Using (5.3.14) and (5.3.15), we would obtain

$$0 = \langle (V^\varepsilon \theta_0)', h_0' \rangle_0 = \langle (V^\varepsilon \theta_0)', \Lambda\theta_0 \rangle_0 \geqq \gamma_4 \|\theta_0\|_2^2 .$$

This would be a contradiction since $\theta_0 \neq 0$. Thus, (*ii*) is shown and this completes the proof.

§ 6. The Fourier method

6.1. *Introductory remarks*

Let us consider again the problem $(\mathscr{P}_\omega^0)$ given by

$$u_{tt}(t, x) - u_{xx}(t, x) = g(t, x), \quad (t, x) \in Q = \mathbf{R} \times (0, \pi), \tag{6.1.1}$$

$$u(t, 0) = u(t, \pi) = 0, \quad t \in \mathbf{R}, \tag{6.1.2}$$

$$u(t + \omega, x) = u(t, x), \quad (t, x) \in Q . \tag{6.1.3}$$

By a solution of the problem $(\mathscr{P}_\omega^0)$ we mean a function $u \in \mathscr{U} = H_\omega^2(Q) \cap \dot{H}_\omega^1(Q)$ that satisfies (6.1.1) almost everywhere.

By II: Remark 1.3.3 and II: Theorem 1.3.4, a function $u \in H_\omega^0(Q)$ belongs to $\mathscr{U}$ if and only if

$$\sum_{j,k} (j^4 + k^4) |u_{j,k}|^2 < +\infty , \tag{6.1.4}$$

where

$$u_{j,k} = \left(\frac{2}{\pi\omega}\right)^{1/2} \int_0^\pi \int_0^\omega u(t, x)\, e^{-ij\nu t} \sin kx \, dt\, dx ,$$

$\nu = 2\pi/\omega$, and $\sum_{j,k} = \sum_{j=-\infty}^{+\infty} \sum_{k=1}^{\infty}$. Hence, u can be written in the form

$$u(t, x) = (2/\pi\omega)^{1/2} \sum_{j,k} u_{j,k}\, e^{ij\nu t} \sin kx . \tag{6.1.5}$$

The square root of the series in (6.1.4) defines a norm $\|\cdot\|_{\mathscr{U}}$ in $\mathscr{U}$, which is equivalent to $\|\cdot\|_{H_\omega{}^2(Q)}$.

Suppose that $g \in \mathscr{G} = H_\omega^\beta(\mathbf{R}; {}^0H^{r-\beta}(0, \pi))$ with fixed $r \in \mathbf{R}^+$ and $\beta \in [0, r]$. By II: Remark 1.3.3 and II: Theorem 1.3.4, a function $g \in H_\omega^0(Q)$ belongs to $\mathscr{G}$ if and only if

$$\sum_{j,k} (1 + |j|)^{2\beta} k^{2(r-\beta)} |g_{j,k}|^2 < +\infty , \tag{6.1.6}$$

where

$$g_{j,k} = \left(\frac{2}{\pi\omega}\right)^{1/2} \int_0^\pi \int_0^\omega g(t, x)\, e^{-ij\nu t} \sin kx \, dt \, dx \, .$$

Hence, g can be written in the form

$$g(t, x) = (2/\pi\omega)^{1/2} \sum_{j,k} g_{j,k}\, e^{ij\nu t} \sin kx \, . \tag{6.1.7}$$

The square root of the series in (6.1.6) defines an equivalent norm $\|\cdot\|_{\mathscr{G}}$ in $H^\beta_\omega(\mathbf{R}; {}^0H^{r-\beta}(0, \pi))$. Substituting (6.1.5) and (6.1.7) in (6.1.1) we obtain

$$(-j^2\nu^2 + k^2)\, u_{j,k} = g_{j,k}\, , \quad j \in \mathbf{Z}\, , \quad k \in \mathbf{N}\, . \tag{6.1.8}$$

Now it is convenient to distinguish two cases according to the number-theoretical character of $\omega : \omega = 2\pi p/q$, $p, q \in \mathbf{N}$, and $\omega = 2\pi\alpha$, α irrational.

6.2. *The case $\omega = 2\pi p/q$*

If $\omega = 2\pi p/q$ with $p, q \in \mathbf{N}$ relatively prime, that is, $\nu = q/p$, then the coefficient of $u_{j,k}$ in (6.1.8) is zero if and only if $|j| = lp$, $k = lq$ for some $l \in \mathbf{N}$. Hence a solution u of $(\mathscr{P}^0_\omega)$ exists only if

$$g_{lp,lq} = 0 \text{ and } g_{-lp,lq} = 0 \quad \text{for} \quad \text{every } l \in \mathbf{N}\, . \tag{6.2.1}$$

(Of course, $g_{lp,lq} = 0$ implies that $g_{-lp,lq} = 0$ and vice versa.) We put

$$S_1 = \{(j, k);\, |j| = lp\, , \quad k = lq\, , \quad l \in \mathbf{N}\}\, ,$$

which is usually called the set of singular pairs,

$$S_2 = (\mathbf{Z} \times \mathbf{N}) \setminus S_1\, ,$$

$$\Sigma_1 = \{(j, k);\, 2\nu|j| < k\}\, , \quad \Sigma_2 = \{(j, k);\, k \leqq 2\nu|j| \leqq 4k\}$$

and

$$\Sigma_3 = \{(j, k);\, 2k < \nu|j|\}\, .$$

For the sake of simplicity we introduce two closed subspaces of $H^0_\omega(Q)$ as follows. The subspaces $\mathscr{V}$ and $\mathscr{W}$ of $H^0_\omega(Q)$ consist of all functions v and w that can be written in the form

$$v(t, x) = \sum_{(j,k)\in S_1} v_{j,k}\, e^{ij\nu t} \sin kx\, ,$$

and

$$w(t, x) = \sum_{(j,k)\in S_2} w_{j,k}\, e^{ij\nu t} \sin kx\, ,$$

respectively. Evidently, $H^0_\omega(Q)$ is the direct sum of $\mathscr{V}$ and $\mathscr{W}$. Finally, the operator L is defined by $\mathscr{D}(L) = \mathscr{U}$ and $Lu = u_{tt} - u_{xx}$. Now when we take $g \in H^\beta_\omega(\mathbf{R}; {}^0H^{1-\beta}(0, \pi)) \cap \mathscr{W}$, $\beta \in [0, 1]$, we shall show that the function w given by

$$w(t, x) = (\Lambda g)(t, x) = (2/\pi\omega)^{1/2} \sum_{(j,k)\in S_2} (-j^2 v^2 + k^2)^{-1} g_{j,k}\, e^{ijvt} \sin kx \tag{6.2.2}$$

belongs to $\mathscr{U}$ and that $Lw = g$. This follows from the inequality

$$\|w\|^2_{\mathscr{U}} = \sum_{(j,k)\in S_2} (j^4 + k^4)(-v^2 j^2 + k^2)^{-2} |g_{j,k}|^2 \leqq c\|g\|^2_{\mathscr{G}}, \tag{6.2.3}$$

where c is independent of g. This inequality is evident since the numbers given by

$$m_{j,k} = (j^4 + k^4)(-v^2 j^2 + k^2)^{-2} (1 + |j|)^{-2\beta} k^{-2(1-\beta)}$$

satisfy

$$\sup \{m_{j,k}; (j, k) \in S_2\} < +\infty . \tag{6.2.4}$$

Indeed, $|k - v|j|| \geqq k/2$ for $(j, k) \in S_2 \cap \Sigma_1$, therefore, $m_{j,k} \leqq 4(1 + (2v)^{-4})$. Next, $|k - v|j|| = p^{-1}|kp - |j|\, q| \geqq p^{-1}$ for all $(j, k) \in S_2$, and this shows immediately that $m_{j,k} \leqq p^2(2v)^{2\beta}((2/v)^4 + 1)$ for $(j, k) \in S_2 \cap \Sigma_2$. Finally, for $(j, k) \in S_2 \cap \Sigma_3$ we have $v|j| - k| \geqq v|j|/2$, which yields $m_{j,k} \leqq 4v^{-4}(1 + (v/2)^4)$. Thus, (6.2.3) is satisfied and we can sum up the result in the following theorem.

THEOREM 6.2.1 *Given $\omega = 2\pi p/q$, p, $q \in \mathbf{N}$ relatively prime, and $g \in H^\beta_\omega(\mathbf{R}; {}^0H^{1-\beta}(0, \pi))$, $\beta \in [0, 1]$, the problem $(\mathscr{P}^0_\omega)$ has a solution if and only if*

$$g_{j,k} = 0 \quad \textit{for all} \quad (j, k) \in S_1 .$$

If these conditions are satisfied, then every solution u of $(\mathscr{P}^0_\omega)$ can be written in the form $u = v + w$, where v is any function from $\mathscr{U} \cap \mathscr{V}$, $w = \Lambda g$ is determined by (6.2.2), *and*

$$\Lambda \in \mathscr{L}(H^\beta_\omega(\mathbf{R}; {}^0H^{1-\beta}(0, \pi)) \cap \mathscr{W}, \mathscr{U} \cap \mathscr{W}) .$$

Remark 6.2.1 Let us define an operator $P : H^0_\omega(Q) \to \mathscr{V}$ by

$$(Ph)(t, x) = (2/\pi\omega)^{1/2} \sum_{(j,k)\in S_1} h_{j,k}\, e^{ijvt} \sin kx .$$

Then Theorem 6.2.1 shows that for every $g \in H^\beta_\omega(\mathbf{R}; {}^0H^{1-\beta}(0, \pi))$, $\beta \in [0, 1]$, satisfying $Pg = 0$ there exists a unique solution $w \in \mathscr{U} \cap \mathscr{W}$ of the problem $(\mathscr{P}^0_\omega)$.

Remark 6.2.2 Observe that the condition $g \in H^\beta_\omega(\mathbf{R}; {}^0H^{1-\beta}(0, \pi))$ cannot be weakened to $g \in H^\beta_\omega(\mathbf{R}; {}^0H^\gamma(0, \pi))$ with $\beta + \gamma < 1$. For $\omega = 2\pi$ it suffices to choose

$$g(t, x) = \sum_{\substack{j \in \mathbf{Z} \\ j \neq 0}} j^{-3/2} e^{ijt} \sin(|j| + 1)\, x$$

and procede similarly as in IV: Remark 1.1.3.

Remark 6.2.3 Poliščuk in [49] and [50], Ptašnik in [53] and [54] and in the joint works [51] and [52] these authors investigate the existence of periodic solutions of linear hyperbolic equations of higher order. In their approach number-theoretical methods and Fourier series are extensively used.

6.3. *The case $\omega = 2\pi\alpha$*

If $\omega = 2\pi\alpha$, α irrational, that is, $v = \alpha^{-1}$, then the coefficient of $u_{j,k}$ in (6.1.8) is always non-zero, but it may be very close to zero, depending on the number-theoretical character of α. Hence, let us suppose that

$$|\alpha - p/q| \geqq cq^{-r} \quad \text{for all} \quad p, q \in \mathbf{N}\,, \tag{6.3.1}$$

with a constant $c > 0$ for a fixed $r \geqq 2$. Given a function $g \in \mathcal{G} = H^\beta_\omega(\mathbf{R}; {}^0H^{r-\beta}(0, \pi))$ with $\beta \in [0, r]$, we can write a solution u of the problem $(\mathcal{P}^0_\omega)$ in the form

$$u(t, x) = (\Lambda g)(t, x) =$$
$$= \left(\frac{2}{\pi\omega}\right)^{1/2} \sum_{j,k} (-j^2\alpha^{-2} + k^2)^{-1}\, g_{j,k}\, e^{ijvt} \sin kx\,. \tag{6.3.2}$$

This function belongs to $\mathcal{U}$ because

$$\sum_{j,k} (j^4 + k^4)(-j^2\alpha^{-2} + k^2)^{-2}\, |g_{j,k}|^2 \leqq c\|g\|^2_{\mathcal{G}} \tag{6.3.3}$$

with c independent of g. This inequality is a direct consequence of the relation

$$\sup\{(j^4 + k^4)(-j^2\alpha^{-2} + k^2)^{-2}(1 + |j|)^{-2\beta}\, k^{-2(r-\beta)};$$
$$(j, k) \in \mathbf{Z} \times \mathbf{N}\} < +\infty\,,$$

which can be proved like the estimate (6.2.4). The decisive role is played again by the estimate on Σ_2:

$$(j^4 + k^4)(|j|\,\alpha^{-1} + k)^{-2}\, \alpha^2 k^{-2} |\alpha - |j|\, k^{-1}|^{-2} (1 + |j|)^{-2\beta} k^{-2(r-\beta)} \leqq$$
$$\leqq (1 + 16\alpha^4)\, \alpha^2\, c^{-2} (\alpha/2)^{-2\beta} < +\infty\,.$$

This leads to the next result.

THEOREM 6.3.1 *Given $\omega = 2\pi\alpha$, α irrational satisfying (6.3.1) and $g \in H^{\beta}_{\omega}(\mathbf{R}; {}^0H^{r-\beta}(0, \pi))$, $\beta \in [0, r]$; then the problem $(\mathscr{P}^0_{\omega})$ has a unique solution $u = \Lambda g \in \mathscr{U}$, where $\Lambda \in \mathscr{L}(H^{\beta}_{\omega}(\mathbf{R}; {}^0H^{r-\beta}(0, \pi)), \mathscr{U})$.*

Remark 6.3.1 It has already been pointed out by BOREL in [5] that there exist irrational numbers α such that the problem $(\mathscr{P}^0_{2\pi\alpha})$ need not have a solution in the class of analytic functions, if g is analytic. In [46] NOVÁK clarified the situation by proving the following proposition:

Let $\delta = \delta(\alpha)$ be the supremum of all γ for which the inequality $|q\alpha^{-1} - n| < q^{-\gamma}$ has infinitely many solutions in positive integers q and n. Let m be a positive integer. Then

(i) for every $g \in H^{m,0}_{2\pi\alpha}(Q)$ there is a generalized solution of the problem $(\mathscr{P}^0_{2\pi\alpha})$, provided that $\delta < m + 1$;

(ii) there exists a function $g \in H^{m,0}_{2\pi\alpha}(Q)$ such that the problem $(\mathscr{P}^0_{2\pi\alpha})$ does not have any generalized solution for $\delta > m + 1$. (For $\delta(\alpha) = m + 1$ solutions may or may not exist).

6.4. *The weakly non-linear case*

Let us deal with the problem $(\mathscr{P}^{\varepsilon}_{\omega})$ described by

$$u_{tt}(t, x) - u_{xx}(t, x) = \varepsilon\, F(u, \varepsilon)\,(t, x)\,, \quad (t, x) \in Q\,, \tag{6.4.1}$$

$$u(t, 0) = u(t, \pi) = 0\,, \quad t \in \mathbf{R}\,, \tag{6.4.2}$$

$$u(t + \omega, x) = u(t, x)\,, \quad (t, x) \in Q\,, \tag{6.4.3}$$

where $F(u, \varepsilon)$ maps $\mathscr{U} \times [-\varepsilon_0, \varepsilon_0]$ into $H^0_{\omega}(Q)$, $\varepsilon_0 > 0$. A function $u \in C([-\varepsilon^*, \varepsilon^*]; \mathscr{U})$, $\varepsilon_0 \geqq \varepsilon^* > 0$ is called a solution of the problem $(\mathscr{P}^{\varepsilon}_{\omega})$ if $u(\varepsilon)$ satisfies (6.4.1) for every $\varepsilon \in [-\varepsilon^*, \varepsilon^*]$.

In this section we prove two theorems concerning the cases $\omega = 2\pi p/q$ and $\omega = 2\pi\alpha$.

THEOREM 6.4.1 *Let $\omega = 2\pi p/q$, $p, q \in \mathbf{N}$, p, q relatively prime, and suppose that:*

(i) the equation

$$P\, F(v_0, 0) = 0$$

(or, what is the same, the system

$$\int_0^{\omega}\int_0^{\pi} F(v_0, 0)\,(t, x)\, e^{-ijvt} \sin kx \, dx\, dt = 0\,, \quad (j, k) \in S_1)$$

has a solution $v_0^* \in \mathscr{U} \cap \mathscr{V}$ (the operator P is defined in Remark 6.2.1);

(*ii*) *$F(u, \varepsilon)$ is a continuous mapping from $B(v_0^*, \varrho; \mathscr{U}) \times [-\varepsilon_0, \varepsilon_0]$ into $\mathscr{G} = H_\omega^\beta(\mathbf{R}; {}^0H^{r-\beta}(0, \pi))$, $\beta \in [0, r]$, $r \geqq 1$, and $F(u, \varepsilon)$ has a continuous G-derivative with respect to u in a neighbourhood of $(v_0^*, 0)$;*

(*iii*) $[P\, F_u'(v_0^*, 0)]^{-1} \in \mathscr{L}(H_\omega^\beta(\mathbf{R}; {}^0H^{r-\beta}(0, \pi)) \cap \mathscr{V}, \mathscr{U} \cap \mathscr{V})$.

Then the problem $(\mathscr{P}_\omega^\varepsilon)$ has a locally unique solution $u^*(\varepsilon) \in \mathscr{U}$ such that $u^*(0) = v_0^*$.

Proof. We verify easily that if two continuous functions $v^*(\varepsilon)$ and $w^*(\varepsilon)$ from $[-\varepsilon^*, \varepsilon^*]$ into $\mathscr{U} \cap \mathscr{V}$ and $\mathscr{U} \cap \mathscr{W}$ satisfy

$$G_1(v, w, \varepsilon) \equiv P\, F(v + w, \varepsilon) = 0\,, \tag{6.4.4}$$

$$G_2(v, w, \varepsilon) \equiv w - \varepsilon\, \Lambda(I - P)\, F(v + w, \varepsilon) = 0 \tag{6.4.5}$$

(Λ is defined by (6.2.2)), then the function $u^*(\varepsilon) = v^*(\varepsilon) + w^*(\varepsilon)$ is a solution of the problem. We obtain the functions $v^*(\varepsilon)$ and $w^*(\varepsilon)$ by applying I: Theorem 3.4.2. For when we put $p = \varepsilon$, $x = (v, w)$, $G = (G_1, G_2)$, $B_1 = (\mathscr{U} \cap \mathscr{V}) \times (\mathscr{U} \cap \mathscr{W})$, $B_2 = \mathbf{R}$, $B = (H_\omega^\beta(\mathbf{R}; {}^0H^{r-\beta}(0, \pi)) \cap \mathscr{V}) \times (\mathscr{U} \cap \mathscr{W})$, $p_0 = 0$ and $x_0 = (v_0^*, 0)$, we find that the assumption (2) of this theorem follows from (*i*) and the assumption (4) from (*iii*). This completes the proof.

Remark 6.4.1 If $F(u, \varepsilon)(t, x) = f(t, x, u, u_t, u_x)$, we find easily that F satisfies the assumption (*ii*) of the preceding theorem only if it depends on u_t and u_x linearly. To avoid this restriction we must look for a smoother solution, for example, in

$$\mathscr{U} = H_\omega^3(\mathbf{R}; H^0(0, \pi)) \cap H_\omega^2(\mathbf{R}; \dot{H}^1(0, \pi)) \cap H_\omega^1(\mathbf{R}; {}^0H^2(0, \pi))$$

with $\mathscr{G} = H_\omega^2(\mathbf{R}; H^0(0, \pi))$, provided that f is ω-periodic in t and continuous together with all its derivatives

$$D_t^{\delta_0} D_u^{\delta_1} D_{u_t}^{\delta_2} D_{u_x}^{\delta_3} f\,, \quad \delta_0 + \delta_1 + \delta_2 + \delta_3 \leqq 3\,, \quad \delta_0 \leqq 2\,,$$

on $\mathbf{R} \times [0, \pi] \times \mathbf{R}^3$. Note that $u_t \in C_\omega(\bar{Q})$ follows from $u_t \in {}^0H_\omega^2(Q)$ by I: Theorem 2.7.2; on the other hand,

$$u_x(t, x) = \sum_{j,k} k u_{j,k}\, e^{ij\nu t} \cos kx \in C(\bar{Q})$$

as

$$\sum_{j,k} |k u_{j,k}| \leqq \{\sum_{j,k} (j^2 + 1)^{-1} k^{-2}\}^{1/2} \{\sum_{j,k} (j^2 + 1)\, k^4 |u_{j,k}|^2\}^{1/2} < +\infty$$

since

$$\{\sum_{j,k} (j^2 + 1)\, k^4 |u_{j,k}|^2\}^{1/2} \leqq \|u\|_{\mathscr{U}}$$

and

$$\sum_{j,k} (j^2 + 1)^{-1} k^{-2} < \infty .$$

Hence

$$F'_u(u) v = D_u f(\cdot, \cdot, u, u_t, u_x) v + D_{u_t} f(\cdot, \cdot, u, u_t, u_x) v_t +$$

$$+ D_{u_x} f(\cdot, \cdot, u, u_t, u_x) v_x$$

and $F'_u(u) \in \mathscr{L}(\mathscr{U}, H^2_\omega(\mathbf{R}; H^0(0, \pi)))$

follows rather easily.

THEOREM 6.4.2 *Let* $\omega = 2\pi\alpha$, α *irrational, and suppose that:*
(*i*) α *satisfies* (6.3.1);
(*ii*) $F(u, \varepsilon)$ *is a continuous mapping from* $B(0, \varrho; \mathscr{U}) \times [-\varepsilon_0, \varepsilon_0]$ *into*

$\mathscr{G} = H^\beta_\omega(\mathbf{R}; {}^0H^{r-\beta}(0, \pi))$, $\beta \in [0, r]$;

(*iii*) $\|F(u_1, \varepsilon) - F(u_2, \varepsilon)\|_{\mathscr{G}} \leqq \lambda \|u_1 - u_2\|_{\mathscr{U}}$

for $u_1, u_2 \in B(0, \varrho; \mathscr{U})$ and $\varepsilon \in [-\varepsilon_0, \varepsilon_0]$.
Then the problem $(\mathscr{P}^\varepsilon_{2\pi\alpha})$ *has a locally unique solution* $u^*(\varepsilon) \in \mathscr{U}$ *such that* $u^*(0) = 0$.

Proof. Let Λ be the operator defined in (6.3.2). If we find a solution $u^*(\varepsilon)$ of the equation

$$u = \varepsilon \Lambda\, F(u, \varepsilon) , \tag{6.4.6}$$

then $u^*(\varepsilon)$ is a solution of $(\mathscr{P}^\varepsilon_{2\pi\alpha})$. Applying I: Corollary 3.4.1 to (6.4.6), we obtain the required solution.

Remark 6.4.2 Suppose that α satisfies (6.3.1) and that $F(u, \varepsilon)(t, x) = f(t, x, u, u_t, u_x)$. Since $r \geqq 2$, f obviously cannot depend on the derivatives u_t and u_x. Then $F(u, \varepsilon)(t, x) = f(t, x, u)$ satisfies the assumptions (*ii*) and (*iii*) of the preceding theorem if $r = \beta = 2$ and the second derivatives of f with respect to t and u are Lipschitz continuous in u.

Remark 6.4.3 MAWHIN in [36] studies the existence of generalized solutions of the problem

$$u_{tt}(t, x) - \alpha^2 u_{xx}(t, x) = f(t, x, u) , \quad (t, x) \in \mathbf{R}^2 ,$$

$$u(t + 2\pi, x) = u(t, x + 2\pi) = u(t, x) , \quad (t, x) \in \mathbf{R}^2 ,$$

supposing that α is an appropriate number and that

$$a \leqq (f(t, x, u) - f(t, x, v))/(u - v) \leqq b\,,$$

where a and b are suitable numbers depending on α.

Remark 6.4.4 SOKOLOV [65]–[66] used the s-Fourier method to study the problem $(\mathscr{P}^{\varepsilon}_{\omega})$ with $\omega = 2\pi\alpha$, $\alpha = \frac{1}{2}\sqrt{n}$, $n \neq m^2$, n, $m \in \mathbf{N}$, and obtained perhaps the first positive result of this kind.

Remark 6.4.5 In [47], making use of the t-Fourier method, PEŠL deals with the problem given by

$$u_{tt}(t, x) - u_{xx}(t, x) = g(t, x) + \varepsilon\, F(u, \varepsilon)\,(t, x)\,, \quad (t, x) \in \mathbf{R} \times (0, \pi)\,,$$

$$u(t + \omega, x) = u(t, x)\,, \quad (t, x) \in \mathbf{R} \times (0, \pi)\,,$$

(where $\omega = 2\pi p/q$, p, $q \in \mathbf{N}$ or $\omega = 2\pi\alpha$, α irrational) and by

$$u(t, 0) = h_0(t) + \varepsilon\, X_0(u, \varepsilon)\,(t)\,,$$

$$u(t, \pi) = h_1(t) + \varepsilon\, X_1(u, \varepsilon)\,(t)$$

or

$$u_x(t, 0) + \sigma_0\, u(t, 0) = h_0(t) + \varepsilon\, X_0(u, \varepsilon)\,(t)\,,$$

$$u_x(t, \pi) + \sigma_1\, u(t, \pi) = h_1(t) + \varepsilon\, X_1(u, \varepsilon)\,(t)$$

or

$$u(t, 0) = h_0(t) + \varepsilon\, X_0(u, \varepsilon)\,(t)\,,$$

$$u_x(t, \pi) + \sigma\, u(t, \pi) = h_1(t) + \varepsilon\, X_1(u, \varepsilon)\,(t)\,,$$

respectively. He looks for a classical solution

$$u \in C([0, \pi]; H^3_\omega(\mathbf{R})) \cap C^1([0, \pi]; H^2_\omega(\mathbf{R})) \cap C^2([0, \pi]; H^1_\omega(\mathbf{R}))\,.$$

Remark 6.4.6 STUART [71] proposes to investigate the problem

$$u_{tt} - u_{xx} = \varepsilon(u^3 + f(x) \sin t)\,, \quad t \in \mathbf{R}\,, \quad x \in (0, \pi)\,,$$

$$u(t, 0) = u(t, \pi) = 0\,, \quad t \in \mathbf{R}\,,$$

$$u(t + \omega, x) = u(t, x)\,, \quad t \in \mathbf{R}\,, \quad x \in (0, \pi)\,,$$

applying the t-Fourier method, but his hint how to solve the bifurcation equations is not sufficiently clear to the authors.

Problem 6.4.1 Making use of t-Fourier method derive theorems analogous to Theorems 6.4.1 and 6.4.2.

6.5. *The (t, s)-Fourier method; the Newton and combined boundary conditions*

In this section we treat the problem $(\mathscr{P}_\omega^{(1)})$ given by

$$u_{tt}(t, x) - u_{xx}(t, x) = g(t, x), \quad (t, x) \in Q, \tag{6.5.1}$$

$$u_x(t, 0) + \sigma_0 u(t, 0) = 0,$$

$$u_x(t, \pi) + \sigma_1 u(t, \pi) = 0, \quad t \in \mathbf{R}, \tag{6.5.2}$$

$$u(t + \omega, x) = u(t, x), \quad (t, x) \in Q. \tag{6.5.3}$$

We deal with this problem more briefly and formally than before. We hope that our introductory comments will enable the reader who is acquainted with the (t, s)-Fourier method through the preceding sections, to provide proofs of the theorems in this section.

Let E be the operator defined by

$$\mathscr{D}(E) = \{v \in H^2(0, \pi); v'(0) + \sigma_0 v(0) = v'(\pi) + \sigma_1 v(\pi) = 0\}$$

and

$$Ev = -v'' + cv,$$

where $c > 0$ is such that E is positive.

By II: Example 1.3.2, we know that there exists a set $\hat{\mathbf{Z}} \subset \{-2, -1, 0\}$ such that the system of eigenfunctions $\{v_k\}$ and eigenvalues $\{\lambda_k\}$ of $E - cI$ can be indexed by $k \in \mathbf{Z}_1 = \hat{\mathbf{Z}} \cup \mathbf{N}$, and λ_k is positive, negative, or zero according as k is positive, negative, or zero.

By a solution of the problem $(\mathscr{P}_\omega^{(1)})$ we mean a function $u \in H_\omega^2(\mathbf{R}; H^0(0, \pi)) \cap H_\omega^1(\mathbf{R}; \mathscr{D}(E^{1/2})) \cap H^0(\mathbf{R}; \mathscr{D}(E))$ satisfying (6.5.1).

Let $\nu = 2\pi/\omega$. We suppose that the function g can be written in the form

$$g(t, x) = \sum_{j \in \mathbf{Z}} \sum_{k \in \mathbf{Z}_1} g_{j,k} \, e^{ij\nu t} \, v_k(x).$$

We put $S_1 = \{(j, k) \in \mathbf{Z} \times \mathbf{Z}_1; -\nu^2 j^2 + \lambda_k = 0\}$ and $S_2 = (\mathbf{Z} \times \mathbf{Z}_1) \setminus S_1$.

If $\sigma_0 = \sigma_1$ and $\omega = 2\pi p/q$, $p, q \in \mathbf{N}$ relatively prime, then
$S_1 = \{(j, k); |j| = lp, \quad k = lq, \quad l \in \mathbf{Z}^+\} \cap (\mathbf{Z} \times \mathbf{Z}_1)$.

THEOREM 6.5.1 *Let $\sigma_0 = \sigma_1 \equiv \sigma$, $\omega = 2\pi p/q$, $p, q \in \mathbf{N}$ relatively prime, and $g \in H_\omega^\beta(\mathbf{R}; \mathscr{D}(E^{(1-\beta)/2}))$, $\beta \in [0, 1]$.*
Then the problem $(\mathscr{P}_{2\pi p/q}^{(1)})$ has a solution only if $g_{lp,lq} = g_{-lp,lq} = 0$ for all $l \in \mathbf{N}$ and $g_{0,0} = 0$, provided that $\sigma = 0$.

If these conditions are satisfied, then every solution u of $\left(\mathscr{P}^{(1)}_{2\pi p/q}\right)$ *is given by*

$$u(t, x) = \sum_{l \in \mathbf{Z}^+ \cap \mathbf{Z}_1} \operatorname{Re}\left(u_l\, e^{ilqt}\right) v_{lq}(x) +$$

$$+ \sum_{(j,k) \in S_2} \left(-j^2 q^2/p^2 + \lambda_k\right)^{-1} g_{j,k}\, e^{ijqt/p}\, v_k(x),$$

where the $u_l \in \mathbf{C}$ *are arbitrary but such that*

$$\sum_{l \in \mathbf{N}} l^4 |u_l|^2 < +\infty .$$

THEOREM 6.5.2 *Suppose that* $\sigma_0 = \sigma_1 \equiv \sigma$, $\omega = 2\pi\alpha$, α *is irrational, and that*

$$|\alpha - m/n| \geqq c_1 n^{-r} \quad \text{for all} \quad m, n \in \mathbf{N},$$

$c_1 > 0$, *and* $g \in H^\beta_\omega(\mathbf{R}; \mathscr{D}(E^{(r-\beta)/2}))$, $\beta \in [0, r]$.
Then the following two assertions hold:

(*i*) $\sigma \neq 0$. *The problem* $(\mathscr{P}^{(1)}_\omega)$ *has a unique solution given by*

$$u(t, x) = \sum_{j \in \mathbf{Z}} \sum_{k \in \mathbf{Z}_1} \left(-j^2 \alpha^{-2} + \lambda_k\right)^{-1} g_{j,k}\, e^{ijt/\alpha}\, v_k(x).$$

(*ii*) $\sigma = 0$. *The problem* $(\mathscr{P}^{(1)}_\omega)$ *has a solution only if* $g_{0,0} = 0$. *If this condition is satisfied, then every solution u of* $(\mathscr{P}^{(1)}_\omega)$ *is given by*

$$u(t, x) = \sum_{j \in \mathbf{Z}} \sum_{k \in \mathbf{N}} \left(-j^2 \alpha^{-2} + k^2\right)^{-1} g_{j,k}\, e^{ijt/\alpha}\, v_k(x) + u_{0,0},$$

where $u_{0,0} \in \mathbf{R}$ *is arbitrary.*

If $\sigma_0 \neq \sigma_1$ and $\omega = 2\pi p/q$, we find that S_1 has at most $2p - 2$ elements. This follows readily from the fact that the λ_k for $k \in \mathbf{N}$ are determined by the equation

$$\cot \pi \sqrt{\lambda} = \frac{\sigma_0 \sigma_1 + \lambda}{(\sigma_1 - \sigma_0) \sqrt{\lambda}}$$

(see II: (1.3.4)), in which $\cot(\pi q j/p)$ assumes $p - 1$ distinct finite values for $j \in \mathbf{Z}$, and $\lambda^{-1/2}(\sigma_0\sigma_1 + \lambda)(\sigma_1 - \sigma_0)^{-1}$ is either non-zero and has at most one extreme value or is monotone.

THEOREM 6.5.3 *Let* $\sigma_0 \neq \sigma_1$, $\omega = 2\pi p/q$, $p, q \in \mathbf{N}$ *relatively prime, and* $g \in H^\beta_\omega(\mathbf{R}; \mathscr{D}(E^{1-\beta/2}))$, $\beta \in [0, 2]$.
Then the problem $\left(\mathscr{P}^{(1)}_{2\pi p/q}\right)$ *has a solution only if*

$$g_{j,k} = 0 \quad \text{for all} \quad (j, k) \in S_1 .$$

If these conditions are satisfied, then the solution u of $(\mathscr{P}^{(1)}_{2\pi p/q})$ *is given by*

$$u(t, x) = \sum_{(j,k)\in S_1} u_{j,k}\, e^{iqjt/p}\, v_k(x) +$$

$$+ \sum_{(j,k)\in S_2} g_{j,k}(-j^2q^2/p^2 + \lambda_k)^{-1}\, e^{iqjt/p}\, v_k(x)\,,$$

where the $u_{j,k} \in \mathbf{C}$ *are arbitrary but such that* $u_{j,k} = \bar{u}_{-j,k}$.

Remark 6.5.1 The case $\sigma_1 \neq \sigma_2$ and $\omega = 2\pi\alpha$, α irrational, is fairly difficult, and we are not able to investigate it in a satisfactory way.

Problem 6.5.1 Derive analogous theorems for the problem given by (6.5.1), (6.5.3) and

$$u_x(t, 0) + \sigma_0\, u(t, 0) = 0\,, \quad u(t, \pi) = 0\,, \quad t \in \mathbf{R}\,.$$

Problem 6.5.2 Formulate the corresponding theorems for the existence of solutions to the weakly non-linear problem described by

$$u_{tt}(t, x) - u_{xx}(t, x) = \varepsilon\, F(u, \varepsilon)\,(t, x)\,, \quad (t, x) \in Q\,,$$

(6.5.2), and (6.5.3).

6.6. *Problems with variable coefficients and problems in several variables*

Investigating the existence of periodic solutions of equations with several space variables or with non-constant coefficients, we find that the following two features are important: the number-theoretical character and the asymptotical behaviour of the eigenvalues of the corresponding eigenvalue problems and the choice of spaces in which it is natural to look for solutions.

As an illustration we investigate the existence of a solution to a problem that occurs in the study of equations with Δ_2 in the rotationally symmetrical case. Let us consider the system

$$u_{tt}(t, r) - u_{rr}(t, r) - r^{-1}\, u_r(t, r) = g(t, r)\,, \quad (t, r) \in \mathbf{R} \times (0, 1)\,, \tag{6.6.1}$$

$$u(t, 1) = 0\,, \quad |u(t, 0)| < +\infty\,, \quad t \in \mathbf{R}\,, \tag{6.6.2}$$

$$u(t + \omega, r) = u(t, r)\,, \quad (t, r) \in \mathbf{R} \times (0, 1) \tag{6.6.3}$$

for a particular value of the period ω.

We denote by $H^0(0, 1; r)$ the space of all real-valued measurable functions f on $(0, 1)$ for which

$$\|f\|_{H^0(0,1;r)} = \left(\int_0^1 r\, f^2(r)\, \mathrm{d}r\right)^{1/2} < +\infty\,.$$

By $H^1(0, 1; r)$ we denote the space of all functions f on $(0, 1]$ which are absolutely continuous on $[\delta, 1]$ for every $\delta \in (0, 1)$, $f(1) = 0$, and

$$\|f\|_{H^1(0,1;r)} = \left(\int_0^1 r(f'(r))^2 \, dr\right)^{1/2} < +\infty .$$

Finally, we set

$$E_1 f(r) = f''(r) + \frac{1}{r} f'(r) , \quad r \in (0, 1) .$$

We now denote by $H_+^2(0, 1)$ the space of all functions $f \in H^1(0, 1; r)$ for which f' is an absolutely continuous function on $[\delta, 1]$ for every $\delta \in (0, 1)$ and

$$\|f\|_{H^2{}_+(0,1)} = (\|f\|^2_{H^1(0,1;r)} + \|E_1 f\|^2_{H^0(0,1;r)})^{1/2} < +\infty .$$

Then $H_+^2(0, 1)$ equipped with the norm $\|\cdot\|_{H^2{}_+(0,1)}$ is a Hilbert space. The eigenvalue problem associated with (6.6.1)–(6.6.2) has the form

$$E_1 v = -\lambda v , \quad |v(0)| < +\infty , \quad v(1) = 0 .$$

Its eigenfunctions v_k and eigenvalues λ_k are given by

$$v_k(r) = \alpha_k J_0(\varrho_k r) , \quad \lambda_k = \varrho_k^2 ,$$

where J_0 is the Bessel function of order 0, ϱ_k is the sequence of positive roots of J_0, and the α_k are such that the v_k form an orthonormal sequence in $H^0(0, 1; r)$. A function f is an element of $H^0(0, 1; r)$ if and only if the numbers

$$f_k = \int_0^1 r f(r) \, v_k(r) \, dr$$

satisfy $\sum_{k=1}^{\infty} |f_k|^2 < \infty$. Moreover, $f \in H_+^2(0, 1)$ if and only if $\sum_{k=1}^{\infty} |f_k|^2 \varrho_k^4 < +\infty$.

We choose $\omega = 2(2q + 1)/p$. Then

$$(2\pi j/\omega)^2 - \lambda_k = (2\pi|j|/\omega + \varrho_k)(2\pi|j|/\omega - \varrho_k) .$$

Since the roots ϱ_k of the Bessel function of order 0 satisfy the relation

$$\varrho_k = \pi(k - 1/4 + O(1/k)) ,$$

we have

$$2\pi|j|/\omega - \varrho_k = \pi \left| \frac{|j| \, p - k(2q + 1)}{2q + 1} + \frac{1}{4} + O\left(\frac{1}{k}\right) \right| \geqq c > 0$$

for all sufficiently large k. Hence the following theorem holds.

THEOREM 6.6.1 *Let* $\omega = 2(2q + 1)/p$, p, $q \in \mathbf{N}$. *Let* $g \in H_\omega^0(\mathbf{R}; H^1(0, 1; r))$ *be such that* $g_{j,k} = 0$ *for every* (j, k) *satisfying*

$$2\pi|j| - \omega\varrho_k = 0 . \tag{6.6.4}$$

Then the problem (6.6.1)–(6.6.3) *has a solution* $u \in H^2_\omega(\mathbf{R}; H^0(0, 1; r)) \cap$ $\cap H^1_\omega(\mathbf{R}; H^1(0, 1; r)) \cap H^0_\omega(\mathbf{R}; H^2_+(0, 1))$.

The reader sees that we have been forced to use the spaces generally called Sobolev weight spaces, the theory of which is outside the scope of this book. The determination of the number of (j, k) satisfying (6.6.4) is connected with the question (not yet solved, as far as we know) of whether the Bessel function of order zero has two roots ϱ_k and ϱ_l for which ϱ_k/ϱ_l is rational.

Remark 6.6.1 Taking account of the above problem, the reader may find the following result due to DE SIMON [62] to be of great interest:

Let $\Omega \subset \mathbf{R}^n$ be open and bounded, and let $m > n - 1$. Then for almost all $\omega > 0$ the following assertion is valid:

If $g \in H^{m,0}_\omega(\mathbf{R} \times \Omega)$, then there exists a unique $u \in H^0_\omega(\mathbf{R} \times \Omega)$ that is a generalized solution of the equation

$$u_{tt} - \Delta_n u = g \quad \text{on} \quad \mathbf{R} \times \Omega$$

with

$$u = 0 \quad \text{on} \quad \mathbf{R} \times \partial\Omega \, .$$

His proof is based on Theorem 14.6 of II: [1].

Remark 6.6.2 In [1] ACQUISTAPACE proves that the problem

$$u_{tt} - \alpha^2 [a(x)\, u_x]_x = \varepsilon f(t, x, u) \, , \quad (t, x) \in \mathbf{R} \times (0, \pi) \, ,$$

$$u(t, 0) = u(t, \pi) = 0 \, , \quad t \in \mathbf{R} \, ,$$

has a unique generalized 2π-periodic solution for every sufficiently small ε and for α belonging to a dense uncountable set of measure zero. The function f is supposed to be 2π-periodic in t and uniformly Lipschitz continuous in u.

§ 7. The wave equation in an unbounded domain

7.1. *A general existence theorem*

In this paragraph we investigate the existence of a solution of the problem $(\mathscr{P}^\varepsilon_\omega)$ given by

$$u_{tt}(t, x) - u_{xx}(t, x) = \varepsilon F(u, \varepsilon)(t, x) \, , \quad (t, x) \in \mathbf{R}^2 \, , \tag{7.1.1}$$

$$u(t, x) = u(t + \omega, x) \, , \quad (t, x) \in \mathbf{R}^2 \, , \tag{7.1.2}$$

where

$$F(u, \varepsilon)(t, x) = f(t, x, u(t, x), u_t(t, x), u_x(t, x), \varepsilon). \tag{7.1.3}$$

Here f is supposed to be ω-periodic in t and at least continuous. The following lemma is evident.

LEMMA 7.1.1 *Suppose that $u \in C^2_{\text{loc},\omega}(\mathbf{R}^2)$ satisfies*

$$u_{tt}(t, x) - u_{xx}(t, x) = 0 \quad \textit{for all} \quad (t, x) \in \mathbf{R}^2 .$$

Then there exist $p, q \in C^2_\omega(\mathbf{R})$ and $c \in \mathbf{R}$ such that

$$u(t, x) = p(t + x) + q(t - x) + 2cx$$

for all $(t, x) \in \mathbf{R}^2$.

Now we turn to the non-linear case. We define a Banach space $\mathscr{U}$ as the set of all functions $u \in C^2_{\text{loc},\omega}(\mathbf{R}^2)$ for which the values of

$$\sup \{|u(t, x)| (1 + |x|)^{-1}; (t, x) \in \mathbf{R}^2\},$$

$$\sup \{|D_t^\alpha D_x^\beta u(t, x)|; \alpha, \beta \in \mathbf{Z}^+, 1 \leqq \alpha + \beta \leqq 2, (t, x) \in \mathbf{R}^2\}$$

are finite. The larger of these two values defines the norm $\|u\|_{\mathscr{U}}$. Next, let $\gamma > 0$. We define a space $\mathscr{G}$ as the set of all functions $g \in C^{1,0}_{\text{loc},\omega}(\mathbf{R}^2)$ for which

$$\sup \{(|g(t, x)| + |g_t(t, x)|) (1 + |x|)^{1+\gamma}; (t, x) \in \mathbf{R}^2\}$$

is finite. This value defines the norm $\|g\|_{\mathscr{G}}$. The following lemma can be proved directly.

LEMMA 7.1.2 *The operator P given by*

$$(Pg)(t, x) = -\frac{1}{2}\int_0^x \int_{t-x+\xi}^{t+x-\xi} g(\tau, \xi)\, d\tau\, d\xi, \quad (t, x) \in \mathbf{R}^2$$

is an element of $\mathscr{L}(\mathscr{G}, \mathscr{U})$.

Finally, we put

$$Z(p, q, c)(t, x) = p(t + x) + q(t - x) + 2cx$$

for $p, q \in C^2_\omega(\mathbf{R})$ and $c \in \mathbf{R}$.

THEOREM 7.1.1 *Let $\omega, \gamma, \varrho, \varepsilon_0$ be positive numbers and $\bar{u} \in \mathscr{U}$. Suppose that:*

(i) if $u \in B(\bar{u}, \varrho; \mathscr{U})$ and $\varepsilon \in [-\varepsilon_0, \varepsilon_0]$, then $F(u, \varepsilon) \in \mathscr{G}$;

(ii) there is a $\lambda > 0$ such that

$$\|F(u, \varepsilon) - F(v, \varepsilon)\|_{\mathscr{G}} \leqq \lambda \|u - v\|_{\mathscr{U}}$$

for all $u, v \in B(\bar{u}, \varrho; \mathscr{U})$ and $\varepsilon \in [-\varepsilon_0, \varepsilon_0]$.

(*iii*) *$F(u, \varepsilon)$ is a continuous mapping from $B(\bar{u}, \varrho; \mathscr{U}) \times [-\varepsilon_0, \varepsilon_0]$ to $\mathscr{G}$. Then for $p, q \in C^2_\omega(\mathbf{R})$ and $c \in \mathbf{R}$ such that $Z(p, q, c) \in B(\bar{u}, \varrho'; \mathscr{U})$, $0 < \varrho' < \varrho$, there is an $\varepsilon^* \in (0, \varepsilon_0]$ and a function $u \in C([-\varepsilon^*, \varepsilon^*], \mathscr{U})$ such that $u(\varepsilon)$, $\varepsilon \in [-\varepsilon^*, \varepsilon^*]$, satisfies (7.1.1)–(7.1.2) and $u(0) = Z(p, q, c)$.*

A *proof* can be easily obtained by I: Theorem 3.4.1, applied to $u = Z(p, q, c) + \varepsilon P\, F(u, \varepsilon)$.

7.2. *Applications*

In this part we describe conditions that ensure that the operator $F(u, \varepsilon)$ defined in (7.1.3) fulfils the assumptions (*i*), (*ii*), and (*iii*) of Theorem 7.1.1.

We say that a function $\hat{f} = \hat{f}(t, x, u_0, u_1, u_2, \varepsilon)$ defined on $\mathbf{R}^5 \times [-\varepsilon_0, \varepsilon_0]$ satisfies the condition $\mathsf{B}^i(r)$, $i = 1, 2$ if $\hat{f}$ is ω-periodic in t and

$$\sup \{|\hat{f}(t, x, u_0, u_1, u_2, \varepsilon)|\,(1 + |x|)^{i+\gamma};\ t \in \mathbf{R},\ x \in \mathbf{R},$$
$$|u_0| \leqq (1 + |x|)\, r, \quad |u_1| \leqq r, \quad |u_2| \leqq r, \quad |\varepsilon| \leqq \varepsilon_0\} < +\infty .$$

We say that a function $\hat{f} = \hat{f}(t, x, u_0, u_1, u_2, \varepsilon)$ defined on $\mathbf{R}^5 \times [-\varepsilon_0, \varepsilon_0]$ satisfies the condition $\mathsf{B}^3(r)$ if $\hat{f}$ is ω-periodic in t and

$$\limsup_{\varepsilon \to \bar{\varepsilon}} \{|\hat{f}(t, x, u_0, u_1, u_2, \varepsilon) - \hat{f}(t, x, u_0, u_1, u_2, \bar{\varepsilon})|\,(1 + |x|)^{1+\gamma};$$
$$t \in \mathbf{R},\ x \in \mathbf{R},\ |u_0| \leqq (1 + |x|)\, r,\ |u_1| \leqq r,\ |u_2| \leqq r\} = 0$$

for every $\bar{\varepsilon} \in [-\varepsilon_0, \varepsilon_0]$.

THEOREM 7.2.1 *Suppose that $f = f(t, x, u_0, u_1, u_2, \varepsilon)$ is defined on $\mathbf{R}^5 \times [-\varepsilon_0, \varepsilon_0]$ and $r > 0$. Let $D^\alpha_t D^{\beta_0}_{u_0} D^{\beta_1}_{u_1} D^{\beta_2}_{u_2} f$ satisfy $\mathsf{B}^1(r)$ for all $\alpha, \beta_0, \beta_1, \beta_2 \in \mathbf{Z}^+$ with*

$$\alpha + \beta_0 + \beta_1 + \beta_2 \leqq 2 \quad \text{and} \quad \alpha \leqq 1 ;$$

and $\mathsf{B}^2(r)$ for all $\alpha, \beta_0, \beta_1, \beta_2 \in \mathbf{Z}^+$ with

$$\alpha + \beta_0 + \beta_1 + \beta_2 \leqq 2 , \quad \alpha \leqq 1 \quad \text{and} \quad \beta_0 \geqq 1 ;$$

and $\mathsf{B}^3(r)$ for all $\alpha, \beta_0, \beta_1, \beta_2 \in \mathbf{Z}^+$ with

$$\alpha + \beta_0 + \beta_1 + \beta_2 \leqq 1 .$$

Then $F(u, \varepsilon)$ satisfies the assumptions (i), (ii), and (iii) of Theorem 7.1.1 provided that $\|\bar{u}\|_{\mathscr{U}} + \varrho \leqq r$.

This can be proved by a straightforward but lengthy calculation. This theorem is applied in the next two examples.

Example 7.2.1 Let

$$f(t, x, u_0, u_1, u_2) = \sum_{\substack{k,m,n \geq 0 \\ k+m+n \leq s}} b_{k,m,n}(t, x)\, u_0^k u_1^m u_2^n ,$$

with $b_{k,m,n} \in C_\omega^{1,0}(\mathbf{R}^2)$ and $s \in \mathbf{N}$. Let $K > 0$ be such that

$$(|b_{k,m,n}(t, x)| + |D_t^1\, b_{k,m,n}(t, x)|)\,(1 + |x|)^{k+1+\gamma} \leqq K$$

for every $(t, x) \in \mathbf{R}^2$ and all k, m, n.
Then the operator $F(u, \varepsilon)$ fulfils the assumptions (i), (ii), and (iii) of Theorem 7.1.1.

Example 7.2.2 Let

$$f(t, x, u_0, u_1, u_2) = b(t) \exp\left(-\alpha|x| + a_0 u_0\right) \hat{f}(u_1, u_2) ,$$

with $b \in C_\omega^1(\mathbf{R})$, $\hat{f} \in C_{\text{loc}}^2(\mathbf{R}^2)$, and $\alpha > 0$.
Then the operator $F(u, \varepsilon)$ fulfils the assumptions (i), (ii), and (iii) of Theorem 7.1.1, provided that $|a_0|\,(\|\bar{u}\|_{\mathscr{U}} + \varrho) < \alpha$.

Remark 7.2.1 Similar results can be found in the paper [22] by HERRMANN and ŠTĚDRÝ.

§ 8. Supplements and comments on non-autonomous hyperbolic equations

8.1. *The adjoint problem method*

Let us deal with the problem $(\mathscr{P}_\omega)$ given by

$$u_{tt} + a(t, x)\, u_{xx} + b(t, x)\, u_x + c(t, x)\, u = g(t, x) ,$$

$$(t, x) \in Q \equiv \mathbf{R} \times (0, \pi) , \tag{8.1.1}$$

$$u(0, x) - u(\omega, x) = 0 ,$$

$$u_t(0, x) - u_t(\omega, x) = 0 , \quad x \in [0, \pi] , \tag{8.1.2}$$

$$M(t)\,(u(t, 0), u_x(t, 0))' + N(t)\,(u(t, \pi), u_x(t, \pi))' =$$

$$= (h_0(t), h_1(t))' , \quad t \in \mathbf{R} , \tag{8.1.3}$$

where $a \in C_\omega^{0,2}(\bar{Q})$, $a(t, x) < 0$, $b \in C_\omega^{0,1}(\bar{Q})$, $c \in C_\omega^0(\bar{Q})$, $g \in C_\omega^0(\bar{Q})$, h_0, $h_1 \in C_\omega^0(\mathbf{R})$, M and N are 2×2-matrices belonging to $C_\omega^0(\mathbf{R})$ and are such that rank $(M, N) = 2$. By II: Sec. 1.2, we find easily that the problem $(\mathscr{P}^*_\omega)$ formally adjoint to the homogeneous problem $(\mathscr{P}_\omega)$ is given by

$$v_{tt} + (a(t,x)\,v)_{xx} - (b(t,x)\,v)_x + c(t,x)\,v = 0\,, \quad (t,x) \in Q\,, \tag{8.1.4}$$

$$v(0,x) - v(\omega,x) = 0\,,$$

$$v_t(0,x) - v_t(\omega,x) = 0\,, \quad x \in [0,\pi]\,, \tag{8.1.5}$$

$$(v(t,0), v_x(t,0))\,\tilde{P}(t) + (v(t,\pi), v_x(t,\pi))\,\tilde{Q}(t) = 0\,, \quad t \in \mathbf{R}\,, \tag{8.1.6}$$

where the 2×2-matrices $\tilde{P}$ and $\tilde{Q}$ satisfy

$$-M(t)\,B^{-1}(t,0)\,\tilde{P}(t) + N(t)\,B^{-1}(t,\pi)\,\tilde{Q}(t) = 0\,, \quad t \in \mathbf{R}\,, \tag{8.1.7}$$

the matrix $B^{-1}(t,x)$ has the form

$$B^{-1}(t,x) = \frac{1}{a^2(t,x)} \begin{pmatrix} 0 & -a(t,x) \\ a(t,x) & b(t,x) - a_x(t,x) \end{pmatrix}$$

and rank $(\tilde{P}, \tilde{Q}) = 2$. Further, suppose that the 2×2-matrices $P_c(t)$ and $Q_c(t)$ defining the complementary adjoint boundary forms are given by the relation

$$-M(t)\,B^{-1}(t,0)\,P_c(t) + N(t)\,B^{-1}(t,\pi)\,Q_c(t) = I\,. \tag{8.1.8}$$

THEOREM 8.1.1 *The problem $(\mathscr{P}_\omega)$ has a solution only if*

$$\int_0^\omega \int_0^\pi v(\tau,\xi)\,g(\tau,\xi)\,\mathrm{d}\xi\,\mathrm{d}\tau = \int_0^\omega \{(v(\tau,0), v_x(\tau,0))\,P_c(\tau) +$$

$$+ (v(\tau,\pi), v_x(\tau,\pi))\,Q_c(\tau)\}\,(h_0(\tau)\,,\; h_1(\tau))'\,\mathrm{d}\tau \tag{8.1.9}$$

for every solution v of $(\mathscr{P}^*_\omega)$.

Let us denote by $(\mathscr{P}^{(1)}_{2\pi})$ the problem given by (8.1.1)–(8.1.3) with $a = -1$, $b = c = 0$, and

$$M = \begin{pmatrix} 1 & 0 \\ 0 & 0 \end{pmatrix}, \quad N = \begin{pmatrix} 0 & 0 \\ 1 & 0 \end{pmatrix}$$

(so that $u(t,0) = h_0(t)$ and $u(t,\pi) = h_1(t)$).

THEOREM 8.1.2 *Let $(\mathscr{P}^{(1)}_{2\pi})$ be the problem with $g \in C^{(\alpha,1-\alpha)}_{2\pi}(\bar{Q})$, $\alpha = 0$ or 1, and $h_0, h_1 \in C^2_{2\pi}(\mathbf{R})$.*

Then $(\mathscr{P}_{2\pi}^{(1)})$ *has a classical solution if and only if*

$$\int_0^{2\pi}\int_0^{\pi} g(\tau, \xi)\,(s(\xi + \tau) - s(-\xi + \tau))\,d\xi\,d\tau =$$

$$= 2\int_0^{2\pi} (-h_0(\tau)\,s'(\tau) + h_1(\tau)\,s'(\tau + \pi))\,d\tau \tag{8.1.10}$$

for every $s \in C_{2\pi}^2(\mathbf{R})$.

Proof. The necessity of (8.1.10) follows at once from Theorem 8.1.1 if we take into account that

$$P_c = \begin{pmatrix} 0 & 0 \\ -1 & 0 \end{pmatrix} \quad \text{and} \quad Q_c = \begin{pmatrix} 0 & 0 \\ 0 & 1 \end{pmatrix}.$$

The sufficiency is a consequence of the equivalence of (8.1.10) and (2.2.2). For when we use some obvious changes of variables and the Fubini theorem, we have

$$\int_0^{2\pi}\int_0^{\pi} g(\tau, \xi)\,(s(\xi + \tau) - s(-\xi + \tau))\,d\xi\,d\tau =$$

$$= \int_0^{2\pi}\left(\int_0^{\pi} (g(\tau - \xi, \xi) - g(\tau + \xi, \xi))\,d\xi\right) s(\tau)\,d\tau\,,$$

and

$$\int_0^{2\pi} (-h_0(\tau)\,s'(\tau) + h_1(\tau)\,s'(\tau + \pi))\,d\tau =$$

$$= \int_0^{2\pi} s(\tau)\,(h_0'(\tau) - h_1'(\tau + \pi))\,d\tau\,.$$

These relations yield (2.2.2) immediately, because s is an arbitrary element of $C_{2\pi}^2(\mathbf{R})$.

Let us denote by $(\mathscr{P}_{2\pi}^{(2)})$ the problem given by (8.1.1)–(8.1.3) with $a = -1$, $b = c = 0$, and

$$M(t) = \begin{pmatrix} \sigma_0(t) & 1 \\ 0 & 0 \end{pmatrix}, \quad N(t) = \begin{pmatrix} 0 & 0 \\ \sigma_1(t) & 1 \end{pmatrix}$$

(so that $u_x(t, 0) + \sigma_0(t)\,u(t, 0) = h_0(t)$ and $u_x(t, \pi) + \sigma_1(t)\,u(t, \pi) = h_1(t)$).

Theorem 8.1.3 *Let* $(\mathscr{P}_{2\pi}^{(2)})$ *be the problem with* $g \in C_{2\pi}^{(\alpha, 2-\alpha)}(\bar{Q})$, $\alpha = 0, 1,$ *or* 2, *and* $\sigma_0, \sigma_1, h_0, h_1 \in C_{2\pi}^2(\mathbf{R})$. *Suppose that*

$$\sigma_1(t + \pi) - \sigma_0(t) \neq 0 \quad \textit{for all} \quad t \in \mathbf{R}\,.$$

Then $(\mathscr{P}_{2\pi}^{(2)})$ *always has a classical solution, provided that*

$$\int_0^{2\pi} \sigma_1(t+\pi)\,\sigma_0(t)\,(\sigma_1(t+\pi)-\sigma_0(t))^{-1}\,dt \neq 2\,.$$

If the integral has the value 2, then there is a solution if and only if

$$\int_0^{2\pi}\int_0^{\pi} g(\tau,\xi)\,v(\tau,\xi)\,d\xi\,d\tau =$$

$$= \int_0^{2\pi} (h_0(\tau)\,v(\tau,0) - h_1(\tau)\,v(\tau,\pi))\,d\tau\,, \tag{8.1.11}$$

where

$$v(t,x) = a(x+t) + a(-x+t) + \int_{-x+t}^{x+t} b(\eta)\,d\eta$$

with

$$a(t) = \sigma_1(t+\pi)\,(\sigma_1(t+\pi)-\sigma_0(t))^{-1}\,,$$

$$b(t) = -\sigma_1(t+\pi)\,\sigma_0(t)\,(\sigma_1(t+\pi)-\sigma_0(t))^{-1}\,.$$

Proof. The first part of the assertion follows readily from Theorem 4.2.2, while the second part is obtained from Theorem 4.2.3, by virtue of the equivalence of (8.1.11) and (4.2.7).

THEOREM 8.1.4 *Let* $(\mathscr{P}_{2\pi}^{(2)})$ *be the problem with* $g \in C_{2\pi}^{(\alpha,1-\alpha)}(\bar{Q})$, $\alpha = 0$ *or* 1, *and* $\sigma_0, \sigma_1, h_0, h_1 \in C_{2\pi}^1(\mathbf{R})$. *Suppose that*

$$\sigma_1(t+\pi) - \sigma_0(t) = 0 \quad \textit{for all} \quad t \in \mathbf{R}\,.$$

Then $(\mathscr{P}_{2\pi}^{(2)})$ *has a solution if and only if*

$$\int_0^{2\pi}\int_0^{\pi} g(\tau,\xi)\,v(\tau,\xi)\,d\xi\,d\tau =$$

$$= \int_0^{2\pi} (h_0(\tau)\,v(\tau,0) - h_1(\tau)\,v(\tau,\pi))\,d\tau$$

for every function v of the form

$$v(t,x) = a(x+t) + a(-x+t) - \int_{-x+t}^{x+t} \sigma_0(\tau)\,a(\tau)\,d\tau\,,$$

where $a \in C_{2\pi}^2(\mathbf{R})$ *satisfies*

$$\int_0^{2\pi} \sigma_0(\tau)\,a(\tau)\,d\tau = 0\,.$$

The *proof* follows readily from Theorems 8.1.1 and 4.2.1.

Problem 8.1.1 Formulate similar theorems for $\omega = 2\pi p/q$.

Problem 8.1.2 Formulate similar theorems for the combined boundary conditions

$$u_x(t, 0) + \sigma(t)\, u(t, 0) = h_0(t)$$

$$u(t, \pi) = h_1(t)$$

for $\omega = 2\pi$ or $2\pi p/q$.

Remark 8.1.1 If we wished to obtain the sufficiency of conditions in Theorems 8.1.2–8.1.4 directly with help of I: Theorem 1.5.1 (or its counterpart in Banach spaces, see e.g. I: [39], Sec. 4.6), we would have to work in spaces in which the considered differential operators are densely defined, closed, and have closed range (and, of course, the adjoint operators should be taken in the sense of functional analysis). For instance, the operator L, $Lu = u_{tt} - u_{xx}$, considered in $H^0((0, 2\pi) \times (0, \pi))$ with

$$\mathscr{D}(L) = \Big\{u \in H^0((0, 2\pi) \times (0, \pi));\ \sum_{\substack{j\in\mathbf{Z}\\ k\in\mathbf{N}}} (j^2 - k^2)^2\, |u_{j,k}|^2 < +\infty\,,$$

$$u_{j,k} = \frac{\sqrt{2}}{\pi} \int_0^{2\pi} \int_0^{\pi} u(t, x)\, e^{-ijt} \sin kx \,\mathrm{d}x\, dt\Big\}$$

has all these required properties.

8.2. *Averaging methods*

In [33] (and in a preliminary form in [32]) KURZWEIL investigates the problem given by

$$u_{tt}(t, x) - u_{xx}(t, x) = \varepsilon f(t, x, u, u_t, u_x)\,, \quad (t, x) \in \mathbf{R}^2\,, \tag{8.2.1}$$

$$u(t, x) = -u(t, -x) = u(t, x + 2\pi)\,, \quad (t, x) \in \mathbf{R}^2\,, \tag{8.2.2}$$

and by the condition of 2π-periodicity in t. Writing

$$u(t, x) = S(x + t, \varepsilon t) - S(-x + t, \varepsilon t)\,,$$

$$u_t(t, x) = s(x + t, \varepsilon t) - s(-x + t, \varepsilon t)\,,$$

where S and s are such that

$$S_\xi(\xi, \tau) = s(\xi, \tau) \quad \text{and} \quad \int_0^{2\pi} S(\xi, \tau)\, \mathrm{d}\xi = 0\,,$$

he reduces (8.2.1) to

$$s_\tau(\xi, \tau) = \tfrac{1}{2}f(\tau/\varepsilon, \xi - \tau/\varepsilon, S(\xi, \tau) - S(-\xi + 2\tau/\varepsilon, \tau), \\ s(\xi, \tau) - s(-\xi + 2\tau/\varepsilon, \tau), s(\xi, \tau) + s(-\xi + 2\tau/\varepsilon, \tau)) . \quad (8.2.3)$$

The existence of a solution to (8.2.3) that is 2π-periodic in ξ and $2\pi\varepsilon$-periodic in τ implies the existence of a 2π-periodic solution to (8.2.1)–(8.2.2). Under certain assumptions on f the solution to (8.2.3) can be found by applying the general theory on the existence of exponentially stable integral manifolds based on the investigation of the averaged equation, which in this case has the form

$$s_\tau(\xi, \tau) = \frac{1}{4\pi}\int_0^{2\pi} f(\sigma, \xi - \sigma, S(\xi, \tau) - S(-\xi + 2\sigma, \tau), \\ s(\xi, \tau) - s(-\xi + 2\sigma, \tau), s(\xi, \tau) + s(-\xi + 2\sigma, \tau))\, d\sigma .$$

In particular, the existence of 2π-periodic solutions is proved for $f = -u_t^3 + h(x + t) - h(-x + t)$ or $f = -u^2u_t + h(x + t) - h(-x + t)$ or $f = d_1u_t^3 + d_2u_x^2u_t + d_3u_t + h(x + t) - h(-x + t)$ $(d_1 < 0,\ 3d_1 \leqq d_2 < -d_1,\ d_3 \leqq 0)$. The author observes that to any of these right-hand sides finitely many terms of the form $u^{m_1}u_t^{m_2}u_x^{m_3}$ $(m_1 + m_2$ positive and even) can be added.

Another approach based on the averaging principle is described in [16] by Fink, Hall, and Hausrath (see VII: 5.5). Here the authors examine in particular the equation

$$u_{tt}(t, x) - u_{xx}(t, x) = \varepsilon(u^2u_t + h(t, x))$$

with the Dirichlet boundary conditions. Further progress in this approach was achieved by Hall in [21].

Fetisov in [14] deals with the problem of the existence of 2π-periodic solutions to the problem

$$u_{tt} - u_{xx} = \varepsilon(f_0(t, x) + p(t, x, u)\,u + q(t, x, u)\,u_t + r(t, x, u)\,u_x) ,$$

$$(t, x) \in \mathbf{R} \times (0, \pi) ,$$

$$u(t, 0) = u(t, \pi) = 0 , \quad t \in \mathbf{R} .$$

Applying the averaging method he shows that this problem has a solution for every sufficiently small ε, provided that

$$\int_0^\pi (f_0(t + \xi, \xi) + f_0(t - \xi, \xi))\, d\xi = 0 , \quad t \in \mathbf{R} ,$$

and the spectrum of an operator formed by means of the functions p, q, and r does not contain any purely imaginary number.

8.3. *Comments on other papers*

A number of authors have examined the equation (or various generalizations of it)

$$u_{xy} = F(u)\,, \tag{8.3.1}$$

requiring u either to be periodic in x on a sufficiently narrow strip $-y_0 \leqq y \leqq y_0$ (see [1], [2], [3], [10], [11], [12], [20], [22], [23], [24], [33], [55], [56], [62] of the "Bibliography of papers on related topics") or, to be periodic in both variables with the same period.

Let us mention in more detail some results concerning the latter problem. If we make the change of variables $x = t + \xi$, $y = t - \xi$, then the existence of a solution of (8.3.1) that is ω-periodic in x and y implies the existence of a solution that is ω-periodic in t and ξ of the equation

$$u_{tt} - u_{\xi\xi} = \tilde{F}(u)\,,$$

which has been the subject of our discussion in the preceding paragraphs.

Using the Banach fixed-point theorem (I: Theorem 3.4.1) AZIZ and HORAK prove in [4] (and AZIZ in somewhat less general form in [3]) that the equation

$$u_{xy} + a(x, y)\, u_x + b(x, y)\, u_y + c(x, y)\, u = f(x, y, u, u_x, u_y)$$

has a solution that is ω-periodic in x and y, provided that the functions a, b, and c are sufficiently smooth, certain expressions formed from them are sufficiently small, and f has a sufficiently small Lipschitz constant.

In [9]–[11] CESARI proves that the equations

$$u_{xy} = \varepsilon[\psi(x, y) + Cu + \psi_1(y)\, u_x + \psi_2(x)\, u_y] + \varepsilon^2\, g(x, y, u, u_x, u_y)\,,$$

$$u_{xy} = \psi(x, y) + Cu + \psi_1(y)\, u_x + \psi_2(x)\, u_y + \varepsilon\, g(x, y, u, u_x, u_y)$$

have a solution that is ω-periodic in x and y, making use of the (t, s)-Fourier method and of the Schauder fixed-point theorem (see I: Theorem 3.2.2). His results are presented in a more "geometrical" form in [17] by HALE.

In [48] the problem

$$u_x(x, y) = F(x, y, u, v)\;, \quad v_y(x, y) = G(x, y, u, v),$$

$$u(x, y) = u(x + \omega, y) = u(x, y + \omega)\,,$$

$$v(x, y) = v(x + \omega, y) = v(x, y + \omega)\,,$$

(which generalizes the problem studied above) is studied by PETROVANU.

8.4. *Comments on papers of an applied character*

STOKER [72] investigates the problem of non-linear periodic vibrations of a string with fixed end-points. He was perhaps the first outside the Soviet Union to point out the necessity of studying such problems.

CHEN [12] deals with the interaction of longitudinal and transverse waves in dispersive non-linear elastic media.

COLLINS [13] investigates the equation $\varrho u_{tt} = (T(u_x))_x$ with the boundary conditions $u(t, 0) = 0$, $u(t, 1) = f \cos \omega t$, which describes longitudinal oscillations in simple non-linear models of a bar, and transverse vibrations of a non-linear string.

MILLMAN and KELLER [37] treat a model equation for small periodic vibrations of a uniform string fixed at one end, and harmonically driven at the other under the action of a non-linear restoring force.

KVAL'VASSER and SAMARIN [34] study the possibility of a periodic motion in a string whose both ends move along some curves $x = f_j(t)$, f_j periodic, $j = 1, 2$.

MIRANKER [38] is concerned with periodic voltage oscillations in transmission lines with periodic driving voltages, where a diode is represented as a non-linear capacitance and is placed in shunt in that line.

VERMA [81] deals with a model equation for periodic vibrations of a uniform square membrane on which a non-linear restoring force and a periodic force are acting, and which is fixed at all four edges.

SIBGATULLIN [61] is concerned with periodic transverse vibrations of an unbounded elastic layer excited by a harmonic tangential force, provided that the velocity of the longitudinal waves is much greater than that of the transverse ones.

JOHNSON [25] studies the motion of a flexible inextensible pendulum with a point mass attached to its free end.

JENTSCH [23]–[24] investigates linear thermoelastic vibrations of inhomogeneous bodies.

§ 9. Comments on autonomous hyperbolic equations

9.1. *Comments on periodic solutions of boundary-value problems*

Using the method described in Sec. 8.2, KURZWEIL [97] succeeded in proving the existence of an $\omega(\varepsilon)$-periodic solution of the problem

$$u_{tt} - u_{xx} = -\varepsilon\{[h(\tfrac{1}{2}(u_x + u_t)) - h(\tfrac{1}{2}(u_x - u_t))] \cos 2x - u_t\}\,,$$

$$(t, x) \in \mathbf{R} \times (0, \pi)\,, \quad u(t, 0) = u(t, \pi) = 0\,, \quad t \in \mathbf{R}\,,$$

where $h = h(\lambda)$ is odd and sufficiently smooth, $h'(\lambda)$ is positive, $h''(\lambda)$ is negative for $\lambda > 0$, $h'(0) > 2$ and $h'(\lambda) \to 0$ with $\lambda \to \infty$.

By a method based on the theory of eigenvalues for quasilinear elliptic operators, POHOŽAEV [103] finds conditions under which there exists a periodic solution of the problem

$$P(u, D_t u, \ldots, D_t^s u) + \sum_{|\alpha| \leqq m} (-1)^{|\alpha|} D^\alpha F_\alpha(x, D^\gamma u) = 0 \quad \text{in} \quad \mathbf{R} \times \Omega,$$

$$D^\beta u \mid \partial\Omega = 0 \quad \text{for} \quad |\beta| \leqq m - 1,$$

where $F_\alpha(x, D^\gamma u) = \partial F(x, D^\gamma u)/\partial(D^\alpha u)$, $|\alpha|$, $|\gamma| \leqq m$, $\Omega \subset \mathbf{R}^n$ and the function P satisfies rather complicated hypotheses.

BERGER [82] tries to prove the existence of countably many distinct periodic solutions of the problem

$$p(x)\, u_{tt} - \Delta_n u + f(x, u) = 0 \quad \text{in} \quad \mathbf{R} \times \Omega,$$

$$u \mid \partial\Omega = 0$$

(Ω is a bounded domain in $\mathbf{R}^n$), under certain assumptions on f. However, at some decisive points the proof is not clear.

FINK, HALL, and HAUSRATH [87] investigate the existence of non-regular 2π-periodic solutions of the system

$$y_{1t} = y_{2x} + \varepsilon(y_1 - y_1^3), \quad (t, x) \in \mathbf{R}^2,$$

$$y_{2t} = y_{1x}, \quad (t, x) \in \mathbf{R}^2,$$

$$y_1(t, x) = y_1(t, x + 2\pi), \quad y_2(t, x) = y_2(t, x + 2\pi), \quad (t, x) \in \mathbf{R}^2.$$

Investigating the asymptotic behaviour of the solutions of the problem

$$u_{tt} - u_{xx} = \varepsilon(u_t - u_t^3), \quad (t, x) \in \mathbf{R} \times (0, \pi),$$

$$u(t, 0) = u(t, \pi) = 0, \quad t \in \mathbf{R},$$

HALL [93] also proves the existence of infinitely many 2π-periodic solutions. The same author in [94] deals with the problem

$$u_t + u_x = \varepsilon(Pu - h(Pu) + f(t, x)), \quad (t, x) \in \mathbf{R}^2.$$

$$u(t, x) = u(t, x + 2\pi), \quad (t, x) \in \mathbf{R}^2,$$

$$u(0, x) = u_0(x), \quad x \in \mathbf{R},$$

where u_0 is 2π-periodic and P is the projection onto the odd part of u. The result obtained makes it possible to prove the existence of 2π-periodic solutions to the equations $u_{tt} - u_{xx} = \varepsilon(u_t - u_t^3)$ and $u_{tt} - u_{xx} = \varepsilon(1 - u^2)\, u_t$ with the Dirichlet boundary conditions

ŠTĚDRÝ and VEJVODA in [105] deal with the existence to solutions to the problem

$$u_{tt}(t, x) - u_{xx}(t, x) = \varepsilon f(x, u, u_t, u_x), \quad t \in \mathbf{R}, \quad x \in [0, \pi], \tag{9.1.1}$$

$$u(t, 0) = u(t, \pi) = 0, \quad t \in \mathbf{R}, \tag{9.1.2}$$

$$u(t + \omega(\varepsilon), x) = u(t, x), \quad t \in \mathbf{R}, \quad x \in [0, \pi],$$

where $\omega(\varepsilon) = 2\pi + \varepsilon v$. In the first part of the paper they derive a necessary condition for the existence of a smooth $\omega(\varepsilon)$-periodic solution, and describe certain classes of functions f for which the branches of smooth solutions, that are continuous in the parameter ε, can bifurcate at $\varepsilon = 0$ only from the zero function. In the second part, which is similar to [87], a sufficient condition for the existence of non-smooth 2π-periodic solutions of (9.1.1)–(9.1.2) is given. This condition is shown to be satisfied for each of the following three types of non-linearity: $f = -\alpha u + \beta u^3$, $f = (-\gamma + u^2) u_t$, and $f = -\alpha u_t + \beta u_t^3$.

MELROSE and PEMBERTON [98] investigate the existence of solutions of the problem

$$u_{tt} - u_{xx} + \varepsilon u = \delta f(u), \quad t \in \mathbf{R}, \quad x \in [0, \pi],$$

$$u(t, 0) = u(t, \pi) = 0, \quad t \in \mathbf{R},$$

$$u(t + \omega(\varepsilon), x) = u(t, x), \quad t \in \mathbf{R}, \quad x \in [0, \pi],$$

where $\omega(\varepsilon) = 2\pi/(k^2 + \varepsilon)^{1/2}$ with some $k \in \mathbf{N}$. Under some assumptions on f they show that this problem has a solution for ε from a certain set (that depends on k and has 0 as an accumulation point), and for every δ sufficiently close to 0.

RABINOWITZ [104] proves the existence and regularity of non-trivial solutions to the problem

$$u_{tt}(t, x) - u_{xx}(t, x) + f(u) = 0, \quad t \in \mathbf{R}, \quad x \in [0, \pi],$$

$$u(t, 0) = u(t, \pi) = 0, \quad t \in \mathbf{R},$$

$$u(t + 2\pi p/q, x) = u(t, x), \quad t \in \mathbf{R}, \quad x \in [0, \pi],$$

where $p, q \in \mathbf{N}$ and f is sufficiently smooth, strictly monotone and satisfies the following two conditions: (a) $f(z) = o(z)$ at $z = 0$, (b) there is a $\theta \in (0, 1/2)$ such that

$$\int_0^z f(s)\, \mathrm{d}s \leqq \theta z f(z)$$

for all z with sufficiently large $|z|$ (for example, $f(u) = u^3$ satisfies these conditions). The variational method developed in this paper can also be applied to the non-autonomous case.

VEJVODA and ŠTĚDRÝ in [106] investigate the problem

$$u_{tt}(t, x) - u_{xx}(t, x) = \varepsilon \left\{ \left(\alpha - \beta \int_0^\pi u^2(t, \xi) \, d\xi \right) u(t, x) + \delta f(x, u) \right\},$$

$$t \in \mathbf{R}, \quad x \in [0, \pi],$$

$$u(t, 0) = u(t, \pi) = 0, \quad t \in \mathbf{R},$$

$$u(t + 2(\pi - \varepsilon v), x) = u(t, x), \quad t \in \mathbf{R}, \quad x \in [0, \pi],$$

for α, β and v positive, fixed, and v sufficiently small. The result, very similar to that in [98], can be described as follows: For every ε from a certain set, which has 0 as an accumulation point, and for every δ with $|\delta|$ sufficiently small this problem has a solution, provided that f is sufficiently smooth.

GREENBERG [92], Sec. 3 investigates the existence of smooth time-periodic solutions of the problem

$$u_{tt} = (\sigma(u_x))_x, \quad t \in \mathbf{R}, \quad x \in [0, \pi],$$

$$u(t, 0) = u(t, \pi) = 0, \quad t \in \mathbf{R}$$

under the assumption that the function σ is of the form

$$\sigma(\xi) = \xi^3 (1 + \sum_{k=1}^{\infty} \sigma_k \xi^{2k}).$$

FILIMONOV [86] deals with the equation $u_{tt} - k|u_x|^n u_{xx} = 0$ (k and n are parameters) and some related equations. He proves the existence of solutions that are periodic in either one or both variables.

9.2. *Comments on periodic solutions in unbounded domains*

In [84] and in the Appendix to [83] BERGER proves that the problem

$$u_{tt} = \Delta_3 u - m^2 u + k(|x|) |u|^\sigma u$$

has countably many stationary solutions (that is, $u_n(t, x) = \exp(i \lambda_n t) v_n(x)$ with $v_n \to 0$ exponentially, as $x \to \infty$) provided that k is continuous, $0 < k_1 \leqq k(|x|) \leqq k_2 < +\infty$, and $m \neq 0$, $0 < \sigma < 4$.

PETIAU in [100] and [101] looks for periodic solutions of the equation

$$u_{tt} - \Delta_3 u + \alpha u + \beta u^3 = 0, \quad t \in \mathbf{R}, \quad x \in \mathbf{R}^3,$$

in the form $u(t, x_1, x_2, x_3) = \varphi(ct - k_1x_1 - k_2x_2 - k_3x_3)$ and finds conditions on α and β under which such a solution (given in terms of Jacobi's elliptic functions) may exist. (Obviously such a solution has the form of a travelling wave whereas in this book we are mostly interested in periodic solutions in the form of a standing wave.)

FLEISHMAN [89] proceeds along the same lines, investigating the equation

$$u_{tt} - u_{xx} + \alpha u + \beta u^3 = 0\,, \quad (t, x) \in \mathbf{R}^2\,.$$

Using the qualitative theory of ordinary differential equations, PISARENKO [102] looks for periodic solutions (in the form of a travelling wave) to the local field theory equation

$$u_{tt} - \Delta_3 u + m^2 u = \varepsilon \sum_{k=1}^{N} a_k u^k\,, \quad (t, x) \in \mathbf{R} \times \mathbf{R}^3\,,$$

as well as to the corresponding non-local field theory equation.

9.3. *Comments on papers of an applied character*

A number of mathematicians and physicists have sought periodic solutions to autonomous wave or Klein-Gordon equations as a formal power series in a small parameter ε.

For example, VITT [107] (who we think, was the first to be interested in periodic solutions of a non-linear problem described by partial differential equations) looks for an $\omega(\varepsilon)$-periodic solution to the problem

$$u_{tt} - u_{xx} = -\varepsilon\, u_t\,, \quad (t, x) \in \mathbf{R} \times (0, \pi)\,,$$

$$u_x(t, 0) = \varepsilon\, c(1 - u^2(t, 0))\, u_t(t, 0)\,,$$

$$\beta\, u_{tt}(t, \pi) + u_x(t, \pi) = -\varepsilon\beta\, u_t(t, \pi)\,, \quad t \in \mathbf{R}\,,$$

which describes a homogeneous electrical line.

ŽABOTINSKIĬ [108] treats the problem

$$\varrho(x)\, u_{tt} - \sum_{j=1}^{3} (p\, u_{x_j})_{x_j} + q(x)\, u = \varepsilon f(x, u, \varepsilon)\,, \quad (t, x) \in \mathbf{R} \times \Omega\,,$$

$$u_{tt} - \Lambda u = \varepsilon\, F(x, u, \varepsilon)\,, \quad (t, x) \in \mathbf{R} \times \partial\Omega\,,$$

$$u(t + \omega(\varepsilon), x) = u(t, x)\,, \quad (t, x) \in \mathbf{R} \times \Omega\,,$$

where Λ is a linear operator. The proposed procedure is purely formal.

FINK, HALL, and KHALILI in [88] consider the problem given by the equation

$$u_{tt}(t, x) - u_{xx}(t, x) = F(u)(t, x), \quad (t, x) \in \mathbf{R} \times (0, \pi),$$

where $F(u)$ is either u^3 or $\alpha(u + \beta u^3)$ or $-M^2 \sin u$, with the boundary conditions

$$u(t, 0) = u(t, \pi) = 0, \quad t \in \mathbf{R},$$

and the periodicity condition

$$u(t + \omega, x) = u(t, x), \quad (t, x) \in \mathbf{R} \times [0, \pi].$$

MILLMAN and KELLER [99] deal with the problem

$$\omega^2 u_{tt} - u_{xx} + u = \varepsilon f(\omega u_t), \quad (t, x) \in \mathbf{R} \times (0, \pi),$$
$$u(t, 0) = u(t, \pi) = 0, \quad t \in \mathbf{R},$$
$$u(t + 2\pi, x) = u(t, x), \quad (t, x) \in \mathbf{R} \times (0, \pi),$$

where $f(\zeta) = \zeta - \frac{1}{3}\zeta^3$ and ω is an undetermined quantity. If u and ω are written in the form

$$u = u_0 + \varepsilon u_1 + \ldots, \quad \omega = \omega_0 + \varepsilon \omega_1 + \ldots$$

with $u_0 = a \cos t \sin nx$ and $\omega_0 = (1 + n^2)^{1/2}$, it turns out that the function u_1 has to satisfy the equation

$$(1 + n^2) u_{1tt} - u_{1xx} + u_1 = -2\omega_0 \omega_1 u_{0tt} + f(\omega_0 u_{0t}). \tag{9.3.1}$$

Taking into account the fact that the right-hand side is to be perpendicular to $\sin t \sin nx$ and $\cos t \sin nx$, the authors obtain the values of a and ω_1. [Let us note (see, for example, the adjoint problem method) that the right-hand side of (9.3.1) has to be orthogonal to all solutions of the problem

$$(1 + n^2) v_{tt} - v_{xx} + v = 0, \quad (t, x) \in \mathbf{R} \times (0, \pi),$$
$$v(t, 0) = v(t, \pi) = 0, \quad t \in \mathbf{R},$$
$$v(t + 2\pi, x) = v(t, x), \quad (t, x) \in \mathbf{R} \times (0, \pi).$$

This problem has infinitely many solutions, because

$$-(1 + n^2) j^2 + k^2 + 1 = 0 \tag{9.3.2}$$

is a Pell equation. To avoid the presence of infinitely many conditions on only the two constants a and ω_1, we must take

$$u_0(t, x) = \sum_{(j,k) \in S} u_{j,k}\, e^{ijt} \sin kx,$$

where S is the set of "singular" pairs (j, k) satisfying (9.3.2). This step is essential if further approximations u_j, $j = 2, 3, \ldots$ are to be calculated.]

The same drawback can be found in the paper [96] by KELLER and TING, where the problem

$$u_{tt} - u_{xx} = f(u)\,, \quad (t, x) \in \mathbf{R} \times (0, \pi)\,,$$

$$u(t, 0) = u(t, \pi) = 0\,, \quad t \in \mathbf{R}\,,$$

$$u(t + \omega, x) = u(t, x)\,, \quad (t, x) \in \mathbf{R} \times [0, \pi]$$

is studied. Again, the fact that the equation

$$(k_0^2 - f'(0))\, j^2 - j_0^2 k^2 + j_0^2 f'(0) = 0$$

can have solutions other than $j = j_0$, $k = k_0$ is not taken into account.

GATIGNOL in [91] asserts that an equation of the form

$$u_{tt} - Eu + V'(u) = 0\,,$$

where E is an elliptic operator with constant coefficients and V is a positive convex function with $V(0) = 0$, has a periodic solution in the form of a travelling wave.

FRIEDLANDER [90] and KELLER [95] study periodic oscillations of a bowed string, whose motion is described by the equations

$$c^{-2}\, u_{tt} - u_{xx} = 0\,, \quad x \in [-b, 0) \cup (0, a]\,,$$

$$T\{u_x(t, 0+) - u_x(t, 0-)\} = -F(u_t(t, 0) - b)\,,$$

where T and b are constants and F is a given non-linear function.

BUSLAEV [85] investigates the existence of a periodic function u satisfying

$$u_{tt} - \Delta_n\, u = F(u)\,, \quad x \in \mathbf{R}^n$$

and tending rapidly to 0 as $|x| \to \infty$.

Chapter VI

The beam equation and related problems

§ 1. The equations of a beam and of a thin plate

1.1. *Equations without damping*

We shall deal with the problem $(\mathscr{P}_\omega)$ given by

$$u_{tt}(t, x) + E\, u(t, x) + c\, u(t, x) = g(t, x)\,, \quad (t, x) \in Q = \mathbf{R} \times \Omega\,, \quad (1.1.1)$$

$$L_j\, u(t, x) = 0\,, \quad (t, x) \in \mathbf{R} \times \partial\Omega\,, \quad j = 0, 1\,, \quad (1.1.2)$$

$$u(t + \omega, x) - u(t, x) = 0\,, \quad (t, x) \in Q\,, \quad (1.1.3)$$

where $\Omega \subset \mathbf{R}^n$ is a bounded domain, $c \in \mathbf{R}$, g is a function ω-periodic in t, the operator E is properly elliptic in Ω,

$$E\, v(x) = \sum_{|\alpha|, |\beta| \leqq 2} (-1)^{|\alpha|}\, D^\alpha(a_{\alpha\beta}(x)\, D^\beta\, v(x))\,, \quad x \in \Omega\,,$$

$$a_{\alpha\beta}(x) = a_{\beta\alpha}(x)\,, \quad |\alpha|, |\beta| \leqq 2\,, \quad x \in \Omega\,, \quad (1.1.4)$$

and the boundary operators L_j $(j = 0, 1)$ of the form

$$L_j\, v(x) = \sum_{|k| \leqq m_j} b_{jk}(x)\, D^k\, v(x)\,, \quad m_j \leqq 3\,, \quad x \in \partial\Omega$$

are E-admissible (see p. 33).

We assume that Ω is sufficiently regular, the $a_{\alpha\beta}$ and b_{jk} are sufficiently smooth and such that the operator $E : \mathscr{D}(E) \subset H^0(\Omega) \to H^0(\Omega)$,

$$\mathscr{D}(E) = \{v \in H^4(\Omega);\, L_j\, v = 0, j = 0, 1\}\,,$$

is self-adjoint and positive definite.

Under these assumptions we know from II: Theorem 1.3.1 that there is a non-decreasing sequence $\{\lambda_k\}_{k=1}^{\infty}$ of eigenvalues of E with ∞ as the only limit point, that all the eigenvalues are positive and of finite multiplicities, and that the corresponding system of eigenfunctions $\{v_k\}_{k=1}^{\infty}$ forms an orthonormal base in $H^0(\Omega)$.

By a solution to $(\mathscr{P}_\omega)$ we mean a function

$$u \in \mathscr{U} = H^0_\omega(\mathbf{R}; \mathscr{D}(E)) \cap H^{2,0}_\omega(Q)$$

that satisfies (1.1.1) almost everywhere in Q. As a subspace of $H^{2,4}_\omega(Q)$, $\mathscr{U}$ is equipped with the norm $\|\cdot\|_{H_\omega^{2,4}(Q)}$. However, as usual, another equivalent norm on $\mathscr{U}$ is of interest for our purposes, namely,

$$\|u\|^2_{\mathscr{U}} = \sum_{j,k} (j^4 + \lambda_k^2) \, |u_{jk}|^2 , \tag{1.1.5}$$

where

$$u_{jk} = \omega^{-1/2} \int_0^\omega \int_\Omega u(\tau, \xi) \, e^{-ij\nu\tau} \, v_k(\xi) \, d\xi \, d\tau , \quad (\nu = 2\pi/\omega)$$

(see II: Theorem 1.3.4).

The function g is assumed to belong to the space

$$\mathscr{G}(\gamma) = H^\alpha_\omega(\mathbf{R}; \mathscr{D}(E^{\gamma + (1-\alpha)/2})) ,$$

where $\gamma \geqq 0$ will be prescribed later and $\alpha \in [0, 2\gamma + 1]$ is arbitrary but fixed. This means, by II: Theorem 1.3.4, that

$$\|g\|^2_{\mathscr{G}(\gamma)} \sim \sum_{j,k} (1 + |j|^{2\alpha}) \, \lambda_k^{2\gamma + 1 - \alpha} \, |g_{jk}|^2 < \infty , \tag{1.1.6}$$

where

$$g_{jk} = \omega^{-1/2} \int_0^\omega \int_\Omega g(\tau, \xi) \, e^{-ij\nu\tau} \, v_k(\xi) \, d\xi \, d\tau .$$

We apply the (t, s)-Fourier method, that is, we use the expansions

$$u(t, x) = \omega^{-1/2} \sum_{j,k} u_{jk} \, e^{ij\nu t} \, v_k(x) , \tag{1.1.7}$$

$$g(t, x) = \omega^{-1/2} \sum_{j,k} g_{jk} \, e^{ij\nu t} \, v_k(x) , \tag{1.1.8}$$

which yield

$$(-\nu^2 j^2 + \lambda_k + c) \, u_{jk} = g_{jk} , \quad (j, k) \in \mathbf{Z} \times \mathbf{N} .$$

We put

$$S_1 = \{(j, k) \in \mathbf{Z} \times \mathbf{N}; \ -\nu^2 j^2 + \lambda_k + c = 0\} \tag{1.1.9}$$

and denote by S_2 the complement of S_1 in $\mathbf{Z} \times \mathbf{N}$. If $S_1 \neq \emptyset$ then $(\mathscr{P}_\omega)$ has a solution only if $g_{jk} = 0$ for all $(j, k) \in S_1$. For $(j, k) \in S_2$ we have $|u_{jk}|^2 = (-v^2j^2 + \lambda_k + c)^{-2} |g_{jk}|^2$. Writing

$$a_{jk} = (j^4 + \lambda_k^2)(-v^2j^2 + \lambda_k + c)^{-2},$$

we find by the same argument as in the proof of IV: Lemma 1.1.1 that $\{a_{jk}; (j, k) \in \Sigma_0 \cup \Sigma_1 \cup \Sigma_3\}$ is bounded. (For the notation see p. 132.) Further investigation (on Σ_2) depends essentially on the behaviour of the eigenvalues λ_k near infinity and on their connection with the numbers $v^2j^2 - c$. We make the following assumption:

there exist constants $m > 0$ and $\varkappa \geqq 0$ such that

$$|(\lambda_k + c)^{1/2} - v|j|| \geqq m\, \lambda_k^{-\varkappa}, \quad (j, k) \in S_2, \quad \lambda_k + c > 0. \tag{1.1.10}$$

Using (1.1.10) and IV: (1.1.10), we obtain for $(j, k) \in \Sigma_2 \cap S_2$

$$a_{jk} \leqq C \frac{(\Theta^4(\lambda_k + c)^2 + \lambda_k^2)\, \lambda_k^{2\varkappa}}{m^2(\Theta + 1)^2 (\lambda_k + c)} \leqq C\, \lambda_k^{2\varkappa+1} \leqq C|j|^{2s}\, \lambda_k^{2\varkappa+1-s},$$

where $s \in [0, 2\varkappa + 1]$ is arbitrary and C denotes various immaterial constants, independent of j, k, $\varkappa$, and s. Since $\bigcup_{l=0}^{3} \Sigma_l = (\mathbf{Z} \setminus \{0\}) \times \mathbf{N}$ and $\{a_{0k}; k \in \mathbf{N}, (0, k) \notin S_1\}$ is bounded, we conclude that

$$a_{jk} \leqq C(1 + |j|^{2s}\, \lambda_k^{2\varkappa+1-s}), \quad (j, k) \in S_2.$$

Thus, the series in (1.1.5) converges, provided that $g \in \mathscr{G}(\gamma)$, where $\gamma = \varkappa$ an $g_{jk} = 0$ for all $(j, k) \in S_1$.

Before summarizing our results let us denote (in the case $S_1 \neq \emptyset$) by $\mathscr{V}$ the set of functions $v \in H^0_\omega(Q)$ whose Fourier expansions are of the form

$$v(t, x) = \omega^{-1/2} \sum_{(j,k)\in S_1} v_{jk}\, e^{ijvt}\, v_k(x),$$

and by P the orthogonal projection of $H^0_\omega(Q)$ onto $\mathscr{V}$.

THEOREM 1.1.1 *Suppose that* (1.1.10) *holds and that* $g \in \mathscr{G}(\varkappa)$. *If* S_1 *given by* (1.1.9) *is empty, then* $(\mathscr{P}_\omega)$ *has a unique solution* $u \in \mathscr{U}$, $u = \Lambda g$, *and* $\Lambda \in \mathscr{L}(\mathscr{G}(\varkappa), \mathscr{U})$.

THEOREM 1.1.2 *Suppose that* (1.1.10) *holds and that* $g \in \mathscr{G}(\varkappa)$. *Let* $S_1 \neq \emptyset$. *Then* $(\mathscr{P}_\omega)$ *has a solution* $u \in \mathscr{U}$ *if and only if*

$$Pg = 0. \tag{1.1.11}$$

If (1.1.11) *is satisfied, then the solutions can be written in the form* $u = v + w$, *where* $v \in \mathscr{U} \cap \mathscr{V}$ *is arbitrary,* $w = \Lambda g$, *and* $\Lambda \in \mathscr{L}(\mathscr{G}(\varkappa) \cap \mathscr{V}^{\perp}, \mathscr{U} \cap \mathscr{V}^{\perp})$.

Remark 1.1.1 Our investigation includes as a special case

(α) *the beam equation*: $\Omega = (0, \pi)$, $E + c = D_x^4$; for example, the following boundary conditions are admissible:

$$u(t, 0) = u_x(t, 0) = u(t, \pi) = u_x(t, \pi) = 0$$

or

$$u(t, 0) = u_{xx}(t, 0) = u(t, \pi) = u_{xx}(t, \pi) = 0$$

or

$$u(t, 0) = u_x(t, 0) = u_{xx}(t, \pi) = u_{xxx}(t, \pi) = 0\,;$$

(β) *the equation of a thin plate*: $\Omega \subset \mathbf{R}^2$ is a bounded domain, $E + c = \Delta^2$, and the L_j are, for example, of the Dirichlet type, that is, $u = \partial u/\partial v = 0$ on $\mathbf{R} \times \partial\Omega$.

Remark 1.1.2 The operator E is of order 4. Of course, the same theory (with only formal changes) can be developed for operators of order $2m$, $m \in \mathbf{N}$, as well as for abstract operators (see VII: [40]).

Problem 1.1.1 Prove that $u \in \mathscr{U}$ implies that $u_t \in H_\omega^0(\mathbf{R}; \mathscr{D}(E^{1/2}))$.

Problem 1.1.2 Investigate the corresponding weakly non-linear case.

Hint: If $S_1 \neq \emptyset$, make the following assumptions:

(a) the equation $P\,F(v_0, 0) = 0$ has a solution $v_0 = v_0^* \in \mathscr{U} \cap \mathscr{V}$;

(b) for some $\varrho > 0$ the operator $F : B(v_0^*, \varrho; \mathscr{U}) \times [-\varepsilon_0, \varepsilon_0] \to \mathscr{G}(\varkappa)$ is continuous and has a continuous G-derivative F'_u;

(c) $[P\,F'_u(v_0^*, 0)]^{-1} \in \mathscr{L}(\mathscr{G}(\varkappa) \cap \mathscr{V}, \mathscr{U} \cap \mathscr{V})$.

See V: Theorems 6.4.1 and 6.4.2.

As an example of the above theory, let us investigate in more detail the problem $(\mathscr{P}_\omega^1)$ given by

$$u_{tt}(t, x) + u_{xxxx}(t, x) = g(t, x)\,, \quad (t, x) \in \mathbf{R} \times (0, \pi)\,, \tag{1.1.12}$$

$$u(t, 0) = u_{xx}(t, 0) = u(t, \pi) = u_{xx}(t, \pi) = 0\,, \quad t \in \mathbf{R}\,, \tag{1.1.13}$$

$$u(t + \omega, x) - u(t, x) = 0\,, \quad (t, x) \in \mathbf{R} \times (0, \pi) \tag{1.1.14}$$

with $\omega = 2\pi p/q$, $p, q \in \mathbf{N}$ relatively prime.

Here $E = D_x^4$, $\mathscr{D}(E) = {}^0H^4(0, \pi)$ and the corresponding eigenvalues and eigenfunctions are

$$\lambda_k = k^4 \quad \text{and} \quad v_k(x) = (2/\pi)^{1/2} \sin kx\,, \quad k \in \mathbf{N}\,,$$

respectively. The space $\mathscr{U}$ can be obtained as the closure of

$$\{u \in C_\omega^\infty(\mathbf{R} \times [0, \pi]);\, u(t, 0) = u_{xx}(t, 0) = u(t, \pi) = u_{xx}(t, \pi) = 0,\, t \in \mathbf{R}\}$$

in $H_\omega^{2,4}(\mathbf{R} \times (0, \pi))$.

First, let us determine the set S_1, that is, the set of all $(j, k) \in \mathbf{Z} \times \mathbf{N}$ such that $-|j|\, q/p + k^2 = 0$. For $(j, k) \in S_1$ we have $k^2 p = |j|\, q$ which means that $|j|\, q$ is divisible by p, and since p, q are relatively prime, there exists a $j' \in \mathbf{N}$ such that $|j| = j'p$, hence $k^2 = j'q$. Writing $q = q_1 q_2^2$, where q_1 is square-free, we see that q_2 divides k, and setting $k = k'q_2$, $k' \in \mathbf{N}$, we easily find that q_1 divides k', $k' = nq_1$, $n \in \mathbf{N}$. Hence, $k = nq_1q_2$, therefore, $|j| = n^2pq_1$, $n \in \mathbf{N}$. On the other hand, if k and j satisfy the above relations, then $(j, k) \in S_1$. So we have proved the following result.

LEMMA 1.1.1 *Let $p, q \in \mathbf{N}$ be relatively prime, $q = q_1q_2^2$, q_1 square-free. Then*

$$S_1 = \{(j, k) \in \mathbf{Z} \times \mathbf{N};\, -|j|\, q/p + k^2 = 0\} =$$
$$= \{(j, k) \in \mathbf{Z} \times \mathbf{N};\, |j| = n^2pq_1,\, k = nq_1q_2,\, n \in \mathbf{N}\}\,.$$

For $(j, k) \in S_2 = (\mathbf{Z} \times \mathbf{N}) \setminus S_1$ we have evidently

$$|k^2 - |j|\, q/p| \geqq 1/p\,,$$

that is, (1.1.10) (with $c = 0$) holds for $\varkappa = 0$.

Thus, the assertion of Theorem 1.1.2 is valid, provided that $g \in \mathscr{G}(0)$. The condition (1.1.11) is equivalent to

$$\int_0^\omega \int_0^\pi g(\tau, \xi)\, e^{\pm in^2q_1{}^2q_2{}^2\tau} \sin nq_1q_2\xi \,\mathrm{d}\xi\,\mathrm{d}\tau = 0\,, \quad n \in \mathbf{N}\,. \tag{1.1.15}$$

Let us mention two cases (analogous to those discussed in V: Sec. 1.3), in which the conditions (1.1.15) are automatically satisfied. This happens if either

(a) $p = 2p' - 1$, $q = 2q'$, and $g(t, x) = g(t, \pi - x)$

or

(b) $p = 2p'$, $q = 2q' - 1$, and $g(t + \omega/2, x) = -g(t, x)$.

Problem 1.1.3 Prove that $u \in H_\omega^0(\mathbf{R};\, {}^0H^4(0, \pi)) \cap H_\omega^{2,0}(Q)$ implies that $u \in C_\omega^{1,2}(\bar{Q})$, $u_{tx}, u_{xxx} \in L_{6,\omega}(Q)$ and $u_{tt}, u_{txx}, u_{xxxx} \in H_\omega^0(Q)$ $(Q = \mathbf{R} \times (0, \pi))$.

1.2. *Equations with a damping term*

Now we investigate the problem $(\mathscr{P}_\omega)$ given by

$$u_{tt}(t, x) + a\, u_t(t, x) + E\, u(t, x) + c\, u(t, x) = g(t, x)\,,$$
$$(t, x) \in Q = \mathbf{R} \times \Omega \tag{1.2.1}$$

and (1.1.2), (1.1.3).

The assumptions introduced in Sec. 1.1 remain unchanged and $a \in \mathbf{R}$, $a \neq 0$. We keep the previous notation.

By a solution we mean a function $u \in \mathscr{U}$ satisfying (1.2.1) almost everywhere in Q. The solutions are sought again in the form (1.1.7). The presence of the damping term au_t makes the investigation much easier in comparison with the problem we studied before, because we need not take into account the behaviour of the eigenvalues λ_k at ∞ and their connections with the numbers $v^2 j^2 - c$.

The Fourier coefficients of a solution have to satisfy the relation

$$(-v^2 j^2 + \lambda_k + c + avji)\, u_{jk} = g_{jk}\,.$$

The set S_1 of $(j, k) \in \mathbf{Z} \times \mathbf{N}$, in which the coefficient of u_{jk} vanishes, is empty except when $-c$ is an eigenvalue of E. In this case $S_1 = \{(0, k);\, k \in \mathbf{N},\, \lambda_k + c = 0\}$ contains precisely r elements, where r is the multiplicity of $-c$.

For $(j, k) \in S_2 = (\mathbf{Z} \times \mathbf{N}) \setminus S_1$ we have $|u_{jk}|^2 = ((-v^2 j^2 + \lambda_k + c)^2 + a^2 v^2 j^2)^{-1}\, |g_{jk}|^2$. As an immediate consequence of IV: Lemma 1.1.1 (p. 132) we obtain

$$(j^4 + \lambda_k^2)\,((-v^2 j^2 + \lambda_k + c)^2 + a^2 v^2 j^2)^{-1} \leqq C(1 + |j|^{2s}\, \lambda_k^{1-s})\,,$$
$$(j, k) \in S_2\,,$$

where $s \in [0, 1]$ is arbitrary and $C > 0$ is a constant independent of j, k, and s.

We have proved the following two theorems.

THEOREM 1.2.1 *Let* $g \in \mathscr{G}(0) = H^\alpha_\omega(\mathbf{R}; \mathscr{D}(E^{(1-\alpha)/2}))$, *where* $\alpha \in [0, 1]$ *is arbitrary but fixed. Let* $c \neq -\lambda_k$ *for all* $k \in \mathbf{N}$.
Then the problem $(\mathscr{P}_\omega)$ *has a unique solution* $u \in \mathscr{U}$, $u = \Lambda g$, *and* $\Lambda \in \mathscr{L}(\mathscr{G}(0), \mathscr{U})$.

THEOREM 1.2.2 *Let* $g \in \mathscr{G}(0)$. *Let* $-c$ *be an eigenvalue of* E.
Then the problem $(\mathscr{P}_\omega)$ *has a solution* $u \in \mathscr{U}$ *if and only if* $Pg = 0$. *If this condition is satisfied, the solutions can be written in the form* $u = v + w$, *where* $v \in \mathscr{U} \cap \mathscr{V}$ *is arbitrary,* $w = \Lambda g$, *and* $\Lambda \in \mathscr{L}(\mathscr{G}(0) \cap \mathscr{V}^\perp, \mathscr{U} \cap \mathscr{V}^\perp)$.

Example 1.2.1 Let $\Omega = (0, \pi)$, $E = D_x^4$, $c = 0$ and let the boundary conditions be of the form (1.1.13).
Then the problem given by (1.2.1), (1.1.13)–(1.1.14) has a unique solution $u \in H_\omega^0(\mathbf{R}; {}^0H^4(0, \pi)) \cap H_\omega^{2,0}(Q)$ provided that $g \in H_\omega^{1,0}(Q)$ or $g \in H_\omega^0(\mathbf{R}; {}^0H^2(0, \pi))$.

Problem 1.2.1 Investigate the corresponding weakly non-linear problem.
Hint: see III: Sec. 1.3.

Problem 1.2.2 (*The Rayleigh equation*). Study the problem given by

$$-\mu\, u_{ttxx}(t, x) + u_{tt}(t, x) + u_{xxxx}(t, x) + a\, u_t(t, x) = g(t, x)\,,$$
$$(t, x) \in \mathbf{R} \times (0, \pi)$$

and (1.1.13)–(1.1.14), where $a \in \mathbf{R}$, $a \neq 0$, and $\mu > 0$ (a small parameter). Prove that for every $g \in H_\omega^{3,0}(Q)$ (or $g \in H_\omega^0(\mathbf{R}; {}^0H^3(0, \pi))$) there exists a unique solution u in the space $\{u, u_{xx} \in H_\omega^2(Q) \cap \dot{H}_\omega^1(Q)\}$. Moreover, u depends continuously on g.

Hint: Use the (t, s)-Fourier method. When estimating

$$(j^4k^4 + k^8)\,[(\mu k^2 + 1)^2\,(-\nu|j| + \xi(k))^2\,(\nu|j| + \xi(k))^2 + a^2\nu^2 j^2]^{-1}$$

where $\xi(k) = k^2(\mu k^2 + 1)^{-1/2}$, decompose $\mathbf{Z} \times \mathbf{N}$ as in the proof of IV: Lemma 1.1.1. The only difference is that $|j| = \nu^{-1}\,\xi(k)$ plays the role of the "dangerous curve" instead of $|j| = \nu^{-1}(\lambda_k + c)^{1/2}$. Show that $c_0 k \leqq \xi(k) \leqq \mu^{-1/2}k$, $k \in \mathbf{N}$, with a suitable $c_0 \in (0, \mu^{-1/2})$.

Problem 1.2.3 Let $g \in H_\omega^{3,0}(Q)$, let u^μ be a solution of the problem 1.2.2, and let u^0 be a solution of the problem in Example 1.2.1. Study the behaviour of $u^\mu - u^0$ in a suitable norm for $\mu \to 0+$. Prove, for example, that

$$\|u^\mu - u^0\|_{H_\omega{}^2(Q)} \leqq \text{const. } \mu \|g\|_{H_\omega{}^{3,0}(Q)}\,.$$

Hint: see IV: § 3.

Problem 1.2.4 (*The Timoshenko equation*). Study the problem given by

$$\mu_1\mu_2\, u_{tttt}(t, x) - (\mu_1 + \mu_2)\, u_{ttxx}(t, x) + u_{tt}(t, x) +$$
$$+ u_{xxxx}(t, x) + a(u_t(t, x) + \mu_2(\mu_1\, u_{ttt}(t, x) - u_{txx}(t, x))) =$$
$$= g(t, x) + \mu_2(\mu_1\, g_{tt}(t, x) - g_{xx}(t, x))\,, \quad (t, x) \in \mathbf{R} \times (0, \pi)$$

and (1.1.13)–(1.1.14). Here $a \in \mathbf{R}$, $a \neq 0$, $\mu_1 > 0$, $\mu_2 > 0$ (small parameters). Prove that for every $g \in H_\omega^0(\mathbf{R}; {}^0H^2(0, \pi)) \cap H_\omega^3(\mathbf{R}; H^0(0, \pi))$ (or $g \in H_\omega^0(\mathbf{R};$

${}^{0}H^3(0, \pi)) \cap H^2_\omega(\mathbf{R}; H^0(0, \pi)))$ there exists a unique solution (depending continuously on g) $u \in H^4_\omega(Q) \cap \{u, u_{xx} \in \dot{H}^1_\omega(Q)\}$. Investigate the behaviour of this solution as $\mu_2 \to 0+$, in comparison with the solution of the Rayleigh equation.

Problem 1.2.5 Study the smoothness properties of functions in the spaces introduced in the preceding problems.

§ 2. Supplements and comments

2.1. *The adjoint problem method*

Let $(\mathscr{P}_\omega)$ be the problem

$$u_{tt}(t, x) + u_{xxxx}(t, x) = g(t, x), \quad (t, x) \in \mathbf{R} \times (0, \pi) \tag{2.1.1}$$

with the boundary conditions

$$M(t)\, \mathrm{J}_4\, u(t, 0) + N(t)\, \mathrm{J}_4\, u(t, \pi) = h(t), \quad t \in \mathbf{R}, \tag{2.1.2}$$

$$u(0, x) - u(\omega, x) = 0,$$
$$u_t(0, x) - u_t(\omega, x) = 0, \quad x \in [0, \pi], \tag{2.1.3}$$

where

$$\mathrm{J}_4\, u(t, x) = (u(t, x), u_x(t, x), u_{xx}(t, x), u_{xxx}(t, x))',$$

M and N are 4×4 -matrices whose elements belong to $C_\omega(\mathbf{R})$, rank $(M, N) = 4$ for all $t \in \mathbf{R}$, $h = (h_0, h_1, h_2, h_3)'$, $h_0, \ldots, h_3 \in C_\omega(\mathbf{R})$, and $g \in C_\omega(\mathbf{R} \times [0, \pi])$.

Following the procedure in III: Sec. 4.1 and using the results of II: Sec. 1.2, we find that the formally adjoint problem $(\mathscr{P}^*_\omega)$ is given by

$$v_{tt}(t, x) + v_{xxxx}(t, x) = 0, \quad (t, x) \in \mathbf{R} \times (0, \pi),$$
$$\mathrm{J}_4\, v(t, 0)'\, P(t) + \mathrm{J}_4\, v(t, \pi)'\, Q(t) = 0, \quad t \in \mathbf{R},$$
$$v(0, x) - v(\omega, x) = 0,$$
$$v_t(0, x) - v_t(\omega, x) = 0, \quad x \in [0, \pi],$$

where the 4×4 -matrices P and Q are determined by

$$-M(t)\, B^{-1}\, P(t) + N(t)\, B^{-1}\, Q(t) = 0,$$
$$\text{rank } (P', Q') = 4, \quad t \in \mathbf{R}$$

while B is given by II: (1.2.3).

The matrices P_c and Q_c defining the complementary adjoint (spatial) forms are determined by

$$-M(t)\,B^{-1}\,P_c(t) + N(t)\,B^{-1}\,Q_c(t) = I\,,$$

$$\operatorname{rank}\,(P_c^{\backprime}, Q_c^{\backprime}) = 4\,, \quad t \in \mathbf{R}\,.$$

In analogy to II: Theorem 1.2.1 and III: Theorem 4.1.1 we obtain the following lemma.

LEMMA 2.1.1 *The problem $(\mathscr{P}_\omega)$ has a solution only if*

$$\int_0^\omega \int_0^\pi v(\tau, \xi)\, g(\tau, \xi)\, d\xi\, d\tau - \int_0^\omega (J_4\, v(\tau, 0)^{\backprime}\, P_c(\tau) +$$

$$+ J_4\, v(\tau, \pi)^{\backprime}\, Q_c(\tau))\, h(\tau)\, d\tau = 0$$

for every solution v of the problem $(\mathscr{P}_\omega^*)$.

2.2. *Results of Krylová, Vejvoda, Kopáčková, Solov'ev, Karimov, Mitrjakov and Filip. (Results based on Fourier methods)*

In [15] (and in a shortened form in [16]) KRYLOVÁ and VEJVODA investigate the existence of a classical solution to the problem $(\mathscr{P}_\omega^1)$ given by (1.1.12)–(1.1.14) (and of the corresponding weakly non-linear problem). The function g is assumed to belong to $C_\omega(\mathbf{R};\, {}^0H^3(0, \pi))$ and the solutions are sought in the space $C_\omega(\mathbf{R};\, {}^0H^5(0, \pi)) \cap C_\omega^1(\mathbf{R};\, {}^0H^3(0, \pi)) \cap C_\omega^2(\mathbf{R};\, {}^0H^1(0, \pi))$.

Assuming that $\varphi \in {}^0H^5(0, \pi)$, $\psi \in {}^0H^3(0, \pi)$, the authors write the solution to the associated initial boundary-value problem in the form

$$u(t, x) = (2\pi)^{-1/2} \left\{ \int_{(-\infty,+\infty)}^{\rightarrow} (\cos \lambda^2 + \sin \lambda^2)\, \varphi_e(x - 2t^{1/2}\,\lambda)\, d\lambda - \right.$$

$$- \int_{(-\infty,+\infty)}^{\rightarrow} (\cos \lambda^2 - \sin \lambda^2)\, \Psi(x - 2t^{1/2}\,\lambda)\, d\lambda -$$

$$\left. - \int_0^t \int_{(-\infty,+\infty)}^{\rightarrow} (\cos \lambda^2 - \sin \lambda^2)\, G(\tau, x - 2(t - \tau)^{1/2}\,\lambda)\, d\lambda\, d\tau \right\},$$

where

$$\Psi(x) = \int_0^x \int_0^\xi \psi_e(\eta)\, d\eta\, d\xi - \frac{x}{2\pi} \int_0^{2\pi} \int_0^\xi \psi_e(\eta)\, d\eta\, d\xi\,,$$

$$G(t, x) = \int_0^x \int_0^\xi g(t, \eta)\, d\eta\, d\xi - \frac{x}{2\pi} \int_0^{2\pi} \int_0^\xi g(t, \eta)\, d\eta\, d\xi\,,$$

(the integrals are meant in the generalized Abel sense, that is,

$$\int_{(-\infty,+\infty)}^{\rightarrow} f(\xi)\,\mathrm{d}\xi = \lim_{\delta\to 0+} \int_{-\infty}^{+\infty} \exp(-\delta\xi^2) f(\xi)\,\mathrm{d}\xi \Big),$$

or in the form

$$u(t, x) = \sum_{k=1}^{\infty} (\varphi_k \cos k^2 t + \psi_k k^{-2} \sin k^2 t +$$

$$+ k^{-2} \int_0^t \sin k^2(t - \tau)\, g_k(\tau)\,\mathrm{d}\tau) \sin kx\,,$$

where φ_k, ψ_k, and $g_k(t)$ are the Fourier coefficients of $\varphi(x)$, $\psi(x)$, and $g(t, x)$, respectively, relative to the system $\{\sin kx\}_{k=1}^{\infty}$. Then the Poincaré method is applied; the cases $\omega = 2\pi p/q$, $p, q \in \mathbf{N}$ and $\omega = 2\pi\alpha$, α irrational, are treated separately. It is shown that for $\omega = 2\pi p/q$, $p, q \in \mathbf{N}$ relatively prime, the conditions (1.1.15) can be written equivalently as follows:

$$\sum_{j=1}^{q} \int_0^{\omega} \int_{(-\infty,+\infty)}^{\rightarrow} (\cos \lambda^2 \mp \sin \lambda^2)\, g(\tau, x - 2(\omega - \tau)^{1/2} \lambda)\,\mathrm{d}\lambda\,\mathrm{d}\tau = 0\,.$$

On the other hand, if $\omega = 2\pi\alpha$, α irrational, the existence of a solution is established, provided that there are constants $m > 0$, $\varrho \in \mathbf{N}$, $\varrho \geqq 3$, such that

$$|\alpha - lk^{-2}| \geqq mk^{-\varrho}\,, \quad l, k \in \mathbf{N}\,,$$

and $g \in C_\omega(\mathbf{R}; {}^0H^{\varrho+1}(0, \pi))$ (hence only the cases $\varrho = 3$ and 4 are of interest in the weakly non-linear problem with a substitution operator).

Applying the t-Fourier method, Kopáčková in [11] proves that the problem $(\mathscr{P}_\omega^1)$ has a classical solution $u \in C^4([0, \pi]; H_\omega^1(\mathbf{R})) \cap C([0, \pi]; H_\omega^3(\mathbf{R}))$, provided that there are constants $m > 0$, $\varrho \geqq 0$, such that

$$\inf_{k\in\mathbf{N}} |(\nu|j|)^{1/2} - k| \geqq m|j|^{-\varrho}$$

for all $j \in \mathbf{Z} \setminus Z_1$, $Z_1 = \{j \in \mathbf{Z};\ (\nu|j|)^{1/2} \in \mathbf{N}\}$ $(\nu = 2\pi/\omega)$, while $g \in C([0, \pi]; H_\omega^r(\mathbf{R}))$, where $r \in \mathbf{N}$, $r \geqq \varrho + 3/2$, and

$$\int_0^{\pi} \int_0^{\omega} g(\tau, \xi) \sin [(\nu|j|)^{1/2} \xi]\, e^{ij\nu\tau}\,\mathrm{d}\tau\,\mathrm{d}\xi = 0\,, \quad j \in Z_1 \quad \text{if} \quad Z_1 \neq \emptyset\,.$$

Using the s-Fourier method, Solov'ev [25], Karimov [8]–[10], Mitrjakov [19], and Krylová and Vejvoda [15] study the problem

$$u_{tt}(t, x) + u_{xxxx}(t, x) = \varepsilon F(u, \varepsilon)(t, x)\,,$$

(1.1.13) – (1.1.14) with $\omega = 2\pi(2p' - 1)/2q'$, $p', q' \in \mathbf{N}$ and $F(u, \varepsilon)(t, x) = F(u, \varepsilon)(t, \pi - x)$, under various assumptions on F. The most general result is in [15] (except for the papers [9] and [10] which allow ε to be arbitrarily large; however, the proofs are incomplete).

The results of [15] are applied in the paper [12] by KOPÁČKOVÁ and VEJVODA, where the equation

$$u_{tt}(t, x) + u_{xxxx}(t, x) = g(t, x) + \varepsilon(f(t, x) + u_{xx}(t, x) \int_0^\pi u_x^2(t, \xi)\, d\xi + \varepsilon F(u)(t, x))$$

is investigated.

Using the (t, s)-Fourier method, FILIP [3] deals with periodic solutions of the equation

$$u_{tt}(t, x, y) + \Delta^2 u(t, x, y) = g(t, x, y), \quad (t, x, y) \in \mathbf{R} \times \Omega,$$

which describes simply supported plates, where $\Omega \subset \mathbf{R}^2$ is a rectangle. The corresponding weakly non-linear problem is also studied.

2.3. *Results of Kurzweil, Hall, Petzeltová and Nakao. (Results based on monotonicity of perturbations)*

A number of papers are devoted to problems with a non-linear perturbation that is monotone with respect to u or u_t. In what follows, ε is a small parameter and all functions in question are supposed to be 2π-periodic in t.

KURZWEIL [18] asserts (without proof) that the problem

$$u_{tt} + u_{xxxx} = \varepsilon(-g(x, u)\, u_t + h(t, x)),$$

(1.1.13)–(1.1.14) has an exponentially stable 2π-periodic solution, provided that $g(x, u) \geqq \text{const.} > 0$.

HALL [4] deals with solutions that are 2π-periodic in both t and x of the equation

$$\Box_p u \equiv u_{tt} + (-1)^p D_x^{2p} u = \varepsilon f(t, x, u), \quad (p \in \mathbf{N}), \tag{2.3.1}$$

under monotonicity and growth conditions on f. Weak solutions from L_r-spaces ($r \in [2, \infty)$) are looked for and the general result applied to the equation

$$\Box_p u = \varepsilon(|u|^{r-1} \operatorname{sgn} u - g(t, x)).$$

In [5], the same author studies more regular solutions of (2.3.1) with f satisfying the strong monotonicity condition $f_u(t, x, u) \geqq \text{const.} > 0$, the growth condition $\|f_t(t, x, u)\|_{L_2} \leqq \text{const.}\,(1 + |u|)$, and then the equation

$$\Box_p u = \varepsilon(\alpha\, u_t + f(t, x, u)), \quad (\alpha > 0)$$

where the only assumptions on f and f_u are smallness with respect to α. (For some further details see V: Sec. 2.5.)

In [6] he examines the existence of a solution that is ω_0-periodic in t and ω_j-periodic in x_j ($j = 1, \ldots, n$) (belonging to L_r) of the equation

$$c^2 u_{tt} + (-1)^p \Delta_n^{2p} u = b|u|^{r-1} \operatorname{sgn} u + g(t, x),$$

where $x = (x_1, \ldots, x_n)$, $p \in \mathbf{N}$, $r \in (2, \infty)$, b, $c \in \mathbf{R}$. He makes the following assumptions:

(a) for each j, $v_j^2/(c\, v_0)^{1/p}$ is rational ($v_j = 2\pi/\omega_j$);
(b) $r \in (2, r_0)$, where $r_0 = \infty$ if $n \leqq 2p$ and $r_0 = 2n/(n - 2p)$ if $n > 2p$;
(c) either b or

$$\int_0^{\omega_0} \int_0^{\omega_1} \cdots \int_0^{\omega_n} |g(\tau, \xi_1, \ldots, \xi_n)|^{r/(r-1)} \, d\tau \, d\xi_1 \ldots d\xi_n$$

is sufficiently small.

Petzeltová [24] investigates first the existence of solutions in $H_{2\pi}^{m,2m}(\mathbf{R} \times (0, \pi))$ ($m \in \mathbf{N}$, $m \geqq 2$) of the problem

$$\Box_2 u \equiv u_{tt} + u_{xxxx} = \varepsilon f(t, x, u, u_t)$$

and (1.1.13)–(1.1.14). Here $f_{u_t} \geqq \gamma > 0$, and a certain additional relation between γ and f_u is assumed to hold (see also V: Theorem 2.6.2). Then she treats the equation

$$\Box_2 u = \varepsilon(\alpha\, u_t + f(t, x, u)), \quad (\alpha > 0),$$

where f is small with respect to α. Finally, she deals with the n-dimensional problem

$$u_{tt} + \Delta_n^2 u = \varepsilon f(t, x, u) \quad (n = 1, 2, 3)$$

under the assumption that $f_u \geqq \text{const.} > 0$.

Nakao [23] extending his results V: [44] studies solutions that are regular 2π-periodic in t and x of the equation

$$\Box_p u = f(t, x, u),$$

where f satisfies certain monotonicity and growth conditions.

2.4. *Results of Štědrý, Krylová and Krejčí.* (*Problems with damping*)

Štědrý in [27] proves the existence of solutions that are 2π-periodic in t of the problem

$$u_{tt} + u_{xxxx} + a\, u_t = \varepsilon f(t, x, u, u_t, u_x, u_{xx}, u_{xxxx})$$

or

$$u_{tt} + u_{xxxx} + a\,u_t = \varepsilon f(t, x, u, u_t, u_x, u_{tt}, u_{xx})$$

with the boundary condition (1.1.13), provided that f is smooth, 2π-periodic in t, $a \neq 0$, and $|\varepsilon|$ is sufficiently small. He uses the method developed by RABINOWITZ in IV: [33].

KRYLOVÁ [14] applies her general result (quoted in IV: Sec. 5.1) to the strongly non-linear equation

$$u_{tt} + \Delta_2^2 u + c\,u + u_t + u_t|u_t| = g\,,$$

with $c \geqq 0$ and various boundary conditions.

KREJČÍ [13] investigates solutions that are 2π-periodic in t and x of the equations

$$u_{tt} + u_{xxxx} + a\,u_t = \beta(u) + g$$

and

$$u_{tt} + u_{xxxx} + a\,u_t = \mu\,u^+ - \nu\,u^- + \varphi(u) + g \qquad (a \neq 0)$$

following MAWHIN IV: [22] and FUČÍK and MAWHIN IV: [13], respectively. The latter problem is much more complicated than the corresponding one for the telegraph equation.

The papers IV: [4], [7], [37] and [40] include results that can also be applied to the beam equation with damping.

§ 3. The dynamic von Kármán equations of thin plates involving rotational inertia and damping

3.1. *Preliminaries*

Let $\Omega \subset \mathbf{R}^2$ be a bounded (sufficiently regular) domain. Given two functions u and v defined on Ω, we put

$$[u, v] = u_{xx}\,v_{yy} + u_{yy}\,v_{xx} - 2u_{xy}\,v_{xy}\,. \tag{3.1.1}$$

We mention the basic properties of the bracket $[u, v]$ (see LIONS II: [13]):

(α) the mapping $u, v \to [u, v]$ is bilinear and continuous from $\dot{H}^2(\Omega) \times \dot{H}^2(\Omega)$ into $H^{-2}(\Omega)$;

(β) the trilinear form $u, v, w \to \langle [u, v], w\rangle$ is continuous and symmetric on $\dot{H}^2(\Omega)$ ($\langle\cdot,\cdot\rangle$ denotes the duality between $\dot{H}^2(\Omega)$ and its dual $H^{-2}(\Omega)$).

The problem in question consists in finding a pair of functions u_1 and u_2, defined on $\mathbf{R} \times \Omega$ and satisfying

$$u_1'' - b\Delta u_1'' + a_1\Delta^2 u_1 + d_1 u_1' - d_2\Delta u_1' - [u_1, u_2] = g \quad \text{on} \quad \mathbf{R} \times \Omega, \tag{3.1.2}$$

$$a_2\Delta^2 u_2 + [u_1, u_1] = 0 \quad \text{on} \quad \mathbf{R} \times \Omega, \tag{3.1.3}$$

$$u_1 = \frac{\partial u_1}{\partial \nu} = 0,$$

$$u_2 = \frac{\partial u_2}{\partial \nu} = 0 \quad \text{on} \quad \mathbf{R} \times \partial\Omega, \tag{3.1.4}$$

and

$$u_1(t + \omega, x, y) - u_1(t, x, y) = 0, \quad (t, x, y) \in \mathbf{R} \times \Omega. \tag{3.1.5}$$

Here a_1, a_2, b, d_1 and d_2 are positive constants, $u_1' = \partial u_1/\partial t$, $u_1'' = \partial^2 u_1/\partial t^2$, Δ^2 is the biharmonic operator, $\Delta^2 = \partial^4/\partial x^4 + 2\partial^4/\partial x^2\,\partial y^2 + \partial^4/\partial y^4$, $\partial/\partial\nu$ is the outward normal derivative to $\partial\Omega$, and g is a function on $\mathbf{R} \times \Omega$ that is ω-periodic in t.

The corresponding initial boundary-value problem given by (3.1.2)–(3.1.4) and by

$$u_1(0, x, y) = u_{01}(x, y),$$
$$u_1'(0, x, y) = u_{11}(x, y), \quad (x, y) \in \Omega \tag{3.1.6}$$

is studied in II: [13], pp. 43–53 (with $d_1 = d_2 = 0$). Under the assumptions

$$u_{01} \in \dot{H}^2(\Omega), \quad u_{11} \in \dot{H}^1(\Omega),$$
$$g \in H^0((0, \omega) \times \Omega),$$

the existence and uniqueness of a weak solution

$$u_1 \in L_\infty(0, \omega; \dot{H}^2(\Omega)) \cap W^1_\infty(0, \omega; \dot{H}^1(\Omega)),$$
$$u_2 \in L_\infty(0, \omega; \dot{H}^2(\Omega)) \tag{3.1.7}$$

is established. (Note that the identities (3.1.6) make sense because it can be shown that $u_1'' \in L_\infty(0, \omega; H^0(\Omega))$ and, consequently, u_1' is equal (almost everywhere in $[0, \omega]$) to a function in $C([0, \omega]; H^0(\Omega))$, whereas u_1 is equal to a function in $C([0, \omega]; \dot{H}^1(\Omega))$, see, for example, I: [3], p. 19.) The initial boundary-value problem remains uniquely soluble for $d_1 > 0$, $d_2 > 0$. Hence the condition (3.1.5) can be replaced by

$$u_1(0, x, y) = u_1(\omega, x, y),$$
$$u_1'(0, x, y) = u_1'(\omega, x, y), \quad (x, y) \in \Omega. \tag{3.1.5'}$$

Following II: [13], we look for a weak solution (u_1, u_2) of the problem $(\mathscr{P}_\omega)$ given by (3.1.2)–(3.1.4), (3.1.5') that fulfils (3.1.7),

$$u_1(0) = u_1(\omega)\,,$$

$$\int_0^\omega (-\langle u_1', \psi' \rangle - b\langle \operatorname{grad} u_1', \operatorname{grad} \psi' \rangle +$$

$$+ a_1\langle \Delta u_1, \Delta\psi \rangle + d_1\langle u_1', \psi \rangle +$$

$$+ d_2\langle \operatorname{grad} u_1', \operatorname{grad} \psi \rangle - \langle [u_1, u_2], \psi \rangle - \langle g, \psi \rangle)\,\mathrm{d}t = 0 \tag{3.1.8}$$

for all $\psi \in L_\infty(0, \omega; \dot{H}^2(\Omega))$ such that $\psi' \in L_\infty(0, \omega; \dot{H}^1(\Omega))$, $\psi(0) = \psi(\omega)$ and

$$a_2\langle \Delta u_2, \Delta\varphi \rangle + \langle [u_1, u_1], \varphi \rangle = 0 \tag{3.1.9}$$

for almost all t and for all $\varphi \in \dot{H}^2(\Omega)$. Throughout this paragraph $\langle \cdot, \cdot \rangle$ denotes the scalar product in $H^0(\Omega)$ ($\langle \operatorname{grad} v, \operatorname{grad} w \rangle = \langle v_x, w_x \rangle + \langle v_y, w_y \rangle$) or the duality between $\dot{H}^2(\Omega)$ and $H^{-2}(\Omega)$.

Since the problem in question is strongly non-linear, the approach via the Fourier method is no longer convenient. We apply the method of Faedo-Galerkin. A sequence $\{u_{1m}\}$ of approximate solutions satisfying certain suitable systems of ordinary differential equations is constructed together with a sequence $\{u_{2m}\}$. After deriving a priori estimates and using the compactness argument we then pass to limits in suitable topologies. The limit functions u_1 and u_2 are verified to be a solution of the problem, the periodicity of u_1 is guaranteed by the ω-periodicity of each approximate solution (by means of the Brouwer fixed-point theorem).

The investigation of strongly non-linear problems lies beyond the scope of this book, and for this reason we proceed more briefly in comparison with the methods explained above. We adopt a notation close to that in II: [13].

In what follows we need the following *Gronwall lemma* which is proved, for example, in II: [3], p. 124.

LEMMA 3.1.1 *Let* $T > 0$. *Let*

(*a*) $a \in L_1(0, T)$, $a(t) \geqq 0$;

(*b*) $\beta \in L_1(0, T)$, $b_0 \in \mathbf{R}$, $b(t) = b_0 + \int_0^t \beta(\tau)\,\mathrm{d}\tau$;

(*c*) $u \in L_\infty(0, T)$.

Then the inequality

$$u(t) \leqq b(t) + \int_0^t a(\tau)\, u(\tau)\, \mathrm{d}\tau\,, \quad t \in [0, T]$$

implies that

$$u(t) \leqq b_0 \exp\left(\int_0^t a(\tau)\, \mathrm{d}\tau\right) + \int_0^t \beta(\tau) \exp\left(\int_\tau^t a(\sigma)\, \mathrm{d}\sigma\right) \mathrm{d}\tau\,,$$

$$t \in [0, T]\,.$$

Before proceeding further, let us denote by G the inverse operator to Δ^2 in Ω with the Dirichlet boundary conditions. Then (3.1.3) is equivalent to

$$u_2 = -a_2^{-1}\, G([u_1, u_1])$$

and this enables us to eliminate u_2 from (3.1.2). We obtain

$$u_1'' - b\Delta\, u_1'' + a_1 \Delta^2\, u_1 + d_1\, u_1' - d_2 \Delta\, u_1' +$$
$$+ a_2^{-1}[u_1, G([u_1, u_1])] = g\,. \tag{3.1.2'}$$

Let us remark here that from the theory of elliptic equations the operator G is known to belong to $\mathscr{L}(H^{-2}(\Omega), \dot{H}^2(\Omega))$ and to satisfy $\langle u, Gu \rangle \geqq 0$, $u \in H^{-2}(\Omega)$.

3.2. *The existence theorem*

We prove the following theorem.

THEOREM 3.2.1 *Let $\Omega \subset \mathbf{R}^2$ be a bounded sufficiently regular domain. Let $g \in H^0_\omega(\mathbf{R} \times \Omega)$, let a_1, a_2, b, d_1, and d_2 be positive constants.*
Then the problem $(\mathscr{P}_\omega)$ has a weak solution.

Proof.

(1) *Approximate solutions*

Since the space $\dot{H}^2(\Omega)$ is separable, there exists a sequence $\{w_k\}_{k\in\mathbf{N}}$ with the following properties:

$w_k \in \dot{H}^2(\Omega)$, $k \in \mathbf{N}$ (we may choose $w_k \in C_0^\infty(\Omega)$);

$w_1, \ldots, w_m$ are linearly independent for every $m \in \mathbf{N}$;

$\mathrm{cl\,lin}\, \{w_k\}_{k\in\mathbf{N}} = \dot{H}^2(\Omega)$.

We look for an approximate solution (of order m) of the initial boundary-value problem (3.1.2)–(3.1.4), (3.1.6) in the form

$$u_{1m}(t) = \sum_{k=1}^{m} q_{km}(t)\, w_k\,;$$

the q_{km} are determined by

$$\begin{aligned}
&\langle u''_{1m}(t), w_j\rangle + b\langle \operatorname{grad} u''_{1m}(t), \operatorname{grad} w_j\rangle + \\
&+ a_1\langle \Delta\, u_{1m}(t), \Delta\, w_j\rangle + d_1\langle u'_{1m}(t), w_j\rangle + \\
&+ d_2\langle \operatorname{grad} u'_{1m}(t), \operatorname{grad} w_j\rangle + a_2^{-1}\langle [u_{1m}(t), G([u_{1m}(t), u_{1m}(t)])], w_j\rangle = \\
&= \langle g(t), w_j\rangle\,, \quad 1 \leqq j \leqq m
\end{aligned} \tag{3.2.1}$$

and

$$\begin{aligned}
u_{1m}(0) &= u_{01m}\,, \\
u'_{1m}(0) &= u_{11m}\,,
\end{aligned} \tag{3.2.2}$$

where u_{01m} and u_{11m} are chosen arbitrarily from $\operatorname{lin}\{w_1, \ldots, w_m\}$.

Standard theorems on non-linear systems of ordinary differential equations guarantee the existence of a solution to the problem (3.2.1)–(3.2.2) on an interval $[0, t_m]$, $t_m > 0$ (note that the Gram determinant $\det(\langle w_k, w_j\rangle + b\langle \operatorname{grad} w_k, \operatorname{grad} w_j\rangle) \neq 0$); the a priori estimates given below guarantee the existence on a large interval $[0, \omega]$.

(2) *Energy inequality*

We multiply (3.2.1) by $q'_{jm}(t) + \delta\, q_{jm}(t)$, $\delta > 0$, and sum up for $j = 1, \ldots, m$. Making use of the symmetry of $[\cdot,\cdot]$ we obtain easily ($|\cdot|$ stands for the norm in $H^0(\Omega)$, $|\cdot|_k$ for the norm in $H^k(\Omega)$, $k = 1, 2$)

$$\begin{aligned}
&\frac{\mathrm{d}}{\mathrm{d}t}\big(|u'_{1m}(t)|^2 + b|\operatorname{grad} u'_{1m}(t)|^2 + a_1|\Delta\, u_{1m}(t)|^2 + \\
&+ (2a_2)^{-1}\langle [u_{1m}(t), u_{1m}(t)],\ G([u_{1m}(t), u_{1m}(t)]\rangle + \\
&+ 2\delta\langle u'_{1m}(t), u_{1m}(t)\rangle + 2\delta b\langle \operatorname{grad} u'_{1m}(t), \operatorname{grad} u_{1m}(t)\rangle\big) + \\
&+ 2\delta|u'_{1m}(t)|^2 + 2\delta b|\operatorname{grad} u'_{1m}(t)|^2 + 2\delta a_1|\Delta\, u_{1m}(t)|^2 + \\
&+ \delta a_2^{-1}\langle [u_{1m}(t), u_{1m}(t)],\ G([u_{1m}(t), u_{1m}(t)])\rangle + 4\delta^2\langle u'_{1m}(t), u_{1m}(t)\rangle + \\
&+ 4\delta^2 b\langle \operatorname{grad} u'_{1m}(t), \operatorname{grad} u_{1m}(t)\rangle = \\
&= 2\langle g(t) - (d_1 - 2\delta)\, u'_{1m}(t), u'_{1m}(t) + \delta\, u_{1m}(t)\rangle - \\
&- 2(d_2 - 2\delta b)\,\langle \operatorname{grad} u'_{1m}(t), \operatorname{grad} u'_{1m}(t) + \delta \operatorname{grad} u_{1m}(t)\rangle - \\
&- \delta a_2^{-1}\langle [u_{1m}(t), u_{1m}(t)],\ G([u_{1m}(t), u_{1m}(t)])\rangle \equiv \gamma_m(t)\,.
\end{aligned} \tag{3.2.3}$$

For $u \in \dot{H}^2(\Omega)$, $v \in \dot{H}^1(\Omega)$ and $\delta > 0$ put

$$F(u, v) = |v|^2 + b|\text{grad } v|^2 + a_1|\Delta u|^2 +$$

$$+ (2a_2)^{-1} \langle [u, u], G([u, u])\rangle + 2\delta\langle v, u\rangle + 2\delta b\langle \text{grad } v, \text{grad } u\rangle$$

and

$$E_m(t) = F(u_{1m}(t), u'_{1m}(t)) .$$

Then (3.2.3) yields

$$E_m(t) = E_m(0)\, e^{-2\delta t} + \int_0^t e^{-2\delta(t-\tau)}\, \gamma_m(\tau)\, d\tau . \tag{3.2.4}$$

Using the equivalence $|v|_2 \sim |\Delta v|$, $v \in \dot{H}^2(\Omega)$, and the inequality $-\alpha x^2 + \beta xy \leqq (4\alpha)^{-1} \beta^2 y^2$, $x, y \in \mathbf{R}$ $(\alpha > 0, \beta \in \mathbf{R})$, we obtain easily that for sufficiently small $\delta > 0$ there exist constants $c_1 > 0$, $0 < c_2 < 1$ independent of m such that

$$\gamma_m(t) \leqq c_1|g(t)|^2 + 2\delta c_2\, E_m(t) . \tag{3.2.5}$$

Applying Lemma 3.1.1 we get from (3.2.4)–(3.2.5)

$$E_m(t) \leqq E_m(0)\, e^{-2\delta(1-c_2)} + c_1 \int_0^t e^{-2\delta(1-c_2)(t-\tau)}\, |g(\tau)|^2\, d\tau . \tag{3.2.6}$$

(3) *A priori estimates*

For sufficiently small $\delta > 0$ there exist constants $c_3 > 0$ and $c_4 > 0$ such that

$$c_3(|u|_2^2 + |v|_1^2) \leqq F(u, v) \leqq c_4(|u|_2^2 + |v|_1^2 + |u|_2^4) ,$$

$$u \in \dot{H}^2(\Omega) , \quad v \in \dot{H}^1(\Omega) . \tag{3.2.7}$$

Hence, if the sequences $|u_{11m}|_1$ and $|u_{01m}|_2$ are bounded, then (3.2.6)–(3.2.7) imply that

$$|u'_{1m}(t)|_1 \leqq C , \quad |u_{1m}(t)|_2 \leqq C , \quad t \in [0, \omega] ,$$

where $C > 0$ is independent of m.

Thus, $t_m = \omega$ and

$$u_{1m} \text{ is bounded in } L_\infty(0, \omega; \dot{H}^2(\Omega)) , \tag{3.2.8}$$

$$u'_{1m} \text{ is bounded in } L_\infty(0, \omega; \dot{H}^1(\Omega)) . \tag{3.2.9}$$

(4) *Existence of periodic approximate solutions*

By virtue of the inequality (3.2.6) there exists a $\varrho > 0$ independent of m such that the translation operator

$$T(u_{01m}, u_{11m}) = (u_{1m}(\omega), u'_{1m}(\omega))$$

defined on the finite-dimensional space $H_m = \text{lin}\{w_1, \ldots, w_m\} \times \text{lin}\{w_1, \ldots, w_m\}$ maps the set

$$M_{\delta,m} = \{(u, v) \in H_m;\ F(u, v) \leqq \varrho\}$$

into itself.

It can be verified that for sufficiently small $\delta > 0$ the set $M_{\delta,m}$ is a compact subset of H_m (uniformly bounded with respect to m) and homeomorphic to $B(0, 1; \mathbf{R}^{2m})$. The latter fact follows from the implication: $(u, v) \in M_{\delta,m} \Rightarrow$ $\Rightarrow s(u, v) \in M_{\delta,m} \setminus \partial M_{\delta,m}$, $s \in [0, 1)$.

Since T is continuous, we get as an easy consequence of I: Theorem 3.2.1 that T has a fixed point in $M_{\delta,m}$, that is, there exists a pair $(u_{01m}, u_{11m}) \in M_{\delta,m}$ such that

$$T(u_{01m}, u_{11m}) = (u_{01m}, u_{11m})\,.$$

Henceforth, u_{1m} will denote a solution of (3.2.1)–(3.2.2) with the initial data (u_{01m}, u_{11m}) just found, which means that u_{1m} is ω-periodic.

From the part (3) of the proof we know that (3.2.8)–(3.2.9) hold. If we define

$$u_{2m} = -a_2^{-1}\, G([u_{1m}, u_{1m}])$$

we get

$$u_{2m} \text{ is bounded in } L_\infty(0, \omega;\ \dot{H}^2(\Omega)). \tag{3.2.10}$$

(5) *Passing to the limit*

The spaces $L_\infty(0, \omega; \dot{H}^2(\Omega))$ and $L_\infty(0, \omega; \dot{H}^1(\Omega))$ are the duals to the separable spaces $L_1(0, \omega; H^{-2}(\Omega))$ and $L_1(0, \omega; H^{-1}(\Omega))$, respectively. Therefore, by Theorem 3.3 from I: [35], p. 196, there exist $u_i \in L_\infty(0, \omega; \dot{H}^2(\Omega))$, $i = 1, 2$, $\chi(= u'_1) \in L_\infty(0, \omega; \dot{H}^1(\Omega))$ and subsequences $u_{i\mu}$ such that

$$\int_0^\omega \langle u_{i\mu}(t), v(t)\rangle\, dt \to \int_0^\omega \langle u_i(t), v(t)\rangle\, dt$$

$$\text{for all} \quad v \in L_1(0, \omega; H^{-2}(\Omega))\,, \quad i = 1, 2\,, \tag{3.2.11}$$

$$\int_0^\omega \langle u'_{1\mu}(t), v(t)\rangle\, dt \to \int_0^\omega \langle \chi(t), v(t)\rangle\, dt$$

$$\text{for all} \quad v \in L_1(0, \omega; H^{-1}(\Omega))\,. \tag{3.2.12}$$

Evidently, u_{1m} is bounded in $H^0(0, \omega; H^1(\Omega))$ and u'_{1m} is bounded in $H^0(0, \omega; H^0(\Omega))$, thus, u_{1m} is bounded in $H^1(Q)$, $Q = (0, \omega) \times \Omega$. By I: Theorem 2.7.3, $H^1(Q)$ is compactly embedded in $H^0(Q)$, hence we may assume that $u_{1\mu}$ satisfies, apart from (3.2.11) and (3.2.12), also

$$u_{1\mu} \to u_1 \quad \text{in} \quad H^0(Q) \,. \tag{3.2.13}$$

Since $u_{1\mu}(0) = u_{1\mu}(\omega)$ we see that $u_1(0) = u_1(\omega)$. The relations (3.2.11)–(3.2.13) enable us to prove that (u_1, u_2) satisfy (3.1.8)–(3.1.9) (we refer to II: [13] for details). Hence the pair (u_1, u_2) of the limit functions is a weak solution to the problem $(\mathscr{P}_\omega)$.

Remark 3.2.1 Morozov in [21] and [22] investigates the same problem making use of a Ljapunov function and of the theory of dissipative differential systems due to Pliss. He asserts (without proof) that in the case of a circular, symmetrically loaded plate there exists a unique stable ω-periodic weak solution, provided that the damping constants d_1 and d_2 are sufficiently large. In [20] using the same methods he treats the problem with $b = d_2 = 0$.

3.3. *Comments on papers of an applied character*

Vorovič and Solop in [29] prove the existence of an ω-periodic solution to non-linear equations of anisotropic, inhomogeneous flat shells of variable thickness, involving damping. The theory of dissipative differential systems is applied.

There are many other papers dealing with the theory of non-linear vibrations of shells, plates, beams, etc. However, in most cases the investigation consists only in formal calculations, for example in [1], [2], [7], [17], [26] and [28], the last two concerning autonomous problems.

Chapter VII

The abstract equations

§ 1. The t-Fourier method: Preliminaries

1.1. *The setting of the problem*

Let H be a fixed Hilbert space with a scalar product $\langle .,. \rangle$ and let $\mathscr{M}$ be the set of all polynomials with complex coefficients of the form

$$M(z) = z^n + a_1 z^{n-1} + \ldots + a_{n-1} z .$$

With any such polynomial we associate the differential operator

$$M\left(\frac{\mathrm{d}}{\mathrm{d}t}\right) = \frac{\mathrm{d}^n}{\mathrm{d}t^n} + a_1 \frac{\mathrm{d}^{n-1}}{\mathrm{d}t^{n-1}} + \ldots + a_{n-1} \frac{\mathrm{d}}{\mathrm{d}t} .$$

The set of all linear operators $A : \mathscr{D}(A) \subset H \to H$ is denoted by $\mathscr{L}^+(H, H)$.

In this chapter we study generalized 2π-periodic solutions (here we restrict ourselves to the case $\omega = 2\pi$ for the sake of simplicity) of the equation

$$M\left(\frac{\mathrm{d}}{\mathrm{d}t}\right) u(t) + A\, u(t) = g(t) , \tag{1.1.1}$$

where $M \in \mathscr{M}$, $A \in \mathscr{L}^+(H, H)$, $u : \mathbf{R} \to H$, $g : \mathbf{R} \to H$, $t \in \mathbf{R}$. The precise definition of this concept will be given later.

The linear problem (1.1.1) is of great importance in solving the corresponding weakly non-linear problem

$$M\left(\frac{\mathrm{d}}{\mathrm{d}t}\right) u(t) + A\, u(t) = \varepsilon f(t, u(t)) , \quad \varepsilon \in \mathbf{R} , \quad f : \mathbf{R} \times H \to H . \tag{1.1.2}$$

Suitable conditions for the solubility of problem (1.1.2) are that the linear operator

$$M\left(\frac{\mathrm{d}}{\mathrm{d}t}\right) + A$$

and its range in a suitably chosen function space are closed.

When we assume that A is normal, then $M(\mathrm{d}/\mathrm{d}t) + A$ is closed and has a closed range if and only if certain relations between the set $\{M(ij); j \in \mathbf{Z}\}$ and the spectrum $\sigma(A)$ hold. These relations represent the main result of this chapter, because they show that the conditions for the existence of a 2π-periodic solution of the problem (1.1.1) can be described as relations between two number sets.

1.2. *Auxiliary notions and results*

PROPOSITION 1.2.1 (*see* I: [16], *pp.* 267–277). *Let* $A \in \mathscr{L}^+(H, H)$. *If* A *is normal, then* $\mathscr{N}(A)$ *is closed,* $\mathscr{R}(A)$ *is dense in* $[\mathscr{N}(A)]^\perp$, A *is one-to-one on* $[\mathscr{N}(A)]^\perp$ *and* $\mathscr{R}(A) = \{Ax; x \in \mathscr{D}(A) \cap [\mathscr{N}(A)]^\perp\}$.

PROPOSITION 1.2.2 *Let* $A \in \mathscr{L}^+(H, H)$. *If* A *is normal, then the following two statements are equivalent*:

(a) $\mathscr{R}(A)$ *is closed*,

(b) *there exists an* $a > 0$ *such that* $\|Ax\| \geqq a \|x\|$ *for every* $x \in \mathscr{D}(A) \cap$ $\cap [\mathscr{N}(A)]^\perp$.

Proof. Use Proposition 1.2.1 and I: Theorem 1.2.1 (the Banach closed graph theorem).

PROPOSITION 1.2.3 *Let* $A \in \mathscr{L}^+(H, H)$, $\lambda \in \mathbf{C}$ *and* $a > 0$. *If* A *is normal, then the following two statements are equivalent*:

(*a*) $\{\mu \in \mathbf{C}; 0 < |\mu - \lambda| < a\} \subset \varrho(A)$,

(b) $\|\lambda x - Ax\| \geqq a\|x\|$ *for all* $x \in [\mathscr{N}(\lambda - A)]^\perp$.

Proof. The implication (a) $\Rightarrow$ (b) follows from the properties of spectral decomposition of normal operators. The implication (b) $\Rightarrow$ (a) is an easy consequence of Propositions 1.2.1 and 1.2.2 (see also I: [16], pp. 178–180).

The operator $A \in \mathscr{L}^+(H, H)$ is called *recompact* if for every $x_k \in \mathscr{D}(A)$ for which Ax_k is a bounded sequence there exists a compact sequence $y_k \in \mathscr{D}(A)$ such that $x_k - y_k \in \mathscr{N}(A)$.

If A is one-to-one, then clearly A is recompact if and only if A^{-1} is compact.

PROPOSITION 1.2.4 *Let* $A \in \mathscr{L}^+(H, H)$. *If* A *is normal and recompact, then* $\sigma_p(A) = \sigma(A)$, *and* $\sigma_p(A)$ *has no finite accumulation point.*

Proof. Let us denote by $\hat{A}$ the restriction of A to $\mathscr{D}(A) \cap [\mathscr{N}(A)]^\perp$. Using Proposition 1.2.1 we obtain easily that $\hat{A}^{-1}$ is a compact operator in the subspace $[\mathscr{N}(A)]^\perp$. Now we apply the Riesz-Schauder theorem to $\hat{A}$ (see I: [16], p. 186).

§ 2. The t-Fourier method: Spectral properties of periodic operators

2.1. *Main results*

For $M \in \mathcal{M}$ and $A \in \mathcal{L}^+(H, H)$ we define a *periodic evolution operator* $\mathsf{L}(M, A)$ (or briefly L) in the following way: $u \in \mathcal{D}(\mathsf{L}(M, A))$ if and only if $u \in H^0_{2\pi}(\mathbf{R}; H)$ and there exists a $g \in H^0_{2\pi}(\mathbf{R}; H)$ such that for every $\varphi \in$ $\in C^\infty_{2\pi}(\mathbf{R}; \mathbf{C})$:

$$\text{(a)} \int_0^{2\pi} \varphi(\tau)\, u(\tau)\, \mathrm{d}\tau \in \mathcal{D}(A)\,,$$

$$\text{(b)} \int_0^{2\pi} \left[M\left(-\frac{\mathrm{d}}{\mathrm{d}\tau}\right)\varphi(\tau)\right] u(\tau)\, \mathrm{d}\tau + A \int_0^{2\pi} \varphi(\tau)\, u(\tau)\, \mathrm{d}\tau =$$

$$= \int_0^{2\pi} \varphi(\tau)\, g(\tau)\, \mathrm{d}\tau\,.$$

The function g, which is clearly uniquely determined, is the value of $\mathsf{L}(M, A)$ at u:

$$\mathsf{L}(M, A)\, u = g\,.$$

We emphasize that in what follows $\mathsf{L}(M, A)$ is always regarded as an operator from $\mathcal{L}^+(H^0_{2\pi}(\mathbf{R}; H), H^0_{2\pi}(\mathbf{R}; H))$.

Lemma 2.1.1 *Let $M \in \mathcal{M}$ and $A \in \mathcal{L}^+(H, H)$. If A is closed, then*

$$\text{(a)} \int_0^{2\pi} e^{-ij\tau}\, u(\tau)\, \mathrm{d}\tau \in \mathcal{D}(A)\,,$$

$$\text{(b)} \ (M(ij) + A) \int_0^{2\pi} e^{-ij\tau}\, u(\tau)\, \mathrm{d}\tau = \int_0^{2\pi} e^{-ij\tau}\, g(\tau)\, \mathrm{d}\tau$$

for every $u \in \mathcal{D}(\mathsf{L}(M, A))$, $g \in H^0_{2\pi}(\mathbf{R}; H)$, such that $\mathsf{L}(M, A)\, u = g$ and for every $j \in \mathbf{Z}$.

Lemma 2.1.2 *Let $M \in \mathcal{M}$ and $A \in \mathcal{L}^+(H, H)$. If A is closed, then $\mathsf{L}(M, A)$ belongs to $\mathcal{L}^+(H^0_{2\pi}(\mathbf{R}; H), H^0_{2\pi}(\mathbf{R}; H))$ and is closed.*

Next, let $r \in \mathbf{Z}^+$. We say that $\mathsf{L}(M, A)$ is *smooth of degree* r if every function $u \in \mathcal{D}(\mathsf{L}(M, A)) \cap [\mathcal{N}(\mathsf{L}(M, A))]^\perp$ belongs to $H^r_{2\pi}(\mathbf{R}; H)$.

THEOREM 2.1.1 *Let* $A \in \mathscr{L}^+(H, H)$, $M \in \mathscr{M}$ *and* $r \in \mathbf{Z}^+$. *If* A *is closed, then the following two assertions are equivalent*:

(a) $L(M, A)$ *has a closed range and is smooth of degree* r;

(b) *there exists a constant* $\gamma > 0$ *such that*
$\|(M(ij) + A)\,x\| \geqq \gamma(1 + |j|^r)\,\|x\|$
for every $j \in \mathbf{Z}$ *and every* $x \in \mathscr{D}(A) \cap [\mathscr{N}(M(ij) + A)]^\perp$.

Proof. Before coming to the proof, let us make some preliminary observations. We shall write L instead of $L(M, A)$ in the whole proof.
Lemma 2.1.2 immediately implies that

$$\mathscr{N}(L) \text{ is a closed subspace of } H^0_{2\pi}(\mathbf{R}; H)\,. \tag{2.1.1}$$

On the other hand, we see from (2.1.1) that

$$\text{for every } g \in \mathscr{R}(L) \text{ there is a unique } u \in \mathscr{D}(L) \cap [\mathscr{N}(L)]^\perp \text{ such that } Lu = g\,. \tag{2.1.2}$$

Next, we denote by $\hat{L}$ the restriction of L to $\mathscr{D}(L) \cap [\mathscr{N}(L)]^\perp$. Clearly, by (2.1.2) and by Lemma 2.1.2

$$\hat{L} \text{ is one-to-one and closed}\,, \tag{2.1.3}$$

$$\mathscr{R}(\hat{L}) = \mathscr{R}(L)\,. \tag{2.1.4}$$

After these preparatory remarks we proceed with the actual proof.

(a) $\Rightarrow$ (b): We see easily from (a) and from (2.1.1)–(2.1.4) that

$$\hat{L} \text{ belongs to } \mathscr{L}^+(H^r_{2\pi}(\mathbf{R}; H), H^0_{2\pi}(\mathbf{R}, H)) \text{ and is closed}\,. \tag{2.1.5}$$

Since $\mathscr{R}(L)$ is closed by assumption (a), we obtain from (2.1.3) and (2.1.5) by means of I: Theorem 1.2.1 (the Banach closed graph theorem) that $\hat{L}^{-1} \in \mathscr{L}(\mathscr{R}(L), H^r_{2\pi}(\mathbf{R}; H))$, that is, there is a constant $c > 0$ such that

$$\|\hat{L}^{-1}g\|_{H_{2\pi}{}^r(\mathbf{R};H)} \leqq c^{-1}\|g\|_{H_{2\pi}{}^0(\mathbf{R};H)} \tag{2.1.6}$$

for every $g \in \mathscr{R}(\hat{L}) = \mathscr{R}(L)$.

However, from (2.1.2), we know that (2.1.6) is equivalent to

$$\|Lu\|_{H_{2\pi}{}^0(\mathbf{R};H)} \geqq c\|u\|_{H_{2\pi}{}^r(\mathbf{R};H)} \tag{2.1.7}$$

for every $u \in \mathscr{D}(L) \cap [\mathscr{N}(L)]^\perp$.

We now choose a fixed $j \in \mathbf{Z}$ and an $x \in [\mathscr{N}(M(ij) + A)]^\perp$. Let $z(t) = e^{ijt}x$ for $t \in \mathbf{R}$. It is easy to verify that $z \in \mathscr{D}(L)$ and $L\,z(t) = e^{ijt}(M(ij)\,x + Ax)$ for every $t \in \mathbf{R}$. But, by Lemma 2.1.1, $z \in [\mathscr{N}(L)]^\perp$. Moreover, it is clear that $z \in H^r_{2\pi}(\mathbf{R}; H)$.

Using (2.1.7), we obtain

$$\|Lz\|_{H_{2\pi}{}^0(\mathbf{R};H)} \geqq c\|z\|_{H_{2\pi}{}^r(\mathbf{R};H)} . \tag{2.1.8}$$

Since $(1 + \xi^2)^{1/2} \geqq 2^{-1/2} (1 + |\xi|)$ for $\xi \in \mathbf{R}$, it follows easily from (2.1.8) that

$$\begin{aligned} &\|(M(ij) + A) x\| \geqq (2\pi)^{1/2} c(1 + j^2 + \ldots + j^{2r})^{1/2} \|x\| \geqq \\ &\geqq (2\pi)^{1/2} c(1 + j^{2r})^{1/2} \|x\| \geqq \pi^{1/2} c(1 + |j|^r) \|x\| . \end{aligned} \tag{2.1.9}$$

Now it suffices to put $\gamma = \pi^{1/2}c$, and we have (b).

(b) $\Rightarrow$ (a): We choose $u \in \mathscr{D}(L) \cap [\mathscr{N}(L)]^\perp$ and write $Lu = g$. Next, we write

$$u_j = \frac{1}{2\pi} \int_0^{2\pi} e^{-ij\tau} u(\tau) \, d\tau , \quad g_j = \frac{1}{2\pi} \int_0^{2\pi} e^{-ij\tau} g(\tau) \, d\tau , \quad j \in \mathbf{Z} .$$

It follows from Lemma 2.1.1 that $u_j \in [\mathscr{N}(M(ij) + A)]^\perp$ for every $j \in \mathbf{Z}$. (The argument is similar as in the first part of the proof with $z(t) = e^{ijt}x$ for arbitrary $x \in [\mathscr{N}(M(ij) + A)]^\perp$). Let us remember that $(1 + \xi^r) \geqq 2^{-r}(1 + \xi)^r$ for $\xi \geqq 0$. Using (b), we see that for every $j \in \mathbf{Z}$,

$$\begin{aligned} &\|(M(ij) + A) u_j\| \geqq \gamma(1 + |j|^r) \|u_j\| \geqq \\ &\geqq 2^{-r} \gamma[(1 + |j|)^{2r}]^{1/2} \|u_j\| \geqq \\ &\geqq 2^{-r} \gamma(1 + j^2 + \ldots + j^{2r})^{1/2} \|u_j\| . \end{aligned} \tag{2.1.10}$$

Taking $\varphi(t) = e^{-ijt}$, we obtain, by the definition of L, that

$$g_j = (M(ij) + A) u_j ,$$

and when (2.1.10) is taken into account, we can write

$$\begin{aligned} &\|Lu\|_{H_{2\pi}{}^0(\mathbf{R};H)} = \|g\|_{H_{2\pi}{}^0(\mathbf{R};H)} = (2\pi \sum_{j\in\mathbf{Z}} \|g_j\|^2)^{1/2} = \\ &= (2\pi \sum_{j\in\mathbf{Z}} \|(M(ij) + A) u_j\|^2)^{1/2} \geqq \\ &\geqq (2\pi)^{1/2} 2^{-r} \gamma(\sum_{j\in\mathbf{Z}} (1 + j^2 + \ldots + j^{2r}) \|u_j\|^2)^{1/2} . \end{aligned} \tag{2.1.11}$$

Since u was chosen arbitrarily in $\mathscr{D}(L) \cap [\mathscr{N}(L)]^\perp$ we see easily from (2.1.11) that

$$\begin{aligned} &u \in H^r_{2\pi}(\mathbf{R}; H) \quad \text{and} \quad \|Lu\|_{H_{2\pi}{}^0(\mathbf{R};H)} \geqq (2\pi)^{1/2} 2^{-r} \gamma\|u\|_{H_{2\pi}{}^r(\mathbf{R};H)} \\ &\text{for every} \quad u \in \mathscr{D}(L) \cap [\mathscr{N}(L)]^\perp . \end{aligned} \tag{2.1.12}$$

However, (2.1.12) shows that L is smooth of degree r. Moreover, it follows from (2.1.12) that $\|\hat{L}u\|_{H_{2\pi}{}^0(\mathbf{R};H)} \geqq (2\pi)^{1/2} 2^{-r} \gamma\|u\|_{H_{2\pi}{}^0(\mathbf{R};H)}$. Using this fact we obtain easily from (2.1.3) and (2.1.4) that $\mathscr{R}(L)$ is a closed subspace of $H^0_{2\pi}(\mathbf{R}; H)$, and this completes the proof.

THEOREM 2.1.2 *Let $A \in \mathscr{L}^+(H, H)$, $M \in \mathscr{M}$* and $r \in \mathbf{Z}^+$. *If A is normal, then the following two assertions are equivalent:*

(a) *$\mathsf{L}(M, A)$ has a closed range and is smooth of degree r;*

(b) *there is a constant $\gamma > 0$ such that*
$\{\lambda \in \mathbf{C}; 0 < |\lambda - M(ij)| < \gamma(1 + |j|^r)\} \cap \sigma(A) = \emptyset$ *for every* $j \in \mathbf{Z}$.

Proof. This is an immediate consequence of Proposition 1.2.3 and Theorem 2.1.1.

THEOREM 2.1.3 *Let $A \in \mathscr{L}^+(H, H)$, $M \in \mathscr{M}$ and $r \in \mathbf{Z}^+$. If A is normal and recompact, then the following two assertions are equivalent:*

(a) *$\mathsf{L}(M, A)$ has a closed range and is smooth of degree r;*

(b) *there is a constant $\gamma > 0$ such that, for sufficiently large $j \in \mathbf{Z}$ ($|j| \geqq j_0$),*
$\{\lambda \in \mathbf{C}; 0 < |\lambda - M(ij)| < \gamma(1 + |j|^r)\} \cap \sigma_p(A) = \emptyset$.

The *proof* uses Proposition 1.2.4 and Theorem 2.1.2.

THEOREM 2.1.4 *Let $A \in \mathscr{L}^+(H, H)$ and $M \in \mathscr{M}$. If A is normal, then the subspaces $\mathscr{N}(\mathsf{L}(M, A))$ and $\mathscr{R}(\mathsf{L}(M, A))$ are mutually orthogonal, and*

$$\mathscr{N}(\mathsf{L}(M, A)) + \mathrm{cl}\, \mathscr{R}(\mathsf{L}(M, A)) = H^0_{2\pi}(\mathbf{R}; H).$$

Proof. Firstly, let $v \in \mathscr{N}(\mathsf{L}(M, A))$ and $g \in \mathscr{R}(\mathsf{L}(M, A))$. We can find a $u \in \mathscr{D}(\mathsf{L}(M, A))$ such that $\mathsf{L}(M, A)\, u = g$.

By Lemma 2.1.1 we have

$$(M(ij) + A) \int_0^{2\pi} e^{-ij\tau}\, v(\tau)\, \mathrm{d}\tau = 0\,, \quad j \in \mathbf{Z}\,, \tag{2.1.13}$$

$$(M(ij) + A) \int_0^{2\pi} e^{-ij\tau}\, u(\tau)\, \mathrm{d}\tau = \int_0^{2\pi} e^{-ij\tau}\, g(\tau)\, \mathrm{d}\tau\,, \quad j \in \mathbf{Z}\,. \tag{2.1.14}$$

Since the operators $M(ij) + A, j \in \mathbf{Z}$, are obviously normal in view of the supposed normality of A, we can apply Proposition 1.2.1. Then we obtain from (2.1.13) and (2.1.14) that for every $j \in \mathbf{Z}$

$$\left\langle \int_0^{2\pi} e^{-ij\tau}\, v(\tau)\, \mathrm{d}\tau\,,\ \int_0^{2\pi} e^{-ij\tau}\, g(\tau)\, \mathrm{d}\tau \right\rangle = 0\,. \tag{2.1.15}$$

But from (2.1.15) it easily follows that $\langle v, g\rangle_{H_{2\pi}{}^0(\mathbf{R};H)} = 0$, which proves the orthogonality.

Secondly, let $u \in H^0_{2\pi}(\mathbf{R}; H)$. We denote by P_j, $j \in \mathbf{Z}$, the orthogonal projections onto $\mathscr{N}(M(ij) + A)$ which is possible since $\mathscr{N}(M(ij) + A)$ is closed according to Proposition 1.2.1.

Putting for $j \in \mathbf{Z}$

$$v_j = P_j \left(\frac{1}{2\pi} \int_0^{2\pi} e^{-ij\tau} u(\tau) \, d\tau \right),$$

$$w_j = (I - P_j) \left(\frac{1}{2\pi} \int_0^{2\pi} e^{-ij\tau} u(\tau) \, d\tau \right), \tag{2.1.16}$$

we can prove easily that

$$\sum_{j \in \mathbf{Z}} \|v_j\|^2 < \infty, \quad \sum_{j \in \mathbf{Z}} \|w_j\|^2 < \infty. \tag{2.1.17}$$

It then follows immediately, from definition (2.1.16), by means of Proposition 1.2.1 that for $j \in \mathbf{Z}$

$$v_j \in \mathcal{N}(M(ij) + A), \quad w_j \in \operatorname{cl} \mathcal{R}(M(ij) + A). \tag{2.1.18}$$

By an easy calculation we obtain from (2.1.18) that for $j \in \mathbf{Z}$

$$v_j e^{ijt} \in \mathcal{N}(\mathsf{L}(M, A)), \quad w_j e^{ijt} \in \operatorname{cl} \mathcal{R}(\mathsf{L}(M, A)). \tag{2.1.19}$$

Now we define in accordance with (2.1.17)

$$v = \sum_{j \in \mathbf{Z}} v_j e^{ijt}, \quad w = \sum_{j \in \mathbf{Z}} w_j e^{ijt},$$

taking the sums in the sense of $H^0_{2\pi}(\mathbf{R}; H)$.

We see from (2.1.16) and (2.1.19) that $u = v + w$, where $v \in \operatorname{cl} \mathcal{N}(\mathsf{L}(M, A))$, $w \in \operatorname{cl} \mathcal{R}(\mathsf{L}(M, A))$. Since by Lemma 2.1.2, $\mathcal{N}(\mathsf{L}(M, A))$ is closed, the result follows, and the proof is complete.

COROLLARY 2.1.1 *Let M be a polynomial of odd degree and let A be a self-adjoint and recompact operator. Then the operator $\mathsf{L}(M, A)$ has a closed range and is smooth of the same degree as M.*

Proof. We easily see that if s is the degree of M, then

$$\frac{|\operatorname{Im}(M(ij))|}{1 + |j|^s} \to 1 \quad \text{as} \quad |j| \to \infty.$$

Therefore, for sufficiently large $j \in \mathbf{Z}$ ($|j| \geqq j_0$)

$$|\operatorname{Im}(M(ij))| \geqq 2^{-1}(1 + |j|^s).$$

Since A is assumed to be self-adjoint, we see that

$$\sigma(A) \subset \mathbf{R}.$$

Obviously, for sufficiently large $j \in \mathbf{Z}$,

$$\{\lambda \in \mathbf{C}; |\lambda - M(ij)| < 2^{-1}(1 + |j|^s)\} \subset \mathbf{C} \setminus \mathbf{R},$$

and hence also

$$\{\lambda \in \mathbf{C}; 0 < |\lambda - M(ij)| < 2^{-1}(1 + |j|^s)\} \cap \sigma_p(A) = \emptyset.$$

Since we assume A to be recompact, we can use Theorem 2.1.3 with $r = s$ and $\gamma = 2^{-1}$, and this completes the proof.

COROLLARY 2.1.2 *Let M be a polynomial of even degree and iA be a self-adjoint and recompact operator. Then the operator $\mathsf{L}(M, A)$ has a closed range and is smooth of the same degree as M.*

Proof. Again, let s be the degree of M. We establish easily that

$$\frac{|\mathrm{Re}\,(M(ij))|}{1 + |j|^s} \to 1 \quad \text{as} \quad |j| \to \infty.$$

Consequently, for sufficiently large $j \in \mathbf{Z}$ $(|j| \geqq j_0)$ we have

$$|\mathrm{Re}\,(M(ij))| \geqq 2^{-1}(1 + |j|^s).$$

Since iA is self-adjoint, we have

$$\sigma(A) \subset i\mathbf{R}.$$

Now we can proceed as in Corollary 2.1.1 and easily arrive at the assertion.

2.2. *Examples*

In this section we choose $H = H^0(0, \pi; \mathbf{C})$.

We define an operator $\Delta \in \mathscr{L}^+(H, H)$ as follows:

$$\varphi \in \mathscr{D}(\Delta) \Leftrightarrow \varphi \in H^2(0, \pi; \mathbf{C}) \cap \dot{H}^1(0, \pi; \mathbf{C})$$

and

$$\Delta\, \varphi(x) = \frac{\mathrm{d}^2}{\mathrm{d}x^2}\, \varphi(x) \quad \text{for every} \quad \varphi \in \mathscr{D}(\Delta).$$

Now it is not difficult to prove that Δ is self-adjoint (see II: Example 1.1.1) and recompact. Its spectrum is

$$\sigma(\Delta) = \sigma_p(\Delta) = \{-k^2; k \in \mathbf{N}\}.$$

Example 2.2.1 (the heat equation). Let $A = -\Delta$ and $M(\xi) = \xi$.

It follows immediately from Theorem 2.1.3 that $L(M, A)$ has a closed range and is smooth of degree 1.

Moreover, since $\mathcal{N}(L(M, A)) = \{0\}$ we see easily that the range of $L(M, A)$ is the whole space $H^0_{2\pi}(\mathbf{R}; H)$.

Example 2.2.2 (the Schrödinger equation). Let $A = -i\Delta$ and $M(\xi) = \xi$.

It follows from Theorem 2.1.3 that $L(M, A)$ has a closed range and is smooth of degree zero.

Example 2.2.3 (the telegraph equation). Let $A = -\Delta$ and $M(\xi) = \xi^2 + a\xi$ with $a \neq 0$.

Then $M(ij) = -j^2 + aij$. Obviously,

$$\frac{|\mathrm{Im}\,(M(ij))|}{|j|} = |a| > 0$$

for every $j \in \mathbf{Z}$, $j \neq 0$, and taking into account the fact that $\sigma(A) \subset \mathbf{R}$ we see easily that for every $j \in \mathbf{Z}$, $j \neq 0$,

$$\{\lambda \in \mathbf{C};\ |\lambda - M(ij)| < |a|\,|j|\} \cap \sigma(A) = \emptyset\,.$$

Hence, by Theorem 2.1.3, $L(M, A)$ has a closed range and is smooth of degree 1.

The range of $L(M, A)$ is again the whole space $H^0_{2\pi}(\mathbf{R}; H)$ (see Example 2.2.1).

Example 2.2.4 (the wave equation). Let $A = -\Delta$ and $M(\xi) = \xi^2$.

Then $M(ij) = -j^2$ for every $j \in \mathbf{Z}$.

If $|j| \neq k$, then $|k^2 - j^2| = |k - |j||\,(k + |j|) \geqq k + |j| \geqq |j|$. Hence,

$$\inf\{|-j^2 + k^2|;\ k \in \mathbf{N},\ j^2 \neq k^2\} \geqq |j|,\ j \in \mathbf{Z}\,.$$

We know that $\sigma_p(A) = \sigma(A) = \{k^2;\ k \in \mathbf{N}\}$, hence we can apply Theorem 2.1.3. Consequently, $L(M, A)$ has a closed range and is smooth of degree 1.

Example 2.2.5 (the beam equation). Let $A = \Delta^2$ and $M(\xi) = \xi^2$.

Then, as in Example 2.2.4, $M(ij) = -j^2$ for every $j \in \mathbf{Z}$.

Moreover, $\sigma_p(A) = \sigma(A) = \{k^4;\ k \in \mathbf{N}\}$.

Now using an argument as in Example 2.2.4, we deduce from Theorem 2.1.3 that $L(M, A)$ has a closed range and is smooth of degree 1.

§ 3. The t-Fourier method: Weakly non-linear problems

3.1. *General scheme*

LEMMA 3.1.1 *Let* $T \in \mathscr{L}(H, H)$. *If* $\langle Tx, x \rangle \geqq d\|x\|^2$ *for every* $x \in H$ *with* $d > 0$, *then* T *is one-to-one and* $\mathscr{R}(T) = H$.

Proof. T is evidently one-to-one. It is easy to see that $\mathscr{R}(T)$ is closed. For, let $Tx_k \to y$. Then $\|Tx_{k_1} - Tx_{k_2}\| \geqq d\|x_{k_1} - x_{k_2}\|$; consequently x_k is a Cauchy sequence, hence convergent. Let $x_k \to x$. Since $T \in \mathscr{L}(H, H)$, we conclude that $Tx = y$, that is, $y \in \mathscr{R}(T)$, which was to be proved.

Finally, let $\mathscr{R}(T) \neq H$. Then there exists an $x \in H$, $x \neq 0$ such that $\langle Tz, x \rangle = 0$ for every $z \in H$. Consequently, $\langle Tx, x \rangle = 0$, which contradicts the assumption that $\langle Tx, x \rangle \geqq d\|x\|^2 > 0$.

LEMMA 3.1.2 *Let* $S : H \to H$. *If* S *is continuous*, $S(H)$ *is an open subset of* H, *and if there is a* $d > 0$ *such that*

$$\|S(x_1) - S(x_2)\| \geqq d\|x_1 - x_2\|$$

for every $x_1, x_2 \in H$, *then* S *is a one-to-one mapping of* H *onto itself.*

Proof. Again it is evident that S is one-to-one. It suffices to prove that $S(H)$ is closed.

Suppose that $S(H)$ is not closed. Then there is a sequence $x_k \in H$ such that $Sx_k \to y$ and $y \notin S(H)$. Now, by assumption,

$$\|S(x_{k_1}) - S(x_{k_2})\| \geqq d\|x_{k_1} - x_{k_2}\|,$$

which implies that x_k is a Cauchy sequence, hence convergent. Let $x_k \to x$.

By the continuity of S, $Sx = y$, which contradicts the assumption that $y \notin S(H)$. Hence $S(H) = H$, which completes the proof.

LEMMA 3.1.3 *Let* $S : H \to H$. *If* S *is F-differentiable on* H, $S' : H \to \mathscr{L}(H, H)$ *is continuous, and if for every* $x \in H$ *the operator* $S'(x)$ *is one-to-one and* $\mathscr{R}(S'(x)) = H$, *then* $S(H)$ *is an open subset of* H.

Proof. We use first I: Theorem 1.2.1 (the Banach closed graph theorem) to prove that $S'(x)^{-1} \in \mathscr{L}(H, H)$ for every $x \in H$, and then I: Theorem 3.4.2 (the implicit function theorem).

THEOREM 3.1.1 *Let* $L \in \mathscr{L}^+(H, H)$ *and* $F : H \to H$. *If*

the operator L *is closed*, (3.1.1)

the range $\mathscr{R}(L)$ *is closed*, (3.1.2)

the subspaces $\mathcal{N}(L)$ *and* $\mathcal{R}(L)$ *are orthogonal to each other,* (3.1.3)

$$\mathcal{N}(L) + \mathcal{R}(L) = H, \tag{3.1.4}$$

$$\|F(x_1) - F(x_2)\| \leqq c\|x_1 - x_2\| \text{ for every } x_1, x_2 \in H \text{ with } c \geqq 0, \tag{3.1.5}$$

the function F *is F-differentiable on the whole of* H, (3.1.6)

the function $F' : H \to \mathcal{L}(H, H)$ *is continuous,* (3.1.7)

$$\langle F'(u)\, h, h\rangle \geqq d\|h\|^2 \text{ for all } u, h \in H \text{ with } d > 0, \tag{3.1.8}$$

then there exist an $\varepsilon_0 > 0$ *and a unique function* $u : [-\varepsilon_0, \varepsilon_0] \to H$ *such that*

$$F(u(0)) \in \mathcal{R}(L), \quad u(\varepsilon) \in \mathcal{D}(L),$$

$$Lu(\varepsilon) = \varepsilon\, F(u(\varepsilon))$$

provided that $|\varepsilon| \leqq \varepsilon_0$. *Moreover,* $u \in C([-\varepsilon_0, \varepsilon_0]; H)$.

Proof. Let us write, for the sake of simplicity, $N = \mathcal{N}(L)$ and $R = \mathcal{R}(L)$. By (3.1.1)–(3.1.4), these subspaces N and R give an orthogonal decomposition of H into closed subspaces.

Next, let us denote by $\hat{L}$ the restriction of L to R. It is clear from (3.1.1)–(3.1.4) (by virtue of I: Theorem 1.2.1) that

$\hat{L}$ is a one-to-one mapping from R onto R, (3.1.9)

$\hat{L}^{-1}$ is bounded. (3.1.10)

Let P be the orthogonal projection of H onto N. Obviously,

$$\mathcal{N}(P) = R, \quad \mathcal{R}(P) = N. \tag{3.1.11}$$

We define an auxiliary system of mappings by

$$\Phi_w(v) = P\, F(v + w), \quad v \in N, \quad w \in R. \tag{3.1.12}$$

It is easy to see that $\Phi_w : N \to N$ for every $w \in R$.
It follows from (3.1.5) that

$$\|\Phi_w(v_1) - \Phi_w(v_2)\| \leqq c\|v_1 - v_2\| \quad \text{for every} \quad w \in R, \quad v_1, v_2 \in N. \tag{3.1.13}$$

Further, by (3.1.6) and (3.1.7),

for every $w \in R$ the function Φ_w is F-differentiable on the whole of N, (3.1.14)

$(\Phi_w)' : N \to \mathcal{L}(N, N)$ is continuous for every $w \in R$. (3.1.15)

Since the projection P is orthogonal and $\mathscr{R}(P) = N$, (3.1.8) implies that

$$\langle (\Phi_w)'(v)\, g, g\rangle = \langle P\, F'(v + w)\, g, g\rangle = \langle F'(v + w)\, g, Pg\rangle =$$
$$= \langle F'(v + w)\, g, g\rangle \geqq d\|g\|^2 \quad \text{for every} \quad v, g \in N \quad \text{and} \quad w \in R\,. \tag{3.1.16}$$

Using Lemma 3.1.1 we obtain from (3.1.14) and (3.1.16) that

$$\mathscr{R}((\Phi_w)'(v)) = N \quad \text{for every} \quad w \in R \quad \text{and} \quad v \in N\,. \tag{3.1.17}$$

By Lemma 3.1.3, we conclude from (3.1.14), (3.1.15), and (3.1.17) that

$$\Phi_w(N) \text{ is an open subset of } N \text{ for every } w \in R\,. \tag{3.1.18}$$

Further, it follows from (3.1.14)–(3.1.16) that

$$\|\Phi_w(v_1) - \Phi_w(v_2)\|\, \|v_1 - v_2\| \geqq$$
$$\geqq \langle \Phi_w(v_1) - \Phi_w(v_2), v_1 - v_2\rangle =$$
$$= \left\langle \int_0^1 (\Phi_w)'(\tau\, v_1 + (1 - \tau)\, v_2)\,(v_1 - v_2)\,\mathrm{d}\tau, v_1 - v_2 \right\rangle =$$
$$= \int_0^1 \langle (\Phi_w)'(\tau\, v_1 + (1 - \tau)\, v_2)\,(v_1 - v_2), v_1 - v_2\rangle\,\mathrm{d}\tau \geqq d\|v_1 - v_2\|^2$$

for $w \in R$ and $v_1, v_2 \in N$, hence

$$\|\Phi_w(v_1) - \Phi_w(v_2)\| \geqq d\|v_1 - v_2\| \quad \text{for all} \quad w \in R\,, \quad v_1, v_2 \in N\,. \tag{3.1.19}$$

Now we can apply Lemma 3.1.2 with $S = \Phi_w$, $w \in R$, and $H = N$, and we find from (3.1.14), (3.1.18), and (3.1.19) that

$$\Phi_w \text{ for every } w \in R, \text{ is a one-to-one mapping of } N \text{ onto itself}\,. \tag{3.1.20}$$

It follows immediately from (3.1.20) that there exists a mapping $\Theta : R \to H$ such that

$$\mathscr{R}(\Theta) \subset N\,, \tag{3.1.21}$$
$$\Phi_w(\Theta(w)) = 0 \quad \text{for every} \quad w \in R\,. \tag{3.1.22}$$

It follows from (3.1.22) that

$$\Phi_{w_1}(\Theta(w_1)) - \Phi_{w_1}(\Theta(w_2)) = \Phi_{w_2}(\Theta(w_2)) - \Phi_{w_1}(\Theta(w_2)) \tag{3.1.23}$$

for every $w_1, w_2 \in R$.

From (3.1.5) and (3.1.12) we obtain that

$$\|\Phi_{w_1}(v) - \Phi_{w_2}(v)\| \leqq c\|w_1 - w_2\| \quad \text{for every} \quad v \in N \quad \text{and} \quad w_1, w_2 \in R\,. \tag{3.1.24}$$

Now (3.1.19), (3.1.23) and (3.1.24) imply that

$$\|\Theta(w_1) - \Theta(w_2)\| \leqq d^{-1}\| \Phi_{w_1}(\Theta(w_1)) - \Phi_{w_1}(\Theta(w_2))\| =$$
$$= d^{-1}\|\Phi_{w_2}(\Theta(w_2)) - \Phi_{w_1}(\Theta(w_2))\| \leqq c\, d^{-1}\|w_1 - w_2\|$$
$$\text{for all} \quad w_1, w_2 \in R\,. \tag{3.1.25}$$

It follows from (3.1.5), (3.1.9), (3.1.10), (3.1.11) and (3.1.25) that

$$\hat{L}^{-1}(I - P)\, F(\Theta(w) + w) \in R \quad \text{for every} \quad w \in R\,, \tag{3.1.26}$$

$$\|\hat{L}^{-1}(I - P)\, F(\Theta(w_1) + w_1) - \hat{L}^{-1}(I - P)\, F(\Theta(w_2) + w_2)\| \leqq$$
$$\leqq \|\hat{L}^{-1}\|\, c\|(\Theta(w_1) + w_1) - (\Theta(w_2) + w_2)\| \leqq$$
$$\leqq \|\hat{L}^{-1}\|\, c(c\, d^{-1} + 1)\, \|w_1 - w_2\| = \|\hat{L}^{-1}\|\, (c^2 + cd)\, d^{-1}\, \|w_1 - w_2\|$$
$$\text{for every} \quad w_1, w_2 \in R\,. \tag{3.1.27}$$

Let

$$\varepsilon_0 = 2^{-1}\|\hat{L}^{-1}\|^{-1}\, d(1 + c^2 + cd)^{-1}\,. \tag{3.1.28}$$

Obviously,

$$\varepsilon_0 > 0\,. \tag{3.1.29}$$

Further, (3.1.27) and (3.1.28) imply that

$$\|\varepsilon\, \hat{L}^{-1}(I - P)\, F(\Theta(w_1) + w_1) - \varepsilon\, \hat{L}^{-1}(I - P)\, F(\Theta(w_2) + w_2)\| \leqq$$
$$\leqq 2^{-1}\|w_1 - w_2\| \quad \text{for every} \quad |\varepsilon| \leqq \varepsilon_0 \quad \text{and} \quad w_1, w_2 \in R\,. \tag{3.1.30}$$

Using I: Theorem 3.1.1 (the Banach contraction principle) we conclude from (3.1.26) and (3.1.30) that there exists a unique function $w : [-\varepsilon_0, \varepsilon_0] \to H$ such that

$$w(\varepsilon) \in R \quad \text{for every} \quad |\varepsilon| \leqq \varepsilon_0\,, \tag{3.1.31}$$

$$w(\varepsilon) = \varepsilon\, \hat{L}^{-1}(I - P)\, F(\Theta(w(\varepsilon)) + w(\varepsilon)) \quad \text{for every} \quad |\varepsilon| \leqq \varepsilon_0\,. \tag{3.1.32}$$

Now we can rewrite (3.1.32) in the form

$$w(\varepsilon) \in \mathscr{D}(L) \quad \text{for every} \quad |\varepsilon| \leqq \varepsilon_0\,, \tag{3.1.33}$$

$$Lw(\varepsilon) = \varepsilon(I - P)\, F(\Theta(w(\varepsilon)) + w(\varepsilon)) \quad \text{for every} \quad |\varepsilon| \leqq \varepsilon_0\,. \tag{3.1.34}$$

On the other hand, by (3.1.12) and (3.1.22), we have

$$P\, F(\Theta(w(\varepsilon)) + w(\varepsilon)) = 0 \quad \text{for every} \quad |\varepsilon| \leqq \varepsilon_0\,. \tag{3.1.35}$$

Hence, by (3.1.34) and (3.1.35),

$$Lw(\varepsilon) = \varepsilon\, F(\Theta(w(\varepsilon)) + w(\varepsilon)) \quad \text{for every} \quad |\varepsilon| \leqq \varepsilon_0\,. \tag{3.1.36}$$

Let us now define

$$u(\varepsilon) = w(\varepsilon) + \Theta(w(\varepsilon)) \quad \text{for} \quad |\varepsilon| \leqq \varepsilon_0 . \tag{3.1.37}$$

By (3.1.21), (3.1.33) and (3.1.36) we have

$$u(\varepsilon) \in \mathscr{D}(L) \quad \text{for every} \quad |\varepsilon| \leqq \varepsilon_0 , \tag{3.1.38}$$

$$Lu(\varepsilon) = Lw(\varepsilon) = \varepsilon\, F(u(\varepsilon)) \quad \text{for every} \quad |\varepsilon| \leqq \varepsilon_0 . \tag{3.1.39}$$

The first statement in our theorem follows easily from (3.1.12), (3.1.21), (3.1.22), (3.1.29), (3.1.38) and (3.1.39). The last statement on continuity is obtained by an easy and simple refinement of our argument.

3.2. *A special problem*

Let $f : \mathbf{R} \times H \to H$. We associate this function with the operator F by the following rule:

$$F(u)\,(t) = f(t, u(t)) ,$$

provided that $t \in \mathbf{R}$ and $u : \mathbf{R} \to H$.

LEMMA 3.2.1 *Let* $f : \mathbf{R} \times H \to H$. *If*

$$f \text{ is continuous on } \mathbf{R} \times H , \tag{3.2.1}$$

$$f(\cdot, x) \text{ is } 2\pi\text{-periodic for every } x \in H , \tag{3.2.2}$$

$$\|f(t, x_1) - f(t, x_2)\| \leqq c\|x_1 - x_2\| \quad \text{for every} \quad t \in \mathbf{R} ,$$

$$x_1, x_2 \in H \quad \text{with} \quad c \geqq 0 , \tag{3.2.3}$$

$$f(t, \cdot) \text{ is F-differentiable on } H \text{ for every } t \in \mathbf{R} , \tag{3.2.4}$$

$$f'_x(t, x) : \mathbf{R} \times H \to \mathscr{L}(H, H) \text{ is continuous} , \tag{3.2.5}$$

$$\langle f'_x(t, x)\, y, y\rangle \geqq d\|y\|^2 \quad \text{for every} \quad t \in \mathbf{R}, \quad x \in H , \quad \text{and} \quad y \in H$$

$$\text{with} \quad d > 0 , \tag{3.2.6}$$

then F *has the following properties*:

$$F \text{ maps } H^0_{2\pi}(\mathbf{R}; H) \text{ into itself} , \tag{3.2.7}$$

$$\|F(u_1) - F(u_2)\|_{H_{2\pi}{}^0(\mathbf{R};H)} \leqq c\|u_1 - u_2\|_{H_{2\pi}{}^0(\mathbf{R};H)}$$

$$\text{for every } u_1, \ u_2 \in H^0_{2\pi}\, (\mathbf{R}; H), \tag{3.2.8}$$

$$F \text{ is F-differentiable on } H^0_{2\pi}(\mathbf{R}; H) , \tag{3.2.9}$$

$$F' : H^0_{2\pi}(\mathbf{R}; H) \to \mathscr{L}(H^0_{2\pi}(\mathbf{R}; H), H^0_{2\pi}(\mathbf{R}; H)) \quad \textit{is continuous}\,, \tag{3.2.10}$$

$$\langle F'(u)\, g, g\rangle_{H_{2\pi}{}^0(\mathbf{R};H)} \geqq d\|g\|^2_{H_{2\pi}{}^0(\mathbf{R};H)} \quad \textit{for all} \quad u, g \in H^0_{2\pi}(\mathbf{R}; H)\,. \tag{3.2.11}$$

The *proof* of this lemma is elementary and is left to the reader.

THEOREM 3.2.1 *Let* $M \in \mathscr{M}$, $A \in \mathscr{L}^+(H, H)$, *and* $f : \mathbf{R} \times H \to H$. *If*

$$A \textit{ is normal}\,, \tag{3.2.12}$$

$$\{\lambda \in \mathbf{C};\, 0 < |\lambda - M(ij)| < \gamma\} \cap \sigma(A) = \emptyset \quad \textit{for every} \quad j \in \mathbf{Z}$$
$$\textit{with} \quad \gamma > 0\,, \tag{3.2.13}$$

$$f \textit{ satisfies } (3.2.1)-(3.2.6)\,, \tag{3.2.14}$$

then there is a constant $\varepsilon_0 > 0$ *and a unique function* $u \in C([-\varepsilon_0, \varepsilon_0];$ $H^0_{2\pi}(\mathbf{R}; H))$ *such that for every* $\varphi \in C^0_{2\pi}(\mathbf{R}; \mathbf{C})$ *and* $|\varepsilon| \leqq \varepsilon_0$,

$$\int_0^{2\pi} \varphi(\tau)\, u(\varepsilon)(\tau)\, \mathrm{d}\tau \in \mathscr{D}(A)\,,$$

$$\int_0^{2\pi} \left[M\left(-\frac{\mathrm{d}}{\mathrm{d}\tau}\right)\varphi(\tau)\right] u(\varepsilon)(\tau)\, \mathrm{d}\tau + A\int_0^{2\pi} \varphi(\tau)\, u(\varepsilon)(\tau)\, \mathrm{d}\tau =$$

$$= \varepsilon\int_0^{2\pi} \varphi(\tau)\, f(\tau, u(\varepsilon)(\tau))\, \mathrm{d}\tau\,.$$

Proof. It follows from Propositions 1.2.2, 1.2.3, Theorems 2.1.2, 2.1.4, and Lemma 3.2.1 that all the assumptions of Theorem 3.1.1 are satified. Therefore, applying this theorem we obviously obtain the required result.

Remark 3.2.1 The preceding theorem shows how to utilize the results on linear differential operators (§ 2), and in particular, why we need to know that the range of such operators is closed. Nevertheless, this theorem is far from being the most general one. In particular, it yields only generalized solutions and does not discuss their regularity. Further, it does not utilize the smoothness of the linear part (see § 2). Moreover, a great variety of methods can be used in finding fixed points instead of the Banach contraction principle which was used above. We have tried only to offer a very simple and special informative model.

§ 4. Comments on papers using direct methods

4.1. *Results of Lions and Magenes, and Ton*

In [48] (Vol. 1. pp. 279–282), LIONS and MAGENES make use of the following isomorphism theorem:

THEOREM 4.1.1 *Let $\mathscr{V}$ and $\mathscr{H}$ be Hilbert spaces, with $\mathscr{V}$ densely embedded in $\mathscr{H}$. (If $\mathscr{H}$ is identified with its dual, then it is densely embedded in $\mathscr{V}^*$, hence $\mathscr{V} \subset \mathscr{H} \subset \mathscr{V}^*$.) Let $G(t)$ $(t \geqq 0)$ be a strongly continuous bounded semigroup of operators in $\mathscr{V}^*$ with the generator Λ. Suppose further, that G also forms a strongly continuous bounded semigroup in $\mathscr{V}$ (hence also in $\mathscr{H}$) and that $\|G(t)\|_{\mathscr{L}(\mathscr{H},\mathscr{H})} \leqq 1$ for $t \geqq 0$. Denoting by $\mathscr{D}(\Lambda; \mathscr{V}^*)$ (or $\mathscr{D}(\Lambda; \mathscr{H})$, or $\mathscr{D}(\Lambda; \mathscr{V})$) the domain of the generator Λ in the space $\mathscr{V}^*$ (or in $\mathscr{H}$, or in $\mathscr{V}$), we equip $\mathscr{D}(\Lambda; \mathscr{V}^*)$ with the norm $\|u\|_{\mathscr{D}(\Lambda;\mathscr{V}^*)} = (\|u\|^2_{\mathscr{V}^*} + \|\Lambda u\|^2_{\mathscr{V}^*})^{1/2}$. Finally, let $M \in \mathscr{L}(\mathscr{V}, \mathscr{V}^*)$ be such that* $\operatorname{Re}\langle Mu, u\rangle \geqq c\|u\|^2_{\mathscr{V}}$ $(c > 0)$ *for all $u \in \mathscr{V}$. Then $\Lambda + M$ is an isomorphism between $\mathscr{V} \cap \mathscr{D}(\Lambda; \mathscr{V}^*)$ and $\mathscr{V}^*$.*

Applying the above result to the equation

$$u'(t) + A(t)\,u(t) = g(t) \tag{4.1.1}$$

the authors put $\mathscr{V} = L_2(0, \omega; V)$, $\mathscr{H} = L_2(0, \omega; H)$; hence $\mathscr{V}^* = L_2(0, \omega; V^*)$, where $V \subset H \subset V^*$ are Hilbert spaces. Further, the semigroup G is defined by

$$(G(t)\,u)\,(s) = \begin{cases} u(s - t + \omega) & \text{for} \quad 0 < s < t < \omega\,, \\ u(s - t) & \text{for} \quad 0 < t < s < \omega\,; \end{cases}$$

hence, $\mathscr{D}(\Lambda; \mathscr{V}^*) = \{u \in \mathscr{V}^*; u' \in \mathscr{V}^*,\ u(0) = u(\omega)\}$. Next, M is given by $(Mu)(t) = A(t)\,u(t)$, where $A(t) \in \mathscr{L}(V, V^*)$ for $t \in [0, \omega]$, and for $u, v \in V$

$$\begin{aligned} &\text{(a)}\ |\langle A(t)\,u, v\rangle| \leqq c_1\|u\|_V\,\|v\|_V\,; \\ &\text{(b)}\ \operatorname{Re}\langle A(t)\,u, u\rangle \geqq c_2\|u\|^2_V \quad (c_2 > 0)\,; \\ &\text{(c)}\ \text{the function } t \to \langle A(t)\,u, v\rangle \text{ is measurable}\,. \end{aligned} \tag{4.1.2}$$

Then Theorem 4.1.1 yields the following result.

THEOREM 4.1.2 *Let (4.1.2) be satisfied and let $g \in L_2(0, \omega; V^*)$. Then the equation (4.1.1) has a unique solution $u \in L_2(0, \omega; V) \cap H^1(0, \omega; V^*)$ with $u(0) = u(\omega)$.*

(This theorem includes the result of LIONS [45], pp. 50–51.)

Applying another isomorphism theorem the authors prove that the equation (4.1.1) has a unique solution $u \in \mathscr{V}$ for any $g \in (\mathscr{V} \cap \mathscr{D}(\Lambda; \mathscr{V}^*))^*$. Then combining the two results and making use of the interpolation theory, they show that the operator $\Lambda + M$ (defined above) represents an isomorphism between a certain space $\Phi^{1/2}$ and $(\Phi^{1/2})^*$. Here $\Phi^{1/2}$ can be described as the space of functions u that can be expanded as Fourier series $u = \sum_{n\in\mathbf{Z}} u_n \, e^{(2\pi/\omega)int}$ with $\sum_{n\in\mathbf{Z}} (\|u_n\|_V^2 + (1 + |n|) \, \|u_n\|_H^2) < \infty$.

In [48] (Vol. 3, pp. 171–174), the same authors are concerned with the problem of regularization of solutions of the equation (4.1.1) in certain spaces of infinitely differentiable functions.

In [47] (Chap. 2, Sec. 7.4), Lions deals with the equation (4.1.1) in non-Hilbert spaces with a non-linear operator A independent of t. He derives, for example, the following theorem.

THEOREM 4.1.3 *Let H be a Hilbert space, V a separable Banach space, $V \subset \subset H \subset V^*$, and $1 < p < \infty$. Further, let A be an operator from V into V^* satisfying the conditions:*

(1) *A is hemicontinuous and monotone;*
(2) $\|A(u)\|_{V^*} \leqq c_1 \|u\|_V^{p-1}$ for $u \in V$;
(3) $\langle A(u), u\rangle \geqq c_2 \|u\|_V^p \; (c_2 > 0)$.

Then for any $g \in L_{p'}(0, \omega; V^*)$ there exists a unique solution u of the equation (4.1.1) for which $u \in L_{p'}(0, \omega; V) \cap W_{p'}^1(0, \omega; V^*)$ and $u(0) = u(\omega)$. (The given equation is replaced by a corresponding difference equation to which the theorem on monotone operators is applied.)

The result is then applied to the equation

$$u_t - \sum_{j=1}^{n} (|u_{x_j}|^{p-2} \, u_{x_j})_{x_j} = g$$

$(x \in \Omega \subset \mathbf{R}^n$, Ω bounded), $u = 0$ on $\partial\Omega$. In Chap. 3, Sec. 2, the same equation (and in Chap. 4 Sec. 7.2 the Navier-Stokes equations) is dealt with by the method of "elliptic regularization" from Chap. 3 Sec. 1.3, 1.4. The following result is an example of an abstract theorem based on this method:

THEOREM 4.1.4 *Let $\mathscr{V}$ be a reflexive Banach space, strongly convex together with its dual $\mathscr{V}^*$. Let $L: \mathscr{D}(L) \subset \mathscr{V} \to \mathscr{V}^*$ be a maximal monotone linear operator. Finally, let $A : \mathscr{V} \to \mathscr{V}^*$ be a pseudomonotone coercive operator. Then for any $g \in \mathscr{V}^*$ there exists a $u \in \mathscr{D}(L)$ such that $Lu + A(u) = g$.*

Roughly speaking, the method consists of solving the modified equation $\varepsilon B(u_\varepsilon) + Lu_\varepsilon + A(u_\varepsilon) = g$ (B being suitably chosen) and in finding some a priori estimates for u_ε which allow us to pass to the limit as $\varepsilon \to 0$. The method seems to have been applied to periodic problems for the first time in [46], where the parabolic equation $u_t - \Delta u = g$ is replaced by the elliptic equation $-\varepsilon(u_\varepsilon)_{tt} + (u_\varepsilon)_t - \Delta u_\varepsilon = g$ $(\varepsilon > 0)$.

Another method (a combination of the Faedo-Galerkin and Poincaré methods) is suggested in Chap. 4, Sec. 6.1 – 6.2 and is used in the case of the Navier-Stokes equations.

The method of elliptic regularization has also been used by TON in [68] as applied to the non-linear heat and the Navier-Stokes equations.

4.2. *Results of Taam and Cend*

TAAM in [65] treats the equation

$$u'(t) + A\,u(t) = F(t, u(t), \varepsilon)\,, \quad t \in \mathbf{R}\,, \quad \varepsilon \in \mathbf{R} \tag{4.2.1}$$

with a linear unbounded operator A in an appropriate Banach space B. His main assumption is that A is closed, densely defined, and that there exist positive numbers α and β such that for any real $\lambda > -\alpha$ the operator $(\lambda + A)^{-1}$ exists, is bounded and everywhere defined, and that $\|(\lambda + A)^{-n}\| \leqq \beta(\lambda + \alpha)^{-n}$ for any $n \in \mathbf{N}$ (in other words, $-A$ generates an exponentially decreasing strongly continuous semigroup of linear operators). Here $F(t, u, \varepsilon)$ is supposed to lie in $\mathscr{D}(A)$. His method is an immediate generalization of a standard method from the theory of ordinary differential equations, namely, carrying over (4.2.1) to the equivalent form

$$u(t) = \int_{-\infty}^{0} \exp(\tau A)\, F(t + \tau, u(t + \tau), \varepsilon)\, \mathrm{d}\tau \tag{4.2.2}$$

and making use of the Banach contraction principle (I: Theorem 3.1.1) in the space $C_\omega(\mathbf{R}; B)$. The author investigates not only periodic but also almost periodic, bounded and compact solutions.

In [66] the same author requires Hölder continuity of F in t in an appropriate sense and studies the equation

$$\frac{\partial u}{\partial t} = \frac{\partial^2 u}{\partial x^2} + a(x)\frac{\partial u}{\partial x} + b(x)\,u + M(u) + f(t, x, u)\,, \quad t \in R\,, \quad x \in R\,.$$

In [67] he weakens the assumption on F and deals (among other things) with the existence of periodic solutions to the equation

$$\frac{\partial u}{\partial t} = a(x)\frac{\partial^2 u}{\partial x^2} + b(x)\frac{\partial u}{\partial x} + c(x)\,u + \sum_{j=0}^{n} d_j(t, x)\,u^j\,, \quad t \in \mathbf{R}\,, \quad x \in \mathbf{R}\,.$$

Equation (4.2.1) is handled in a similar way by CEND, in [16], who also makes (4.2.2) the point of departure. He works in a Hilbert space, A is assumed to be self-adjoint, strictly positive, and to have a compact inverse. Assumptions on the non-linear operator F are chosen in such a way that they enable the author to prove the existence of a generalized ω-periodic solution by means of the Schauder fixed-point theorem.

4.3. *Results of Da Prato and Barbu*

In [21] and [22], DA PRATO deals with the existence of ω-periodic solutions to the equation

$$\lambda u - \varepsilon_n u^{(n)}(t) - B(t)\, u(t) - F(t, u(t)) = g(t) \tag{4.3.1}$$

where $\lambda > 0$, $\varepsilon_n = -1$ for odd n and $\varepsilon_n = (-1)^{n/2+1}$ for even n and $B(t)$ is a linear operator. In [22] F is a non-linear operator and $n = 1$ or 2, while in [21] $F \equiv 0$. The proofs in both papers are based on general theorems concerning properties of a sum of operators (one of which is always the operator of differentiation with respect to t) from certain special classes of operators acting on a space of ω-periodic functions (for example, maximally dissipative ones or generators of contraction semigroups).

A theorem due to BARBU [3] for an abstract first-order equation is based on similar ideas.

4.4. *Results of Gajewski, Gröger and Zacharias*

In [34] pp. 213–215 these authors investigate (using I: Theorem 3.3.2) the existence of a periodic solution to the equation

$$u'(t) + (A\, u)\,(t) = f\,, \tag{4.4.1}$$

where A is a (not necessarily pointwise) hemicontinuous, monotone and coercive operator. (If A is strictly monotone, then the periodic solution is unique.) On pp. 241–244 a similar result is proved for the equation

$$u''(t) + (A\, u')\,(t) + B\, u(t) = f\,,$$

where A satisfies the same assumptions and B is a (pointwise) linear bounded self-adjoint and positive definite operator.

In [29]–[33] GAJEWSKI and GRÖGER deal mainly with the approximability of periodic solutions to the first-order equation of a certain type. GRÖGER [37] treats the same problem for a certain second-order equation and in [38] he establishes the regularity of periodic solutions of (4.4.1).

4.5. *Results of Sova*

SOVA [60] investigates the equation

$$C\,u''(t) + B\,u'(t) + A\,u(t) = g(t) + \varepsilon\,\Gamma(u(t)),$$

(if $C = 0$, then $B = I$) with help of the t-Fourier method. He looks for 2π-periodic solutions in spaces of continuous functions, or of functions of bounded semivariation (introduced by the author), or of L_2-integrable functions with values in a given Banach space. The existence of a solution depends on certain spectral properties. More precisely, it is assumed that

$$\sum_{k \in \mathbf{Z}} \|(-k^2 C + ikB + A)^{-1}\| < \infty,$$

or

$$\sup_{k \in \mathbf{Z}} \|(-k^2 C + ikB + A)^{-1}\| < \infty,$$

or

$$\sum_{k \in \mathbf{Z}} \|(-k^2 C + ikB + A)^{-1}\|^2 < \infty.$$

4.6. *Results of Dubinskiĭ*

DUBINSKIĬ [24]–[25] proves the existence of 2π-periodic solutions of the abstract equations of the type

$$\sum_{j=0}^{s} A_j\, u^{(j)}(t) = h(t), \quad A_s = I,$$

where $A_0, A_1, \ldots, A_{s-1}$ are closed, generally unbounded, mutually commuting operators (in a Hilbert space H) whose joint spectrum satisfies a certain condition. In the proof he uses an explicit formula which is based on the spectral resolution for $i(\mathrm{d}/\mathrm{d}t)$ in $H^0_{2\pi}(\mathbf{R}; H)$. In [26] some higher order elliptic-parabolic equations are treated in an abstract operator setting.

4.7. *Results of Herrmann*

HERRMANN in [40] treats ω-periodic solutions of abstract first- and second-order differential equations. He makes the following assumptions:

(1) V and H are Hilbert spaces and V is densely and compactly embedded in H;

(2) $((\cdot, \cdot))$ is a sesquilinear form on $V \times V$ and is continuous, V-elliptic (that is $|((v, v))| \geqq \gamma \|v\|_V^2$, $v \in V$, $(\gamma > 0)$) and symmetric.

Let $\mathfrak{a} : V \to V^*$ be an isomorphism determined by $((u, v)) = \langle \mathfrak{a}\, u, v \rangle$, $u, v \in V$. The author examines the properties of the operators

$$L_1\, u = u'(t) + \mathfrak{a}\, u(t) + c\, u(t)\,,$$

$$L_2\, u = u''(t) + a\, u'(t) + \mathfrak{a}\, u(t) + c\, u(t)\,,$$

and

$$L_3\, u = u''(t) + \mathfrak{a}\, u(t) + c\, u(t)$$

$(a, c \in \mathbf{R},\ a \neq 0)$ defined on $\mathscr{D}(L_1) = H^0_\omega(\mathbf{R}; V) \cap H^1_\omega(\mathbf{R}; V^*)$ and $\mathscr{D}(L_2) = \mathscr{D}(L_3) = H^0_\omega(\mathbf{R}; V) \cap H^1_\omega(\mathbf{R}; H) \cap H^2_\omega(\mathbf{R}; V^*)$, respectively.

Regularity properties are also studied (by means of the t-Fourier method): Let $\mathscr{D}(A) = \mathfrak{a}^{-1}H$, $A = \mathfrak{a} \mid \mathscr{D}(A)$, let M be a Hilbert space that is densely embedded in H. It is supposed that there is a $\Theta \in [0, 1)$ such that $\mathscr{D}(A)$ is embedded in $[M, H]_\Theta$ ($[\cdot, \cdot]_\Theta$ stands for the interpolation space between M and H with the exponent Θ). Let $s_0 = 2(1 - \Theta)^{-1}$ and $M_s = [M, H]_{(s_0 - s)/s_0}$, $s \in [0, s_0]$. Hence $M_0 = H$, $M_{s_0} = M$ and $\mathscr{D}(A)$ is embedded in M_2. It is also supposed that $A^{-1} \in \mathscr{L}(M_{s_0 - 2}, M_{s_0})$. (In particular, one can choose $M_{s_0} = \mathscr{D}(A^{s_0/2})$.) Under these assumptions the author proves, for example, that in the non-critical case

$$L_1^{-1} \in \mathscr{L}(H^0_\omega(\mathbf{R}; M_{s-2}) \cap H^{s/2-1}_\omega(\mathbf{R}; H),\ H^0_\omega(\mathbf{R}; M_s) \cap H^{s/2}_\omega(\mathbf{R}; H))\,.$$

When applying this result to partial differential equations, for example, to III: (1.3.1), the right-hand sides are not required to belong (for almost all t) to domains of some powers of the operator E ($= A$) (which leads to superfluous conditions on their behaviour on $\partial\Omega$) as is necessary when we use directly the (t, s)-Fourier method (see II: Remark 2.4.3). In fact, it suffices to choose $M_{s_0} = H^{s_0}(\Omega)$ (the choice $M_{s_0} = \mathscr{D}(A^{s_0/2})$ leads to results that can be obtained by the (t, s)-Fourier method). Similar regularity results are valid for the operators L_2 and L_3.

4.8. *Results of Straškraba*

STRAŠKRABA [61] deals with the equation

$$P\left(\frac{\mathrm{d}}{\mathrm{d}t}, A_1, \ldots, A_r\right) u(t) = g(t)\,, \quad t \in \mathbf{R} \tag{4.8.1}$$

with generally unbounded, mutually commuting operators $A_1, \ldots, A_r$ in a Banach space B, where P is a polynomial of $(r + 1)$ variables of order m and n_k

($k = 1, \ldots, r$) in $\mathrm{d}/\mathrm{d}t$ and A_k, respectively, with complex coefficients. It is assumed that each of the operators A_j generates a strongly continuous group $\{T_j(s_j)\}_{s_j \in \mathbf{R}}$ of linear bounded operators in B, that the roots $\lambda = \lambda_j(\sigma)$, $j = 1, \ldots, m$ of $P(\lambda, \mathrm{i}\sigma_1, \ldots, \mathrm{i}\sigma_r)$ satisfy

$$\max_{1 \leqq j \leqq m} \operatorname{Re} \lambda_j(\sigma) \leqq -d(1 + |\sigma|)^\alpha, \quad \sigma \in S = \{ z \in C^r ; \quad b_j^- \leqq \operatorname{Im} z_j) \leqq b_j^+ ,$$

$j = 1, \ldots, r\}$ for some constants $d > 0$, $\alpha \in \mathbf{R}$ and that the coefficient $Q_m(\mathrm{i}\sigma)$ of λ^m does not vanish for $\sigma \in S$. Here $b_j^\pm > a_j^\pm$, where the $a_j^\pm \geqq 0$ are such that $\|T_j(s_j)\| \leqq \text{const.}\, e^{a_j^+ s_j}$ for $s_j \geqq 0$ and $\|T_j(s_j)\| \leqq \text{const.}\, e^{-a_j^- s_j}$ for $s_j < 0$, ($j = 1, \ldots, m$). The right hand side g is supposed to belong to $C_\omega(\mathbf{R}; \bigcap_{j=1}^{r} \mathscr{D}(A_j^{p_j}))$ with $p_j \in \mathbf{N}$ sufficiently large. It is proved that there exists a unique solution of (4.8.1) in $\bigcap_{j=0}^{m} C_\omega^j(\mathbf{R}; \bigcap_{k=1}^{r} \mathscr{D}(A_k^{n_k}))$. Following general ideas of HERSH, II: [8], the author constructs an auxiliary equation

$$P\left(\frac{\partial}{\partial t}, D_s\right) v(t, s) = \delta(s)\, g(t), \quad t \in \mathbf{R}, \quad s \in \mathbf{R}^r$$

and looks for an ω-periodic solution with help of the Fourier transform.

4.9. *Results of Mokeĭčev and Kopáčková*

MOKEĬČEV [51] investigates the existence of a solution $u \in H_\Phi^2$, 2π-periodic in all variables x_j, $j = 1, 2, \ldots, n$, of the equation

$$R(D, x) = \sum_{m \in \Phi} c_m(x)\, D^m u(x) = g(x) \quad (0 \leqq x_j \leqq 2\pi, j = 1, 2, \ldots, n),$$

where Φ is a finite set of multi-indices and

$$H_\Phi^2 = \{u(x) = \sum_{p \in \mathbf{Z}^n} u_p e^{i<p,x>} ; \sum_{p \in \mathbf{Z}^n} |u_p|^2 \, |(ip)^m|^2 < \infty \quad \text{for all} \quad m \in \Phi\}.$$

The basic property defined for $R(D, x)$ is the so-called property (E), which reads as follows:

(E) If the series $\sum_{p \in \mathbf{Z}^n} u_p\, R(ip, x)\, e^{i\langle p,x\rangle}$ is convergent in L_2, then

$$\sum_{\substack{p \in \mathbf{Z}^n \\ R(ip,x) \not\equiv 0}} |u_p|^2 \, |(ip)^m|^2 < \infty \quad \text{for every } m \in \Phi.$$

Among other results the following two theorems are derived:

(1) If $R(D, x)$ has the property (E), then it has a closed range.

(2) The operator $R(D, x)$ (with constant coefficients) has the property (E) if and only if it has a closed range.

Let us note that Mokeĭčev's method is essentially the (t, s)-Fourier method.

In [52] the results of the preceding paper are generalized in the sense that the periodicity conditions are replaced by a system of additional linear conditions. The method, the assumptions, and the assertions are similar to those mentioned above.

KOPÁČKOVÁ in [41] derives an existence theorem for 2π-periodic solutions of the equation

$$P_1\left(\frac{\partial}{\partial t}\right)\frac{\partial^{2n}u}{\partial x^{2n}}(t, x) + P_2\left(\frac{\partial}{\partial t}\right)u(t, x) = g(t, x), \quad x \in [0, a],$$

where P_1 and P_2 are polynomials, with the boundary conditions

$$\frac{\partial^{2k}u}{\partial x^{2k}}(t, 0) = \frac{\partial^{2k}u}{\partial x^{2k}}(t, a) = 0, \quad k = 0, 1, \ldots, n - 1.$$

The t-Fourier method is applied to obtain the result.

4.10. *Results of Crandall and Rabinowitz*

CRANDALL and RABINOWITZ [20] give a generalization of the Hopf bifurcation theorem to the case of infinite-dimensional spaces. They look for $2\pi\varrho$-periodic solutions of the equation $u'(t) + Lu(t) + f(u(t), \mu) = 0$. Under some intricate, but natural assumptions they show that there exists a triple $(u(\varepsilon), \mu(\varepsilon), \varrho(\varepsilon))$ (depending regularly on the parameter $|\varepsilon| < \varepsilon_0$) that solves the given problem.

4.11. *Results of Dezin*

DEZIN in [23] studies the problem

$$u'(t) - A\,u(t) = g(t)$$
$$\mu\,u(0) - u(\omega) = h \quad (\mu \neq 0) \tag{4.11.1}$$

in a Banach space B. (For $\mu = 1$ and $h = 0$ the problem becomes periodic.) Some very weak conditions on A are found under which the problem has a solution (among others, A need not be a generator of a semigroup of operators). With this aim the author develops a sophisticated functional calculus that is applicable to the given problem.

4.12. *Results of Comincioli and Gaultier*

In [17] and [18] COMINCIOLI proves the existence and uniqueness of a weak periodic solution to the equation

$$u'(t) + A(t)\,u(t) + B(t)\,u(t - \alpha(t)) + \lambda\,u(t) = g(t)$$

for $\operatorname{Re}\lambda \geqq \lambda_0$, under various assumptions on the linear operators $A(t)$ and $B(t)$ and on the functions g and α.

A more general problem is treated by GAULTIER in [35], where he sketches an existence and uniqueness proof for an ω-periodic solution of the equation

$$u'(t) + A(t)\,u(t) + (Mu)(t) = g(t)\,,$$

where M is a linear operator in $L_\infty(0, \omega; H)$ (this means, in particular, that the equation may include operators with time displacement).

4.13. *Results of Borisovič*

BORISOVIČ in [7] states theorems on the existence of an ω-periodic solution of the system

$$\varepsilon\,u'(t) + A(t)\,u(t) = f(t, u, v, \varepsilon)\,,$$
$$v'(t) + B(t)\,v(t) = g(t, u, v, \varepsilon)$$

under various hypotheses on $A(t)$, $B(t)$, f, and g, provided that the limit system ($\varepsilon = 0$) has a sufficiently smooth solution. He applies the Schauder and Banach fixed-point theorems to suitably constructed non-linear integral equations.

§ 5. Comments on papers using indirect methods

5.1. *Results of Browder*

In [15] (briefly in [14] and in a preliminary version in [12] and [13]) BROWDER considers the equation

$$u' = F(t, u) \tag{5.1.1}$$

in a reflexive Banach space B. He supposes that $F(t, \varphi) \in B$ for fixed t is defined for $\varphi \in \mathscr{D}_t(F) \subset B$, where $\mathscr{D}_t(F)$ is dense in B, and that F is ω-periodic in t, that is, $\mathscr{D}_{t+\omega}(F) = \mathscr{D}_t(F)$ and $F(t + \omega, \varphi) = F(t, \varphi)$ for $\varphi \in \mathscr{D}_t(F)$. Making use of the Poincaré method, he first solves the initial problem given by (5.1.1) and

$$u(0) = \varphi \tag{5.1.1'}$$

(we denote its solution by $U(\varphi, t)$) and then searches for a fixed point of the translation operator $U(\varphi, \omega)$. To ensure the existence of such a fixed point he uses an appropriate Ljapunov function for (5.1.1). A continuous real-valued function V on B is called a Ljapunov function for (5.1.1) if

(1) V is convex, $V(\varphi) \geqq 0$ for $\varphi \in B$, and $V(\varphi) = 0 \Leftrightarrow \varphi = 0$;

(2) there exists a demicontinuous mapping $T: B \to B^*$ such that $V(\varphi_1) - V(\varphi_2) \geqq \langle T(\varphi_2), \varphi_1 - \varphi_2 \rangle$ for $\varphi_1, \varphi_2 \in B$ (the role of T is often played by the differential or subdifferential of V);

(3) there exists an $r > 0$ such that $\langle T(\varphi), F(t, \varphi) \rangle \leqq 0$ for any t and any $\varphi \in \mathscr{D}_t(F)$ with $\|\varphi\| \geqq r$;

(4) $\langle T(\varphi_1 - \varphi_2), F(t, \varphi_1) - F(t, \varphi_2) \rangle \leqq 0$ for any t and any $\varphi_1, \varphi_2 \in \mathscr{D}_t(F)$.

To be able to apply I: Theorem 3.3.1 (due to himself) the author assumes, in addition, that

(a) if $V(\varphi_n) \to 0$ for a sequence $\varphi_n \in B$, then $\varphi_n \to 0$;

(b) level sets of V are bounded and uniformly convex, that is for each $r > 0$, $\varepsilon > 0$ there exists an $r_1 < r$ such that if $V(\varphi_1) \leqq r$, $V(\varphi_2) \leqq r$, $V(\varphi_1 - \varphi_2) \geqq \varepsilon$, then $V((\varphi_1 + \varphi_2)/2) \leqq r_1$. (5.1.2)

By a strict solution of (5.1.1) the author means a continuous function $u : \mathbf{R}^+ \to B$ such that

(1) u is continuously differentiable for $t > 0$;
(2) $u(t) \in \mathscr{D}_t(F)$ for $t > 0$;
(3) $u'(t) = F(t, u(t))$ for $t > 0$.

By a strong solution the author means a function $u : \mathbf{R}^+ \to B$ for which there exists a sequence u_n of strict solutions such that $\sup_{t \in \mathbf{R}^+} \|u(t) - u_n(t)\|_B \to 0$ as $n \to \infty$.

Provided that there exists a dense set in B of initial data for which the equation (5.1.1) is strictly solvable and a Ljapunov function for (5.1.1) satisfying (5.1.2a), he proves that for any $\varphi \in B$ there exists a unique strong solution u of (5.1.1) with $u(0) = \varphi$. Hence the translation operator $U(\varphi, \omega)$ is uniquely defined and the following two relations hold:

$$V(U(\varphi, \omega)) \leqq r \quad \text{if} \quad V(\varphi) \leqq r \quad (r > 0), \quad V(U(\varphi_1, \omega) - U(\varphi_2, \omega)) \leqq \leqq V(\varphi_1 - \varphi_2), \quad (\varphi_1, \varphi_2 \in B).$$

Then by virtue of the fixed-point theorem mentioned above the author proves the following theorem:

THEOREM 5.1.1 *Suppose that there are strict solutions of* (5.1.1) *for a dense set of initial data. Suppose also that a Ljapunov function for* (5.1.1) *exists and satisfies* (5.1.2). *Then there is an* ω*-periodic strong solution of* (5.1.1).

Other existence theorems on ω-periodic solutions for weak or mild or generalized solutions are also stated.

Problem 5.1.1 Given the problem

$$\begin{aligned} &u_t = u_{xx} - c\,u + g(t, x) \equiv F(t, u)\,, \\ &(c \geqq 0,\ g \in C([0, \omega];\ L_2(0, \pi)))\,, \\ &u(t, 0) = u(t, \pi) = 0\,, \end{aligned} \tag{5.1.3}$$

show that the functional

$$V(\varphi) = \frac{1}{2}\int_0^\pi \varphi^2(x)\,\mathrm{d}x\,, \quad (\varphi \in L_2(0, \pi))$$

is a Ljapunov function for (5.1.3) satisfying (5.1.2) if $B = L_2(0, \pi)$ and $\mathscr{D}_t(F) = H^2(0, \pi) \cap \dot{H}^1(0, \pi)$.

Problem 5.1.2 Given the problem

$$w_t \equiv \begin{pmatrix} u_t \\ v_t \end{pmatrix} = \begin{pmatrix} v - \delta u \\ u_{xx} - (a - \delta)\,v + (a - \delta)\,\delta\,u + g \end{pmatrix} \equiv F(t, w) \tag{5.1.4}$$

($a > 0$, $\delta > 0$, $g \in C([0, \omega];\ L_2(0, \pi))$), show that the functional

$$V(\chi) = \frac{1}{2}\left(\int_0^\pi \varphi_x^2(x)\,\mathrm{d}x + \int_0^\pi \psi^2(x)\,\mathrm{d}x\right)$$

$$(\chi = (\varphi, \psi),\ \varphi \in \dot{H}^1(0, \pi),\ \psi \in L_2(0, \pi))$$

is for sufficiently small δ a Ljapunov function for (5.1.4) satisfying (5.1.2) if $B = \dot{H}^1(0, \pi) \times L_2(0, \pi)$ and $\mathscr{D}_t(F) = (H^2 \cap \dot{H}^1) \times \dot{H}^1$. (Note that the system (5.1.4) is for any δ "equivalent" to the equation $u_{tt} = u_{xx} - a\,u_t + g$.)

5.2. *Results of Brézis, Benilan, Biroli, Crandall, Pazy, Pavel and Prouse*

BRÉZIS in [9] proves the following theorem:

THEOREM 5.2.1 *Let* A *be a maximal monotone multivalued operator in a Hilbert space* H *and let* $g \in C([0, \omega];\ H)$ *with* $g' \in L_1(0, \omega;\ H)$ *and* $g(0) =$

$= g(\omega)$. *Then for any* $\lambda > 0$ *there is a unique function* $u \in W^1_\infty(0, \omega; H)$ *such that* $u(t) \in \mathscr{D}(A)$ *for* $t \in [0, \omega]$ *and*

$$u'(t) + A(u(t)) + \lambda\, u(t) \ni g(t) \quad \textit{for a. e.} \quad t \in (0, \omega),\, u(0) = u(\omega)\,. \quad (5.2.1)$$

First he proves the existence of an ω-periodic solution u of the "regularized" equation $u'_\varepsilon(t) + A_\varepsilon(u_\varepsilon(t)) + \lambda\, u_\varepsilon(t) = g(t)$ $(\varepsilon > 0)$, where $A_\varepsilon = \varepsilon^{-1}(I - (I + \varepsilon A)^{-1})$, and then he shows that u_ε converges to an ω-periodic solution of (5.2.1).

Some other results of almost the same character are proved by BRÉZIS in [10]–[11], by BENILAN and BRÉZIS [5] and by BIROLI [6].

In [19], CRANDALL and PAZY establish the existence of a unique ω-periodic solution to the equation $u'(t) + A(t)\, u(t) \ni 0$, where $\mathscr{D}(A(t))$ does not depend on t, $A(t) - \lambda I$ is, roughly speaking, maximal monotone for some $\lambda > 0$, and $A(t)$ is sufficiently regular in t. Almost the same result is obtained by PAVEL [54].

In [56] (and in a preliminary form in [55]), PROUSE considers (besides other questions) the problem of ω-periodic solutions of the equation $u'(t) + (A_1 + A_3)\, u(t) + BA_2\, u(t) = g(t)$ under rather complicated assumptions, of which the most fundamental are the following:

(1) A_1 and A_2 are permutable unbounded positive linear operators;
(2) A_3 is a linear operator "subordinate" to A_1;
(3) B is a "dissipative" non-linear operator.

The proof that an ω-periodic solution exists is based on the fact that the translation operator satisfies the assumptions of I: Theorem 3.2.3.

5.3. *Results of Vejvoda, Straškraba, Krylová, Sobolevskiĭ and Pogorelenko*

STRAŠKRABA and VEJVODA in [62], (and in a correction [63]) deal with abstract first- and second-order differential equations which include all standard evolution equations of mathematical physics, for example, the heat equation, the Schrödinger equation, the telegraph equation, the wave equation, and the beam equation. In the first part they deal with the equation

$$u'(t) + (A + c)\, u(t) = g(t)\,, \quad t \in \mathbf{R} \quad (5.3.1)$$

by the Poincaré method both in Banach and Hilbert spaces. In the former case $(-A)$ is a generator of a semigroup $\{T(t)\}_{t \in \mathbf{R}^+}$ of linear bounded operators in B, c is a constant, and $g : \mathbf{R} \to B$ is an ω-periodic function. Special attention is paid to the case when A is strictly positive (that is, $\|\lambda I + A\| \leqq c/(1 + |\lambda|)$, $\operatorname{Re} \lambda \geqq 0$, which implies that $(-A)$ is a generator of a holomorphic semigroup $\{T(t)\}_{t \in \mathbf{R}^+}$ in B).

In the latter case (that is, $B = H$, H a Hilbert space with a scalar product $\langle \cdot, \cdot \rangle$), $A : \mathscr{D}(A) \subset H \to H$ is assumed to be positive definite and self-adjoint, and $(-c)$ at most an isolated point of $\sigma(A)$.

If $U(\varphi, t)$ is a solution of (5.3.1) with $u(0) = \varphi$, then the equation $U(\varphi, \omega) - \varphi = 0$ can be solved by means of the spectral resolution for A (see I: [43], pp. 313–315). The following theorem is proved.

THEOREM 5.3.1 *An ω-periodic solution of (5.3.1) exists and is given by*

$$u(t) = \varphi + e^{-ct}\, T(t)\,(I - e^{-c\omega}\, T(\omega))_P^{-1} \int_0^\omega e^{-c\tau}\, T(\tau)\, g(t - \tau)\, \mathrm{d}\tau +$$

$$+ \int_0^t e^{-c(t-\tau)}\, T(t - \tau)\, g(\tau)\, \mathrm{d}\tau\,, \quad t \in \mathbf{R}\,, \quad \varphi \in \mathscr{N}(A + c)$$

if and only if

$$\int_0^\omega \langle g(\tau), \xi \rangle\, \mathrm{d}\tau = 0\,, \quad \xi \in \mathscr{N}(A + c)\,. \tag{5.3.2}$$

Here P is the orthogonal projection onto $\mathscr{N}(A + c)$, and if C is an operator from H to H then we define $C_P = C \mid \mathscr{N}(P)$. The solution is in $C^1_\omega(\mathbf{R}; H) \cap C_\omega(\mathbf{R}; \mathscr{D}(A))$, provided that $g \in C_\omega(\mathbf{R}; \mathscr{D}(A^\eta))$, where $\eta > 0$ is arbitrary (the operator A^η is defined in I: [43], *p.* 260).

If $A = iB$ in (5.3.1), where B is a self-adjoint operator in H and $c = 0$, then the assumption that there exist constants $\gamma > 0$ and $\varrho \geqq 0$ such that

$$\min_{l \in \mathbf{Z}} |\lambda - \nu l| \geqq \frac{\gamma}{|\lambda|^\varrho}\,, \quad \lambda \in M = \sigma(B) \setminus \{k\nu\}_{k \in \mathbf{Z}}\,, \quad (\nu = 2\pi/\omega) \tag{5.3.3}$$

turns out to be reasonable. Then the following assertion holds:

THEOREM 5.3.2 Let $g \in C_\omega(\mathbf{R}; \mathscr{D}(B^{1+\varrho}))$. *An ω-periodic solution $u \in C^1_\omega(\mathbf{R}; H) \cap C_\omega(\mathbf{R}; \mathscr{D}(A))$ of (5.3.1) exists if and only if*

$$\int_0^\omega e^{-ij\nu\tau} \langle g(\tau), \xi_j \rangle\, \mathrm{d}\tau = 0\,, \quad \xi_j \in \mathscr{N}(B + j\nu)\,, \quad j \in \mathbf{Z}\,. \tag{5.3.4}$$

When (5.3.4) is satisfied, every solution of (5.3.1) can be written in the form $u = v + w$, where

$$v(t) = \sum_{j=-\infty}^{\infty} e^{ij\nu t} \varphi_j\,, \quad \varphi_j \in \mathscr{N}(B + j\nu)\,, \sum_{j=-\infty}^{\infty} \varphi_j \in \mathscr{D}(B)\,,$$

and

$$w(t) = T(t)\, C_P^{-1} \int_0^\omega T(\tau)\,(B - \beta)^\varrho\, g(-\tau)\, \mathrm{d}\tau + \int_0^t T(t - \tau)\, g(\tau)\, \mathrm{d}\tau\,,$$

where

$$C = e^{i\pi\varrho} \int_{-\infty}^{\beta} (\beta - \lambda)^{\varrho} (1 - e^{-i\lambda\omega}) \, dE(\lambda) + \int_{\beta}^{\infty} (\lambda - \beta)^{\varrho} (1 - e^{-i\lambda\omega}) \, dE(\lambda),$$

$\beta \in \varrho(B)$ *is arbitrary, P is the orthogonal projection onto* $\mathcal{N}_{\infty} = \sum_{j=-\infty}^{\infty} \mathcal{N}(B + jv)$, $E(\lambda)$ *is the resolution of the identity corresponding to B* (*see* I: [43], *p.* 309).

The second-order equation

$$u''(t) + 2(a + bA) u'(t) + (A + c) u(t) = g(t), \quad t \in \mathbf{R}, \tag{5.3.5}$$

is examined in Hilbert spaces only. It is assumed that A is a positive definite self-adjoint operator in H ($\langle Ax, x \rangle \geqq m\|x\|_H^2$, $x \in H$ $(m > 0)$), $a \geqq 0$, $b \geqq 0$, $c \in \mathbf{R}$. The following results are established:

THEOREM 5.3.3 *Let* $(-c)$ *be at most an isolated point of* $\sigma(A)$, $a + bm > 0$, $g \in C_{\omega}(\mathbf{R}; \mathcal{D}(A^{\varkappa}))$ ($\varkappa = \frac{1}{2}$ *for* $b = 0$, $\varkappa = 1$ for $b > 0$). *Then a solution* $u \in C_{\omega}^2(\mathbf{R}; H) \cap C_{\omega}^1(\mathbf{R}; \mathcal{D}(A^{\varkappa})) \cap C_{\omega}(\mathbf{R}; \mathcal{D}(A))$ *exists if and only if* (5.3.2) *holds.*

THEOREM 5.3.4 Let $a = b = 0$, suppose that there are constants $\gamma > 0$ and $\varrho \geqq 0$ such that

$$\min_{l \in \mathbf{Z}} |(\lambda + c)^{1/2} - lv| \geqq \frac{\gamma}{\lambda^{\varrho}}, \quad \lambda \in \sigma(A) \setminus \{k^2 v^2 - c\}_{k \in \mathbf{Z}^+}$$

and let $g \in C_{\omega}(\mathbf{R}; \mathcal{D}(A^{\varrho + 1/2}))$. Then a solution $u \in C_{\omega}^2(\mathbf{R}; H) \cap C_{\omega}(\mathbf{R}; \mathcal{D}(A))$ exists if and only if

$$\int_0^{\omega} e^{ikv\tau} \langle g(\tau), \xi_k \rangle \, d\tau = 0, \quad \xi_k \in \mathcal{N}(k^2 v^2 - A - c), \quad k \in \mathbf{Z}^+.$$

As in Chapters III–VI, weakly non-linear problems corresponding to those studied above are solved by the Banach contraction principle in the non-critical case and by the implicit function theorem in the critical case.

The special case of the wave equation on an n-dimensional parallelepiped with the Dirichlet boundary conditions is investigated in detail.

In [64] the same authors are concerned with the equations

$$\mu u''(t) + u'(t) + A u(t) = g(t), \quad t \in \mathbf{R}, \quad \mu > 0$$

and

$$\mu u''(t) + u'(t) + A u(t) = \varepsilon F(u)(t), \quad t \in \mathbf{R}, \quad \mu > 0, \quad |\varepsilon| \leqq \varepsilon_0 \ (\varepsilon_0 > 0),$$

and they examine the convergence of the u_μ to u_0 $(\mu \to 0+)$ where u_0 solves the limit equation

$$u'(t) + A\,u(t) = g(t)$$

or

$$u'(t) + A\,u(t) = \varepsilon\,F(u)\,(t)\,.$$

In [44], KRYLOVÁ and VEJVODA describe a general scheme of the Poincaré method for the equation

$$D\,u = g + \varepsilon\,F(u)\,(\varepsilon)$$

with the boundary condition

$$B\,u = 0$$

and the periodicity condition

$$u(t, x) = u(t + \omega, x)\,,$$

where D is a differential operator of order k in D_t, and B is, for example, of the type II: (1.2.6).

SOBOLEVSKIĬ and POGOROLENKO in [59] treat (with proofs indicated) the equation (5.3.5) (with $a = b = c = 0$) under the assumption that g is sufficiently smooth in t. For $\sigma(A^{1/2}) = \sigma_p(A^{1/2}) = \{\lambda_k\}_{k=1}^{\infty}$ they formulate explicitly a sufficient condition for the existence of an ω-periodic solution. In particular, if $\lambda_k = k$ and $\omega = 2\pi p/q$ $(p, q \in \mathbf{N})$ or $\omega = 2\pi\alpha$ (α irrational), they state two theorems, which become after some modifications generalizations of V: Theorems 6.2.1 and 6.3.1 if we set there $\beta = 0$.

5.4. *Results of Simoněnko*

SIMONĚNKO in [57] deals with the existence of $\varepsilon\omega$-periodic solutions (with $\varepsilon > 0$ sufficiently small) of the abstract "parabolic" equation

$$u'(t) = A\,u(t) + F\left(u(t)\,,\,\frac{t}{\varepsilon}\right) \tag{5.4.1}$$

in a Banach space B, where A generates a holomorphic semigroup in B, $-A$ is a positive operator with a compact inverse, and F is ω-periodic in t and in a certain sense subordinate to A. He looks for a solution as a continuous function with values in $\mathscr{D}((-A)^\beta)$ (where $0 < \beta < 1$ is suitably chosen). Provided that the "averaged" equation

$$v'(t) = A\,v(t) + G(v(t))\,, \quad G(v) = \frac{1}{\omega}\int_0^\omega F(v, s)\,\mathrm{d}s \tag{5.4.2}$$

has a steady state solution $v(t) = v_0$ for all t and that the spectrum of the operator $A + G'(v_0)$ does not contain the origin, the author proves that there is a unique $\varepsilon\omega$-periodic solution of (5.4.1) near v_0. (He also states some theorems on exponential stability and instability of the solution.) The interesting idea of the proof can be indicated as follows: Let $H(u_0, \varepsilon, t)$, $\varepsilon > 0$, be the value at t of the solution of (5.4.1) with the initial value u_0, and suppose $H(u_0, 0, t)$ is the value at t of the solution of (5.4.2) with the initial value u_0. Then he shows that, by assumptions imposed on F, the operator H is defined and continuous together with its F-derivative H'_{u_0} on $B(v_0, \varrho; B) \times [0, \varepsilon_0] \times (0, t_0)$, and that there exists a $t_1 \in (0, t_0)$ such that the spectrum of $H'_{u_0}(v_0, 0, t_1)$ does not contain 1. Hence, by the implicit function theorem, the equation

$$u_0 = H(u_0, \varepsilon, t) \tag{5.4.3}$$

has a unique solution in some $B(v_0, \varrho_1; B)$ for $\varepsilon < \varepsilon_1$ and $t \in (t_1 - \delta, t_1 + \delta)$. For sufficiently small ε there is an $n \in \mathbf{N}$ for which $t_\varepsilon = n\varepsilon\omega \in (t_1 - \delta, t_1 + \delta)$, and the solution u_0^* of the equation $u_0 = H(u_0, \varepsilon, t_\varepsilon)$ represents the initial value of a superharmonic solution of (5.4.1). The local uniqueness of solutions of (5.4.3) implies that this solution is the required $\varepsilon\omega$-periodic solution of (5.4.1).

In [58] the results of the above paper are applied to a parabolic equation, to the Navier-Stokes equation and to the problem of convection in a field of rapidly oscillating forces (in the last case some modifications of the procedure are needed).

5.5. *Results of Fink, Hall and Hausrath*

The *two-time method* suggested in [28] by Fink, Hall and Hausrath is designed especially for finding ω-periodic solutions of problems of the type

$$z'(t) = A\, z(t) + \varepsilon(g(z(t)) + h(t)) \tag{5.5.1}$$

in a Banach space B, where A generates a strongly continuous group of operators in B which is ω-periodic (that is, $e^{\omega A} = I$), and h is ω-periodic. By the ω-periodic transformation $z(t) = e^{At}\, u(t)$ we obtain an equation in the standard form

$$u'(t) = \varepsilon\, F(t, u(t))\,, \tag{5.5.2}$$

where $F(t, u) = e^{-At}\, g(e^{At}\, u) + e^{-At}\, h(t)$ is ω-periodic in t. A solution to the equation (5.5.2) with the initial condition

$$u(0) = u_0 \tag{5.5.3}$$

is assumed to be of the form

$$u(t) = w(t, v(\varepsilon t, \varepsilon), \varepsilon)\,, \tag{5.5.4}$$

where w is an ω-periodic solution of the equation

$$w'(s, v_0, \varepsilon) = \varepsilon(F(s, w(s, v_0, \varepsilon)) - \\ - \omega^{-1} \int_0^\omega F(\tau, w(\tau, v_0, \varepsilon))\, \mathrm{d}\tau)\,, \quad w(0, v_0, \varepsilon) = v_0 \tag{5.5.5}$$

and v has to satisfy the equation

$$w'_v(s/\varepsilon, v(s, \varepsilon), \varepsilon)\, v'(s, \varepsilon) = \\ = \omega^{-1} \int_0^\omega F(\tau, w(\tau, v(s, \varepsilon), \varepsilon))\, \mathrm{d}\tau\,, \quad v(0, \varepsilon) = u_0 \tag{5.5.6}$$

in order that u be a solution of (5.5.2), (5.5.3).

Under the hypothesis that F and its first and second F-derivatives with respect to u are sufficiently smooth, the authors prove the following two theorems:

THEOREM 5.5.1 *Suppose that the equation* (5.5.6) *has a stable (asymptotically stable, exponentially stable) constant solution. Then the equation* (5.5.2) *has a stable (asymptotically stable, exponentially stable) ω-periodic solution.*

THEOREM 5.5.2 *Suppose that the equation*

$$G(v) \equiv \int_0^\omega F(s, v)\, \mathrm{d}s = 0 \tag{5.5.7}$$

has a solution $v_0 \in B$ and that $G'(v_0)$ has a bounded inverse. Then for sufficiently small $\varepsilon > 0$ the equation (5.5.2) *has an ω-periodic solution, which is even exponentially stable if the same holds for the zero solution of the equation $v' = G'(v_0)\, v$.*

FINK and HALL [27] investigate the existence of periodic solutions to the evolution equation

$$u' = A\, u + \varepsilon f(t, u, \varepsilon)\,, \tag{5.5.8}$$

when the group e^{At}, describing the solutions to the linear problem $u' = A\, u$, is 2π-periodic in t and f is periodic with the period $\omega(\varepsilon) = 2\pi(1 + \varepsilon\mu)$, $\mu \neq 0$. Denoting by $u(t, p, \varepsilon)$ the solution of (5.5.8) with an initial value $p \in \mathscr{D}(A)$, the authors show that $u(t, p, \varepsilon)$, $\varepsilon \neq 0$, is an $\omega(\varepsilon)$-periodic solution of (5.5.8) if and only if

$$\frac{1}{\varepsilon}\left(I - e^{-2\pi\varepsilon\mu A}\right) p + \int_0^{2\pi(1+\varepsilon\mu)} e^{-As} f(s, u(s, p, \varepsilon))\, ds = 0 .$$

Under certain additional assumptions this operator equation is solved by applying the inverse function theorem, provided that ε is sufficiently close to 0 but $\varepsilon \neq 0$.

5.6. *Results of Amann*

AMANN in [1] deals with the equation

$$u'(t) + A(t)\, u(t) = f(t, u(t)) \tag{5.6.1}$$

in a Banach space. In addition to some rather usual assumptions ($\mathscr{D}(A(t))$ independent of t, $A(t)$ is continuous in t in a certain sense, $A(t)$ generates an analytic semigroup, $A(0)^{-1}$ is compact, f is continuous in a certain sense etc.) he assumes that f satisfies a "Nagumo type" condition and that some a priori estimates hold. Making use of the Poincaré method and looking for an initial value in a suitably chosen space he proves the existence of an ω-periodic solution of (5.6.1). The general theorem is applied to a parabolic system.

5.7. *Other results*

MASUDA [50] investigates the equation

$$u''(t) + 2\gamma\, u'(t) + A\, u(t) + F(t, u) = f(t)\,, \quad t \geqq 0 \tag{5.7.1}$$

where $\gamma > 0$, A is a positive definite self-adjoint operator in a Hilbert space H, F is an operator from $\mathbf{R} \times H$ into H and $f : \mathbf{R}^+ \to H$. In addition to the ω-periodicity in t of F and f the author assumes that

(i) $F(t, 0) = 0$;

(ii) F is small when u is small ($|F(t, u)| = o(|u|)$) and satisfies a Lipschitz condition with respect to certain norms;

(iii) f is small enough.

It is shown that there exists an asymptotically stable ω-periodic solution of (5.7.1). In the proof the author writes the solution of the corresponding Cauchy problem with the help of the spectral resolution of A and then applies the method of FICKEN and FLEISHMANN developed in IV: [9].

KRASNOSELSKIĬ in [42] and KRASNOSELSKIĬ, ZABREĬKO, PUSTYLNIK, SOBOLEVSKIĬ in [43] present general ideas in searching for periodic solutions of equations of the types

$$u'(t) = A(t)\, u(t) + f(t, u(t)),$$
$$u''(t) + A(t)\, u(t) = f(t, u(t), u'(t)),$$
$$u'(t) = f(t, u(t), u(t-h))$$

with a linear unbounded operator $A(t)$ and non-linear operators f in a Banach space B. They show how these problems can be transformed to integral equations in appropriate function spaces and suggest topological methods of solving them.

ZABREĬKO and FETISOV [69] investigate the existence of periodic (and almost periodic) solutions of the equation $u''(t) + A\, u(t) = \varepsilon\, F(t, u(t), u'(t))$ in a Hilbert space H with a selfadjoint positive definite operator A in H, and with $F(t, u, v) = F_0(t) + C(t)\, u + D(t)\, v + G(t, u, v)$ satisfying some further conditions. A version of an averaging principle is used to obtain a solution for small ε as a fixed point of a certain related integral operator.

LOVICAR in [49] proves the existence of a generalized ω-periodic solution of the equation $u''(t) + A^2\, u(t) = F(t, u, u')$ in a Hilbert space H. It is assumed that A is a strictly positive self-adjoint operator in H and that F satisfies, in addition to the usual regularity conditions, some assumptions that guarantee the validity of appropriate a priori estimates. A theorem of Tihonov type is then applied. By means of the general result the existence of an ω-periodic transverse vibration of an extensible beam described by the equation

$$u_{tt} + a\, u_t + u_{xxxx} - b\left(\int_0^1 |u_x(t, \xi)|^2\, d\xi\right) u_{xx} = f$$

is established.

BARBU [4] examines the existence of ω-periodic solutions to the equation $u'(t) \in A\, u(t) + f(t)$ in a Banach space. He finds a fixed point of the translation operator assuming A m-dissipative.

BAILLON and HARAUX [2] deal with the equation $u'(t) + A\, u(t) \ni g(t)$ in a Hilbert space H, where A is a subdifferential of a convex functional in H and g is ω-periodic. They show that the existence of an ω-periodic solution follows from the existence of a bounded one.

HARAUX [39] treats the same problem and proves the existence of an ω-periodic solution provided that $g_0 \equiv \omega^{-1} \int_0^\omega g(t)\, dt$ lies in the interior of

$\overline{\mathscr{R}(A)}$. If g_0 is an extreme point of $\overline{\mathscr{R}(A)}$ then the problem is equivalent to the problem given by $u'(t) = g(t) - g_0$, $u(t) \in A^{-1}(g_0)$. The set of all ω-periodic solutions of the equation is also investigated.

NAGY [53] derives necessary and sufficient conditions on a closed, linear, densely defined operator A, in a Banach space B, for any solution of the equation $u''(t) = A\,u(t)$ to be ω-periodic. He uses the result of GIUSTI [36] who shows that the ω-periodicity of the cosine operator function generated by A is connected with the existence of an analytic continuation of the operator-valued function $(1 - e^{-\omega z})(z^2 I - A)^{-1}$, $(z \in \sqrt{\varrho(A)})$.

Bibliography

Chapter I *Preliminaries from functional analysis*

[1] R. A. Adams: Sobolev spaces. Academic Press, New York—San Francisco—London 1975.

[2] N. I. Ahiezer, I. M. Glazman: Theory of linear operators in Hilbert space, I, II• Frederick Ungar Publishing Co., New York 1966.

[3] V. Barbu: Nonlinear semigroups and differential equations in Banach spaces. Nordhoff International Publishing, Leyden 1976.

[4] O. V. Besov, V. P. Il'in, S. M. Nikol'skiĭ: Integral representations of functions and embedding theorems. (Russian.) Izd. Nauka, Moscow 1975.

[5] M. Biroli: Gli operatori monotoni: teoria ed applicazioni. *Rend. Sem. Mat. Fis. Milano* **42** (1972), 143—228.

[6] N. Bourbaki: Topologie générale, Chap. X — Espaces fonctionnelles. Hermann, Paris 1949.

[7] N. Bourbaki: Espaces véctoriels topologique. Chap. V — Espaces Hilbertiens. Hermann, Paris 1955.

[8] N. Bourbaki: Intégration. Hermann, Paris 1965.

[9] H. Brézis: Operateurs maximaux monotones et semi-groupes de contractions dans les espaces de Hilbert. North-Holland/American Elsevier, Amsterdam—London—New York 1973.

[10] F. E. Browder: Problèmes non-linéaires. Séminaire de mathématiques supérieures — été 1965. Univ. de Montréal 1966.

[11] J. Dieudonné: Foundations of modern analysis. Academic Press, New York—London 1969.

[12] N. Dunford, J. T. Schwartz: Linear operators, I. Interscience publishers, New York—London 1958.

[13] S. Fučík, J. Nečas, J. Souček, V. Souček: Spectral analysis of non-linear operators. Lecture Notes in Mathematics Vol. 346. Springer-Verlag, Berlin—Heidelberg—New York 1973.

[14] H. Gajewski, K. Gröger, K. Zacharias: Nichtlineare Operatorgleichungen und Operatordifferentialgleichungen. Akademie-Verlag, Berlin 1974.

[15] E. Hille, R. S. Phillips: Functional analysis and semi-groups. American Math. Soc., Providence 1957.

[16] T. Kato: Perturbation theory for linear operators. Springer-Verlag, Berlin—Heidelberg—New York 1966.

[17] M. A. Krasnosel'skiĭ: Topological methods in the theory of non-linear integral equations. (Russian.) Gostehizdat, Moscow 1956.

[18] M. A. Krasnosel'skiĭ: Positive solutions of operator equations. (Russian.) Fizmatgiz, Moscow 1967.

[19] M. A. Krasnosel'skiĭ, P. P. Zabreĭko, E. I. Pustyl'nik, P. E. Sobolevskiĭ: Integral operators in spaces of measurable functions. (Russian.) Izd. Nauka, Moscow 1966.

[20] A. Kufner, O. John, S. Fučík: Function spaces. Academia, Prague 1977.

[21] A. Kufner, J. Kadlec: Fourier series. Academia, Prague 1971.

[22] J.-L. Lions, E. Magenes: Problèmes aux limites non homogènes et applications, I. Dunod, Paris 1968.

[23] J. Leray, J. Schauder: Topologie et équations fonctionnelles. *Ann. Sci. École Norm. Sup.* **51** (1934), 45—78.

[24] L. A. Ljusternik, V. I. Sobolev: Elements of functional analysis. (Russian.) Izd. Nauka, Moscow 1965.

[25] J. Moser: A rapidly convergent iteration method and nonlinear partial differential equations I, II. *Ann. Scuola Norm. Sup. Pisa Ser.* III **20** (1966), 265—315, 499—535.

[26] J. Nečas: Les méthodes directes en théorie des équations elliptiques. Academia, Prague 1967.

[27] S. M. Nikol'skiĭ: Approximation of functions of several variables and imbedding theorems. Springer-Verlag, Berlin—Heidelberg—New York 1975.

[28] A. I. Plesner, V. A. Rohlin: Spectral theory of linear operators, II. (Russian.) *Uspehi Mat. Nauk* **1** (1946), 71—191.

[29] G. Prodi, A. Ambrosetti: Analisi non lineare. Scuola normale superiore, Pisa 1973.

[30] P. H. Rabinowitz: Periodic solutions of non-linear hyperbolic partial differential equations. *Comm. Pure Appl. Math.* **20** (1967), 145—205.

[31] M. Reed, B. Simon: Methods of modern mathematical physics I: Functional Analysis. Academic Press, New York—London 1972.

[32] F. Riesz, B. Sz.-Nagy: Leçons d'analyse fonctionnelle. Akadémiai Kiadó, Budapest 1952.

[33] W. Rudin: Real and complex analysis. Mc Graw-Hill, London—New York 1970.

[34] B. N. Sadovskiĭ: On one fixed point principle. (Russian.) *Funkcional. Anal. i Priložen.* **1** (1967), 74—76.

[35] M. Schechter: Principles of functional analysis. Academic Press, New York—London 1971.

[36] J. T. Schwartz: Non-linear functional analysis. Notes on Math. its Appl. Gordon and Breach. Science Publ., New York 1969.

[37] L. Schwartz: Théorie des distributions, I. Hermann, Paris 1957.

[38] M. Sova: Abstract semilinear equations with small non-linearities. *Comment. Math. Univ. Carolinae* **12** (1971), 785—805.

[39] A. E. Taylor: Introduction to functional analysis. John Wiley and Sons, New York 1967.

[40] H. Triebel: Höhere analysis. VEB Deutscher Verlag der Wissenschaften, Berlin 1972.

[41] M. M. Vaĭnberg: Variational methods for the study of non-linear operators. (Russian.) Gostehizdat, Moscow 1956.

[42] A. VOIGT, J. WŁOKA: Hilberträume und elliptische Differentialgleichungen. Bibliographisches Institut, Mannheim—Wien—Zürich 1975.
[43] K. YOSIDA: Functional analysis. Springer-Verlag, Berlin—Göttingen—Heidelberg 1965.

Chapter II *Preliminaries from the theory of differential equations*

[1] S. AGMON: Lectures on elliptic boundary value problems. Van Nostrand, Toronto—New York—London 1965.
[2] L. BERS, F. JOHN, M. SCHECHTER: Partial differential equations. Interscience publishers, New York—London—Sydney 1964.
[3] R. W. CARROLL: Abstract Methods in Partial Differential Equations. Harper Row Publishers, New York-Evanston-London 1969.
[4] R. H. COLE: Theory of ordinary differential equations. Appleton-Century-Crofts, New York 1968.
[5] R. COURANT, D. HILBERT: Methods of mathematical physics. New York 1953, vol. I, II.
[6] A. FRIEDMAN: Partial differential equations. Prentice-Hall, INC, New York 1964.
[7] G. HELLWIG: Differentialoperatoren der mathematischen Physik. Springer-Verlag 1964.
[8] R. HERSH: Explicit Solution of a Class of Higher-Order Abstract Cauchy Problems. *J. Differential Equations* **8** (1970), 570—579.
[9] S. G. KREĬN: Linear differential equations in a Banach space. (Russian.) Nauka, Moscow 1968.
[10] S. G. KREĬN: Linear equations in a Banach space. Translations of mathematical monographs, Amer. Math. Soc. 1972.
[11] O. A. LADYŽENSKAJA: Mixed problem for hyperbolic equations. (Russian.) Gos. Izd. Teh.-Teoret. literatury, Moscow 1953.
[12] O. A. LADYŽENSKAJA: Mathematical questions of the dynamics of a viscous incompressible liquid. (Russian.) Nauka, Moscow 1970.
[13] J. - L. LIONS: Quelques méthodes de résolution des problèmes aux limites non linéaires. Dunod, Paris 1969.
[14] J. - L. LIONS, E. MAGENES: Problèmes aux limites non homogènes et applications. Dunod, Paris 1968, vol. 1.
[15] M. A. NAĬMARK: Linear differential operators. (Russian.) Gos. Izd. Teh.-Teoret. literatury, Moscow 1954.
[16] J. NEČAS: Les méthodes directes en théorie des équations elliptiques. Academia, Prague 1967.
[17] O. PLAAT: Linear difference operators on periodic functions. *Proc. Amer. Math. Soc.* **18** (1967), 257—262.
[18] S. SCHWABIK, M. TVRDÝ, O. VEJVODA: Ordinary differential equations: Boundary value problems and adjoints. Academia, Prague 1979.
[19] O. VEJVODA: The mixed problem and periodic solutions for a linear and weakly non-linear wave equation in one dimension. *Rozpravy Československé Akad. Věd, Řada Mat. Přírod. Věd* **80** (1970), 3, Academia, Prague, 78 pp.
[20] V. S. VLADIMIROV: Equations of mathematical physics. (Russian.) Nauka, Moscow 1971.

[21] G. N. WATSON: A treatise on the theory of Bessel functions. University Press, Cambridge 1922.

[22] D. V. WIDDER: The heat equation. Academic Press, New York—San Francisco—London 1975.

Chapter III *The heat equation*

[1] H. AMANN: Periodic solutions of semi-linear parabolic equations. Non-linear analysis: A collection of papers in honor of Erich H. Rothe. Ed. Cesari-Kannan-Weinberger. Academic Press, New York—San Francisco—London 1978, 1—29. (p. 115)

[2] H. AMANN: Invariant sets and existence theorems for semi-linear parabolic and elliptic systems. *J. Math. Anal. Appl.* **65** (1978), 432—467. (p. 320)

[3] D. W. BANGE: Periodic solutions of a quasi-linear parabolic differential equations. *J. Differential Equations* **17** (1975), 61—72. (p. 114)

[4] D. W. BANGE: An existence theorem for periodic solutions of a nonlinear parabolic boundary value problem. *J. Differential Equations* **24** (1977), 426—436. (p. 114)

[5] M. BIROLI: Sur un lemme de convergence et ses applications aux équations aux dérivées partielles d'évolution non linéaires et non monotones, I, II, III. *Atti Accad. Naz. Lincei Rend. Cl. Sci. Fis. Mat. Natur.* (8) **54** (1973), no. 3, 200—204, no. 4, 511—518, no. 5, 448—451. (p. 120)

[6] H. BRÉZIS, L. NIRENBERG: Characterisation of the ranges of some nonlinear operators and applications to boundary value problems. *Ann. Scuola Norm. Sup. Pisa Ser. IV* **5** (1978), 225—326. (p. 120)

[7] F. E. BROWDER: Problèmes non-linéaires. Séminaire de mathématiques supérieures — été 1965. Univ. de Montréal 1966, 114—122. (p. 122)

[8] JA. V. BYKOV, A. I. GORŠKOV: On periodic solutions of a boundary value problem for a non-linear equation of parabolic type. (Russian.) *Trudy Krasnodar. Politehn. Inst.* **33** (1970), 137—161. (p. 107)

[9] B. M. CHERKAS: On nonlinear diffusion equations. *J. Differential Equations* **11** (1972), 284—298. (p. 123)

[10] J. DEUEL: Nichtlineare parabolische Randwertprobleme mit Unter- und Oberlösungen. ETH Diss. Nr. 5750, Zürich 1976, 37 pp. (p. 116)

[11] J. DEUEL, P. HESS: Nonlinear parabolic boundary value problems with upper and lower solutions. *Israel J. Math.* **29** (1978), 92—104. (p. 116)

[12] S. J. FARLOW: An existence theorem for periodic solution of a parabolic boundary value problem of the second kind. *SIAM J. Appl. Math.* **16** (1968), 1223—1226. (p. 112)

[13] S. J. FARLOW: Periodic solutions of nonlinear boundary value problems of the second kind. *Portugal. Math.* **32** (1973), 25—37. (p. 112)

[14] P. FIFE: Solutions of parabolic boundary problems existing for all time. *Arch. Rational Mech. Anal.* **16** (1964), 155—186. (p. 117)

[15] P. FIFE: Solutions of parabolic boundary problems existing for quasi-linear parabolic equations of second order. *Funkcial. Ekvac.* **9** (1966), 129—138. (p. 117)

[16] R. GAINES, W. WALTER: Periodic solutions to nonlinear parabolic differential equations. *Rocky Mountain J. Math.* **7** (1977), 297—312. (p. 118)

[17] JU. D. GOR'KOV: On periodic solutions of parabolic equations. (Russian.) *Differencial'nye Uravnenija* **2** (1966), 943—952. (p. 123)

[18] K. Gröger: Evolution equations modelling adsorption and diffusion in a fixed bed column (unpublished manuscript).
[19] J. Havlová: Periodic solutions of non-linear parabolic equations. Proc. Fourth Conf. Non-lin. Oscill., Prague 1967. Academia, Prague 1968, 169—172. (p. 99)
[20] L. N. Howard, N. Kopell: Slowly varying waves and shock structures in reaction-diffusion equations. *Studies in Appl. Math.* **56** (1977), 95—145. (p. 124)
[21] C. Kallina: Periodicity and stability for linear and quasi-linear parabolic equations. *SIAM J. Appl. Math.* **18** (1970), 601—613. (p. 123)
[22] D. Ch. Karimov: On periodic solutions of non-linear differential equations of parabolic type. (Russian.) *Dokl. Akad. Nauk SSSR* **25** (1939), 3—6. (p. 100)
[23] D. Ch. Karimov: On periodic solutions of non-linear equations of parabolic type. (Russian.) *Dokl. Akad. Nauk SSSR* **28** (1940), 404—407. (p. 100)
[24] D. Ch. Karimov: On periodic solutions to non-linear differential equations of parabolic type. (Russian.) *Dokl. Akad. Nauk SSSR* **46** (1945), 175—178. (p. 100)
[25] D. Ch. Karimov: On periodic solutions of non-linear differential equations of parabolic type. (Russian.) *Dokl. Akad. Nauk SSSR* **54** (1946), 293—295. (p. 100)
[26] D. Ch. Karimov: On periodic solutions of non-linear differential parabolic equations. (Russian.) *Dokl. Akad. Nauk SSSR* **56** (1947), 119—122. (p. 100)
[27] D. Ch. Karimov: On periodic solutions of non-linear differential equations of parabolic type. (Russian.) *Dokl. Akad. Nauk SSSR* **58** (1947), 969—972. (p. 100)
[28] D. Ch. Karimov: On periodic solutions of non-linear differential equations of parabolic type. (Russian.) *Trudy Inst. Mat. Meh. UzSSR* **5** (1949), 30—53. (p. 100)
[29] D. Ch. Karimov: On a certain equation of the problem of heat conduction. (Russian.) *Trudy Inst. Mat. Meh. UzSSR* **6** (1950), 67—106. (p. 100)
[30] D. Ch. Karimov: On one equation of parabolic type. (Russian.) *Trudy Inst. Mat. Meh. UzSSR* **8** (1951), 128—137. (p. 100)
[31] V. S. Klimov: Continuous branches of periodic solutions to quasi-linear parabolic equations. (Russian.) *Sibirsk. Mat. Ž.* **15** (1975), 434—438. (p. 122)
[32] V. S. Klimov: Periodic and stationary solutions of quasi-linear parabolic equations. (Russian.) *Sibirsk. Mat. Ž.* **17** (1976), 692—696, 718. (p. 115)
[33] H. Knolle: Periodische Lösungen der 3. Randwertaufgabe einer quasilinearen parabolischen Differentialgleichung. Diss., Gesamthochschule Paderborn, 1977, 78 pp. (p. 119)
[34] N. N. Kočina: On periodic solutions of the Burgers equation. (Russian.) *Prikl. Mat. Meh.* **25** (1961), 1068—1075. (p. 123)
[35] N. N. Kočina: On autooscillations of a liquid of high density in tubes. (Russian.) *Prikl. Mat. Meh.* **27** (1963), 609—617. (p. 124)
[36] N. N. Kočina: On a certain solution of the non-linear diffusion equation. (Russian.) *Prikl. Mat. Meh.* **28** (1964), 699—707. (p. 123)
[37] N. N. Kočina: On periodic régimes of some distributed systems. (Russian.) *Dokl. Akad. Nauk SSSR* **165** (1965), 1015—1018. (p. 123)
[38] N. N. Kočina: On a solution of a diffusion problem with the non-linear boundary condition. (Russian.) *Dokl. Akad. Nauk SSSR* **174** (1967), 305—308. (p. 123)
[39] N. N. Kočina: On some periodic solution to the diffusion equation. (Russian.) *Ž. Prikl. Meh. i Tehn. Fiz.* (1967), no. 1, 77—83. (p. 123)
[40] N. N. Kočina: Periodic solution of the diffusion equation with non-linear boundary condition. (Russian.) *Dokl. Akad. Nauk SSSR* **179** (1968), 1297—1300. (p. 123)

[41] Ju. S. Kolesov: On some criteria for the existence of stable periodic solutions to quasi-linear parabolic equations. (Russian.) *Dokl. Akad. Nauk SSSR* **157** (1964), 1288—1290. (p. 114)

[42] Ju. S. Kolesov: On one criterion of the existence of periodic solutions to parabolic equations. (Russian.) *Dokl. Akad. Nauk SSSR* **170** (1966), 1013—1015. (p. 114)

[43] Ju. S. Kolesov: On periodic solutions of one class of differential equations with hysteresis non-linearity. (Russian.) *Dokl. Akad. Nauk SSSR* **176** (1967), 1240—1243. (p. 123)

[44] Ju. S. Kolesov: Schauder's principle and stability of periodic solutions. (Russian.) *Dokl. Akad. Nauk SSSR* **188** (1969), 1234—1236. (p. 115)

[45] Ju. S. Kolesov: Periodic solutions of relay systems with distributed parameters. (Russian.) *Mat. Sb.* **83** (1970), 349—371. (p. 123)

[46] Ju. S. Kolesov: On a new method of proving the existence of stable periodic solutions. (Russian.) Trudy 5. Mežd. Konf. Nelin. Koleb., Kiev 1969, Vol. 1, Inst. Mat. Akad. Nauk USSR, Kiev 1970, 299—303. (p. 115)

[47] Ju. S. Kolesov: Periodic solutions of the quasi-linear parabolic equations of the second order. (Russian.) *Trudy Moskov. Mat. Obšč.* **21** (1970), 103—134. (p. 114)

[48] M. Kono: Remarks on periodic solution of linear partial differential equations of the second order. *Proc. Japan Acad.* **42** (1966), 5—9. (p. 112)

[49] M. Kopáčková: On periodic solutions of some equations of mathematical physics. *Apl. Mat.* **18** (1973), 33—42. (p. 91)

[50] N. Kopell, L. N. Howard: Plane wave solutions to reaction-diffusion equations. *Studies in Appl. Math.* **52** (1973), 291—328. (p. 124)

[51] M. A. Krasnosel'skiĭ: To the theory of periodic solutions to non-autonomous differential equations. (Russian.) *Uspehi Mat. Nauk* **21** (1966), 53—74. (p. 123)

[52] M. A. Krasnosel'skiĭ, P. E. Sobolevskiĭ: On some non-linear problems for the partial differential equations. Soviet-Amer. Symp. Part. Diff. Eq., Novosibirsk 1963, 129—133. (p. 122)

[53] S. N. Kružkov: Periodic solutions to non-linear equations. (Russian.) *Differencial'nye Uravnenija* **6** (1970), 731—740. (p. 118)

[54] T. Kusano: A remark on a periodic boundary problem of parabolic type. *Proc. Japan Acad.* **42** (1966), 10—12. (p. 118)

[55] T. Kusano: Periodic solutions of the first boundary problem for quasi-linear parabolic equations of second order. *Funkcial. Ekvac.* **9** (1966), 129—137. (p. 118)

[56] T. Kusano: Periodic solutions of parabolic partial differential equations. (Japanese.) *Sûgaku* **18** (1966), 104—106. (Not available to the authors.)

[57] D. Lauerová: A note to the theory of periodic solutions of a parabolic equation. *Apl. Mat.* **25** (1980), (to appear). (p. 104)

[58] A. P. Mal'cev: The convergence and stability of the Rothe method in the search for a periodic solution of a quasi-linear parabolic equation with non-linear boundary conditions. (Russian.) *Izv. Vysš. Učebn. Zaved. Radiofizika* 12 (1969), 415—424. (p. 119)

[59] A. P. Mal'cev: Construction of periodic solutions of boundary value problems for equations of parabolic type by the method of lines. (Russian.) *Izv. Vysš. Učebn. Zaved. Radiofizika* 12 (1969), 1657—1665. (Not available to the authors.)

[60] A. P. Mal'cev: An application of the method of lines to the determination of positive periodic solutions of problems of parabolic type with Stefan-Boltzmann boundary

conditions. (Russian.) *Izv. Vysš. Učebn. Zaved. Radiofizika* 12 (1969), 1666—1674. (Not available to the authors.)

[61] A. P. Mal'cev: The convergence of the Rothe method in the construction of a bounded, almost periodic and periodic solution of a boundary value problem of parabolic type. (Russian.) *Izv. Vysš. Učebn. Zaved. Radiofizika* 15 (1972), 332—339. (p. 119)

[62] A. P. Mal'cev: The construction of a bounded, almost periodic and periodic solution of a problem of parabolic type by the method of lines. (Russian.) *Izv. Vysš. Učebn. Zaved. Radiofizika* 15 (1972), 340—345. (p. 119)

[63] R. S. Minasjan: On a problem of a periodic heat flow in an infinite cylinder. (Russian.) *Dokl. Akad. Nauk Arm. SSR* **48** (1969), 3—8. (p. 91)

[64] A. P. Mitrjakov: On a solution to infinite systems of non-linear integral and integro-differential equations. (Russian.) *Trudy Uzb. Gos. Univ.* **37** (1948), 105—138. (p. 100)

[65] A. P. Mitrjakov: On periodic solutions to systems of non-linear equations of parabolic type. (Russian.) *Trudy Samarkand. Gos. Univ.* **119** (1962), 109—114. (p. 100)

[66] C. Monari: Soluzioni periodiche di una equazione parabolica con un termine di ritardo non lineare. *Ist. Lombardo Accad. Sci. Lett. Rend. A* **103** (1969), 688—703. (p. 123)

[67] C. Monari: Su una equazione parabolica con termine di ritardo non lineare, discontinua rispetto all incognita: soluzioni periodiche. *Ricerche Mat.* **20** (1971), 118—142. (p. 123)

[68] M. Nakao: On boundedness, periodicity, and almost periodicity of solutions of some nonlinear parabolic equations. *J. Differential Equations* **19** (1975), 371—385. (p. 120)

[69] M. Nakao, T. Nanbu: Bounded or almost periodic classical solutions for some nonlinear parabolic equations. *Mem. Fac. Sci. Kyushu Univ. Ser. A* **30** (1976), 191—211. (p. 120)

[70] M. Nakao, T. Nanbu: On the existence of global, bounded, periodic and almost-periodic solutions of nonlinear parabolic equations. *Math. Rep. College General Ed. Kyushu Univ.* **10** (1976), 99—112. (p. 120)

[71] G. Palmieri, F. Porcari: Sull' equazione non lineare del calore, discontinua rispetto all incognita: soluzioni periodiche. *Atti Accad. Sci. Torino Cl. Sci. Fis. Mat. Natur* **103** (1968—69), 969—982. (p. 119)

[72] A. D. Pendjur: On the question of the periodic solutions of neutral relay systems. (Russian.) *Vestnik Jaroslav. Univ.* **13** (1975), 153—158. (p. 123)

[73] G. Prodi: Soluzioni periodiche di equazioni alle derivate parziali di tipo parabolico e non lineari. *Riv. Mat. Univ. Parma* **3** (1952), 193—196. (p. 113)

[74] G. Prodi: Soluzioni periodiche di equazioni alle derivate parziali di tipo parabolico e non lineari. *Riv. Mat. Univ. Parma* **3** (1952), 265—290. (p. 113)

[75] G. Prodi: Problemi al contorno non lineari per equazioni di tipo parabolico non lineari in due variabili-soluzioni periodiche. *Rend. Sem. Mat. Univ. Padova* **23** (1954), 25—85. (p. 114)

[76] P. H. Rabinowitz: Periodic solutions of nonlinear hyperbolic partial differential equations, II. *Comm. Pure Appl. Math.* **22** (1969), 15—40. (p. 112)

[77] T. I. Seidman: Periodic solutions of a nonlinear parabolic equation. *J. Differential Equations* **19** (1975), 242—257. (p. 122)

[78] I. I. ŠMULEV: On periodic solutions to bundary value problems without initial data for quasi-linear parabolic equations. (Russian.) *Dokl. Akad. Nauk SSSR* **139** (1961), 1318—1321. (p. 118)

[79] I. I. ŠMULEV: On periodic solutions to boundary value problems without initial data for quasi-linear parabolic equations. (Russian.) *Dokl. Akad. Nauk SSSR* **141** (1961), 1313—1316. (p. 91)

[80] I. I. ŠMULEV: Periodic solutions of the first boundary value problem for parabolic equations. (Russian.) *Mat. Sb.* **66** (1965), 398—410. (p. 117)

[81] I. I. ŠMULEV: Quasi-periodic and periodic solutions of the problem with oblique derivative for parabolic equations. (Russian.) *Differencial'nye Uravnenija* **5** (1969), 2225—2236. (p. 117)

[82] *V.* ŠŤASTNOVÁ, S. FUČÍK: Note to periodic solvability of the boundary value problem for non-linear heat equation. *Comment. Math. Univ. Carolinae* **18** (1977), 735— 740 (p. 121)

[83] V. ŠŤASTNOVÁ, S. FUČÍK: Weak periodic solutions of the boundary value problem for nonlinear heat equation. *Apl. Mat.* **24** (1979), 284—303. (pp. 76, 121)

[84] V. ŠŤASTNOVÁ, O. VEJVODA: Periodic solutions of the first boundary value problem for a linear and weakly non-linear heat equation. *Apl. Mat.* **13** (1968), 466—477, **14** (1969), 241. (p. 107)

[85] L. - Y. TSAI: Periodic solutions of non-linear parabolic differential equations. *Bull. Inst. Math. Acad. Sinica* **5** (1977), 219—247. (p. 116)

[86] C. VAGHI: Soluzioni limitate, o quasi-periodiche, di un equazione di tipo parabolico non lineare. *Boll. Un. Mat. Ital.* **4** (1968), 559—580. (p. 114)

[87] C. VAGHI: Su un equazione parabolica con termine quadratico nella derivate Z_x. *Ist. Lombardo Accad. Sci. Lett. Rend. A* **104** (1970), 3—23. (p. 114)

[88] C. VAGHI: Soluzione limitate, o quasi-periodiche, dell equazione quasi-lineare del calore. *Rend. Sem. Mat. Fis. Milano* **42** (1972), 25—46. (p. 114)

[89] W. WALTER: The line method for parabolic differential equations. Problems in boundary layer theory and existence of periodic solutions. Lecture Notes in Mathematics Vol. 430. Springer-Verlag, Berlin—Heidelberg—New York 1974, 395—413. (p. 119)

[90] P. ZECCA: Soluzioni periodiche per un equazione d'evoluzione multivoca. *Boll. Un. Mat. Ital.* **15** (1978), 140—146. (p. 120)

The Navier-Stokes equations

[91] R. G. GORDEEV: On the existence of a periodic solution in some tide-dynamics problem. (Russian.) Problemy Mat. Anal. 4, Izd. Leningrad. Univ., Leningrad 1973, 3—9. (p. 129)

[92] G. IOOSS: Bifurcation des solutions périodique de certains problèmes d'évolution. *C. R. Acad. Sci. Paris Sér. A* **273** (1971), 624—627. (p. 127)

[93] G. IOOSS: Existence et stabilité de la solution périodique secondaire intervenant dans les problèmes d'évolution du type Navier-Stokes. *Arch. Rational Mech. Anal.* **47** (1972), 301—329. (p. 127)

[94] G. IOOSS: Bifurcation d'une solution T-périodique vers une solution nT-périodique, pour certains problèmes d'évolution du type Navier-Stokes. *C. R. Acad. Sci. Paris Sér. A* **275** (1972), 935—938. (p. 127)

[95] G. Iooss: Bifurcation of a T-periodic flow towards an nT-periodic flow and their non-linear stabilities. *Arch. Mech. Stos.* **26** (1974), 795—804. (p. 127)

[96] G. Iooss: Sur la deuxième bifurcation d'une solution stationnaire de systèmes du type Navier-Stokes. *Arch. Rational Mech. Anal.* **64** (1977), 339—369.

[97] G. Iooss, D. D. Joseph: Bifurcation and stability of nT-periodic solutions branching from T-periodic solutions at points of resonance. *Arch. Rational Mech. Anal.* **66** (1977), 135—172.

[98] D. D. Joseph: Remarks about bifurcation and stability of quasi-periodic solutions which bifurcate from periodic solutions of the Navier-Stokes equations. Lecture Notes in Mathematics 322, Springer-Verlag, Berlin—Heidelberg—New York 1973, 130—158.

[99] D. D. Joseph, D. H. Sattinger: Bifurcating time periodic solutions and their stability. *Arch. Rational Mech. Anal.* **45** (1972), 79—109. (p. 127)

[100] V. I. Judovič: Periodic motions of a viscous incompressible fluid. (Russian.) *Dokl. Akad. Nauk SSSR* **130** (1960), 1214—1217. (p. 125)

[101] V. I. Judovič: Generation of self-oscillations within a viscous fluid. (Russian.) *Prikl. Mat. Meh.* **35** (1971), 638—655. (p. 127)

[102] V. I. Judovič: Investigation of self-oscillations of a continuum, arising at a loss of stability of the stationary state. (Russian.) *Prikl. Mat. Meh.* **36** (1972), 424—431. (p. 128)

[103] S. Kaniel, M. Shinbrot: A reproductive property of the Navier-Stokes equations. *Arch. Rational Mech. Anal.* **24** (1967), 363—369. (p. 126)

[104] J. - L. Lions: Sur quelques propriétes des solutions d'inéquations variationnelles. *C. R. Acad. Sci. Paris Sér. A* **267** (1968), 631—633. (p. 129)

[105] J. - L. Lions: Quelques méthodes de résolution des problèmes aux limites non linéaires. Dunod, Paris 1969, 328, 482—489. (p. 128)

[106] G. S. Markman: On the origin of the time periodic convection. (Russian.) *Meh. Židk. Gaza* (1971), 109—119. (p. 128)

[107] G. S. Markman: On the origin of the time periodic secondary convection flows. (Russian.) *Meh. Židk. Gaza* (1973), 58—63. (p. 129)

[108] J. E. Marsden, M. McCracken: The Hopf bifurcation and its applications. Springer-Verlag, New York—Heidelberg—Berlin 1976. (p. 127)

[109] G. Morimoto: On existence of periodic weak solutions of the Navier-Stokes equations in regions with periodically moving boundaries. *J. Fac. Sci. Univ. Tokyo Sect. I A Math.* **18** (1972), 499—524. (p. 126)

[110] J. Naumann: Periodic solutions to evolution inequalities of Bingham type in arbitrary dimensions. *Rend. Mat.* **10** (1977), 213—232. (p. 129)

[111] J. Naumann: Periodic solutions to evolution inequalities related to the Bingham problem in two dimensions. *J. Math. Anal. Appl.* **61** (1977), 774—784. (p. 129)

[112] G. Prodi: Qualche risultato riquardo alle equazioni di Navier-Stokes nel caso bidimensionale. *Rend. Sem. Mat. Univ. Padova* **30** (1960), 1—15. (p. 125)

[113] G. Prouse: Soluzioni periodiche dell' equazione di Navier-Stokes. *Atti Accad. Naz. Lincei Rend. Cl. Sci. Fis. Mat. Natur.* **35** (1963), 443—447. (p. 125)

[114] G. Prouse: Su alcuni problemi per l'equazione di Navier-Stokes. Symposia Matematica 7, Academic Press, New York—London 1971, 43—83. (p. 125)

[115] R. Salvi: Sui moti periodici dei fluidi di Bingham. *Ist. Lombardo Accad. Sci. Lett. Rend. A* **108** (1974), 778—788. (p. 129)

[116] D. H. Sattinger: Bifurcation of periodic solutions of the Navier-Stokes equation. *Arch. Rational Mech. Anal.* **41** (1971), 66—80. (p. 127)

[117] D. H. Sattinger: Topics in stability and bifurcation theory. Lecture Notes in Mathematics 309, Springer-Verlag, Berlin—Heidelberg—New York 1973. (p. 127)

[118] J. Serrin: A note on the existence of periodic solutions of the Navier-Stokes equations. *Arch. Rational Mech. Anal.* **3** (1959), 120—122. (p. 125)

[119] G. Shinbrot: Lectures in fluid mechanics. Gordon and Breach, New York 1973, 213—219. (p. 126)

[120] D. A. Silaev: Time periodic solutions of a boundary layer system. (Russian.) *Prikl. Mat. Meh.* **36** (1972), 460—470. (p. 129)

[121] I. I. Šmulev: Periodic solutions of a model equation of a motion of a viscous liquid. (Russian.) *Žurn. Vyčisl. Mat. Fiz.* **7** (1967), 694—697. (p. 128)

[122] A. Takeshita: On the reproductive property of the 2-dimensional Navier-Stokes equations. *J. Fac. Sci. Univ. Tokyo Sect. I A Math.* **16** (1970), 297—311. (p. 126)

[123] B. A. Ton: Periodic solutions of non-linear evolution equations in Banach spaces. *Canad. J. Math.* **23** (1971), 189—196. (pp. 126, 128)

[124] A. Zaretti: Soluzioni periodiche di un problema misto non lineare per le equazioni di Navier-Stokes. *Atti Accad. Naz. Lincei Rend. Cl. Sci. Fis. Mat. Natur.* **51** (1971), 154—161. (p. 125)

Chapter IV *The telegraph equation*

[1] M. Biroli: Sur l'équation des ondes avec un terme non linéaire, monotone dans la fonction inconnue, I, II. *Atti Accad. Naz. Lincei Rend. Cl. Sci. Fis. Mat. Natur.* (8) **53** (1972), no. 5, 359—361, no. 6, 508—515. (p. 172)

[2] M. Biroli: Solutions bornées ou presque périodiques de l'équation non linéaire de la corde vibrante, I, II, III. *Atti Accad. Naz. Lincei Rend. Cl. Sci. Fis. Mat. Natur.* (8) **53** (1972), no. 1—2, 1—8, 9—14, no. 3—4, 229—233. (p. 171)

[3] M. Biroli: Bounded or almost periodic solutions of the non-linear vibrating membrane equation. *Ricerche Mat.* **22** (1973), 190—202. (p. 171)

[4] H. Brézis, L. Nirenberg: Characterisation of the ranges of some nonlinear operators and applications to boundary value problems. *Ann. Scuola Norm. Sup. Pisa Ser. IV* **5** (1978), 225—326. (pp. 172, 280)

[5] F. Buzzetti: Soluzioni periodiche dell' equazione della corda vibrante con termine dissipativo monotono crescente. *Ist. Lombardo Accad. Sci. Lett. Rend. A* **102** (1968), 623—634. (p. 170)

[6] F. Buzzetti: Soluzioni periodiche dell' equazione della corda vibrante con termine dissipativo discontinuo rispetto alla velocità. *Rend. Sem. Mat. Fis. Milano A* **103** (1969), 654—662. (p. 170)

[7] J. C. Clements: Existence theorems for some nonlinear equations of evolution. *Canad. J. Math.* **22** (1970), 726—745, **23** (1971), 1086. (pp. 173, 280)

[8] F. A. Ficken: Periodic solutions of a nonlinear wave equation. *Bull. Amer. Math. Soc.* **60** (1954), 148. (p. 165)

[9] F. A. Ficken, B. A. Fleishman: Initial value problems and time-periodic solutions for a nonlinear wave equation. *Comm. Pure Appl. Math.* **10** (1957), 331—356. (p. 165)

[10] B. A. Fleishman: Periodic solutions of a nonlinear wave equation. *Bull. Amer. Math. Soc.* **59** (1953), 349. (p. 165)

[11] B. A. FLEISHMAN: On the periodic solutions and initial-value problem for a Duffing-type nonlinear wave equation. *Bumblebee Report* no. 209 (1954), 29 pp. (p. 165)

[12] B. A. FLEISHMAN: Periodic solutions of a nonlinear wave equation, II. *Bull. Amer. Math. Soc.* **61** (1955), 289. (p. 165)

[13] S. FUČÍK, J. MAWHIN: Generalized periodic solutions of nonlinear telegraph equations. *Nonlinear Analysis, Theory, Methods & Applications* **2** (1978), 609—617. (p. 172)

[14] J. HAVLOVÁ: Periodic solutions of a non-linear telegraph equation. *Časopis Pěst. Mat.* **90** (1965), 273—289. (p. 165)

[15] O. HORÁČEK: Über die Existenz einer periodischen Lösung der nichtlinearen Wellengleichung in E_2. *Comment. Math. Univ. Carolinae* **10** (1969), 421—424. (p. 173)

[16] O. HORÁČEK: L'existence d'une solution classique périodique d'une équation d'ondes non linéaire dans E_1. *Comment. Math. Univ. Carolinae* **12** (1971), 635—638. (p. 173)

[17] T. KAKITA: Time periodic solutions of some nonlinear evolution equations. *Publ. Res. Inst. Math. Sci. Kyoto Univ.* **9** (1974), 477—492. (p. 174)

[18] M. KOPÁČKOVÁ-SUCHÁ: On the weakly nonlinear wave equation involving a small parameter at the highest derivative. *Czechoslovak Math. J.* **19** (1969), 469—491. (p. 164)

[19] M. KOPÁČKOVÁ: On periodic solutions of some equations of mathematical physics. *Apl. Mat.* **18** (1973), 33—42. (p. 141)

[20] N. KRYLOVÁ: Periodic solutions of hyperbolic partial differential equation with quadratic dissipative term. *Czechoslovak Math. J.* **20** (1970), 375—405. (p. 170)

[21] J. - L. LIONS: Quelques méthodes de résolution des problèmes aux limites non linéaires. Dunod, Paris 1969, 489—505. (p. 171)

[22] J. MAWHIN: Periodic solutions of nonlinear telegraph equations. Dynamical Systems, Bednarek-Cesari Ed., Academic Press, New York 1977, 193—210. (pp. 137, 139, 171)

[23] T. I. MUSUKAEVA, G. K. NAMAZOV: Asymptotic behaviour with respect to a small parameter of periodic solutions of degenerate hyperbolic equations. (Russian.) *Učen. Zap. MV SSO AzerbaĭdžanSSR* (1975), no. 6, 68—73. (p. 164)

[24] M. NAKAO: Bounded, periodic or almost periodic solutions of nonlinear hyperbolic partial differential equations. *J. Differential Equations* **23** (1977), 368—386. (p. 171)

[25] G. PRODI: Qualche risultato sulle soluzioni periodiche di equazioni del tipo iperbolico. Atti V. Congr. Unione Mat. Ital., Pavia—Torino, 1956, 1—3. (p. 169)

[26] G. PRODI: Soluzioni periodiche di equazioni a derivative parziali di tipo iperbolico non lineari. *Ann. Mat. Pura Appl.* **42** (1956), 25—49. (pp. 137, 169)

[27] G. PRODI: Soluzioni periodiche di equazioni di tipo iperbolico non lineari. Atti Convegno Eq. Deriv. Parz., Nervi, 1965, 106—107. (p. 170)

[28] G. PRODI: Soluzioni periodiche dell' equazione delle onde con termine dissipativo non lineare. *Rend. Sem. Mat. Univ. Padova* **36** (1966), 37—49. (p. 170)

[29] G. PROUSE: Soluzioni periodiche dell' equazione delle onde non omogenea con termine dissipativo quadratico. *Ricerche Mat.* **13** (1964), 261—280. (p. 170)

[30] G. PROUSE: Problemi di propagazione per equazioni non lineari della fisica matematica. *Rend. Sem. Mat. Fis. Milano* **36** (1966), 1—19. (p. 170)

[31] P. H. RABINOWITZ: Periodic solutions of nonlinear wave equation. Proc. Inter. Symp. Univ. Puerto Rico 1965, Diff. Eq. Dynamical Syst., Academic Press, New York 1967, 69—74. (p. 168)

[32] P. H. RABINOWITZ: Periodic solutions of nonlinear hyperbolic partial differential equations. *Comm. Pure Appl. Math.* **20** (1967), 145—205. (pp. 137, 167)

[33] P. H. RABINOWITZ: Periodic solutions of nonlinear hyperbolic partial differential equations, II. *Comm. Pure Appl. Math.* **22** (1969), 15—40. (p. 168)

[34] C. O. A. ŞOWUNMI: On the existence and uniqueness of periodic solution of the equation $\varrho\, u_{tt} - \mu\, u_{xx} - K^*\, u_{xx} - \lambda\, u_{xtx} - f = 0$. *Boll. Un. Mat. Ital.* **10** (1974), 724—729. (p. 174)

[35] C. O. A. ŞOWUNMI: On the existence of periodic solutions of the equation $\varrho\, u_{tt} - (\sigma(u_x))_x - \lambda\, u_{xtx} - f = 0$. *Rend. Ist. Mat. Univ. Trieste* **8** (1976), 58—68. (p. 174)

[36] I. STRAŠKRABA: On a singularly perturbed abstract diffusion equation with time periodicity condition. VIII. Int. Konf. Nichtlineare Schwingungen, B. I, 1 Akademie-Verlag, Berlin 1977, 313—316. (p. 154)

[37] M. TOUŠEK: Periodic solutions to partial differential equations of evolution type. (Czech.) Dipl. Thesis, Fac. Math. Phys. of Charles Univ., Prague 1973, 44 pp. (p. 171)

[38] C. VAGHI: Su una equazione iperbolica con coefficienti periodici e termine noto quasi-periodico. *Ist. Lombardo Accad. Sci. Lett. Rend. A* **100** (1966), 155—180. (p. 145)

[39] V. VÍTEK: Periodic solutions of a weakly nonlinear hyperbolic equation in E_2 and E_3. *Apl. Mat.* **19** (1974), 232—245. (p. 167)

[40] W. VON WAHL: Periodic solutions of nonlinear wave-equations with a dissipative term. *Math. Ann.* **190** (1971), 313—322. (pp. 173, 280)

[41] P. ZECCA: Soluzioni periodiche per un'equazione d'evoluzione multivoca. *Boll. Un. Mat. Ital.* **15** (1978), 140—146. (p. 173)

Chapter V *The wave equation*

The non-autonomous wave equation

[1] P. ACQUISTAPACE: Soluzioni periodiche di un'equazione iperbolica non lineare. *Boll. Un. Mat. Ital.* (5) **13-B** (1976), 760—777. (p. 251)

[2] N. A. ARTEM'EV: Periodic solutions of a class of partial differential equations. (Russian.) *Izv. Akad. Nauk SSSR Ser. Mat.* 1 (1937), 15—50. (p. 192)

[3] A. K. AZIZ: Periodic solutions of hyperbolic differential equations. *Proc. Amer. Math. Soc.* **17** (1966), 557—566. (p. 260)

[4] A. K. AZIZ, M. G. HORAK: Periodic solutions of hyperbolic partial differential equations in the large. *SIAM J. Math. Anal* **3** (1972), 176—182. (p. 260)

[5] E. BOREL: Sur les équations aux dérivées partielles à coefficients constants et les fonctions non analytiques. *C. R. Acad. Sci. Paris Sér. A—B.* **121** (1895), 933—935. (p. 243)

[6] H. BRÉZIS: Nonlinear equations at resonance (to appear). (p. 211)

[7] H. BRÉZIS, L. NIRENBERG: Characterizations of the ranges of some nonlinear operators and applications to boundary value problems. *Ann. Scuola Norm. Sup. Pisa Ser. IV,* **5** (1978), 225—326. (p. 210)

[8] H. BRÉZIS, L. NIRENBERG: Forced vibrations for a nonlinear wave equation. *Comm. Pure Appl. Math.* **31** (1978), 1—30. (p. 211)

[9] L. CESARI: Existence in the large of periodic solutions of hyperbolic partial differential equations. *Arch. Rational Mech. Anal.* **20** (1965), 170—190. (p. 260)

[10] L. CESARI: Periodic solutions of nonlinear hyperbolic partial differential equations. Coll. Intern. Centre Nat. Rech. Sci. 148 (1965), 425—437. (p. 260)

[11] L. CESARI: Smoothness properties of periodic solutions in the large of nonlinear hyperbolic differential systems. *Funkcial. Ekvac.* **9** (1966), 325—338. (p. 260)

[12] I. M. CHEN: Interaction of longitudinal waves with transverse waves in dispersive nonlinear elastic media, I. *Quart. Appl. Math.* **29** (1971), 125—133. (p. 261)

[13] W. D. COLLINS: Forced oscillations of system governed by one-dimensional nonlinear wave equation. *Quart. J. Mech. Appl. Math.* **24** (1971), 129—153. (p. 261)

[14] JU. I. FETISOV: The Krylov-Bogoljubov averaging method for the oscillations of a homogeneous string with fixed ends. (Russian.) *Vestnik Jaroslav. Univ.* **12** (1975), 131—154. (p. 259)

[15] J. P. FINK, W. S. HALL: Entrainment of frequency in evolution equations. *J. Differential Equations* **14** (1973), 9—41. (p. 229)

[16] J. P. FINK, W. S. HALL, A. P. HAUSRATH: A convergent two-time method for periodic differential equations. *J. Differential Equations* **15** (1974), 459—498. (p. 259)

[17] J. K. HALE: Periodic solutions of a class of hyperbolic equations containing a small parameter. *Arch. Rational Mech. Anal.* **23** (1967), 380—398. (pp. 184, 193, 260)

[18] W. S. HALL: Periodic solutions of a class of weakly nonlinear evolution equations. *Arch. Rational Mech. Anal.* **39** (1970), 294—322. (p. 210)

[19] W. S. HALL: On the existence of periodic solutions for the equations $D_{tt} u + (-1)^p D_x^{2p} u = \varepsilon f(.,., u)$. *J. Differential Equations* **7** (1970), 509—526. (p. 210)

[20] W. S. HALL: The bifurcation of solutions in Banach spaces. *Trans. Amer. Math. Soc.* **161** (1971), 207—218. (p. 210)

[21] W. S. HALL: The Rayleigh and van der Pol wave equations, some generalizations. Proc. Conf. Diff. Eq. Appl., EQUADIFF IV (1977), Lecture Notes in Mathematics 703, Springer-Verlag, Berlin—Heidelberg—New York 1979, 130—138. (p. 259)

[22] L. HERRMANN, M. ŠTĚDRÝ: Periodic solutions of weakly nonlinear wave equations in unbounded intervals. *Czechoslovak Math. J.* **28** (1978), 343—355. (p. 254)

[23] L. JENTSCH: Über stationäre thermoelastische Schwingungen in inhomogenen Körpern. *Math. Nachr.* **64** (1974), 171—231. (p. 261)

[24] L. JENTSCH: Stationäre thermoelastische Schwingungen in stückweise homogenen Körpern infolge zeitlich periodischer Aussentemperatur. *Math. Nachr.* **69** (1975), 15—37. (p. 261)

[25] H. L. JOHNSON: The existence of a periodic solution of a vibrating hanging string. *SIAM J. Appl. Math.* **16** (1968), 1048—1058. (p. 261)

[26] V. N. KARP. On periodic solutions of a non-linear hyperbolic equation. (Russian.) *Dokl. Akad. Nauk UzSSR* 5 (1953), 8—13. (p. 192)

[27] V. N. KARP: On forced periodic vibrations of strings. On periodic solutions of some non-linear equations of hyperbolic type. (Russian.) Avtoref. diss. kand. fiz. matem. nauk, Uz. univ., Samarkand, 1955. (p. 192)

[28] V. N. KARP: On existence and uniqueness of a periodic solution of a non-linear problem on forced vibrations of a string. (Russian.) *Dokl. Akad. Nauk SSSR* **133** (1960), 515—518, (p. 192)

[29] V. N. KARP: Application of the wave-region method to the solution of the problem of forced non-linear periodic vibrations of a string. (Russian.) *Izv. Učebn. Zaved. Matematika* 6 (1961), 51—59. (p. 192)

[30] V. N. KARP: On existence and uniqueness of a periodic solution of a non-linear equation of hyperbolic type. (Russian.) *Izv. Vysš. Učebn. Zaved. Matematika* 5 (1963), 43—50. (p. 192)

[31] N. KLIMPEROVÁ: Periodic solutions of hyperbolic equations of the first and second order. (Czech.) Dipl. Thesis, Fac. Math. Phys. of Charles Univ., Prague 1978, 36 pp. (p. 193)

[32] J. KURZWEIL: Problems which lead to a generalization of the concept of an ordinary differential equation. Proc. Conf. Diff. Eq. Appl., EQUADIFF I (1962), Academia, Prague 1963, 65—76. (p. 258)

[33] J. KURZWEIL: Exponentially stable integral manifolds, averaging principle and continuous dependence on a parameter. *Czechoslovak Math. J.* **16** (1966), 380—423, 463—492. (p. 258)

[34] V. I. KVAL'VASSER, JU. P. SAMARIN: Quasiperiodic and periodic solutions to problems with moving boundaries for a wave equation in one-dimensional space. (Russian.) *Differencial'nye Uravnenija* **2** (1966), 1541—1543. (p. 261)

[35] H. LOVICAROVÁ: Periodic solutions of a weakly non-linear wave equation in one dimension. *Czechoslovak Math. J.* **19** (1969), 324—342. (pp. 204, 208)

[36] J. MAWHIN: Solutions périodiques d'équations aux dérivées partielles hyperboliques non linéaires. Mélanges ,,Théodore Vogel" — 1978, 301—315. (p. 245)

[37] M. H. MILLMAN, J. B. KELLER: Perturbation theory of nonlinear boundary value problems. *J. Mathematical Phys.* **10** (1969), 342—361. (p. 261)

[38] W. MIRANKER: Periodic solutions of the wave equation with a nonlinear interface condition. *IBM Journal Res. Develop.* **5** (1961), 2—24. (p. 261)

[39] A. P. MITRJAKOV: On a solution to infinite systems of nonlinear integral and integro-differential equations. (Russian.) *Trudy Uzb. Gos. Univ.* **37** (1948), 105—138. (p. 192)

[40] A. P. MITRJAKOV: On periodic solutions of the nonlinear hyperbolic equations. (Russian.) *Trudy Inst. Mat. Meh. Akad. Nauk UzSSR* **7** (1949), 137—149. (p. 192)

[41] E. MUSTAFAZADE: Periodic solutions of a class of non-linear partial differential equations. (Azerbaijani, Russian Summary.) *Trudy Inst. Mat. Meh. Akad. Nauk Azerbaĭdžan. SSR* 1 (1961), 112—136. (p. 192)

[42] M. NAKAO: Remarks on the existence of periodic solutions of nonlinear wave equations. *Mem. Fac. Sci. Kyushu Univ. Ser. A* **27** (1973), 323—334. (p. 210)

[43] M. NAKAO. On the regularity of periodic solutions to a class of nonlinear evolution equations. *Mem. Fac. Sci. Kyushu Univ. Ser. A* **28** (1974), 101—111. (p. 210)

[44] M. NAKAO: On some regularity of periodic solutions of nonlinear wave equations. *J. Differential Equations* **17** (1975), 187—197. (p. 210)

[45] M. NAKAO: Periodic solutions of linear and nonlinear wave equations. *Arch. Rational Mech. Anal.* **62** (1976), 87—98. (p. 210)

[46] B. NOVÁK: Remark on periodic solutions of a linear wave equation in one dimension. *Comment. Math. Univ. Carolinae* **15** (1974), 513—519. (p. 243)

[47] J. PEŠL: Periodic solutions of a weakly nonlinear wave equation in one dimension. *Časopis Pěst. Mat.* **98** (1973), 333—356. (pp. 229, 246)

[48] D. PETROVANU: Periodic solutions of the Tricomi problem. *Michigan Math. J.* **16** (1969), 331—348. (p. 260)

[49] V. N. POLIŠČUK: Periodic boundary value problem for linear hyperbolic equations. (Russian.) Mathematical Methods and Physical-Mechanical Fields (Russian.), Vol. 2. Naukova Dumka, Kiev 1975, 158—160.(p. 242)

[50] V. N. POLIŠČUK: Periodic boundary value problem for hyperbolic equations with variable coefficients. (Russian.) Theoretical and applied questions of algebra and differential equations. (Russian.) Kiev 1976, 60—65. (p. 242)

[51] V. N. Polіščuk, B. I. Ptašnik: On a periodic boundary value problem for hyperbolic operators decomposed into linear factors of the first order with constant coefficients. (Russian.) Mathematical Methods and Physical-Mechanical Fields (Russian.), Vol. 3. Naukova Dumka, Kiev 1976, 6—12. (p. 242)

[52] V. N. Polіščuk, B. I. Ptašnik: On a periodic boundary-value problem for a system of hyperbolic equations with constant coefficients. (Russian.) *Ukrain. Mat. Ž.* **30** (1978), 326—333. (p. 242)

[53] B. I. Ptašnik: Periodic boundary value problem for hyperbolic operator decomposing into the first order linear factors. (Ukrainian.) *Dopovidi Akad. Nauk Ukrain. RSR Ser. A* (1973), 985—989. (p. 242)

[54] B. I. Ptašnik: Periodic boundary value problem for linear hyperbolic equations with constant coeficients. (Russian.) Matematičeskaja fizika 12, Naukova dumka, Kiev 1972, 117—121. (p. 242)

[55] P. H. Rabinowitz: Periodic solutions of a nonlinear nondissipative wave equation. Courant Inst. Math. Sci., New York Univ., 1965, 1—45. (p. 209)

[56] P. H. Rabinowitz: Periodic solutions of a nonlinear wave equation. Proc. Intern. Symp. Univ. Puerto Rico, Diff. Eq. Dynamical Syst., 1965, Academic Press, New York, 1967, 69—74. (p. 209)

[57] P. H. Rabinowitz: Periodic solutions of nonlinear hyperbolic partial differential equations. *Comm. Pure Appl. Math.* **20** (1967), 145—205. (p. 209)

[58] P. H. Rabinowitz: Time periodic solutions of nonlinear wave equations. *Manuscripta Math.* **5** (1971), 165—194. (p. 209)

[59] P. H. Rabinowitz: Some global results for nonlinear eigenvalue problems. *J. Functional Analysis* **7** (1971), 487—513. (p. 209)

[60] P. H. Rabinowitz: Some minimax theorems and applications to nonlinear partial differential equations. Non-linear analysis: A collection of papers in honor of Erich H. Rothe. Ed. Cesari-Kannan-Weinberger. Academic Press, New York—San Francisco—London 1978, 161—177. (pp. 169, 211)

[61] N. R. Sibgatulin: On non-linear transverse vibrations at resonance in elastic layer and in layer of ideal conducting liquid. (Russian.) *Prikl. Mat. Meh.* **36** (1972), 79—87. (p. 261)

[62] L. de Simon: Sull'equazione delle onde con termine noto periodico. *Rend. Ist. Mat. Univ. Trieste* **1** (1969), 150—162. (p. 251)

[63] L. de Simon, G. Torelli: Soluzioni periodiche di equazioni a derivate parziali di tipo iperbolico non lineari. *Rend. Sem. Mat. Univ. Padova* **40** (1968), 380—401. (p. 209)

[64] G. T. Sokolov: On periodic solutions of a class of partial differential equations. (Russian.) *Dokl. Akad. Nauk UzSSR* 12 (1953), 3—7. (p. 192)

[65] G. T. Sokolov: On periodic solutions of a class of equations. (Russian.) *Učen. Zap. Ferganskogo Gos. Ped. Inst.* 7 (1957), 2—5. (p. 246)

[66] G. T. Sokolov: On periodic solution of a wave equation. (Russian.) *Učen. Zap. Ferganskogo Gos. Ped. Inst.* 1 (1965), 17—25. (p. 246)

[67] P. V. Solov'ev: Some remarks on periodic solutions of the non-linear equations of hyperbolic type. (Russian.) *Izv. Akad. Nauk SSSR, Ser. Mat.* 2 (1939), 150—164. (p. 192)

[68] M. Sova: Abstract semi-linear equations with small nonlinearities, I. *Comment. Math. Univ. Carolinae* **12** (1971), 785—805. (p. 210)

[69] M. Sova: Abstract semi-linear equations with small parameter. Proc. Conf. Diff.

Eq. Appl., EQUADIFF III (1972), J. E. Purkyně University, Brno 1973, 71—79. (p. 210)

[70] M. Štědrý: Periodic solutions of a weakly nonlinear wave equation. *Časopis Pěst. Mat.* **102** (1977), 128—143. (p. 235)

[71] F. M. Stewart: Periodic solutions of a nonlinear wave equation. *Bull. Amer. Math. Soc.* **60** (1954), 344. (p. 246)

[72] J. J. Stoker: Periodic oscillations of nonlinear systems with infinitely many degrees of freedom. Actes Coll. Inter. Vibrations Nonlinéaires. Ile de Porquerolles, 1951 Publ. Sci. Tech. Ministère de l'Air, 281 (1953), 61—75. (p. 261)

[73] I. Straškraba: Periodic solutions of a weakly nonlinear wave equation. (Czech.) Dipl. Thesis, Fac. Math. Phys. of Charles Univ., Prague 1970, 61 pp. (p. 193)

[74] G. Torelli: Soluzioni periodiche dell equazione non lineare $u_{tt} - u_{xx} + \varepsilon f(x, t, u) = 0$. *Rend. Ist. Mat. Univ. Trieste* **1** (1969), 123—137. (p. 209)

[75] K. N. Tursunov: Investigation of periodic, almost everywhere solutions of a multidimensional boundary value problem for a class of hyperbolic equations of second order with the non-linear operator right-hand side. (Russian.) *Voprosy vyčisl. prikl. mat.*, **40**, Taškent, 1976, 34—47. (Not available to the authors.)

[76] O. Vejvoda: Nonlinear boundary-value problems for differential equations. Proc. Conf. Diff. Eq. Appl., EQUADIFF I (1962), Academia, Prague 1963, 199—215. (p. 192)

[77] O. Vejvoda: Periodic oscillations of a weakly nonlinear string. (Russian and English.) Trudy Mežd. Symp. Nelin. Koleb., Tom 2, Kiev 1963, 120—122. (p. 192)

[78] O. Vejvoda: Periodic solutions of a linear and weakly nonlinear wave equation in one dimension, I. *Czechoslovak Math. J.* **14** (1964), 341—382. (p. 192)

[79] O. Vejvoda: Periodic solutions of a weakly nonlinear wave equation in E_3 in a spherically symmetrical case. *Apl. Mat.* **14** (1969), 160—167. (p. 193)

[80] O. Vejvoda: The mixed problem and periodic solutions for a linear and weakly nonlinear wave equation in one dimension. *Rozpravy Československé Akad. Věd, Řada Mat. Přírod. Věd* **80** (1970), 3, Academia, Prague, 1—78. (pp. 192, 229)

[81] G. R. Verma: Nonlinear vibrations of beams and membranes. *Z. Angew. Math. Phys.* **23** (1972), 805—814. (p. 261)

Autonomous hyperbolic equations

[82] M. S. Berger: On periodic solutions of nonlinear hyperbolic equations and the calculus of variations. *Bull. Amer. Math. Soc.* **76** (1970). 633—637. (p. 262)

[83] M. S. Berger: Stationary states for a nonlinear wave equation. *J. Mathematical Phys.* **11** (1970), 2906—2912. (p. 264)

[84] M. S. Berger: On the existence and structure of stationary states for a nonlinear Klein-Gordon equation. *J. Functional Analysis* **9** (1972), 249—261. (p. 264)

[85] V. S. Buslaev: Double soliton type solutions of the multidimensional equation $\Box u = F(u)$. (Russian.) *Teoretičeskaja i matematičeskaja fizika*, **31** (1977), No 1, 23—32. (p. 267)

[86] A. M. Filimonov: Periodic solutions of some non-linear partial differential equations. (Russian.) *Differencial'nye uravnenija,* **12** (1976), 2076—2084. (p. 264)

[87] J. P. Fink, W. S. Hall, A. R. Hausrath: Discontinuous periodic solutions for an autonomous nonlinear wave equation. *Proc. Roy. Irish Acad.* **75** A 16 (1975), 195—226. (p. 262)

[88] J. P. Fink, W. S. Hall, S. Khalili: Perturbation expansions for some nonlinear wave equations. *SIAM J. Appl. Math.* **24** (1973), 575—595. (p. 266)

[89] B. A. Fleishman: Progressing waves in an infinite nonlinear string. *Proc. Amer. Math. Soc.* **10** (1959), 329—334. (p. 265)

[90] F. G. Friedlander: On the oscillations of a bowed string. *Math. Proc. Cambridge Philos. Soc.* **49** (1953), 516—530. (p. 267)

[91] P. Gatignol: Sur une méthode asymptotique en théorie des ondes dispersives non linéaires. *J. Mécanique* **11** (1972), 95—117. (p. 267)

[92] J. M. Greenberg: Smooth and time-periodic solutions to the quasi-linear wave equation. *Arch. Rational Mech. Anal.* **60** (1975), 29—50. (p. 264)

[93] W. S. Hall: A Rayleigh wave equation — an analysis. *Nonlinear Analysis, Theory, Methods & Applications* **2** (1978), 129—156. (p. 262)

[94] W. S. Hall: The Rayleigh and van der Pol wave equations; some generalizations. Proc. Conf. Diff. Eq. Appl., EQUADIFF IV (1977), Lecture Notes in Mathematics 703, Springer-Verlag, Berlin—Heidelberg—New York 1979, 130—138. (p. 262)

[95] J. B. Keller: Bowing of violin strings. *Comm. Pure Appl. Math.* **6** (1953), 483—495. (p. 267)

[96] J. B. Keller, L. Ting: Periodic vibrations of systems governed by nonlinear partial differential equations. *Comm. Pure Appl. Math.* **19** (1966), 371—420. (p. 267)

[97] J. Kurzweil: Exponentially stable integral manifolds, averaging principle and continuous dependence on a parameter. *Czechoslovak Math. J.* **16** (1966), 380—423, 463—492. (p. 261)

[98] R. B. Melrose, M. Pemberton: Periodic solutions of certain nonlinear autonomous wave equations. *Math. Proc. Cambridge Philos. Soc.* **78** (1975), 137—143. (p. 263)

[99] M. H. Millman, J. B. Keller: Perturbation theory of nonlinear boundary value problems. *J. Mathematical Phys.* **10** (1969), 342—361. (p. 266)

[100] G. Petiau: Sur des fonctions d'ondes d'un type nouveau, solutions d'équations non linéaires généralisant l'équation des ondes de la mécanique ondulatoire. *C. R. Acad. Sci. Paris. Sér. A-B* **244** (1957), 1890—1893. (p. 264)

[101] G. Petiau: Sur une généralisation non linéaire de la mécanique ondulatoire et les propriétés des fonctions d'ondes correspondantes. *Nuovo Cimento* **9** (1958), 542—568. (p. 264)

[102] V. G. Pisarenko: Problems of relativistic dynamics of many bodies and of the non-linear field theory. (Russian.) Naukova Dumka, Kijev 1974, 378—395. (p. 265)

[103] S. I. Pohožaev: On periodic solutions of some non-linear hyperbolic equations. (Russian.) *Dokl. Akad. Nauk SSSR* **198** (1971), 1274—1277. (p. 262)

[104] P. H. Rabinowitz: Free vibrations for a semi-linear wave equation. *Comm. Pure Appl. Math.* **31** (1978), 31—68. (p. 263)

[105] M. Štědrý, O. Vejvoda: Periodic solutions to weakly non-linear autonomous wave equations. *Czechoslovak Math. J.* **25** (1975), 536—555. (p. 263)

[106] O. Vejvoda, M. Štědrý: Periodic solutions of a weakly non-linear autonomous wave equation. Trudy seminara S. L. Soboleva, **2** (1978), Novosibirsk, 17—36. (p. 264)

[107] V. A. Vitt: Distributed self-oscillating systems. (Russian.) *Žurn. Teoret. Fiz.* **4** (1934), 144—157. (p. 265)

[108] M. E. Žabotinskiĭ: On periodic solutions of nonlinear partial differential equations. (Russian.) *Dokl. Akad. Nauk SSSR* **56** (1947), 469—472. (p. 265)

Chapter VI *The beam equation and related problems*

[1] A. A. Berezovskiĭ, Ju. I. Žariĭ: On increase of the shell statical stability with the help of vibrations. (Russian.) Trudy V. Konf. Nelin. Koleb., Tom 3, Inst. Mat. Akad. Nauk USSR, Kiev 1970, 63—88. (p. 287)

[2] R. Van Dooren: Two mode subharmonic vibrations of ordre 1/9 of a non-linear beam forced by a two mode harmonic load. *J. Sound and Vibration* **41** (1975), 133—142. (p. 287)

[3] P. Filip: Periodic vibrations of rectangular plates. (Czech.) Dipl. thesis, Fac. Math. Phys. of Univ. Charles, Prague 1976, 35 pp. (p. 278)

[4] W. S. Hall: On the existence of periodic solutions for the equations $D_{tt}u + (-1)^p D_x^{2p} u = \varepsilon f(.,.,u)$. *J. Differential Equations* **7** (1970), 509—526. (p. 278)

[5] W. S. Hall: Periodic solutions of a class of weakly non-linear evolution equations. *Arch. Rational Mech. Anal.* **39** (1970), 294—322. (p. 278)

[6] W. S. Hall: The bifurcation of solutions in Banach spaces. *Trans. Amer. Math. Soc.* **161** (1971), 207—217. (p. 279)

[7] W. Heinrich: Über Biegeschwingungen einer flachen rotationssymetrischen Kugelschale. Trudy V. Konf. Nelin. Koleb., Tom 3, Inst. Mat. Akad. Nauk USSR, Kiev 1970, 182—196. (p. 287)

[8] D. Ch. Karimov: On periodic solutions of non-linear equations of the fourth order. (Russian.) *Dokl. Akad. Nauk SSSR* **49** (1945), 618—621. (p. 277)

[9] D. Ch. Karimov: On periodic solutions of non-linear equations of the fourth order. (Russian.) *Dokl. Akad. Nauk SSSR* **57** (1947), 651—653. (p. 277)

[10] D. Ch. Karimov: On periodic solutions of non-linear equations of the fourth order. (Russian.) *Dokl. Akad. Nauk UzSSR* **8** (1949), 3—7. (p. 277)

[11] M. Kopáčková: On periodic solutions of some equations of mathematical physics. *Apl. Mat.* **18** (1973), 33—42. (p. 277)

[12] M. Kopáčková, O. Vejvoda: Periodic vibrations of an extensible beam. *Časopis Pěst. Mat.* **102** (1977), 356—363. (p. 278)

[13] P. Krejčí: Periodic solutions to the equations of mathematical physics. (Czech.) Dipl. thesis, Fac. Math. Phys. of Univ. Charles, Prague 1978, 63 pp. (p. 280)

[14] N. Krylová: Periodic solutions of hyperbolic partial differential equation with quadratic dissipative term. *Czechoslovak Math. J.* **20** (1970), 375—405. (p. 280)

[15] N. Krylová, O. Vejvoda: A linear and weakly nonlinear equation of a beam: the boundary-value problem for free extremities and its periodic solutions. *Czechoslovak Math. J.* **21** (1971), 535—566. (pp. 276, 277)

[16] N. Krylová, O. Vejvoda: Periodic solutions to partial differential equations, especially to a biharmonic wave equation. Symposia Matematica 7, Academic Press, New York—London 1971, 85—96. (p. 276)

[17] I. Kurbanov: Non-linear vibrations of axially-symmetrically deformable circular plates. Analytic methods of investigation of solutions to non-linear differential equations. (Russian.) Inst. Mat. Akad. Nauk USSR, Kiev 1975, 83—87. (p. 287)

[18] J. Kurzweil: Problems which lead to a generalization of the concept of an ordinary differential equation. Proc. Conf. Diff. Eq. Appl., EQUADIFF I (1962), Academia, Prague 1963, 65—76. (p. 278)

[19] A. P. Mitrjakov: On periodic solutions to non-linear partial differential equations of higher order. (Russian.) *Trudy Samarkand. Gos. Univ.* 65 (1956), 31—44. (p. 277)

[20] N. F. MOROZOV: Investigation of non-linear vibration of thin plates involving damping. (Russian.) *Differencial'nye Uravnenija* **3** (1967), 619—635. (p. 287)

[21] N. F. MOROZOV: Non-linear vibrations of thin plates with allowance for rotational inertia. (Russian.) *Dokl. Akad. Nauk SSSR* **176** (1967), 522—525. (p. 287)

[22] N. F. MOROZOV: Non-linear vibrations of thin plates with allowance for rotational inertia. (Russian.) *Differencial'nye Uravnenija* **4** (1968), 932—937. (p. 287)

[23] M. NAKAO: On the regularity of periodic solutions to a class of non-linear evolution equations. *Mem. Fac. Sci. Kyushu Univ. Ser. A* **28** (1974), 101—111. (p. 279)

[24] H. PETZELTOVÁ: Periodic solutions of the equation $u_{tt} + u_{xxxx} = \varepsilon f(\cdot, \cdot, u, u_t)$. *Czechoslovak Math. J.* **23** (1973), 269—285. (p. 279)

[25] P. V. SOLOV'EV: Sur les solutions périodiques de certaines équations nonlinéaires du quatrième ordre. *Dokl. Akad. Nauk SSSR* **25** (1939), 731—734. (p. 277)

[26] W. SZEMPLIŃSKA-STUPNICKA: Periodyczne drgania swobodne belki przy słabonieliniowych warunkach brzegowych. *Rozprawy Inż.* **15** (1967), 93—103. (p. 287)

[27] M. ŠTĚDRÝ: Periodic solutions of nonlinear equations of a beam with damping. (Czech.) Thesis, Math. Inst. Czechoslovak Acad. Sci., Prague 1973, 59 pp. (p. 279)

[28] G. R. VERMA: Nonlinear vibrations of beams and membranes. *J. Appl. Math. Phys.* **23** (1972), 805—814. (p. 287)

[29] I. I. VOROVIČ, S. A. SOLOP: On the existence of periodic solutions in the non-linear theory of vibrations of flat shells with account of damping. (Russian.) *Prikl. Mat. Meh.* **40** (1976), 699—705. (p. 287)

Chapter VII *The abstract equations*

[1] H. AMANN: Invariant sets and existence theorems for semi-linear parabolic and elliptic systems. *J. Math. Anal. Appl.* **65** (1978), 432—467. (p. 320)

[2] J. B. BAILLON, A. HARAUX: Comportement a l'infini pour les équations d'évolution avec forcing périodique. *Arch. Rational Mech. Anal.* **67** (1977), 101—109. (p. 321)

[3] V. BARBU: Nonlinear semigroups and evolution equations. Iaşi 1970, 72—93. (p. 306)

[4] V. BARBU: Nonlinear semigroups and differential equations in Banach spaces. Nordhoff International Publishing, Leyden 1976, 138—139. (p. 321)

[5] P. BENILAN, H. BRÉZIS: Solutions faibles d'équations d'évolution dans les espaces de Hilbert. *Ann. Inst. Fourier, (Grenoble)* **22** (1972), 311—329. (p. 314)

[6] M. BIROLI: Gli operatori monotoni: teoria ed applicazioni. *Rend. Sem. Mat. Fis. Milano* **42** (1972), 143—228. (p. 314)

[7] JU. G. BORISOVIČ: On periodic solutions of differential-operator equations with a small parameter at the derivative. (Russian.) *Dokl. Akad. Nauk SSSR* **148** (1963), No. 2, 255—258. (p. 311)

[8] JU. G. BORISOVIČ, ČAN ZEŇ BONG: On periodic solutions of a parabolic equation with a small parameter. (Russian.) *Voroněž. Gos. Univ. Trudy Sem. Mat. Fak.* 1972, No. 7, 24—29. (Not available to the authors.)

[9] H. BRÉZIS: Problèmes unilatéraux. *J. Math. Pures Appl.* **51** (1972), 128—139. (p. 313)

[10] H. BRÉZIS: Semigroupes non linéaires et applications. Symposia Matematica 7, Academic Press, New York—London 1972, 3—27. (p. 314)

[11] H. BRÉZIS: Operateurs maximaux monotones. North-Holland Publishing Co., Amsterdam—London 1973, 93—97. (p. 314)

[12] F. E. Browder: Existence of periodic solutions for nonlinear equations of evolution. *Proc. Nat. Acad. Sci. U.S.A.* **53** (1965), 1100—1103. (p. 311)

[13] F. E. Browder: Problèmes non-linéaires. Séminaire de mathématiques supérieures — été 1965. Univ. de Montréal 1966, 114—148. (p. 311)

[14] F. E. Browder: Nonlinear functional analysis and nonlinear partial differential equations. Proc. Conf. Differential Equations Their Appl., EQUADIFF II 1966, Slov. Ped. Nakl. Bratislava 1969, 45—64. (p. 311)

[15] F. E. Browder: Periodic solutions of nonlinear equations of evolution in infinite dimensional spaces. Lecture Series in Differential Equations, 1. Ed. A. K. Aziz. Van Nostrand, New York 1969, 71—96. (pp. 28, 311)

[16] L. Cend: Theorem on the existence of a periodic solution to a parabolic equation with a non-linearity. (Russian.) *Vestnik Moskov. Univ.* (1967), 32—35. (p. 305)

[17] V. Comincioli: Problemi periodici relativi a equazioni d'evoluzione paraboliche con termini di ritardo. Risoluzione e approssimazione delle soluzioni mediante uno schema alle differenze finite. *Rend. Ist. Lombardo Ser. A* **104** (1970), 356—382. (p. 311)

[18] V. Comincioli: Ulteriori osservazioni sulle soluzioni del problema periodico per equazioni paraboliche lineari con termini di perturbazione. *Rend. Ist. Lombardo Ser. A* **104** (1970), 726—735. (p. 311)

[19] M. G. Crandall, A. Pazy: Non-linear evolution equations in Banach spaces. *Israel J. Math.* **11** (1972), 57—94. (p. 314)

[20] M. G. Crandall, P. H. Rabinowitz: The Hopf bifurcation theorem in infinite dimensions. *Arch. Rational Mech. Anal.* **67** (1977), 53—72. (p. 310)

[21] G. Da Prato: Weak solutions for linear abstract differential equations. *Advances in Math.* **5** (1970), 181—245. (p. 306)

[22] G. Da Prato: Somme d'applications non-linéaires. Symposia Matematica 7, Academic Press, London—New York 1971, 233—268. (p. 306)

[23] A. A. Dezin: Operator with first time derivative and a non-local boundary condition. (Russian.) *Izv. Akad. Nauk SSSR* **31** (1967), vyp. 1, 61—86. (p. 310)

[24] Ju. A. Dubinskiĭ: Periodic solutions of elliptico-parabolic equations. (Russian.) *Mat. Sb.* **76** (1968), 620—633. (p. 307)

[25] Ju. A. Dubinskiĭ: On some differential-operator equations of a general kind. *Dokl. Akad. Nauk SSSR* **201** (1971), No. 5, 1033—1036. (p. 307)

[26] Ju. A. Dubinskiĭ: On some differential-operational equations of an arbitrary order. (Russian.) *Mat. Sb.* **90** (1973), 3—22. (p. 307)

[27] J. P. Fink, W. S. Hall: Entrainment of frequency in evolution equations. *J. Differential Equations* **14** (1973), 9—41. (p. 319)

[28] J. P. Fink, W. S. Hall, A. R. Hausrath: A convergent two time method for periodic differential equations. *J. Differential Equations* **15** (1974), 459—498. (p. 318)

[29] H. Gajewski, K. Gröger: Ein Iterations-verfahren für Gleichungen mit einem maximal monotonen Lipschitz-stetigen Operator. *Math. Nachr.* **69** (1975), 307—317. (p. 306)

[30] H. Gajewski, K. Gröger: Zur Konvergenz eines Iterations-verfahrens für Evolutionsgleichungen. *Math. Nachr.* **68** (1975), 331—343. (p. 306)

[31] H. Gajewski, K. Gröger: Zur Konvergenz eines Iterations-verfahrens für Gleichungen der Form $Au' + Lu = f$. *Math. Nachr.* **69** (1975), 319—331. (p. 306)

[32] H. Gajewski, K. Gröger: Ein Projektions-Iterationsverfahrens für Evolutionsgleichungen. *Math. Nachr.* **72** (1976), 119—136. (p. 306)

[33] H. GAJEWSKI, K. GRÖGER: Ein Projektions-Iterationsverfahren für Gleichungen der Form $Au' + Lu = f$. *Math. Nachr.* **73** (1976), 249—267. (p. 306)

[34] H. GAJEWSKI, K. GRÖGER, K. ZACHARIAS: Nichtlineare Operator-gleichungen und Operatordifferentialgleichungen. Academie Verlag, Berlin 1974, 213—215, 241—244. (p. 306)

[35] M. GAULTIER: Solutions "faibles" périodiques d'équations d'évolution du premier ordre perturbées. *Comptes Rendus* **272** (1971), 118—120. (p. 311)

[36] E. GIUSTI: Funzioni coseno periodiche. *Boll. Un. Mat. Ital.* **22** (1967), 478—485. (p. 321)

[37] K. GRÖGER: An iteration method for non-linear second order evolution equations. *Comment. Math. Univ. Carolinae* **17** (1976), 575—592. (p. 306)

[38] K. GRÖGER. Regularitätsaussagen für Evolutionsgleichungen mit stark monotonen Operatoren. *Math. Nachr.* **67** (1975), 21—34. (p. 306)

[39] A. HARAUX: Equations d'évolution non linéaires: solutions bornées et périodiques. *Ann. Inst. Fourier (Grenoble)* **28** (1978), 201—220. (p. 321)

[40] L. HERRMANN: Periodic solutions of abstract differential equations: the Fourier method. (Czech.) Thesis, Math. Inst. Czechoslovak Acad. Sci., Prague 1977, 70 pp. (pp. 271, 307)

[41] M. KOPÁČKOVÁ: On periodic solutions of some equations of mathematical physics. *Apl. Mat.* **18** (1973), 33—42. (p. 310)

[42] M. A. KRASNOSELSKIĬ: Translation operator along the trajectories of differential equations. (Russian.) Moscow 1966, 296—321. (p. 321)

[43] M. A. KRASNOSELSKIĬ, P. P. ZABREĬKO, E. I. PUSTYLNIK, P. E. SOBOLEVSKIĬ: Integral operators in spaces of integrable functions. (Russian.) Moscow 1966, 484—486. (p. 321)

[44] N. KRYLOVÁ, O. VEJVODA: Periodic solutions to partial differential equations, especially to a biharmonic wave equation. Symposia Matematica 7, Academic Press, New York—London 1971, 85—96. (p. 317)

[45] J.-L. LIONS: Equations différentielles — opérationnelles et problèmes aux limites. Springer-Verlag, Berlin—Göttingen—Heidelberg 1961, 50—51. (p. 303)

[46] J.-L. LIONS: Sur certaines équations paraboliques non linéaires. *Bull. Soc. Math. France* **93** (1965), 155—176. (p. 304)

[47] J.-L. LIONS: Quelques méthodes de résolution des problèmes aux limites non linéaires. Dunod, Paris 1969, 236—237, 328—329. (p. 304)

[48] J.-L. LIONS, E. MAGENES: Problèmes aux limites non homogènes et applications. Vol. 1, Dunod, Paris 1968, 279—282. Vol. 3, Dunod, Paris 1970, 171—174. (p. 303)

[49] V. LOVICAR: Periodic solutions of non-linear abstract second order equations with dissipative terms. *Časopis Pěst. Mat.* **102** (1977), 364—369. (p. 321)

[50] K. MASUDA: On the existence of periodic solutions of non-linear differential equations. *J. Fac. Sci. Univ. Tokyo Sect. I A Math.* **12** (1966), 247—257. (p. 320)

[51] V. S. MOKEĬČEV: Periodic solutions of partial differential equations. (Russian.) *Izv. Vysš. Učebn. Zaved. Matem.* **130** (1973), No. 3, 77—87. (p. 309)

[52] V. S. MOKEĬČEV: Boundary value problems for partial differential equations. (Russian.) *Izv. Vysš. Učebn. Zaved.* **155** (1975), No. 4, 103—107. (p. 310)

[53] B. NAGY: Cosine operator functions and the abstract Cauchy problem. *Period. Math. Hungar.* **7** (1976), 213—217. (p. 322)

[54] N. PAVEL: Differential equations associated to some non-linear operators on Banach spaces. (Roumanian.) Ed. Acad. Rep. Soc. Rom., Bucureşti 1977, 124—129. (p. 314)

[55] G. PROUSE: Problemi di propagazione per un' equazione funzionale non lineare. Atti Convegno Eq. Deriv. Parz., Bologna 1967, 136—144. (p. 314)

[56] G. PROUSE: Periodic or almost-periodic solutions of a non-linear functional equation, I—IV. *Atti Accad. Naz. Lincei Rend. Cl. Sci. Fis. Mat. Natur.* (8) **42** (1967), 161—167, 281—287, 448—452, 24 (1968), 1—8. (p. 314)

[57] I. B. SIMONENKO: Basing of the averaging principle for abstract parabolic equations. (Russian.) *Mat. Sb.* **31** (1970), 53—61. (p. 317)

[58] I. B. SIMONENKO: Basing of the averaging principle for the problem of convection in a field of rapidly oscillating forces and for other parabolic equations. (Russian.) *Mat. Sb.* **37** (1972), 236—253. (p. 318)

[59] P. E. SOBOLEVSKIĬ, V. A. POGORELENKO: On periodic solutions of hyperbolic equations. (Russian.) Trudy V. Mežd. Konf. Nelin. Koleb., Tom 1, Inst. Mat. Akad. Nauk USSR, Kiev 1970, 530—534. (p. 317)

[60] M. SOVA: Solutions périodiques des équations différentielles opérationnelles: La méthode de développements de Fourier. *Časopis Pěst. Mat.* **93** (1968), 386—421. (p. 307)

[61] I. STRAŠKRABA: Existence and uniqueness of periodic solutions of linear differential equations in a Banach space. (Czech.) Thesis, Math. Inst. Czechoslovak Acad. Sci., Prague 1978, 60 pp. (p. 308) (*To appear in Czechoslovak Math. J.*)

[62] I. STRAŠKRABA, O. VEJVODA: Periodic solutions to abstract differential equations. *Czechoslovak Math. J.* 23 (1973), 635—669. (pp. 100, 314)

[63] I. STRAŠKRABA, O. VEJVODA: Correction to our paper: Periodic solutions to abstract differential equations. *Czechoslovak Math. J.* **27** (1977), 511—513. (p. 314)

[64] I. STRAŠKRABA, O. VEJVODA: Periodic solutions of singular abstract equations. *Czechoslovak Math. J.* **24** (1974), 528—540. (pp. 154, 316)

[65] C. T. TAAM. Stability, periodicity and almost periodicity of solutions of nonlinear differential equations in Banach spaces. *J. Math. and Mech.* **15** (1966), 849—876. (p. 305)

[66] C. T. TAAM: On nonlinear diffusion equations. *J. Differential Equations* **3** (1967), 482—499. (p. 305)

[67] C. T. TAAM: Nonlinear differential equations in Banach spaces, and applications. *Michigan Math. J.* **15** (1968), 177—186. (p. 305)

[68] B. A. TON: Periodic solutions of non-linear evolution equations in Banach spaces. *Canad. J. Math.* **23** (1971), 189—196. (p. 305)

[69] P. P. ZABREĬKO, JU. I. FETISOV: On a method of averaging of Bogoljubov-Krylov to hyperbolic equations. *Vestnik Jaroslav. Univ.* 7 (1974), 150—155. (p. 321)

Papers on related topics

[1] A. K. AZIZ: Periodic solutions of hyperbolic partial differential equations. *Proc. Amer. Math. Soc.* **17** (1966), 557—566.

[2] A. K. AZIZ, A. M. MEYERS: Periodic solutions of hyperbolic partial differential equations in a strip. *Trans. Amer. Math. Soc.* **146** (1969), 167—178.

[3] A. K. AZIZ, S. L. BRODSKY: Periodic solutions of a class of weakly nonlinear hyperbolic partial differential equations. *SIAM J. Math. Anal.* **3** (1972), 300—313.

[4] S. G. Boguslavskiĭ: On one mathematical problem of the propagation of the heat waves. Non-linear boundary-value problems of mathematical physics. (Russian.) Kiev 1973, 41—51.

[5] R. Bouc: Solution stationnaire d'une équation integro-différentielle périodique intervenant en viscoélasticité. EQUADIFF 78, Convegno inter. equaz. diff. ord. ed equaz. funz. Firenze 1978, 341—349.

[6] H. Brézis: Problèmes unilatéraux. *J. Math. Pures Appl.* **51** (1972), 1—168.

[7] H. Brézis, L. Nirenberg: Some first order nonlinear equations on a torus. *Comm. Pure Appl. Math.* **30** (1977), 1—11.

[8] N. Calistru: Periodic solutions of a boundary value problem. *An. Şti. Univ. „Al. I. Cuza" Iaşi Secţ I a Mat.* **23** (1977), 353—366.

[9] G. A. Carpenter: Periodic solutions of nerve impulse equations. *J. Math. Anal. Appl.* **58** (1977), 152—173.

[10] L. Cesari: Periodic solutions of partial differential equation. Symp. Non-lin. Vibrations, Kiev 1961, 440—457.

[11] L. Cesari: Periodic solutions of hyperbolic partial differential equations. Intern. Symp. Non-lin. Diff. Eq. Non-lin. Mech., Academic Press, New York 1963, 33—57.

[12] L. Cesari: A criterion for the existence in a strip of periodic solutions of hyperbolic partial differential equations. *Rend. Circ. Mat. Palermo* **15** (1965), 95—118.

[13] P. Concus: Standing capillary-gravity waves of finite amplitude. *J. Fluid Mech.* **14** (1962), 568—576.

[14] O. E. Cynkova: On the régime of forced vibrations for a non-linear filtration of a liquid in a layer. (Russian.) *Meh. Židk. Gaza*, No. 4 (1974), 62—68.

[15] Ph. Gatignol: Description asymptotique des solutions d'ondes périodiques lentement variables pour l'équation de Korteweg - de Vries. *C. R. Acad. Sci. Paris Sér. A* **274** (1972), 1861—1864.

[16] E. A. Grebenikov, R. Š. Šakibaliev: On the existence of periodic solutions. (Russian.) *Izv. Akad. Nauk Kazah. SSR Ser. Fiz.-Mat.* (1971), 34—37.

[17] E. A. Grebenikov, R. Š. Šakibaliev: On the existence of the asymptotic solutions of a class of partial differential equations. (Russian.) *Differencial'nye Uravnenija* **11** (1975), 505—511.

[18] H. Günzler: Hyperbolische Differentialgleichungen und Klassen fastperiodischer Funktionen. Inaugural-Dissertation der Universität zu München, 1957, 108 pp.

[19] A. Haimovici: Periodic solutions of differential equations with set functions as unknowns. *Rev. Roumaine Math. Pures Appl.* **19** (1974), 161—170.

[20], A. Haimovici: Periodic solutions of hyperbolic partial differential equations. *Ann. Mat. Pura Appl.* **98** (1974), 297—309.

[21] V. V. Harasahal, I. Tažimuratov: On the existence of periodic solutions to systems of linear homogeneous partial differential equations. (Russian.) Nekotoryje voprosy differencnial'nych uravnenij. Izdat. Nauka Kazah. SSR, Alma-Ata 1969, 57—62.

[22] G. Hecquet: Utilisation de la méthode d'Euler-Cauchy pour la démonstration d'un théoreme de L. Cesari. *C. R. Acad. Sci. Paris Sér. A* **273** (1971), 712—715.

[23] G. Hecquet: Contribution à la recherche des solutions périodiques en x_1 de l'équation $u_{x_1 \dots x_n} = f(x_1, \dots, x_n, u, u_{x_1}, \dots, u_{x_n}, u_{x_1 x_2}, \dots, u_{x_2 \dots x_n})$. *C. R. Acad. Sci. Paris Sér. A* **276** (1973), 997—1000.

[24] G. Hecquet: Contribution à la recherche des solutions périodiques en x_1 et x_2 de l'équation $u_{x_1 \dots x_n} = f(x_1, \dots, x_n, u, u_{x_1}, \dots, u_{x_n}, \dots, u_{x_2 \dots x_n})$. *C. R. Acad. Sci. Paris Sér. A* **276** (1973), 1047—1050.

[25] G. HECQUET: Sistemi ai differenziali totali. *Atti Accad. Naz. Lincei Rend. Cl. Sci. Fis. Mat. Natur.* **57** (1974), 40—47.

[26] G. HECQUET: Sur l'éxistence des solutions périodiques en x de l'équation $u_{x^2y^2} = f(x, y, u, u_x, u_y, u_{x^2}, u_{xy}, u_{y^2}, u_{x^2y}, u_{xy^2})$. VII. Internationale Konferenz über Nichtlineare Schwingungen, Band I, 1, Akademie-Verlag, Berlin 1975, 339—345.

[27] G. JUMARIE: Condition d'existence et de stabilité pour les solutions périodiques de certaines équations aux dérivées partielles non linéaires. *C. R. Acad. Sci. Paris Sér. A* **273** (1971), 510—513.

[28] G. JUMARIE: Conditions d'existence et de stabilité pour les solutions périodiques de certaines équations aux dérivées partielles non linéaires. *C. R. Acad. Sci. Paris Sér. A.* **273** (1971), 992—994.

[29] J. B. KELLER: The solitary wave and periodic waves in shallow water. *Comm. Pure Appl. Math.* **1** (1948), 323—339.

[30] J. B. KELLER: Periodic oscillations in a model of thermal convection. *J. Fluid Mech.* **26** (1966), 599—606.

[31] A. A. KOLOKOLOV: Existence of stationary solutions to a non-linear wave equation. (Russian.) *Prikl. Mat. Meh.* **38** (1974), 923—928.

[32] G. MATTEI: Propagazione di onde magnetofluidodinamiche in un plasma comprimibile. *Atti Sem. Mat. Fis. Univ. Modena* **22** (1973), 1—24.

[33] JU. A. MITROPOL'SKIĬ, B. P. TKAČ: Periodic solutions to non-linear systems of partial differential equations of neutral type. (Russian.) *Ukrain. Mat. Ž.* **21** (1969), 475—486.

[34] V. B. MOSEENKOV: Quasi-periodic solutions of a non-linear wave equation with damping. (Russian.) *Ukrain. Mat. Ž.* **29** (1977), 537—541.

[35] V. B. MOSEENKOV: Quasi-periodic solutions of a weakly dissipative non-linear wave equation. (Russian.) *Ukrain. Mat. Ž.* **30** (1978), 254—257.

[36] B. P. MOSEENKOV: Application of the method of accelerated convergence to investigation of a non-linear wave equation. (Russian.) *Ukrain. Mat. Ž.* **30** (1978), 54—62.

[37] T. NAGAI: Periodic solutions for certain time-dependent parabolic variational inequalities. *Hiroshima Math. J.* **5** (1975), 537—549.

[38] M. NAKAO: Periodic solutions of some systems of semilinear partial differential equations in complex domains. *Mem. Fac. Sci. Kyushu Univ. Ser. A* **25** (1971), 318 to 336.

[39] J. NAUMANN: Periodic solutions to a class of second order evolution inequalities. *Boll. Un. Mat. Ital.* **12** (1975), 361—369.

[40] J. NAUMANN: Periodic solutions to evolution inequalities in Hilbert spaces. *Boll. Un. Mat. Ital.* **13** (1976), 686—711.

[41] J. NAUMANN: Periodic solutions to certain evolution inequalities. *Czechoslovak Math. J.* **27** (1977), 424—433.

[42] J. M. OKUNIEWICZ: Quasiperiodic pointwise solutions of the periodic, time-dependent Schrödinger equation. *J. Mathematical Phys.* **15** (1974), 1587—1595.

[43] D. PETROVANU: Solutions périodiques de certaines équations aux dérivées partielles. *Ann. Mat. Pura Appl.* **82** (1969), 83—96.

[44] B. PINI: Su certe questioni di periodicità e asinteticità per i sistemi lineari del primo ordine ai differenziali totali. *Rend. Sem. Mat. Univ. Padova* **20** (1951), 249—277.

[45] Ž. A. SARTABANOV: Application of the truncation method to the construction of a multiperiodic solution to a countable system of partial differential equations. (Russian.) *Izv. Akad. Nauk. Kazah. SSR Ser. Fiz.-Mat.* (1973), 48—56.

[46] Ž. A. SARTABANOV, D. U. UMBETŽANOV: On the construction of a multiperiodic solution to a countable system of partial differential equations. (Russian.) *Izv. Akad. Nauk Kazah. SSR Ser. Fiz.-Mat.* (1972), 61—66.

[47] D. A. SILAEV: Continuation of a T-periodic boundary layer. (Russian.) *Uspehi Mat. Nauk* **27** (1972), 215—216.

[48] R. Š. ŠAKIBALIEV: To the problem of the existence of periodic solutions to a system of partial differential equations. (Russian.) *Izv. Akad. Nauk Kazah. SSR Ser. Fiz.-Mat.* (1972), 74—76.

[49] R. Š. ŠAKIBALIEV: A modified method of K. Siegel for certain systems of non-linear partial differential equations. (Russian.) *Izv. Akad. Nauk Kazah. SSR Ser. Fiz.-Mat.* (1973), 79—81.

[50] R. Š. ŠAKIBALIEV, I. TAŽIMURATOV: On the existence of periodic solutions to a system of partial differential equations of the second order. (Russian.) *Izv. Akad. Nauk Kazah. SSR Ser. Fiz.-Mat.* (1971), 78—81.

[51] I. TADJBAKHSH, J. B. KELLER: Standing surface waves of finite amplitude. *J. Fluid Mech.* **8** (1960), 442—451.

[52] B. P. TKAČ: The periodic solutions of systems of partial differential equations with deviating argument of neutral type. (Russian.) *Trudy Sem. Mat. Fiz. Nelineĭn. Koleban.* **1** (1968), 237—252.

[53] B. P. TKAČ: The solution of an equation of the vibrating type with time delay. (Russian.) *Trudy Sem. Mat. Fiz. Nelineĭn. Koleban.* **1** (1968), 252—258.

[54] B. P. TKAČ: Periodic solutions to systems of partial differential equations of neutral type. (Russian.) *Differencial'nye Uravnenija* **5** (1969), 735—748.

[55] B. P. TKAČ: On the existence of periodic solutions to non-linear partial differential equations of neutral type. (Russian.) *Izv. Vysš. Učebn. Zaved. Matematika* **8** (1970), 112—117.

[56] B. P. TKAČ: On the construction of the periodic solutions of the partial differential equations with deviating argument of the retardation type. (Russian.) Trudy V. Mežd. Konf. Nelineĭn. Koleban., Tom 1, Kiev 1970, 544—550.

[57] B. P. TKAČ: On periodic solutions for a class of a system of equations with partial derivatives of a higher order of neutral type. (Russian.) *Diff.-raznost. Uravnenija, Resp. Mežved. Sb.* **9** (1971).

[58] B. P. TKAČ: Construction of periodic solutions to systems of partial differential equations of a higher order with retarded argument. (Russian.) Matematičeskaja Fizika 9, Naukova dumka, Kiev 1971, 132—141.

[59] B. P. TKAČ: Periodic solutions to systems of partial differential equations with variable time lag. (Russian.) *Non-linear Vibration Problems (Zagadnenia Drgań Nieliniowych)* **16** (1975), 89—96.

[60] B. P. TKAČ: Periodic solutions to systems of two dimensional partial differential equations with variable delay. (Russian.) Matematičeskaja Fizika 19, Naukova dumka, Kiev 1976, 115—120.

[61] B. P. TKAČ: On finding periodic solutions for systems of equations with mixed partial derivatives of neutral type. (Russian.) Kačestvennye metody teorii differencial'nych uravněniĭ s otkl. arg., Kiev 1977, 130—140.

[62] B. P. TKAČ: On periodic solutions of some non-linear systems of equations with mixed partial derivatives of higher orders with retardation. (Russian.) Matematičeskaja Fizika 21, Naukova dumka, Kiev 1977, 105—110.

[63] W. C. TROY: Large amplitude periodic solutions of a system of equations derived from the Hodgkin-Huxley equations. *Arch. Rational Mech. Anal.* **65** (1977), 227—247.

[64] D. U. UMBETŽANOV: On periodic and quasi-periodic solutions holomorphic with respect to a parameter to one type of partial differential equations. (Russian.) *Vestnik Moskov. Univ. Ser. I Mat. Meh.* **3** (1969), 37—43.

[65] D. U. UMBETŽANOV: On almost periodic solutions of a certain system of hyperbolic type. (Russian.) *Izv. Akad. Nauk Kazah. SSR Ser. Fiz.-Mat.* (1971), 55—59.

[66] D. U. UMBETŽANOV: Construction of a multiperiodic solution to a system of quasi-linear partial differential equations. (Russian.) *Izv. Akad. Nauk Kazah. SSR Ser. Fiz.-Mat.* (1971), 73—77.

[67] D. U. UMBETŽANOV: Construction of a multiperiodic solution to a system of partial differential equations with the same principal part. (Russian.) *Differencial'nye Uravnenija* **8** (1972), 1326—1329.

[68] D. U. UMBETŽANOV, A. B. BERŽANOV: On a holomorphic almost-quasi-periodic solution of an integral-partial differential equation. (Russian.) *Izv. Akad. Nauk Kazah. SSR Ser. Fiz.-Mat.* (1977), 61—66.

[69] D. U. UMBETŽANOV, Ž. SARTABANOV: On a quasi-periodic solution to a certain system of partial differential equations with a constant lag. (Russian.) *Izv. Akad. Nauk Kazah. SSR Ser. Fiz.-Mat.* (1976), 48—52.

[70] D. U. UMBETŽANOV, M. B. TULEGENOVA: On multiperiodic solutions to a system of partial differential equations. (Russian.) *Izv. Akad. Nauk Kazah. SSR Ser. Fiz.-Mat.* (1973), 57—61.

[71] G. R. VERMA, J. B. KELLER: Three-dimensional standing surface waves of finite amplitude. *Phys. Fluids* **5** (1962), 52—56.

[72] M. YAMAGUCHI: On the existence of quasiperiodic solutions of nonlinear hyperbolic partial differential equations. *Proc. Japan Acad.* **48** (1972), 320—322.

[73] M. YAMAGUCHI: On the existence of quasiperiodic solutions of nonlinear partial differential equations. *Proc. Fac. Sci. Tokai Univ.* **10** (1974), 1—12.

[74] M. YAMAGUCHI: Note on quasiperiodic solutions of undamped linear wave equations. *Proc. Fac. Sci. Tokai Univ.* **12** (1976), 3—10.

[75] M. YAMAGUCHI: Existence of quasi-periodic solutions of perturbed nonlinear and quasi-linear partial differential equations of standard types. *J. Math. Anal. Appl.* **59** (1977), 15—28.

Theory of stability

[76] J. BARTÁK: Stability and correctness of abstract differential equations in Hilbert spaces. *Czechoslovak Math. J.* **28** (1978), 548—593.

[77] J. BARTÁK, J. NEUSTUPA: Remark to the stability of solutions of the equation of a vibrating beam. (Russian.) *Čas. Pěst. Mat.* **104** (1979), 338—352.

[78] H. KIELHÖFER: Stability and semilinear equations in Hilbert space. *Arch. Rational Mech. Anal.* **57** (1974), 150—165.

[79] J. KURZWEIL: Exponentially stable integral manifolds, averaging principle and continuous dependence on a parameter. *Czechoslovak Math. J.* **16** (1966), 380—423, 463—492.

[80] M. NAKAO: Asymptotic stability for some nonlinear evolution equations of second order with unbounded dissipative terms. *J. Differential Equations* **30** (1978), 54—63.

[81] J. NEUSTUPA: The uniform exponential stability and the uniform stability at constantly acting disturbances of a periodic solution of a wave equation. *Czechoslovak Math. J.* **26** (1976), 388—410.

[82] J. NEUSTUPA: A contribution to the theory of stability of differential equations in Banach space. *Czechoslovak Math. J.* **29** (1979), 27—52.

[83] P. C. PARKS: A stability criterion for panel flutter via the second method of Liapunov. *AIAA J.* (1966), 175—177.

[84] D. H. SATTINGER: Topics in stability and bifurcation theory. Lecture Notes in Mathematics 309 (1973).

[85] J. SERRIN: On the stability of viscous fluid motions. *Arch. Rational Mech. Anal.* **3** (1959), 1—13.

Addenda to the bibliography

to Chapter III

P. H. GOULD: Oscillations in nonlinear parabolic systems. *Bull. Inst. Math. Acad. Sinica* **6** (1978), 107—132.

J. M. GREENBERG: Periodic solutions to a reaction-diffusion equations. *SIAM J. Appl. Math.* **30** (1976), 199—205.

P. HESS: Nonlinear perturbations of linear elliptic and parabolic problems at resonance: existence of multiple solutions. *Ann. Scuola Norm. Sup. Pisa Cl. Sci. Ser. IV* **5** (1978), 527—537.

J. HEUSS: Existenzsätze mit der Linienmethode für parabolische Probleme und periodische Lösungen von $u_t = f(t, x, u, u_x, u_{xx})$. *Manuscripta Math.* **30** (1979), 137—162.

A. A. KEREFOV: Non-local boundary-value problems for parabolic equations. (Russian.) *Differencial'nyje Uravnenija* **15** (1979), 74—78.

H. KIELHÖFER: Hopf bifurcation at multiple eigenvalues. *Arch. Rational Mech. Anal.* **69** (1979), 53—83.

JU. S. KOLESOV, A. D. PENDJUR: The method of adjustment and the Fourier method for a certain nonlinear boundary-value problem of a parabolic type. (Russian.) Issl. po ustoĭčivosti i teorii kolebaniĭ. Jaroslavl' 1976, 3—57.

J. NAUMANN: Periodic solutions to evolution inequalities of a modified Navier-Stokes type. *Boll. Un. Mat. Ital.* (5) **15**-B (1978), 351—369.

M. ÔTANI, Y. YAMADA: On the Navier-Stokes equations in non-cylindrical domains: An approach by the subdifferential operator theory. *J. Fac. Sci. Univ. Tokyo Sect. IA Math.* **25** (1978), 185—204.

D. PETROVANU: Periodic and almost periodic solutions of parabolic equations and E. Rothe's method. *Bul. Inst. Politehn. Iaşi* **23** (1977), 17—22.

O. VEJVODA, M. KOPÁČKOVÁ: Periodic solutions of abstract and partial differential equations with deviation (to appear).

M. ŠTĚDRÝ, O. VEJVODA: Time periodic solutions of a one-dimensional two-phase Stefan problem. *Ann. Mat. Pura Appl.* (to appear).

Y. YAMADA: Periodic solutions of certain nonlinear parabolic differential equations in domains with periodically moving boundaries. *Nagoya Math. J.* **70** (1978), 111—123.

to Chapter IV

A. Haraux: Comportement à l'infini pour certains systèmes dissipatifs non linéaires. Analyse numérique et fonctionnelle, Univ. Paris VI, Lab. Associé 189, 35 pp.

M. Nakao: Bounded, periodic and almost periodic classical solutions of some nonlinear wave equations with a dissipative term. *J. Math. Soc. Japan* **30** (1978), 375—394.

M. Nakao: Periodic solutions of the dissipative wave equation in a time-dependent domain. *J. Differential Equations* **34** (1979), 393—404.

K. N. Tursunov: A study of the periodic (in time) classical solution of a multidimensional boundary-value problem for second order hyperbolic equations with a nonlinear operator right-hand side. (Russian.) *Izv. Akad. Nauk UzSSR, Ser. Fiz.-Mat. Nauk* (1977), no 6, 34—40.

O. Vejvoda, M. Kopáčková: Periodic solutions of abstract and partial differential equations with deviation (to appear).

O. Vejvoda, I. Straškraba: Perturbed nonlinear abstract equations (to appear), 22 pp.

to Chapter V

H. Amann: Saddle points and multiple solutions of differential equations. *Math. Z.* **169** (1979), 127—166.

H. Amann, G. Mancini: Some applications of monotone operator theory to resonance problems (to appear).

H. Amann, E. Zehnder: Nontrivial solutions for a class of nonresonance problems and applications to nonlinear differential equations (to appear), 79 pp.

A. Bahri, H. Brézis: Periodic solutions of a nonlinear wave equation. Anal. num. fonct., Lab. Anal. Num., Univ. Paris 6, 1979, 10 pp.

V. Benci, P. H. Rabinowitz: Critical point theorems for indefinite functionals. *Invent. Math.* **52** (1979), 241—273.

H. Brézis: Cordes vibrantes non linéaires. Sem. Goulaouic-Schwartz 1978—1979, 8 pp.

H. Brézis, J. M. Coron: Periodic solutions of nonlinear wave equations and Hamiltonian systems (to appear), 13 pp.

H. Brézis, J. M. Coron, L. Nirenberg: Free vibrations for a nonlinear wave equation and a theorem of P. Rabinowitz (to appear), 25 pp.

H. Hofer: A multiplicity result for a class of nonlinear problems, with application to a nonlinear wave equation (to appear), 23 pp.

N. Klimperová: Periodic solutions to the inhomogeneous sine-Gordon equation. *Comment. Math. Univ. Carolinae* **20** (1979), 329—333.

G. Mancini: Periodic solutions of some semilinear autonomous wave equations. *Boll. Un. Mat. Ital.* **15**-B (1978), 649—672.

J. Mawhin: Periodic solutions of nonlinear dispersive wave equations. Sem. Math. Appl. Mec., 1^e semestre 1979. Inst. Math. Pure Appl. Univ. Cath. Louvain, 8 pp.

V. N. Poliščuk, V. I. Ptašnik: Periodic solutions of partial differential equations with constant coefficients. (Russian.) *Ukrain. Mat. Ž.* **32** (1980), 239—243.

P. H. Rabinowitz: A priori bounds for a semilinear wave equation. Equadiff IV, Proc., Prague 1977, Lecture Notes Math. 703, 340—347.

M. Štědrý: Time-periodic solutions to a class of first order hyperbolic systems (to appear), 26 pp.

O. Vejvoda, M. Kopáčková: Periodic solutions of abstract and partial differential equations with deviation (to appear).

M. WILLEM: Periodic solutions of wave equations with jumping nonlinearities. Sem. Math. Appl. Mec., 1e semestre 1979, Inst. Math. Pure Appl. Univ. Cath. Louvain, 15 pp.

P. P. ZABREĬKO, JU. I. FETISOV: On an application of the Bogoljubov-Krylov averaging method to hyperbolic equations. (Russian.) *Vestnik Jaroslav. Univ.* Vyp. 7 (1974), 150—155.

to Chapter VI

L. CESARI, R. KANNAN: Existence of solutions of nonlinear hyperbolic equations. *Ann. Scuola Norm. Sup. Pisa Sci. Fis. Mat. Ser. IV* **6** (1979), 573—592.

Ž. N. DMITRIEVA: To the theory of nonlinear vibrations of thin rectangular plates. (Russian.) *Izv. Vysš. Učebn. Zaved. Matematika* 8 (1978), 62—66.

S. FUČÍK: Nonlinear noncoercive problems: Generalized periodic solutions of nonlinear beam equation. 3° Sem. Anal. Funz. Appl., Bari 1978, 52 pp.

V. N. MATVEEV: On vibrations of a beam with a weak nonlinearity. (Russian.) *Vestnik Jaroslav. Univ.* Vyp. 1 (1976), 127—132.

S. A. SOLOP: On the existence of periodic solutions in nonlinear theory of oscillations of Reissner's non-flat shells of revolution with regard to damping. (Russian.) *Prikl. Mat. Meh.* (1980), Vyp. 1, 188—192.

to Chapter VII

L. HERRMANN: Periodic solutions of abstract differential equations: the Fourier method. *Czechoslovak Math. J.* **30** (1980), 177—206.

H. KIELHÖFER: Hopf bifurcation at multiple eigenvalues. *Arch. Rational Mech. Anal.* **69** (1979), 53—83.

H. KIELHÖFER: Generalized Hopf bifurcation in Hilbert space. Mathematische Institute der Julius-Maximilians-Universität Würzburg. Preprint No. 50, 1979, 27 pp.

V. A. POGORELENKO, P. E. SOBOLEVSKIĬ: On periodic solutions to a class of differential equations of second order in a Hilbert space. (Russian.) *Voronež. Gos. Univ. Trudy Mat. Fak.* (1970), vyp. 1, 93—100.

V. A. POGORELENKO, P. E. SOBOLEVSKIĬ: Periodic solutions of quasilinear hyperbolic equations. (Russian.) *Voronež. Gos. Univ. Trudy Mat. Fak.* (1972), vyp. 7, 85—89.

H. SHAW, P. J. MC KENNA: On the structure of the set of solutions to some nonlinear boundary-value problems. *J. Differential Equations* **35** (1980), 183—199.

of papers on related problems

M. S. BORTEJ, V. I. FODČUK: On quasiperiodic solutions of linear partial differential functional equations. (Russian.) *Ukrain. Mat. Ž.* **31** (1979), 237—246.

S. A. GABOV: On the equation of Whitham. (Russian.) *Dokl. Akad. Nauk SSSR* **242** (1978), 983—996.

G. HECQUET: Solutions périodiques d'équations de type hyperbolique. *Ann. Mat. Pura Appl. Ser. IV* **116** (1978), 217—315.

G. HECQUET: Sur la recherche des solutions périodiques en x du système: $u_x = f(x, y, u, v)$, $v_y - A(x)\, v/y = g(x, y, u, v, v_x)$. *C. R. Acad. Sci. Paris Sér. A* **290** (1980), 279—282.

L. HERRMANN, O. VEJVODA: Periodic and quasi-periodic solutions of abstract differential equations (to appear), 12 pp.

T. MATAHASHI, M. TSUTSUMI: On a periodic problem for pseudoparabolic equations of Sobolev-Galpern type. *Math. Japon.* **22** (1978), 535—553.

V. B. MOSEENKOV: Quasiperiodic solutions of a nonlinear hyperbolic equation with damping. (Russian.) *Differencial'nyje Uravnenija* **15** (1979), 695—703.

Y. NISHIURA: Existence of periodic solutions of nonlinear hyperbolic system. *Proc. Japan. Acad. Ser. A* **53** (1977), 190—194.

M. YAMAGUCHI: Existence and stability of bounded and almost periodic solutions of perturbed quasilinear partial differential equations of standard types. *Proc. Fac. Sci. Tokai Univ.* **14** (1978), 45—57.

List of symbols

The numbers denote the pages where the symbols are defined or used for the first time.

$\sim$	equivalence of norms	10
$\langle .,. \rangle$	inner (scalar) product or duality	11
$B(u_0, \varrho; X)$	closed ball	9
$\operatorname{cl} M$, $\overline{M}$	closure	9
∂M	boundary	9
$\operatorname{lin} M$	linear hull	10
$M^{\perp}$	orthogonal complement	11
$\mathscr{L}(B_1, B_2)$	space of bounded linear operators $L: B_1 \to B_2$	10
$\mathscr{L}^{+}(B_1, B_2)$	space of linear operators $L: \mathscr{D}(L) \subset B_1 \to B_2$	288
$\mathscr{D}(A)$	domain of an operator	10
$\mathscr{R}(A)$	range	10
$\mathscr{N}(A)$	null-space	10
$\varrho(A)$	resolvent set	13
$\sigma(A)$	spectrum	13
$\sigma_p(A)$	point spectrum	13
$A \mid M$	restriction	10
A^{-1}	inverse	10
A^{α}	power	15
A^{*}	adjoint	12
	formally adjoint	38
	dual (antidual) of a space	29
A'	transpose of a matrix	36
$\det A$	determinant of a matrix	36

In the sequel let $k \in \mathbf{N}$, $\varkappa \in \mathbf{R}^{+}$, $(\gamma_1, \ldots, \gamma_n) \in (\mathbf{Z}^{+})^n$, $p \in [1, \infty)$ (or $p = \infty$), $\Omega \subset \mathbf{R}^n$, $Q = \mathbf{R} \times \Omega$, $I \subset \mathbf{R}$.

$C_{\mathrm{loc}}(\Omega)$	$C_{\mathrm{loc}}(\Omega; B)$	18
$C^{k}_{\mathrm{loc}}(\Omega)$	$C^{k}_{\mathrm{loc}}(\Omega; B)$	17
$C^{\gamma_1, \ldots, \gamma_n}_{\mathrm{loc}}(\Omega)$	$C^{\gamma_1, \ldots, \gamma_n}_{\mathrm{loc}}(\Omega; B)$	17
$C^{(\gamma_1, \ldots, \gamma_n)}_{\mathrm{loc}}(\Omega)$	$C^{(\gamma_1, \ldots, \gamma_n)}_{\mathrm{loc}}(\Omega; B)$	17
$C(\Omega)$	$C(\Omega; B)$	18
$C^{k}(\Omega)$	$C^{k}(\Omega; B)$	17
$C^{\gamma_1, \ldots, \gamma_n}(\Omega)$	$C^{\gamma_1, \ldots, \gamma_n}(\Omega; B)$	17
$C^{(\gamma_1, \ldots, \gamma_n)}(\Omega)$	$C^{(\gamma_1, \ldots, \gamma_n)}(\Omega; B)$	17
$C^{\infty}(\Omega)$	$C^{\infty}(\Omega; B)$	18
$C^{\infty}_{0}(\Omega)$	$C^{\infty}_{0}(\Omega; B)$	18
$C^{k}(\partial\Omega)$	$C^{k}(\partial\Omega; B)$	24
$C_{\omega}(\mathbf{R})$	$C_{\omega}(\mathbf{R}; B)$	20
$C^{k}_{\omega}(\mathbf{R})$	$C^{k}_{\omega}(\mathbf{R}; B)$	20

Index